Digital Microwave Communication Systems

with selected topics in Mobile Communications

P V Sreekanth

Universities Press

Universities Press (India) Private Limited

Registered Office
3-5-747/1/A & 3-6-754/1 Himayatnagar,
Hyderabad 500 029 (A.P.), INDIA
e-mail: info@universitiespress.com

Distributed by
Orient Blackswan Private Limited

Registered Office
3-6-752 Himayatnagar, Hyderabad 500 029 (A.P.), India

Other Offices
Bangalore, Bhopal, Bhubaneshwar, Chennai,
Ernakulam, Guwahati, Hyderabad, Jaipur, Kolkata,
Lucknow, Mumbai, New Delhi, Noida, Patna

First Published 2003
Reprinted 2009, 2013

ISBN : 978 81 7341 395 8

Typeset by
krishtel eMaging Solutions Pvt. Ltd., Chennai

Printed in India at
B B Press
NOIDA

Published by
Universities Press (India) Private Limited
3-5-747/1/A & 3-6-754/1 Himayatnagar,.
Hyderabad 500 029 (A.P.), INDIA

Dedicated to

My father

Sri P Venkateswarlu
Retd. Principal, AP State Board of Technical Education
who made me an Engineer

and

My Professors

Dr. D C Reddy
Former V.C., Osmania University, Hyderabad
Dr. V V Rao
Sr. Professor, IIT Madras
who taught me Communications Engineering

Contents

Preface

Digital microwave communications system is one of the integral constituents of modern telecommunications. The idea of this book evolved when I was working on the Vijayawada–Renigunta–Madras Digital Microwave Project of the Indian Railways, and training personnel to install, maintain, and handle problems related to the working of long haul digital MW links. I have used the course material of this training program for the core contents of this book; it covers the principles of digital communications systems; digital transmission hierarchies; design aspects of the constituents of a digital MW communications system, viz. radio, MUX, and various other accessories; installation practices; and problems encountered in the working of the systems. Some associated topics in mobile communications have been included to enhance the value of the book for professionals as well as engineering students. The entire contents have been organized in four parts.

Part-I deals with the principles of digital communications system—the constituents, modulation schemes, multiplexing hierarchies and the choice of schemes for digital MW communications systems. I have limited the mathematical analysis of concepts to the extent that is absolutely necessary for understanding the basics of the technology; practical implications of the same follow such discussions. To provide the readers with a clear insight, each constituent of the digital MW communications system, viz. digital MW radio, primary MUX, and higher-order MUX have been described function-wise and module-wise, with the help of one practical model each.

Part-II deals with microwave links. The techniques for synchronization on data link; bit stream integration/segregation; the challenges posed by the trio, namely noise, ISI, and jitter, and methods for handling them; counter-fading techniques; and interfacing analog MW link and digital MW link through trans-MUX, are discussed.

Part-III deals with the outdoor assets of a MW station, namely, waveguides, accessories and antenna. The modes of transmission and their practical significance, and antenna mounting and waveguide hoisting techniques are presented using photographs and illustrations.

Digital MW links act as the backbone of mobile communications systems. They not only provide cell connectivities but also offer tower space for the BTS, etc. Part-IV covers selected topics in mobile communications, viz. GSM, DECT, GPRS, UMTS, and TETRA.

It is hoped that the book would serve as a comprehensive study material for engineering students as well as professionals undergoing training for work on digital microwave systems.

P V Sreekanth

Acknowledgements

I am grateful to Sri P.S. Venkataraman, CSTE/Plg., S.C. Railway, Sri P.K. Srivastava, Senior Professor, Telecom, IRISET and Sri Satish Kumar, SDGM, S.C. Railway, who encouraged me to write a book after the successful commissioning of the Vijayawada–Madras–Renigunta digital MW link of the S.C. Railway.

I am greatly indebted to Sri B. Sambi Reddy CSTE/Projects, S.C.Rly for critically reviewing the manuscript.

I am grateful to Sri R. Sivaramakrishna, CSTE, S.C. Railway for granting permission for publishing the book, and also allowing me to include certain references to S.C. Railway installations in the manuscript.

I am grateful to M/s ITI Ltd., Bangalore, M/s Precision Electronics Ltd., New Delhi and M/s PCL, Chandigarh for allowing reference to some of their equipment in the book.

I am greatly indebted to Universities Press, Hyderabad, for bringing out the book. Interactions with their team of editors helped in giving a professional touch to the manuscript. Although care has been taken to keep the book error-free, suggestions for improvement are welcome.

I am thankful to my wife Kalyani and my son Koustubha Krishna and daughter Sree Durga for their patience and encouragement during the time I spent on this project.

P V Sreekanth

1

Introduction

Students of Electronics and Communications Engineering and postgraduate students of Electronics often find that there is a wide gap between what was learnt as a part of their academic curriculum and the technology pertaining to working systems. Though these courses provide them the foundation in communications theory, practical details of the working systems are generally not covered. In this book I have attempted to provide a unique blend of theoretical and practical coverage on digital microwave communications systems, which will be useful for students as well as working engineers. The practical aspects covered include equipment, interfaces, as well as systems-engineering aspects.

The book is organized in four parts.

Part–I

Part 1 deals with the fundamentals of digital communications system and digital transmission hierarchy. The reader is first introduced to the basic theory necessary for understanding the functioning of a digital communications system. This is followed by a descriptive treatment of various modules that form part of the digital microwave communications system, with specific examples of equipment taken from the infrastructure of the Indian Railways communications network.

Chapter 2 provides an overview of a general communications system. The concept of communication resources and its effective utilization in a digital communications system have been highlighted here.

Chapter 3 is on pulse code modulation (PCM), the basis of any time division multiplexing (TDM) system. The concept of TDM is developed in this chapter, as also the foundation for digital transmission hierarchy.

Different types of modulation schemes used in a digital radio are discussed in Chapter 4. To provide a common and convenient analysis of the noise performance of various modulation schemes, geometric modelling of transmitted signals is developed using Gram–Schmidt orthogonalization procedure.

Chapter 5 provides an overview of digital microwave (MW) communications system. A digital microwave communications system has the same constituents as that of a general communications system, with digital transmission hierarchies duly incorporated. It comprises digital MW radios with antennas, waveguides and accessories, placed over a series of hops which support III-order/IV-order digital multiplexing hierarchy.

No amount of theory or any number of block diagrams can substitute an actual description of a practical system. One model each of a digital MW radio, primary multiplexer (MUX) and higher order MUX are covered in Chapters 6, 7 and 8 respectively, with descriptions of its various constituent modules and their circuitry (included to the extent necessary for understanding the working of the system). As already stated, the examples of equipment considered in this book are taken from the Indian Railways communications network.

Part–II

While Part-I focuses on individual equipment in a digital microwave network, Part-II focuses on problems associated with systems integration of long haul digital MW links and their solution.

In Chapter 9, the two aspects of synchronization, viz., carrier synchronization and symbol synchronization are discussed, and an example of a 34 Mb digital MW radio is presented.

In Chapter 10 the concept of optimum receiver is developed to tackle additive white Gaussian noise (AWGN). Three different schemes of correlative level coding are discussed as effective means of combating ISI (inter symbol interference). Practical applications of these techniques are also presented.

The equalization operations performed in a digital radio to eliminate group delay distortion, IF signal distortion and line distortion are discussed in Chapter 11.

Chapters 12 and 14 deal with the integration of bit streams from lower order MUXes to higher order MUXes. The converse operations are also presented in detail. Integration of data streams (main, wayside and digital service channel) in a digital radio is also discussed. The problems associated with bit stream integration and their counter-measures are presented in Chapter 14.

Chapter 13 discusses various line codes used at different hierarchical interfaces, their merits and demerits. Examples have been provided for complex coding schemes.

Chapter 15 discusses the reliability considerations for long haul digital MW links. The reliability objectives, against the backdrop of overall error propability P_e and the effect of jitter on P_e are discussed. Diversity techniques to counter degradation due to fading and the concept of hitless switching are also presented.

Chapter 16 discusses CCITT recommendations pertaining to network limits for jitter accumulation, measurement of jitter and jitter reduction techniques.

Chapters 17 and 18 provide extensive coverage on TRANS-MUX (transmultiplier) which provides for direct conversion of FDM to TDM channels. Practical problems relating to the functioning of TRANS-MUXes are also highlighted.

Results of experiments conducted on a microwave link commissioned by the South Central Railway (India) are briefly presented in Chapter 19. This will help the reader assess the performance of a practical setup based on the knowledge gained thus far.

An engineer who is thoroughly conversant with all the aspects dealt in Part-II can confidently design, install and maintain a long haul digital MW link.

Part–III

While becoming engrossed in modulation, multiplexing and digital hierarchical interfaces, one should not lose sight of good installation practices. This part focuses on the outdoor assests of a MW station, namely waveguides, antennas and accessories. These components have been described in detail, together with sufficient theoritical support, to enable the reader to appreciate various excitation/feeding arrangements in different modules of the MW radio equipment and their interconnections.

In Chapter 20 a heuristic approach to determine the modes of an electromagnetic wave transmission through waveguides and its practical implications are presented. Chapter 21 is devoted to waveguide components, accessories and installation practices. Chapter 22 covers branching circuit in digital radios and the application of coaxial switches. Chapter 23 presents antenna-mounting techniques with several photographs from a real-life installation.

Part–IV

Part-IV covers some of the latest topics in communications like GSM viz. (global system for mobile communication), satellite system for mobile communication and new data services on digital cellular networks (e.g. HSCSD, TETRA, DECT and UMTS). These networks use digital MW links for some of their interconnections and hence this part has been included to look into these aspects. Synchronous digital hierarchy (SDH), which is suitable for large volumes of data traffic, is also briefly covered.

The topics included in the four parts, together with some thought-provoking exercises provided in the end, will serve as a comprehensive course material for students in their final semester of Communications Engineering, first/second semester of postgraduate courses, and those persuing special diploma courses in this area. It can also be used as training material for engineers and technicians working on digital MW networks.

Part - I

2

Digital Transmission Systems—An Overview

2.1 Elements of Communications System

Communications has become an inherent part of our lives today—we take for granted its overwhelming presence in our environment. Radios, televisions, computers, newspapers and magazines—all media for communicating—provide us with an update on events anywhere in the world and allow us to keep in touch with people in any part of the world. Communications enable provision of navigational information to ships and aircraft and data to the weather bureau for making weather forecasts. Mobile telephony and other wireless communication facilities help the executives to be in touch with the office while on the move. The list of applications is indeed quite large, and growing.

In all the above applications, communications in the most fundamental sense involves implicitly the transmission of information from one point to another in the following sequence:

- Generation of a thought pattern or an image in the mind of an originator (data in the case of computers)
- Description of the image with a certain measure of precision by a set of symbols
- Encoding of symbols to a form suitable for transmission over a selected medium
- Transmission over the selected medium
- Decoding and reproduction of the original symbols
- Recreation of originator's pattern/data

Thus, irrespective of the form of communication, there are three basic elements in every communications system, namely

- transmitter,
- channel,
- receiver.

Figure 2.1 gives a schematic representation of a communications system. The transmitter transforms the message signal produced by the source of information into a form suitable for transmission over the channel. The transmitted signal is distorted due to channel imperfections, and also subjected to interference from noise and signals from other sources. So the output of the channel, i.e. the received signal, is a corrupted version of the transmitted signal. The receiver reconstructs a recognizable form of the original message signal and delivers it to the user.

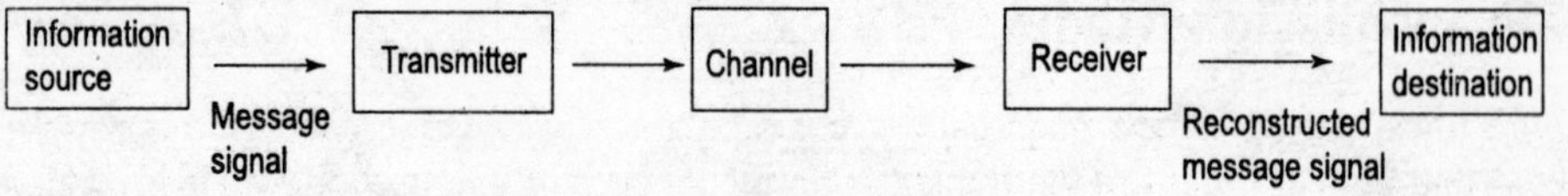

Fig. 2.1 *Schematic representation a communications system.*

2.2 Sources of Information

A source of information is characterized in terms of the signal that carries the information. A signal is a single-valued function of time. Information sources can be classified into four basic types: audio, video, facsimile and computer-generated data.

Audio

Speech: Speech production is well defined by the sequence:

Nabhi - Hruth - Kantha - Rasana.

Nabhi-Hruth: region acts as the sound source which excites '*Kantha*', the vocal tract filter. The vocal tract consists of a tube of non-uniform cross-sectional area, beginning at the glottis (i.e., the opening between the vocal cords) and ending at the lips. As the sound propagates along the vocal tract, the spectrum (i.e. the frequency content) is shaped by the selectivity of the vocal tract. Finally, speech is delivered through the movement of 'Rasana' i.e., tongue. Audio signals lie in the 300–3400 Hz range, with a bandwidth of 4 kHz.

Other than speech: Mechanical, environmental and instrumental sounds. Human ear can discern signals upto 15 kHz.

Video

Television signal: The two-dimensional picture form is converted to a one-dimensional waveform through a spatial sampling process known as raster scanning. The bandwidth of these signals is 4.2 MHz.

Facsimile: Facsimile or FAX, is used for transmitting still pictures over a communication channel. An image sensor is used in the FAX machine to convert the original document to an electrical signal. FAX machines operating at 4.8 kbps and 9.6 kbps are common; machines operating at 64 kbps are also available.

Computer generated data: Computers are a source of information when used for e-mail, file transfer and sharing of resources. Characters of various lower case and upper case letters, numbers, special control symbols, punctuation symbols, etc., are encoded using ASCII (American Standard Code for Information Interchange). Each character is represented by seven bits (i.e., $2^7 = 128$ different characters can be represented in ASCII), and the eighth bit is used as the parity bit. Thus, 1 byte (eight bits) is used for transmission of each character.

2.3 Transmission Media

Open wire lines, paired cables, quad cables, coaxial cables, waveguides, radios (microwave, UHF, VHF, etc.) and optical fibres are the transmission media currently in use today. Their properties and uses are mentioned below.

Open wire lines

These are the original telephone and telegraph transmission lines laid for local and long haul transmission. They have very low loss at voice frequency (for 6|1|1.5 ACSR, it is 0.038 dB/km), and its low attenuation makes long distance communication possible. Crosstalk and noise pickup are minimised by transposing each pair at regular intervals. This medium is referred to as the 'overhead alignment' in Railway telecom parlance. It is becoming obsolete now because of its high maintenance cost (theft is rampant), susceptability to weather and environmental damages, and low channel capacity.

Paired cables

Multipair cable is used extensively in telephone networks for wiring within buildings, from buildings to local office, and from local office to central office.

The number of pairs can be as high as 2000. Loss is typically 0.05 dB/km.

Quad cables

Quad cables have paper-insulated 'quads' (one Tx pair, one Rx pair, i.e., four conductor make a quad), few are poly-insulated (PET) quads as well. Quad cables are used for long distance communication and are laid along railway tracks. Attenuation in a pair is 0.25 dB/km. The cable is screened to reduce induced voltages. The standard sizes are 0+12+2 and 0+17+3 (0+12+2 implies zero carrier quads, twelve paper-insulated quads and two PET quads).

Coaxial cables

A coaxial cable consists of a single wire conductor centred within a cylindrical outer conductor or shield. The two conductors are insulated from each other by a dielectric. The outer conductor can be either braided or solid, depending on the application, and has a proctective jacket over it.

In India, the DOT (Department of Telecommunications) used analog coaxial systems initially and later replaced them with digital coaxial systems, retaining the cable.

System	Channels	Cable dia Inner/Outer	Attenuation at max. frequency	Repeater spacing
Analog, 2.6 MHz	960	1.2/4.4 mm	5.3 dB/km	9.5 km
Analog, 12 MHz	2700	2.6/9.5 mm	8.5 dB/km	4.5 km
Digital, 140 Mbps	1920	2.6/9.5 mm	8.5 dB/km	4.5 km

Normal coaxial cable is used in LANs (local area networks) and can support bit rates upto 100 Mbps in data transmission.

Braided coaxial cable, also called 'leaky coaxial cable' can be used as a radiating element to establish communication links with trains/moving vehicles in tunnels.

Waveguides

A waveguide consists of a hollow conductor or a tube with a rectangular, elliptical or circular cross-section. Waveguides can be rigid or flexible. Electromagnetic waves at microwave frequencies can travel through waveguides. They are used in communications subsystems, for example in connecting a microwave transmitter/receiver to an antenna. Attenuation is as low as 5 dB per 100 m (tower height is normally not more than 100 m), and data rates can be upto 140 Mbps.

In the early 1970s, AT&T announced its WT-4 millimeter waveguide system having a capacity upto 2,40,000 duplex voice circuits, but it was not commercially exploited due to rapid developments in optical fibre technology.

Radio (HF, VHF, UHF, MW)

Radio propagation modes and their corresponding systems are shown in Table 2.1. HF band was used by the Railways* for emergency communications and also for providing a few channels to remote places. It is becoming obsolete now.

Table 2. 1 *Radio propagation modes and characteristics of various radio frequency bands.*

Frequency band	Frequency range	Propagation modes	System/Uses/Characteristics
VLF	3–30 kHz	Earth–ionosphere	Worldwide, military and navigation
LF	30–300 kHz	Surface wave	Stable signal over distances upto 1500 km.
MF	300–3 MHz	Ground/skywave for short/long distances, respectively	Radio broadcasting. Long distance sky-wave signals are subject to fading
HF	3–30 MHz	Skywave, but very limited; short distance ground waves	3–6 MHz: continental; 6–30 MHz: intercontinental land and ship-to-shore communications
UHF	30–300 MHz	Space wave	Close to line-of-sight over short distances. Brodcasting and land mobile
	Above 300 MHz	Scatter wave	Tropospheric scatter over 150–800 km distances
SHF	3–30 GHz	Space wave	The 'workhorse' microwave band. Line-of-sight, terrestrial and satellite relay links
EHF	30–300 GHz	Space	Line-of-sight millimeter waves. Space-to-space links, military use, and possible short distance local distribution

*The 'Railways', wherever mentioned in this book, refers to the Indian Railways.

VHF is the most familiar band for single channel equipment over distances upto 50 km. Single channel duplex sets for control communication, emergency communication, walkie-talkie sets for communication between the train driver and the guard, microwave tower-to-equipment room communication, communication within the yard, etc., are some of the applications of VHF radio by the Railways. VHF simplex sets are quite popular with the Police for communication within a town or a city, and within their patrol parties.

Microwave band (3 GHz to 30 GHz, of which the 7 GHz band and the 18 GHz band are used by the Railways) is the work-horse for inter-railway communications. Normally, in this band the signal strength falls off at a rate of 6 dB per octave, and therefore interference with adjacent carriers will not be a problem. However, fading can be a daunting problem caused due to multipath or inverse path bending. Fading can cause signal variations in excess of 30 dB for a duration of a few seconds to minutes. For frequencies above 10 GHz, fading as a result of attenuation due to rain must be taken into account as well.

Repeaters are spaced at 40 km (7 GHz band) or 10 km (18 GHz band). The 7 GHz and the 18 GHz digital MW radios can take upto 140 Mbps and 274 Mbps per carrier, respectively. The 7 GHz analog microwave radio can take upto 960 channels.

Optical fibre

Graded index fibre with optimized single mode transmission system at 1300 nm wavelength and with repeaters every 30 km are being used for data rates upto 565 Mbps. Optical fibres are also used for providing connectivity between exchanges and in LAN (fibre distributed digital interface (FDDI)). Transmission loss in an optical fibre is typically 0.4 dB per km.

High data rate transmission systems based on SDH (synchronous digital hierarchy) using optical fibre medium at 1550 nm wavelength are being commissioned. Transmission loss at this wavelength is only 0.2 dB per km.

2.4 Modulation Process

In a communications system the source and the destination are separated by a transmission medium. A communications system delivers message signals from an information source in a form that can be detected at the destination. To do so, the transmitter has to modify the message signal to a form that is suitable for the transmission medium. This is achieved through modulation—a process in which some parameter of the carrier wave is varied in accordance with the message signal. The reverse process, known as demodulation, takes place at the receiver. However, due to the presence of noise and distortion in the received signal, the original signal is not reconstructed exactly. The extent of degradation in the overall system performance depends on the medium and the modulation scheme used. There are several modulation schemes possible, and these are classified as shown in Fig. 2.2

2.5 Multiplexing

When huge investments are made on the transmitter, receiver and transmission medium arrangement, it is desirable that we transmit as many channels as possible using the same medium. This is acheived through a process called multiplexing, where several message signals are combined for simultaneous,

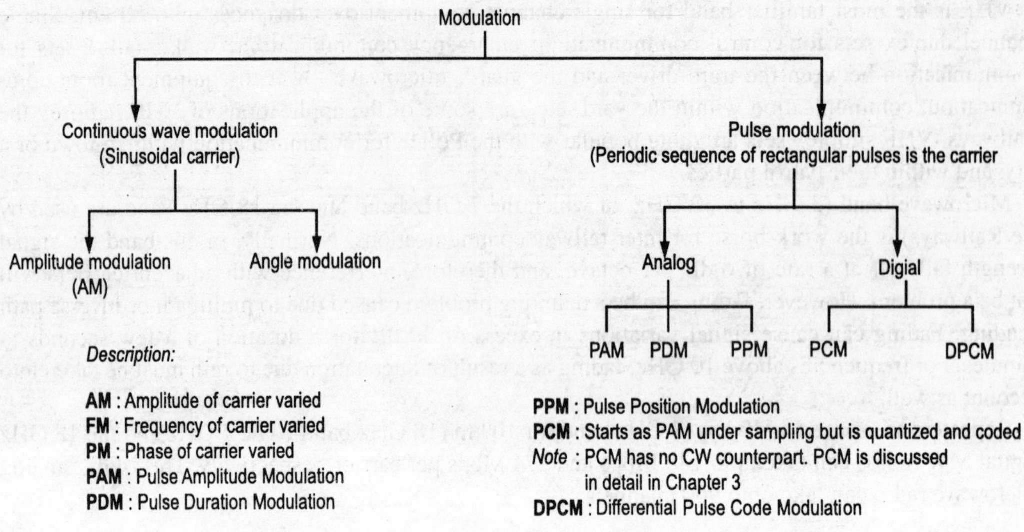

Fig. 2.2 *Classification of modulation methods.*

or near-simultaneous transmission over the same channel. Two of the multiplexing methods generally followed are:

– FDM (frequency division multiplexing),
– TDM (time division multiplexing).

In FDM, through CW (continuous wave) modulations, each message signal is translated to reside in a specific frequency slot inside the passband of the channel by assigning it a distinct carrier frequency. At the receiver a bank of filters is used to separate different modulated signals, and each of the separated signal is then demodulated.

In TDM, pulse modulation is used to position samples of different message signals in non-overlapping time-slots.

In FDM there is a possibility of cross-talk but bandwidth utilisation is good. There is no cross-talk in TDM, but bandwidth requirement is more (as seen later in Chapter 3).

FDM uses CW modulaltion and TDM uses pluse modulation. Thus, different methods of modulation and multiplexing go hand (in hand and must be compatible.

A block diagram of FDM is given in Fig. 2.3(a). Its practical implementation involves many steps of modulaltion and demodulation as illustrated in Fig. 2.3(b). They are:

– Twelve voice channels are combined into a basic group (the nth input modulates a carrier of frequency $F_c = (60 + 4n)$ kHz, where $n = 1, 2, \ldots 12$). The lower sidebands are then selected by a bandpass filter to form a group of twelve channels in the 60–108 kHz band.
– Five basic groups are combined to form 'supergroups' (the nth group modulates a carrier of frequency $F_c = (372 + 48n)$ kHz, where $n = 1, 2, \ldots, 5$). The lower sidebands are selected through bandpass filtering to form a supergroup in the 312–552 kHz, band.

Transmission of sixteen supergroups are possible in the bands 60–300, 312–552 and 564–804 kHz etc. by suitable translations of supergroups.

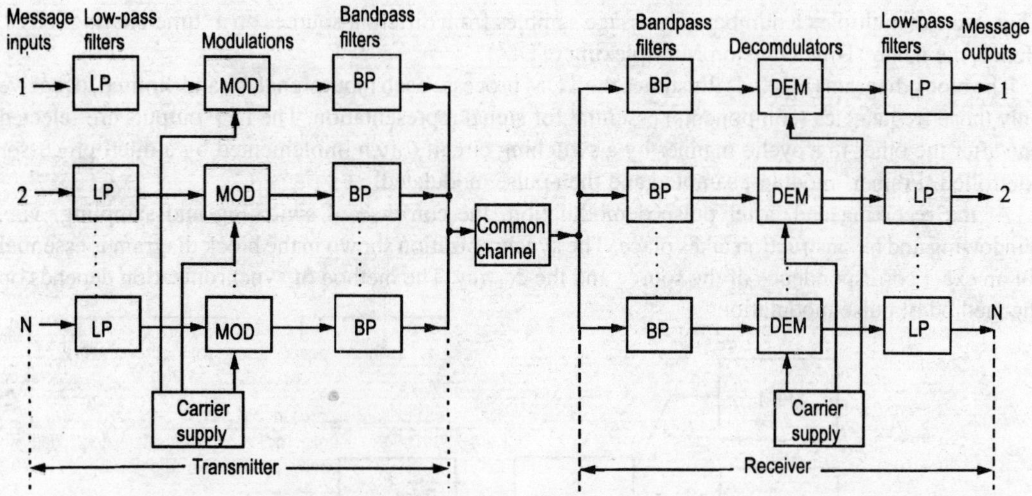

Fig. 2.3(a) *Block diagram of FDM.*

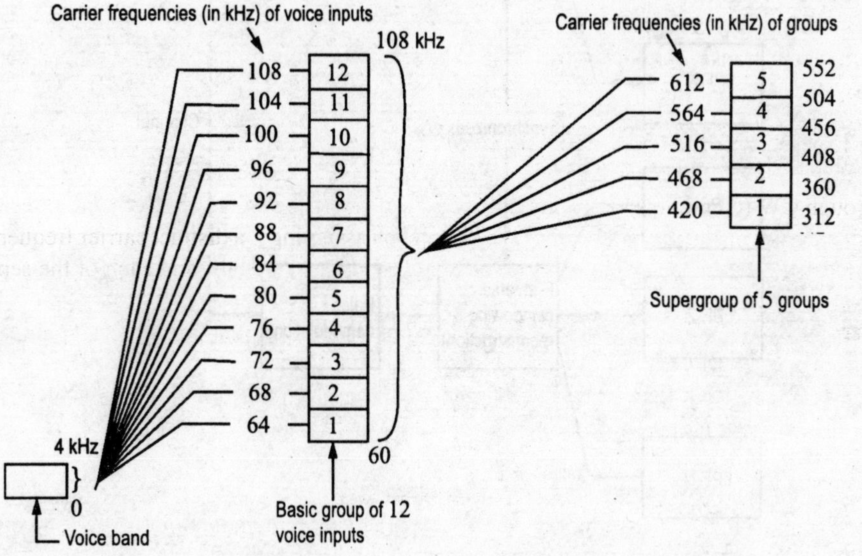

Fig. 2.3(b) *Modulation steps in FDM.*

2.5.1 Time division multiplexing (TDM)

The sampling theorem provides the basis for transmitting information contained in a band-limited signal as a sequence of samples taken at a rate higher than the Nyquist rate (refer Section 3.2 of Chapter 3). That is, the transmission of message samples engages the communication channel only for a fraction of the sampling interval on a periodic basis. So the time interval between adjacent samples can be used for samples of other message sources (as per the data rates feasible in the technology being used).

Thus, we can multiplex a number of message samples from different sources on a 'time-sharing' basis. Hence, the name 'Time Division Multiplexing (TDM)'.

The block diagram (Fig. 2.4) illustrates the TDM process. Each input signal is band-limited to receive only those frequencies (components) essential for signal representation. The LPF outputs are selected one after the other in a cyclic manner by a switching circuit (often implemented by a microprocessor controlled 'Framer' module), sampled and then pulse modulated.

At the receiving end, after pulse demodulation, the converse of switching and sampling, viz., windowing and reconstruction takes place. The synchronization shown in the block diagram is essential for an exact correspondence of the source and the destiny. The method of synchronization depends on the method of pulse modulation.

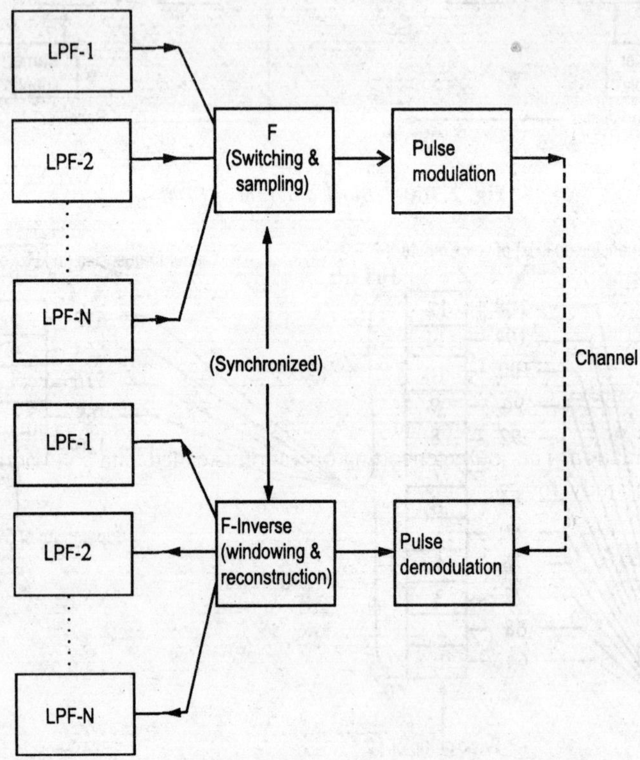

Fig. 2.4 *Block diagram of TDM.*

2.6 Communications Resources

We have seen how the bandwidth (BW) of a transmission media is optimized by the technique of multiplexing. The other parameter of importance is the transmitted carrier power, which is the average power of the transmitted signal. The power output required determines the configuration of the power plant, the ergonomics of equipment room layout, AC plant requirement, etc. Above all, for a given receiver design, the power output of the transmitter signal determines the SNR (ratio of the average signal power to the average noise power) at the receiver, which in turn determines the performance of the receiver. Noise power to some extent is dependent on bandwidth. Here, the transmitter power and the available BW are considered primary resourses of communications.

The objective of communications system design is to use these two primary resources as efficiently as possible. The question that immediately arises is—which modulation scheme provides for their most efficient use? In other words, for a specified channel bandwidth, which modulation scheme requires the smallest SNR for a prescribed level of system performance?

The answer to this question lies in Information Theory.

2.7 Information Theory

Information theory is a multidisciplinary study rooted in communications and mathematics. It is the application of laws of probability and other mathematical branches to the performance analysis of a communications systems. Information theory answers two fundamental questions:

1. What is the fundamental limit on the compression and refinement of information generated by a source?

The answer provided by information theory is the Source Coding Theorem. An information source can be discrete memoryless (digital) or continuous.

According to the source coding theorem, the average number of bits per symbol required to encode the output of a discrete memoryless source with an arbitrarily small probability of decoding error is lower-bound by the 'entropy' (average information content per symbol) of the source.

The source encoding is exact or lossless if the original information source output can be reconstructed without loss of information. The source encoding operation is called data compaction.

To generalize the study for continuous source (analog) also, the concept of 'rate-distortion' is used. Rate distortion is the lower-bind on the rate at which the source output can be transmitted over a noiseless channel and still be reconstructed within the prescribed limit of distortion (digital representation of analog signal is necessarily accompanied by distortion). The conversion from analog to digital is termed 'data compression' in information theory. In data compression there is entropy loss, whereas in data compaction there is no entropy loss (Fig. 2.5).

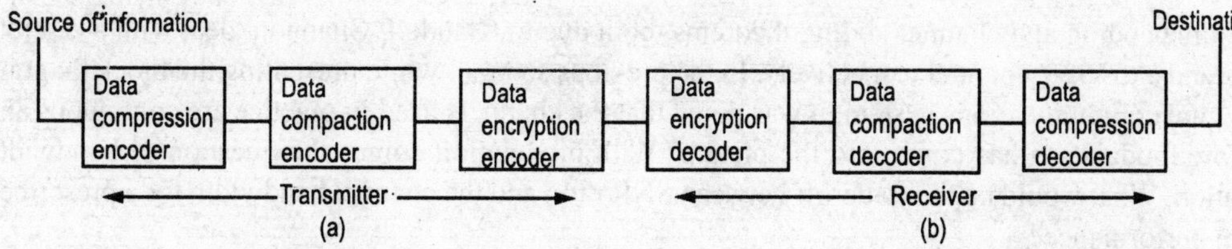

Fig. 2.5 *(a) Source encoding at transmitter, (b) source decoding at receiver.*

Finally, data encryption requires certain additional bits for error correction (not just detection) or confidentiality, as the case may be. At the receiving end a sequence of decoding takes place as shown in Fig. 2.5(*b*).

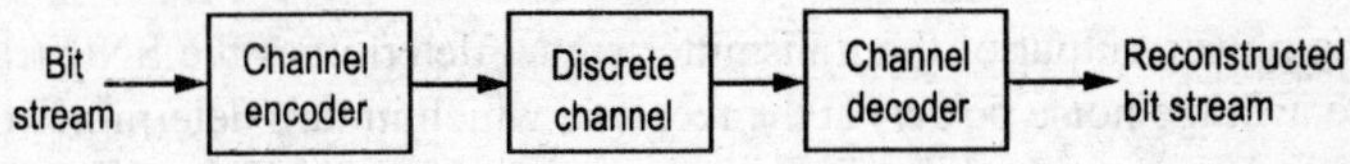

Fig. 2.6 *Channel coding.*

The second question answered by information theory is:

2. What is the fundamental limit on the transmission rate of information over a noisy communication channel?

The answer provided by information theory is the Channel Coding Theorem. Source coding theorem uses theterm entropy whereas channel coding theorem uses the term 'channel capacity'.

According to channel coding theorem, as long as the rate of information transmission is below the channel capacity, it is possible to have error-free transmission over the channel. This is achieved by using a channel encoder in the transmitter and a channel decoder in the receiver (Fig. 2.6).

In other words, the channel coding theorem asserts that error-free transmission is possible over a noisy channel upto a certain rate of transmission. It does not reveal how to design the channel encoder/decoder or how complex their design is. Channel coding theorem has opened up vistas of exploration for error-free communications systems. Many good solutions have been found but the final or the best solution is still unknown.

2.8 Digital Communications System

The channels used for communications in practice are the continuous (analog) type. Bit stream have to be first modulated to transmit them over continuous channels. Thus, combining the operations of source coding, channel coding and modulation, a digital communications system would be as shown in Fig. 2.7.

2.9 Information Capacity

The source coding and channel coding theorems, both due to Claude E Shannon, deal with a discrete source and a discrete channel respectively. In the previous section, while presenting the block diagram of a digital communications system, it was noted that the channels used in practice are continuous and therefore modulation has come into the picture. With modulation comes the question of bandwidth utilization. What would be the trade-off between SNR ratio and the channel bandwidth for a prescribed system performance?

The answer to this question is provided by the Information Capacity Theorem, also due to Claude E Shannon.

The information capacity theorem states that $C = B \log_2(1 + \text{SNR})$ bits per second, where B is the channel bandwidth and C the information capacity. Information capacity means the maximum rate at which information may be transmitted without error through a channel. From this theorem we can see the trade-off between B and SNR for a prescribed information capacity.

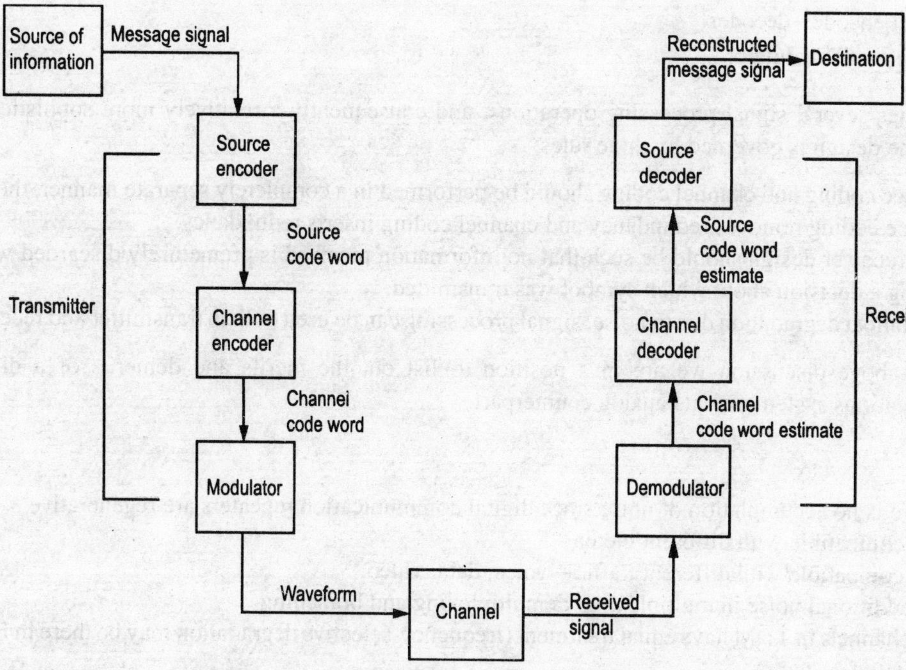

Fig. 2.7 *Block diagram of a digital communications system.*

2.10 Communications System Design

In the design of a communications system, the source of information channel and the destination are given. The design can be for an analog or a digital system. The channel makes available certain resources for information transmission, subject to some limitations. The designer has to make the communications system at an affordable cost such that the system takes the message signals from the source of information, transmits them over the channel, and produces an 'estimate' of it at the receiver which satisfies the user.

Consider the case of an analog communications system. The design is quite simple. The transmitter and the receiver are designed as per the CW modulation scheme chosen. The analog communications system makes some relatively superficial changes to the message signal to prepare it for transmission. There is no deep effort as in the case of a digital communications system to tailor the waveform of the transmitted signal to suit the channel.

A digital communications system is the culmination of an endeavour to find a set of waveforms that closely match the channel characteristics. In this system the functional blocks of the transmitter and the receiver are paired as under:

Source encoder–decoder
Data compression encoder–decoder
Data compaction encoder–decoder
Data encryption encoder–decoder

Channel encoder–decoder
Modulator–demodulator

This implies several signal processing operations, and consequently a relatively more sophisticated design. The design is governed by three rules:

- Source coding and channel coding should be performed in a completely separate manner, though source coding removes redundancy and channel coding inserts redundancy.
- The receiver design should be such that no information received is prematurely discarded while taking a decision about which symbol was transmitted.
- To combat degradation due to noise, signal processing can be used both in transmitter and receiver.

With the above discussion we are in a position to list out the merits and demerits of a digital communications system over its analog counterpart.

Merits

- There is no accumulation of noise since digital communication repeaters are regenerative
- It is compatible with different media
- It is compatible with different traffic—voice, data, video
- No additional noise in multiplexing, demultiplexing and branching
- All channels in TDM have equal treatment (frequency-selective degradation may be there in FDM of analog systems)
- Level fluctuations during transmission have no effect on the primary signal recovered in the receiver (in FDM sophisticated equipments are required to maintain the level constant)
- Novel facilities like cryptography digital processing of various functions are possible

Demerits

- Complex design
- Sudden kill (since it is not affected by fluctuations, the degradation cannot be detected till the sudden kill occurs)
- Bandwidth required is more (but this is offset by the advantages of a digital communications system)

Exercises

1. Identify source encoder, channel encoder and modulator, medium in an optical fibre communications system.
2. Identify source encoder, channel encoder, modulator and medium in a digital MW communications system.
3. In TDM frame, we have a series of 1's and 0's. How then can we get a band-limited signal for transmission?
4. Why are 6 GHz or 7 GHz carriers preferred for terrestrial MW links?

3

Pulse Code Modulation

3.1 Introduction

In Chapter 2 we have distinguished between two families of modulation: continuous wave (CW) modulation and pulse modulation. Again, under 'pulse modulation' we have distinguished bewteen two types: analog pulse modulation and digital pulse modulation. In digital pulse modulation, the message signal is represented in a form that is discrete both in time and amplitude, and thus the transmission is in digital form a sequence of coded pulses.

Signals can be continuous in both frequency and time. Obviously, we cannot make a coded pulse representation of the signal, transmit it and then recontruct the original signal at every instant. Using sampling process an analog signal is converted to a corresponding sequence of samples that are usually spaced uniformaly in time. But to be able to construct the original signal, the sampling rate has to be chosen properly. Sampling theorem which deals with this, is the starting point for discussion on PCM (pulse code modulation).

3.2 Sampling Theorem

Consider a signal $s(t)$ of finite energy which is specified for all time. A segment of the signal $s(t)$ is shown in Fig. 3.1(a).

Let the signal $s(t)$ be sampled instantaneously and at a uniform rate once every T_s second. On such sampling, a sequence of samples spaced T_s seconds is obtained. $s(t)$ is defined for all times, however, only a part of it is shown in Fig. 3.1(a).

The signal sampled as above results in a sequence of pulses. This sequence is infinite. Let the sequence of pulses be denoted by $\{s(nT_s)\}$ where n is an integer, T_s is the sampling period and $(1/T_s) = f_s$ is the sampling rate or the sampling frequency (refer Fig. 3.1(b)).

Consider $\delta(t)$, the unit impulse function

$$\delta(t) = 0 \quad \text{for} \quad t \neq = 0 \quad \text{and} \quad \int_{-\infty}^{\infty} \delta(t)dt = 1, \tag{3.1}$$

i.e., $\delta(t)$ has non-zero value only at $t = 0$ and the area under the function is unity. Then $\delta(t - nT_s)$ (n is an integer) defines the train of unit pulses spaced T_s seconds apart. By sampling $s(t)$, we get the sequence $\{s(nT_s)\}$ which also has samples T_s seconds apart.

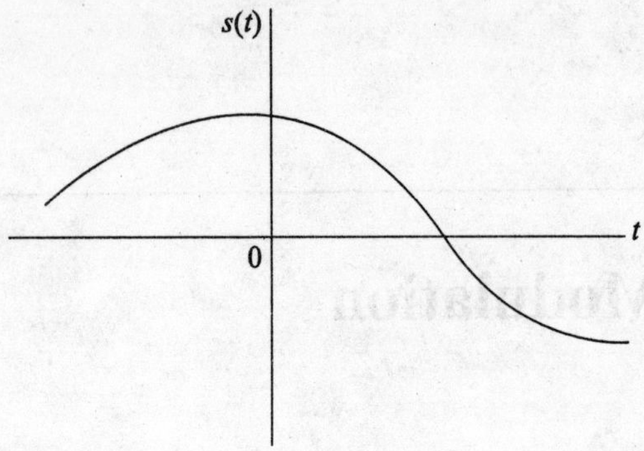

Fig. 3.1(*a*) *A segment of the signal.*

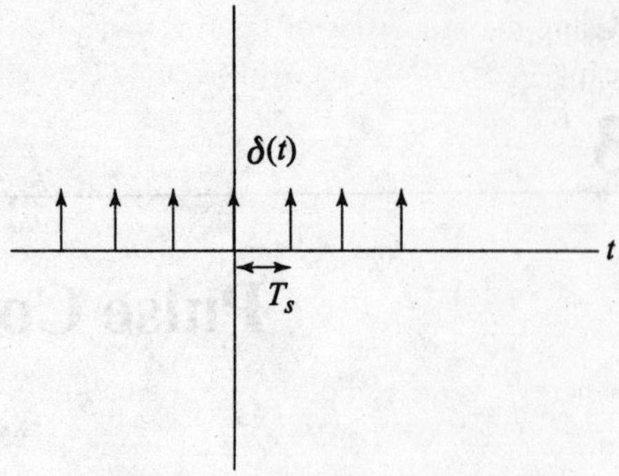

Fig. 3.1(*b*) *Impulse function used for sampling.*

Let $\delta(t - nT_s)$ be weighted by $s(nT_s)$.

$$\text{We get} \quad s_\delta(t) = v = \sum_{n=-\infty}^{\infty} s(nT_s)\delta(t - nT_s). \tag{3.2}$$

$s_\delta(t)$ is the 'ideal sampled signal' corresponding to $s(t)$ (Fig. 3.1(c)).

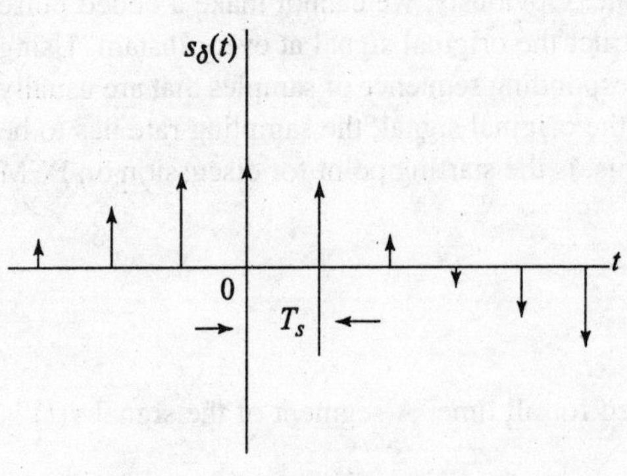

Fig. 3.1(*c*) *Sampled signal.*

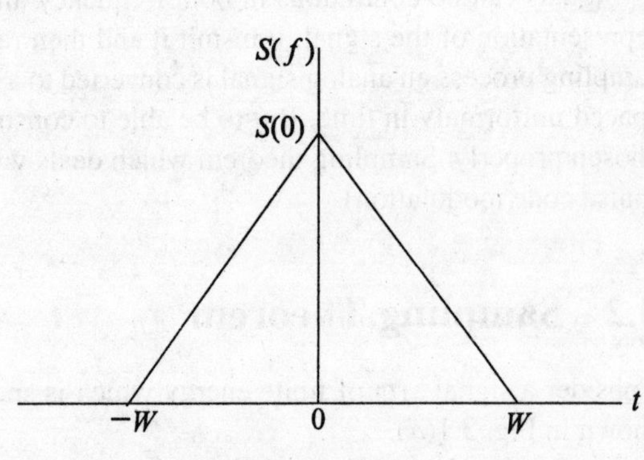

Fig. 3.1(*d*) *Fourier transform of* $s(t)$.

Let $s(t)$ be band-limited with maximum frequency component at W Hz. $S(f)$, the Fourier transform of $s(t)$ shall be as in Fig. 3.1(d). Fourier transform of $s(t)$ by definition is given by

$$S(f) = \int_{-\infty}^{\infty} s(t)\exp(-j2\pi ft)dt,$$

and the inverse transform of $S(f)$ is

$$s(t) = \int_{-\infty}^{\infty} S(f)\exp(j2\pi ft)df.$$

Using the properties of Fourier transform, it can be shown that the Fourier transform $\delta(t - nT_s) = \exp(-j2\pi nfT_s)$. Let $S_\delta(f)$ denote Fourier Transform of $s_\delta(t)$. Then $S_\delta(f)$ is expressed as

$$S_\delta(f) = \sum_{n=-\infty}^{\infty} s(nT_s)\exp(-j2\pi nfT_s), \tag{3.3}$$

$$\text{or} \quad S_\delta(f) = f_s \sum_{m=-\infty}^{\infty} S(f - mf_s), \tag{3.4}$$

where f_s is the sampling rate. Equation (3.4) can also be expressed as

$$S_\delta(f) = f_s S(f) + f_s \sum_{\substack{m=-\infty \\ m \neq 0}}^{m=\infty} S(f - mf_s).$$

Since $s(t)$ is band-limited (to W Hz), by setting $f_s = 2W$ we get the variation of $S_\delta(f)$ with f as shown in Fig. 3.1(e).

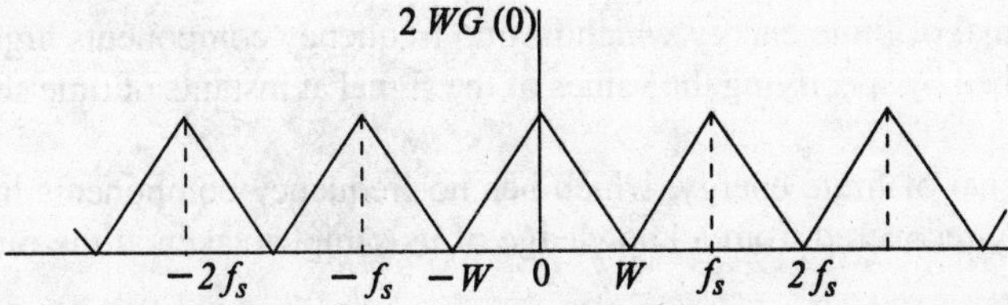

Fig. 3.1(e) *Spectrum of the sampled signal (signal is band-limited to W).*

From Fig. 3.1(e), we can interpret that

$$S(f) = \frac{1}{2W} S_\delta(f), \quad -W < f < W \tag{3.5}$$

$$= \frac{1}{2W} \sum_{n=-\infty}^{\infty} s\left(\frac{n}{2W}\right)\exp\left(\frac{-j\pi nf}{W}\right). \quad -W < f < W \tag{3.6}$$

This means that if the sample values $s(n/2W)$ of a signal $s(t)$ are specified for all time, then the Fourier transform $S(f)$ of the signal is uniquely determined by using the discrete time Fourier transform of equation (3.6). Since $s(t)$ is related to $S(f)$ by inverse Fourier transform, it follows that the signal $s(t)$ is itself uniquely determined by the sample values $s(n/2W)$ for $-\infty < n < \infty$. In other words, the sequence $\{s(n/2W)\}$ has all the information contained in $s(t)$.

Consider the reconstruction of $s(t)$ from the samples $\{s(n/2W)\}$. Substituting (3.6) in the inverse Fourier transform expression for $s(t)$,

$$s(t) = \int_{n=-\infty}^{\infty} S(f)\exp(j2\pi ft)df$$

$$= \int_{-W}^{W} \frac{1}{2W} \sum_{n=-\infty}^{\infty} s\left(\frac{n}{2W}\right)\exp\left(\frac{-j\pi nf}{W}\right)\exp(j2\pi ft)df.$$

Interchanging the order of $\sum$ and $\int$,

$$\begin{aligned} s(t) &= \sum_{n=-\infty}^{\infty} s\left(\frac{n}{2W}\right) \frac{1}{2W} \int_{-W}^{W} \exp\left(j2\pi f\left(t - \frac{n}{2W}\right)\right) df \\ &= \sum_{n=-\infty}^{\infty} s\left(\frac{n}{2W}\right) \frac{\sin(2\pi Wt - n\pi)}{2\pi Wt - n\pi} \\ &= \sum_{n=-\infty}^{\infty} s\left(\frac{n}{2W}\right) \text{sinc}\,(2Wt - n) \qquad -\infty < t < \infty. \end{aligned} \tag{3.7}$$

Equation (3.7) provides the means of reconstructing $s(t)$ from the sequence of samples $s(n/2W)$ using sinc$(2Wt)$ as the interpolation function. Each sample is multiplied by a delayed version of the interpolation function and all the resultant waveforms are added to obtain $s(t)$. Now, we are in a position to state 'Sampling Theorem' for strictly band-limited signals of finite energy as

- a band-limited signal of finite energy which has no frequency components higher than W Hz, is completely described by specifying the values of the signal at instants of time separated by $1/2W$ seconds, and
- a band-limited signal of finite energy, which has no frequency components higher than W Hz, may be completely recovered from a knowledge of its samples taken at the rate of $2W$ samples per second.

Both the parts are to be mentioned in the Sampling Theorem.

The rate $2W$ samples per second for a signal of bandwidth W Hz is called the 'Nyquist rate' and $1/2W$ is called the 'Nyquist interval'. If the sampling rate is less than $2W$, the signal will be subjected to distortion called 'aliasing' (see Fig. 3.1(f)).

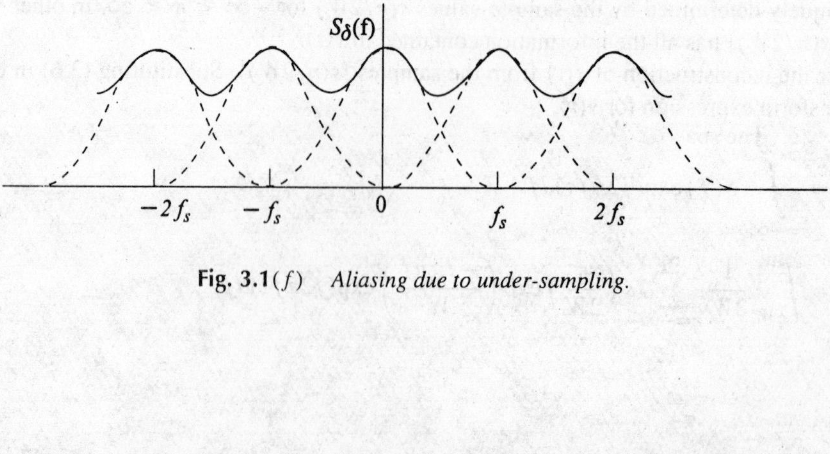

Fig. 3.1(f) *Aliasing due to under-sampling.*

3.3 Quantization Process

3.3.1 What is quantization

A continuous signal such as voice, has a continuous range of amplitudes and therefore its samples have a continuous amplitude range. It is not necessary to transmit the exact amplitudes of the samples. A human sense organ, as a receiver, can detect only finite differences in amplitudes. This implies that if the original continuous signal is approximated by a signal constructed of discrete amplitude levels with sufficiently close spacing, the approximated signal is practically indistinguishable from the original continuous signal.

The process of transforming the sample amplitude $m(nT_s)$ of a message signal $m(t)$ at time $t = nT_s$ into a discrete amplitude $v(nT_s)$ taken from a finite set of possible amplitudes is called amplitude quantization.

3.3.2 Some important terms related to quantization

Memoryless quantization is 'memoryless' and 'instantaneous' if the transformation at time $t = nT_s$ is not affected by earlier or later samples of the message signal. Memoryless and instantaneous quantizations are commonly used.

Decision thresholds: Since we are dealing with memoryless quantization, the time variable nT_s can be dropped and symbol m be used. Signal amplitude m is specified by the index k if it lies in the interval

$$I_k : \{m_k < m \leq m_{k+1}\} \qquad k = 1, 2, \ldots L,$$

where L is the total number of amplitude levels used in the quantizer. The amplitudes m_k, $k = 1, 2, \ldots, L$ are called the 'decision levels' or the decision thresholds (Fig. 3.2(a)).

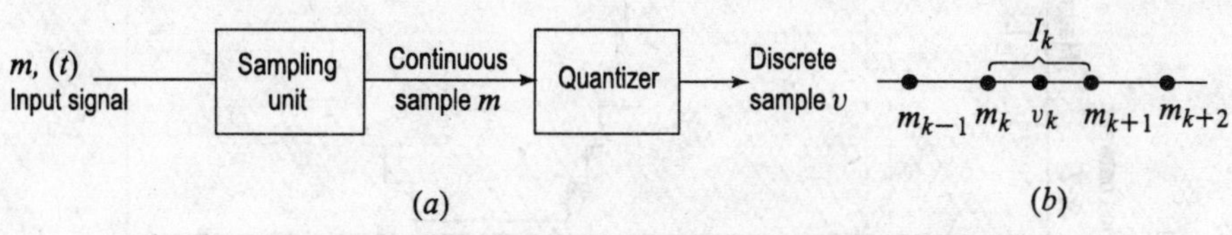

Fig. 3.2 *Sampling and quantization.*

Reconstruction levels: At the quantizer output, the index 'k' is transformed into an amplitude v_k that represents all amplitudes of the interval I_{kj}. The amplitudes $v_k, k = 1, 2, \ldots, L$ are called the representation levels.

Quantum size: The spacing between two adjacent representation levels is called the quantum size or the step size.

Quantizer characteristic: The quantizer output v equals v_k if the input signal sample m belongs to the interval I_k; the mapping $v = g(m)$ is the 'quantization characteristic'. This is a staircase function and can be either mid-tread (Fig. 3.2(c)) or mid-rise (Fig. 3.2(d)).

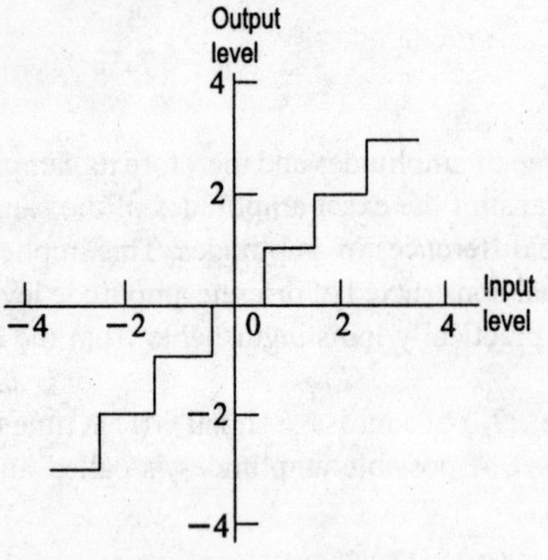

Fig. 3.2(c) *Mid-tread quantization.*

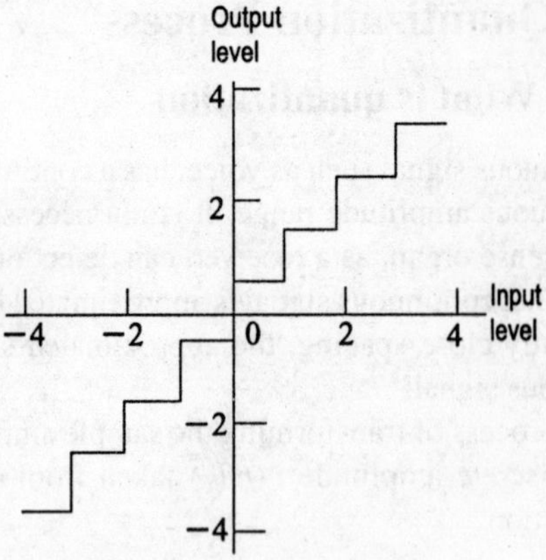

Fig. 3.2(d) *Mid-rise quantization.*

3.3.3 Quantization noise

The difference between the input m and the output v of the quantizer is the error due to approximation in the process of quantization. This error is called the 'quantization noise' (Fig. 3.2(e)). This is a random variable with zero mean.

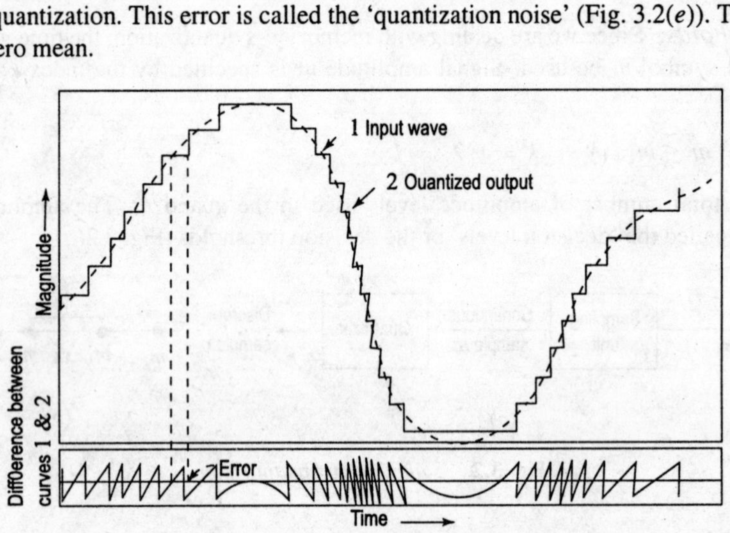

Fig. 3.2(e) *Quantization noise.*

For a uniform mid-rise type quantizer, if L is the total number of representation levels and input m has continuous amplitude in the range $(-m_{\max}, m_{\max})$, the step size is given by

$$\Delta = 2m_{\max}/L. \tag{3.8}$$

The quantization error Q as a random variable is bounded by $-\Delta/2 \leq q \leq \Delta/2$.

Usually, the number of representation level L is large, i.e., Δ is small. Hence, the quantization error Q can be considered as a uniformly distributed random variable. It has interference effects similar to thermal noise. Its probability density function is given by

$$f_Q(q) = Q(q) = \begin{cases} \frac{1}{\Delta} & \frac{-\Delta}{2} < q \le \frac{\Delta}{2} \\ 0 & \text{otherwise.} \end{cases} \tag{3.9}$$

Mean $[Q] = 0$.

$$\text{The variance } \sigma_Q^2 = \int_{-\Delta/2}^{\Delta/2} q^2 f_Q(q)dq = E[Q^2]. \tag{3.10}$$

From (3.9) and (3.10) we have

$$\sigma_Q^2 = \frac{1}{\Delta}\int_{-\Delta/2}^{\Delta/2} q^2 dq = \frac{\Delta^2}{12}. \tag{3.11}$$

If each level is encoded into binary representation using N bits, we have

$$L = 2^N \quad \text{or} \quad N = \log_2^L. \tag{3.12}$$

Substituting (3.12) in (3.11) gives

$$\Delta = \frac{2m_{\max}}{2^N} \tag{3.13}$$

and

$$\sigma_Q^2 = \frac{1}{3}m_{\max}^2 2^{-2N} \tag{3.14}$$

If P is the average power of the message signal, then the output signal-to-(quantization) noise ratio is

$$(\text{SNR})_0 = \frac{P}{\sigma_Q^2} = \left(\frac{3P}{m_{\max}^2}\right)2^{2N}.$$

We have the following important observations from the above discussion:

- SNR increases exponentially by increasing the number of bits per sample
- An increase in N requires a proportionate increase in channel bandwidth
- Use of binary code for the representation of a message signal (as in pulse code modulation) provides a more efficient method than either FM (frequency modulation) or PPM (pulse position modulation) for improved noise performance at the cost of increased channel bandwidth
- FM, PPM are limited by receiver noise whereas binary coded modulation techniques (PCM) are limited by quantization noise
- For a sinusoidal signal of amplitude A_m, $P = (A_m^2/2)$

$$\text{Hence,} \quad \sigma_Q^2 = \frac{1}{3}A_m^2 2^{-2N}$$

$$(\text{SNR})_0 = \frac{A_m^2/2}{\frac{1}{3}A_m^2 2^{-2N}} = \frac{3}{2}(2^{2N})$$

$$10\log(\text{SNR}_{10}) = 1.8 + 6N.$$

3.4 Pulse Code Modulation

3.4.1 Operations in PCM

Having discussed sampling and quantization, we can now turn our attention to PCM (pulse code modulation). In PCM, a message signal is represented by a sequence of coded pulses, which is accomplished by representing the signal in a discrete form in both time and amplitude. The basic operations performed in a PCM system are (Fig. 3.3) given below.

In the transmitter:

- Sampling
- Quantizing
- Encoding

} A/D conversion

In the transmission path:

- Regenerative repetition on every hop (of pre determined length as per Tx media used)

In the receiver

- Regeneration of impaired signals (as in a repeater)
- Decoding
- Reconstruction

Now, the above operations are described in brief.

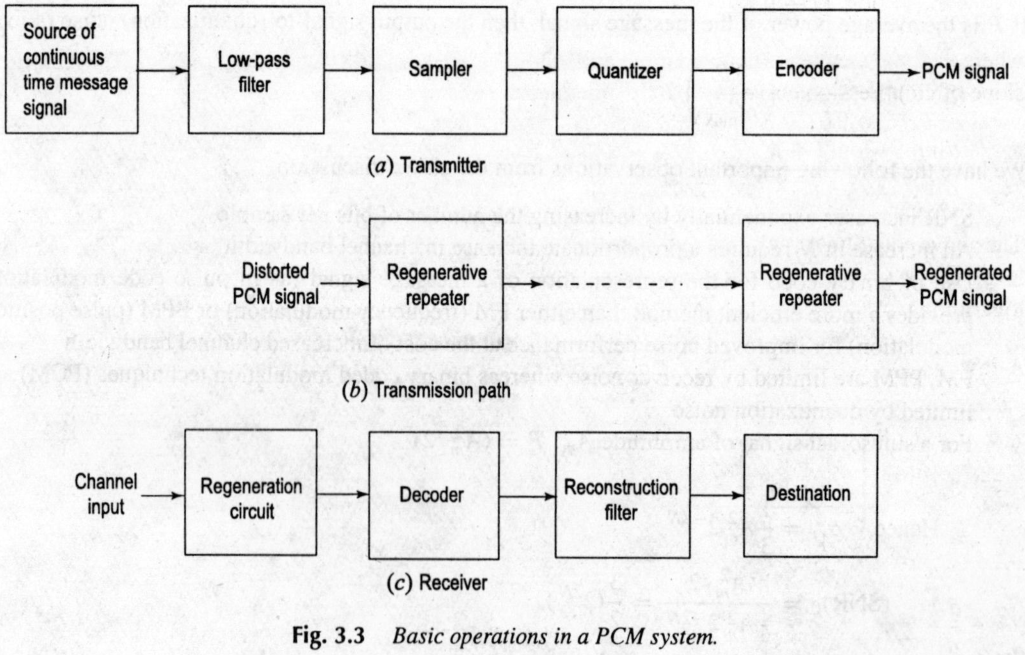

Fig. 3.3 *Basic operations in a PCM system.*

3.4.2 Sampling

The incoming message signal is sampled with a pulse train so as to closely approximate the instantaneous sampling process. (The pulses in the train are not exactly $\delta(t)$ but are rectangular pulses of very low duration compared to the period of sampling.) The sampling rate is greater than twice the highest frequency component of the message signal. A pre-alias filter, which is a low-pass filter, is used at the front end of the sampler to exclude frequencies greater than W before sampling, and the sampling rate is chosen greater than $2W$ (for noise signals $W = 3.4$ kHz and sampling rate $= 8\text{ kHz} > 2W$). By sampling, a limited number of discrete values per second are obtained from the continuously varying message signal.

3.4.3 Quantization—Linear or Non-linear?

By sampling we got a version of the signal discrete in time. With quantization we make it discrete in amplitude as well. In all practical systems, a non-uniform quantizer is used. The reason is simple: The range of voltages covered by voice signals—from the loudest to the feeblest talk—are of the order 1000 to 1. So, more steps (or quantization levels) are needed on the feeble side to protect the message, and this is done by having fewer levels on the louder scale.

Non-uniform quantizer is equivalent to a compressor followed by uniform quantizer. There are two laws of compression generally followed—the μ-law and the A-law.

μ–law This is used in PCM systems of USA, Canada, Japan (Fig. 3.4(a)). μ-law is defined by

$$|\upsilon| = \frac{\log\ (1+\mu|m|)}{\log\ (1+\mu)},$$

where m and υ are the normalized input and output voltages and μ is a positive constant. The reciprocal slope of compression curve $(d|m|/d|\upsilon|)$ defines the quantum steps:

$$\frac{d|m|}{d|\upsilon|} = \frac{\log(1+\mu)}{\mu}(1+\mu|m|).$$

The above equation reveals that the μ-law is neither strictly linear nor strictly logarithmic. For low input levels ($\mu|m| \ll 1$), it is approximately linear, and for high input levels ($\mu|m| \gg 1$), it is approximately logarithmic.

A-law This is used in PCM systems in Europe and India. A-law is defined by (see Fig. 3.4(b))

$$|\upsilon| = \begin{cases} \dfrac{A|m|}{1+\log A} & 0 \le |m| \le \dfrac{1}{A} \\ \dfrac{1+\log A|m|}{1+\log A} & \dfrac{1}{A} \le |m| \le 1, \end{cases}$$

$$\frac{d|m|}{d|\upsilon|} = \begin{cases} \dfrac{1+\log A}{A} & 0 \le |m| \le \frac{1}{A} \\ (1+\log A)|m| & \dfrac{1}{A} \le |m| \le 1. \end{cases}$$

The quantum steps at the lower end are diminished by a factor $A/(1+\log A)$ (25 dB approx in practice) to preserve the small signals. Practical values of A or μ are in the vicinity of 100.

In the receiver, we need to use a complimentary function of compression, i.e., expansion. Ideally, the compression and expansion laws should be exactly inverse so that except for the effect of quantization, the expander output is equal to the compressor input.

The combination of a compressor and an expander is called a 'compander'.

In actual PCM systems the companding circuitry does not produce an exact replica of the non-linear compression curves (Fig. 3.4(a) and Fig. 3.4(b)). It provides a piecewise linear approximation of the desired curve. By using a large-enough number of linear segments, the approximation can approach the true compression curve very closely.

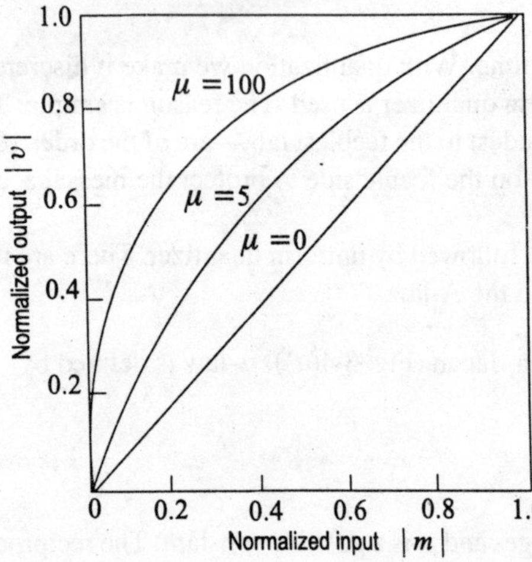

Fig. 3.4(a) *μ–Law.*

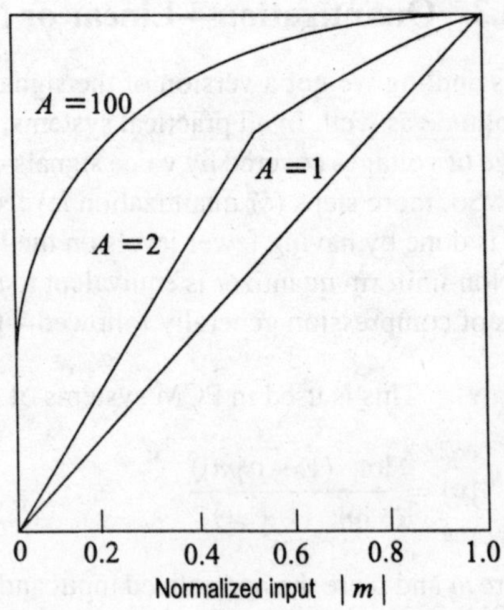

Fig. 3.4(b) *A–Law.*

3.4.4 Encoding

Any plan for representing each of the discrete set of amplitude values as a particular arrangement of the discrete set of events is called a code, and the process of translating the discrete set of sample values to the agreed-upon arrangement (code) is called 'encoding'.

The use of binary code in which each symbol has one of the two distinct values has the following advantages:

- Binary symbol withstands a relatively high level of noise
- It is easy to regenerate binary symbols

Let there be N bits in a code word. Then $2N$ distinct code words are possible. Using an N-bit code, we can therefore represent 2^N distinct discrete amplitude levels.

3.4.5 Representation of binary symbols

There are several methods for the electrical representation of binary symbols 1 and 0. Some of them are described here (Fig. 3.5).

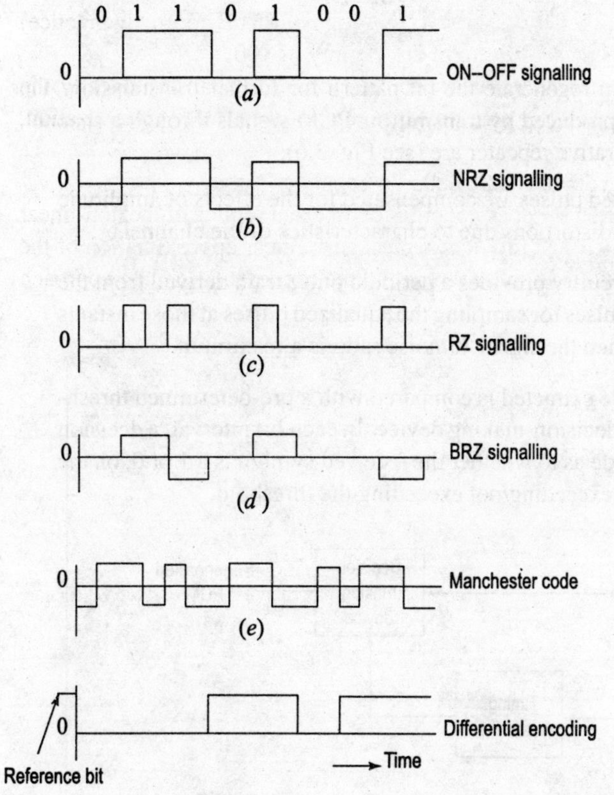

Fig. 3.5 *Different methods of representing binary signals.*

ON–OFF signalling	:	Symbol 1 represented by transmission of a pulse (of fixed amplitude) for the duration of the symbol.
NRZ (non-return to zero) signalling	:	Symbols 1 and 0 are represented by pulses of equal positive and negative amplitudes.
RZ (return to zero) signalling	:	Symbol 1 is represented by a positive rectangular pulse of half-symbol width and symbol 0 is represented by transmitting no pulse.
BRZ (bipolar return to zero) signalling	:	Positive and negative pulses of equal amplitude are used alternately for symbol 1 and no pulse for symbol 0. The power spectrum has no dc component and relatively insignificant low-frequency components.
Split phase (Manchester code)	:	Symbol 1 is represented by a positive pulse followed by a negative pulse (both pulses have equal amplitude and half-symbol width). For symbol 0, the polarities of these two pulses are reversed.
Differential encoding	:	A transition is used to designate symbol 0 and no transition designates symbol 1.

3.4.6 Regeneration

The repeaters on PCM transmission system regenerate the bit pattern for further transmission. This controls the effects of distortion and noise produced by transmitting PCM signals through a channel.

The operations performed at any regenerative repeater are (see Fig. 3.6):

Equalization	:	The received pulses are compensated for the effects of amplitude and phase distortions due to characteristics of the channel.
Timing	:	Timing circuitry provides a periodic pulse train derived from the received pulses for sampling the equalized pulses at those instants of time when the signal-to-noise ratio is a maximum.
Decision making	:	The sample extracted is compared with a pre-determined threshold in the decision-making device. In each bit interval, a decision is then made as to whether the received symbol is a 1 or 0, on the basis of it exceeding/not exceeding the threshold.

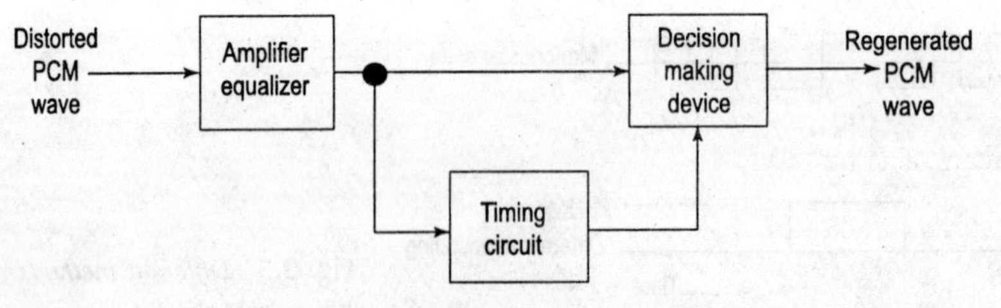

Fig. 3.6 *Block diagram of a regenerative repeater.*

Ideally, except for the delay, the regenerated signal is exactly the same as the signal originally transmitted. But in all practical systems the regenerated signal departs from the original signal for two reasons:

- Presence of channel noise and interference causes the repeater to make wrong decisions occasionally, thereby 'bit errors' come up in the regenerated signal.
- 'Jitter' is introduced into the regenerated pulse position due to deviation of spacing between received pulses from its assigned value. (A detailed discussion on jitter appears later in chapters 13–15.)

3.4.7 Operations at the receiver

The receiver also regenerates the received pulses finally and then groups them into code words/frames which are decoded into a quantized PAM signal. The decoder output is passed through a low-pass reconstruction filter whose cut-off frequency is equal to the message bandwidth W. If the jitter accumulation and bit errors are under control, the received signal has no noise except the initial distortion introduced by the quantization process.

3.5 Multiplexing in PCM Systems

3.5.1 The TDM way

In applications using PCM, it is natural to multiplex different messages by time division. As the number of messages (channels) increase, the time interval allotted to each one gets reduced, i.e. the allowable duration of the codeword representing a single sample is reduced. This reduction cannot go beyond a certain stage, depending on the transmission medium properties. On digital MW radio, data rates upto 140 Mbps are supported. On OFC (optical fibre cable), data rates upto 565 Mbps are supported.

In India, digital hierarchies based on 2 Mbps PCM are followed.

A primary digital MUX (multiplexer) operates at 2048 kbps. In this, 30 channels are combined to form an E1 frame, with 2 channels carrying signalling and synchronization information.

Max. frequency of voice channel	:	4 KHz
Sampling rate (Nyquist)	:	8 KHz
No. of bits per sample	:	8

$\therefore$ per channel, the data rate is 64 kbps.

For 32 channels the data rate is 2048 kpbs or 2 Mbps (1 k in binary is $2^{10} = 1025$, 1 M in binary = 1000 k, hence 2 Mbps = 2048 kpbs).

Four 2 Mbps tributaries are combined to form an 8 Mbps stream and four such 8 Mbps streams form a 34 Mbps stream. The frame structures for 2 Mb, 8 Mb and 34 Mb streams are given in Chapter 7.

Exercises

1. Find signal-to-noise (quantization noise) ratio in dB in a PCM system when the number of levels of representation is (*a*) 32, (*b*) 64, (*c*) 128 (*d*) 256.
2. The signal $S(t) = 8 \sin 2t$ volts is transmitted using an 8-bit binary PCM system. The quantizer is of mid-rise type with step size of 0.5 V. Sketch the resultant PCM wave for one complete cycle of input.
3. Let there be $(n-1)$ regenerative repeaters, i.e., a total of n sequential decisions are made on a binary PCM wave. If p_1 is the probability of erroneous regeneration of any binary symbol by any repeater, show that the probability of the symbol in error at the final stage (p_n) is given by $p_n = 1/2[1 - (1-2p_1)^n)$.

4

Hierarchy of Digital Transmission Systems

4.1 Introduction

In digital transmission, the incoming data stream is modulated onto a carrier with fixed frequency limits imposed by a bandpass channel. The channel can be a microwave radio link, a satellite channel or an optical fibre. In any case, the modulation process involves switching (keying) the amplitude, frequency or phase of a sinusoidal carrier in some fashion in accordance with the incoming data.

The issue of concern is the optimum design of the receiver vis-á-vis the modulation and transmission schemes so as to minimize the probability of error in the presence of noise. Two different cases are considered in the study—coherent receivers and non-coherent receivers. In a coherent receiver, the receiver is phase-locked to the transmitter, i.e., there is phase synchronization between the local oscillator used in the receiver for demodulation and the oscillator providing carriers in the transmitter for modulation. In a non-coherent receiver there is no such phase lock.

4.2 Geometric Representation of Transmitted Signals

To study the effects of additive white Gaussian noise (AWGN) on the transmitted signals, it is convenient to have a geometric model of the transmitted signals. With this model and using vector analysis, decision regions regarding the received signal and the correctness of the decision in the presence of AWGN (or probability of error otherwise) can be easily comprehended.

In a geometric model we have orthogonal axes, and a point on a plane in space is represented as the sum of vectors (of suitable magnitude) along these axes. Similarly, for geometric representation of transmitted signals, we have orthogonal functions. These are called orthonormal basis functions.

Gram–Schmidt orthogonalization procedure helps us in the representation of any set of M energy signals $\{S(t)\}$ as linear combinations of N orthonormal basis functions where $N \leq M$.

4.3 Gram–Schmidt Orthogonalization Procedure

Let the incoming message of a digital transmission system have M states m_i, for $i = 1, 2, \ldots, M$, each of duration T. This is converted to modulated wave $S_i(t)$. The set $\{S_i(t)\}$ is a set of M energy signals. This set $[S_1(t)S_2(t)\ldots S_M(t)]$ is to be represented as a linear combination of N orthonormal functions, where $N \leq M$.

$$S_i(t) = \sum_{j=1}^{N} S_{ij}\Phi_j(t) \begin{cases} 0 \le t \le T \\ i = 1, 2, \dots, M \end{cases}$$

where the coefficients S_{ij} are defined by

$$S_i(t) = \int_0^T S_i(t)\Phi_j(t)dt \begin{cases} i = 1, 2, \dots, M \\ j = 1, 2, \dots, N. \end{cases}$$

$\Phi_1(t), \Phi_2(t) \dots \Phi_N(t)$ are real-valued basis functions called orthonormal basis functions. By 'Orthonormal', we mean

$$\int_0^T \Phi_i(t)\Phi_j(t)dt = \begin{cases} 1 & \text{if } i = j \\ 0 & \text{if } i \neq j. \end{cases}$$

(Remember dot-product of two orthogonal vectors is 0 and that of parallel vectors of unit magnitude is 1.)

We have the following observations:

- The integral = 1 for $i = j$ implies that each basis function is normalized to have unit energy.
- The integral = 0 for $i \neq j$ implies that the basis functions are orthogonal w.r.t each other over the interval $0 \le t \le T$.
- The coefficient S_{ij} is the jth element of the N-dimensional vector S_i.
- The transmission system with input signal having M states (i.e. m_i for $i = 1, 2, \dots, M$) and the modulated energy signals $S_i(t)$ can be modelled as shown in Fig. 4.1
- The scheme for generating $S_i(t)$ using orthonormal basis functions and also the coefficients $\{S_i\}$ can be modelled, as shown in Fig. 4.1(a) and 4.1(b) respectively.
- It can be seen from Fig. 4.1(b) that generation of $\{S_i\}$, i.e. $S_{i1}, S_{i2}, \dots, S_{iN}$ is through a bank of N-product integrators or correlators with a common input and relevant basis functions.
- The concept of correlator or product integrator is vital to the design of a receiver since it is the first stage of detection in the receiver. At the receiver, the received signal is first subjected to product integration with the basis functions in order to find out the components (the constituents of the N-dimensional signal space). An important assumption made here is that the basis functions used in the product integrator of the receiver are in phase (coherent) with the basis functions used at the transmitting end. Hence the reception is termed 'coherent reception'.

We can draw the analogy from vector algebra. If $A\mathbf{x} + B\mathbf{y} + C\mathbf{z}$ is the given vector,

$$(A\mathbf{x} + B\mathbf{y} + C\mathbf{z}) \cdot \mathbf{x} = A \text{ (The magnitude or content of } x \text{ component)},$$
$$(A\mathbf{x} + B\mathbf{y} + C\mathbf{z}) \cdot \mathbf{y} = B \text{ (The magnitude or content of } y \text{ component)},$$
$$(A\mathbf{x} + B\mathbf{y} + C\mathbf{z}) \cdot \mathbf{z} = C \text{ (The magnitude or content of } z \text{ component)},$$

where $\mathbf{x}, \mathbf{y}$ and $\mathbf{z}$ are unit vectors along the x, y and z direction respectively, of a rectangular coordinate system.

Compare the orthonormal basis function series with the conventional Fourier series:

a) Orthonormal basis functions need not be sinusoidal functions.

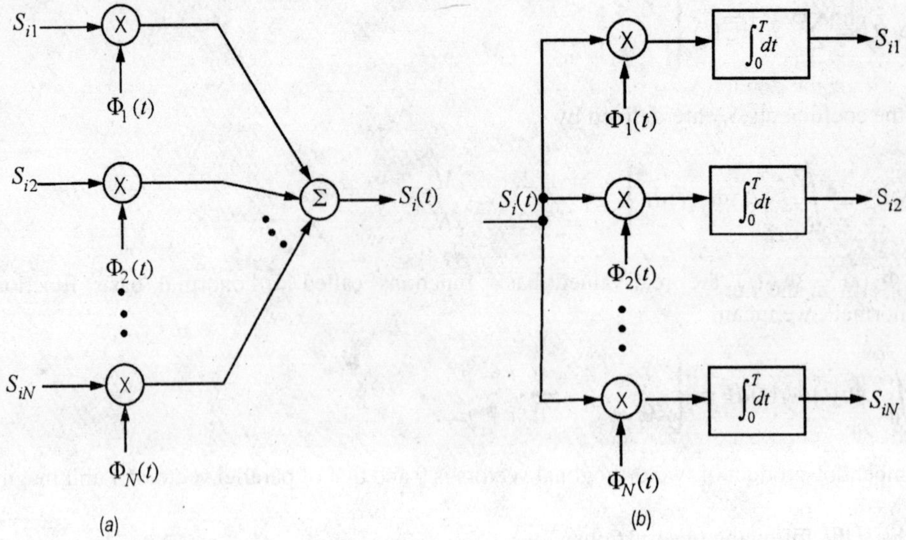

Fig. 4.1(*a*) *Scheme for generating the signal* $S_i(t)$, (*b*) *Scheme for generating the set of coefficients* $\{S_{iN}\}$.

b) Expansion in terms of Fourier series is infinite–we take the first few harmonics for approximation in all practical analysis. But in the case of orthonormal basis functions, there is no such approximation.

Expansion in terms of orthonormal basis functions has been defined as per the Gram–Schmidt orthogonalization procedure. Unless we prove that $\Phi_1, \Phi_2, \ldots, \Phi_n$ are orthonormal, the discussion is not complete.

The proof can be arrived at by taking two basis functions at a time.

Step 1: Define $\Phi_1(t) = \dfrac{S_1(t)}{\sqrt{E_1}}$, (*a*)

where E_1 is energy of the signal $S_1(t)$.

$$S_1(t) = \sqrt{E_1}\,\Phi_1(t) = S_{11}\,\Phi_1(t), \tag{b}$$

where $S_{11} = \sqrt{E_1}$ and $\Phi_1(t)$ has unit energy.

Step 2: Using $S_2(t)$ and by definition,

$$S_{21} = \int_0^T S_2(t)\Phi_1(t)dt. \tag{c}$$

Step 3: Introduce a new function $g_2(t) = S_2(t) - S_{21}\,\Phi_1(t)$. (*d*)

$$\int_0^T g_2(t)\Phi_1(t) = \int_0^T S_2(t)\Phi_1(t) - S_{21}\int_0^T [\Phi_1(t)]^2 dt$$
$$= S_{21} - S_{21} = 0,$$

i.e., $g_2(t)$ and $\Phi_1(t)$ are orthogonal. (*e*)

Step 4: Define $\Phi_2(t) = \dfrac{g_2(t)}{\sqrt{\int_0^T g_2^2(t)dt}}$. (f)

Since the denominator is energy of $g_2(t)$, $\Phi_2(t)$ has unit energy. (g)

Step 5: Substitute (d) in (f):

$$\Phi_2(t) = \frac{S_2(t) - S_{21}\Phi_1(t)}{\sqrt{E_2 - S_{21}^2}}. \quad (h)$$

The first term in the numerator is S_{21} (from (c)). The second term in the numerator = S_{21} (since $\int_0^T \Phi_1^2(t)dt = 1$).

$$\therefore \quad \int_0^T \Phi_1(t)\Phi_2(t)dt = 0. \quad (i)$$

We have the results

$\int_0^T \Phi_1^2(t)dt = 1$ (from (a))

$\int_0^T \Phi_2^2(t)dt = 1$ (from (g))

$\int_0^T \Phi_1(t)\Phi_2(t)dt = 0$ (from (i)).

$\therefore$ $\Phi_1(t)$, $\Phi_2(t)$ are orthonormal.

Similarly, taking other pairs, we can prove that $\Phi_1(t), \Phi_2(t) \ldots \Phi_n(t)$ are orthonormal.

The dimensions N and M depend on one of the two possibilities:

- If $S_1(t), S_2(t) \ldots S_M(t)$ are linearly independent, $N = M$.
- If $S_1(t), S_2(t) \ldots S_M(t)$ are not linearly independent, $N < M$ and $g_i(t) = 0$ for $i > N$.

4.4 Response to Noisy Input at Receiver

$S_i(t)$ is expressed in terms of $\Phi_1(t), \Phi_2(t) \ldots \Phi_N(t)$, or rather, formed in terms of them by product integrators by associating the coefficients $S_{i1}, S_{i2}, \ldots, S_{iN}$ with the basis functions $\Phi_1(t), \Phi_2(t), \ldots \Phi_N(t)$, as already seen. At the receiver, if we derive $S_{i1}, S_{i2}, \ldots, S_{iN}$ from the received signal, we get back the transmitted message. To achive this, we multiply the received signal with $\Phi_1(t), \Phi_2(t) \ldots \Phi_N(t)$ and integrate, thus arriving at the contents of $\Phi_1(t), \Phi_2(t) \ldots \Phi_N(t)$ in the received signal. This process is called 'correlation reception' (recall the analogy with dot product of vector algebra presented in the previous section).

But the transmitted signal is corrupted by AWGN (additive white Gaussian noise), so the received signal is

$$X(t) = S_i(t) + W_i(t) \begin{cases} 0 \le t \le T \\ i = 1, 2, \ldots, M. \end{cases}$$

Output of the jth correlator is

$$X_j = \int_0^T X(t)\Phi_j(t)dt$$
$$= S_{ij} + W_j \qquad j = 1, 2, \ldots, N,$$

where $S_{ij} = \int_0^T S_i(t)\Phi_j(t)dt$.

S_{ij} is the deterministic term which would have come if there was no noise.

$$W_j = \int_0^T W(t)\Phi_j(t)dt.$$

This is a random term since $W(t)$ is the AWGN.

When we sample $X(t)$, the result is a random process, and can be expressed as

$$X^1(t) = X(t) - \sum_{i=1}^{N} X_j\Phi_j(t)$$
$$= S_i(t) + W(t) - \sum_{j=1}^{N}(S_{ij} + W_j)\Phi_j(t)$$
$$= W(t) - \sum_{j=1}^{1} W_j\Phi_j(t),$$

i.e., the sample function $X^1(t)$ depends only on the noise.

The transmitted signal (signal vector), the observed signal and the noise vector are represented in the Euclidean space in Fig. 4.2. The received signal point is randomly placed about the message point, i.e., it lies anywhere inside the Gaussian-distributed noise 'cloud'.

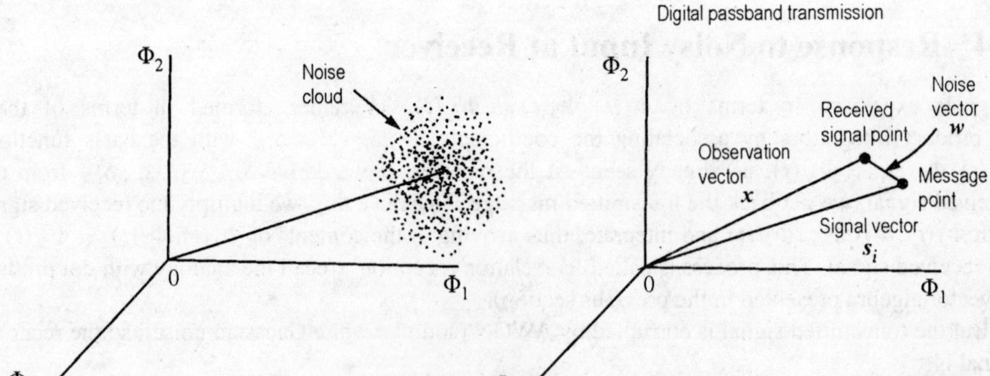

Fig. 4.2 *Effect of noise disturbtion on the location of received signal point.*

4.5 Maximum Likelihood Decision Rule and Decision Zones

Suppose we decide that the message sent is m_i when x is received. Then the probability of error is

$$P_e(m_i, x) = P(m_i \text{ not sent}/x),$$
$$= 1 - P(m_i \text{ sent}/x).$$

On the basis of this equation, the optimum decision rule can be stated as:

$$\text{Set} = m_i \text{ if } \quad P(m_i \quad \text{sent}/x) = P(m_i \quad \text{sent}/x) \text{ for all } k,$$

where $k = 1, 2, \ldots, M$. This is called MAP (maximum *a posteriori* probability) rule.

MAP rule can also be stated as

$$\text{Set} = m_i \text{ if } f_x(x/m_k) \text{ is maximum for } \ k = 1,$$

where f_x is the probability density function of random variable x. Since f_x is always non-negative, $\ln(f_x)$ is a monotonously increasing function. Hence we can state,

$$\text{Set } = m_i \text{ if} \ln[f_x(x/m_k)] \text{ is maximum for } k = i.$$

This is called the maximum likelihood decision rule (ln denotes natural logarithm).

In the signal space representation, the region where this is fulfilled is the decision region in favour of m_i. So, the orthonormal space can be divided into decision regions $Z_i (i = 1, 2, \ldots, M)$. For example, for $m = 4$, $N = 2$, the decision regions are illustrated in Fig. 4.3.

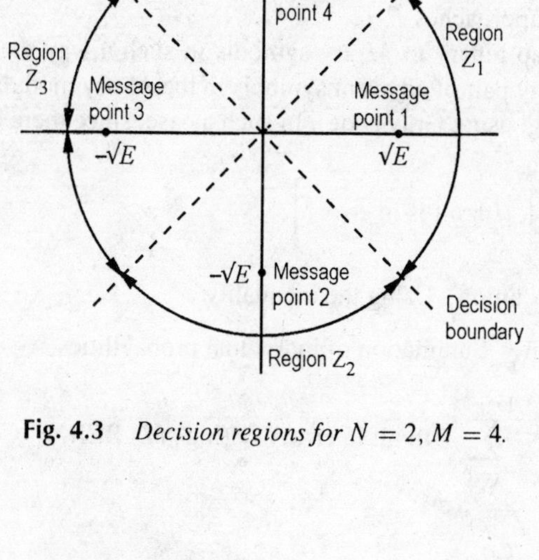

Fig. 4.3 *Decision regions for N = 2, M = 4.*

4.5.1 Probability of error

The average probability of symbol error is given by

$$\begin{aligned} P_e &= \frac{1}{M}\sum_{i=1}^{M} P(X \text{ does not lie in } Z_i/m_i \text{ sent}) \\ &= 1 - \frac{1}{M}\sum_{i=1}^{M} P(X \text{ lies in } Z_i/m_i \text{ sent}) \\ &= 1 - \frac{1}{M}\sum_{i=1}^{M}\int_{Z_i} f_X(x/m_i)dx. \end{aligned}$$

4.5.2 Bit error versus symbol error

Bit error probability and symbol error probability are the figures of merit used in the noise performance of a digital passband transmission system. When messages such as alphanumeric symbols of length $m = \log_2 M$ (i.e., $M = 2^m$) are transmitted, the figure of merit to be considered is the 'probability of symbol error'. When binary data (for example digital computer data) is transmitted, the figure of merit to be considered is the 'probability of bit error' or 'bit error rate (BER)'. Probability of symbol error (P_e) and BER can be related in the following practical situations:

(*a*) $M = 2^k$, where k is an integer. Further, all symbol errors are equally likely. Then, the probability of occurrence of any symbol error $= P_e/(M-1) = P_2/(2^k-1)$. The probability that the ith bit in a symbol is in error is given by

$$\text{BER} = \frac{(2^{k-1}-1)P_e}{2^k-1}.$$

Since there are two cases of symbol error in which this particular ith bit has been changed,

$$\text{BER} = \left[\frac{2^{k-1}}{2^k-1}\right]P_e = \left[\frac{M/2}{M-1}\right]P_e.$$

For large M, BER approaches $P_e/2$.

(*b*) It is possible to map binary to M-ary symbols in such a way that the two binary M-tuples corresponding to any pair of adjacent symbols in the M-ary modulation scheme differ only in one bit position (e.g. using Gray code). In such a case, since there are log M bits per symbol,

$$P_e = P\left[\bigcup_{i=0}^{\log_2 M} \{i\text{th bit is in error}\}\right],$$

where $\cup$ stands for 'union'. Using the inequality

Probability of union $\leq$ Summation of individual probabilities,

$$\text{we get}\quad P_e \leq \sum_{i=1}^{\log_2 M} (i\text{th bit in error}) = (\log_2 M)\,\text{BER}.$$

Also, $P_e \geq P(i\text{th bit in error}) = \text{BER}$. So, it follows that

$$(P_e/\log_2 M) \leq \text{BER} \leq P_e,$$

i.e., BER is bounded by $P_e/\log_2 M$ on the lower side and P_e on the upper side.

4.5.3 Union bound on the probability of error

$$P(A \bigcup B) = P(A) + P(B). \qquad \text{(From probability theory)}$$

$$\therefore \quad P_e(m_i) = \sum_{\substack{i=1 \\ k \neq 1}}^{M} P(A_{ik}) \qquad \text{for } i = 1, 2, \ldots, M,$$

where A_{ik} is the event that the observation sector X is closer to S_k than to S_i when m_i is sent.

Since the decision between S_i and S_k starts midway between S_i and S_k,

$$P(A_{ik}) = \int_{d_{ik}/2}^{\infty} \frac{1}{\sqrt{\pi N_0}} \exp\left(\frac{-u^2}{N_0}\right) du,$$

where d_{ik} = distance between S_i and S_k, and N_0 is the noise spectral density. From the definition of complementary error function (denoted by $\text{erfc}(n)$, refer Appendix 1),

$$\text{erfc}\,(n) = \frac{2}{\sqrt{\pi}} \int_n^{\infty} \exp\left(-z^2\right) dz.$$

$$\therefore \quad P(A_{ik}) = \frac{1}{2}\,\text{erfc}\left(\frac{d_{ik}}{2\sqrt{N_0}}\right)$$

$$P_e = \frac{1}{M} \sum_{i=1}^{M} P_e(m_i)$$

$$\leq \frac{1}{2M} \sum_{i=1}^{M} \sum_{\substack{k=1 \\ k \neq 1}}^{M} \text{erfc}\left(\frac{d_{ik}}{2\sqrt{N_0}}\right).$$

'erfc' has an upper bound

$$\text{erfc}\left(\frac{d_{ik}^2}{2\sqrt{N_0}}\right) \leq \frac{1}{\sqrt{\pi}} \exp\left(\frac{-d_{ik}^2}{2N_0}\right).$$

(Derivation of this is beyond the scope of this book.)

$$P_e \leq \frac{1}{2\sqrt{\pi}} \sum_{\substack{k=1 \\ k \neq 1}}^{M} \exp\left(\frac{-d_{ik}^2}{2N_0}\right) \quad \text{for all } i.$$

Since transmitted signal energy $\gg N_0$, the exponential term with the smallest d_{ik} is the relevant quantity in the above inequality. Hence we have,

$$P_e \leq \frac{M_{\min}}{2\sqrt{\pi}} \exp\left[-\min\left\{\frac{d_{ik}^2}{2N_0}\right\}_{\substack{i,k \\ i \neq k}}\right].$$

4.6 Correlation Receiver

From the above discussions, we have arrived at the structure of a correlation receiver.

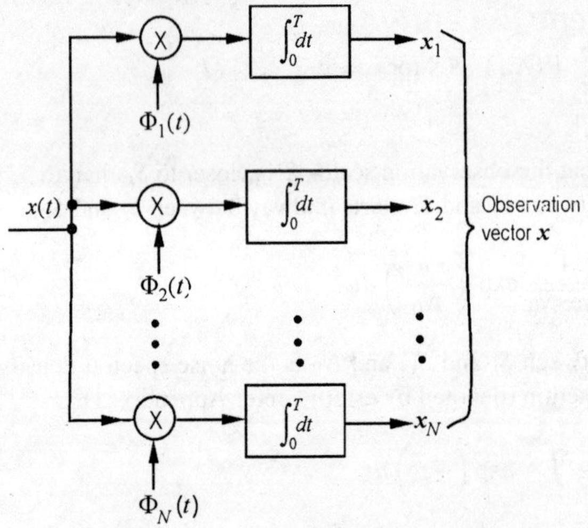

Fig. 4.4(a) *Detector or demodulator.*

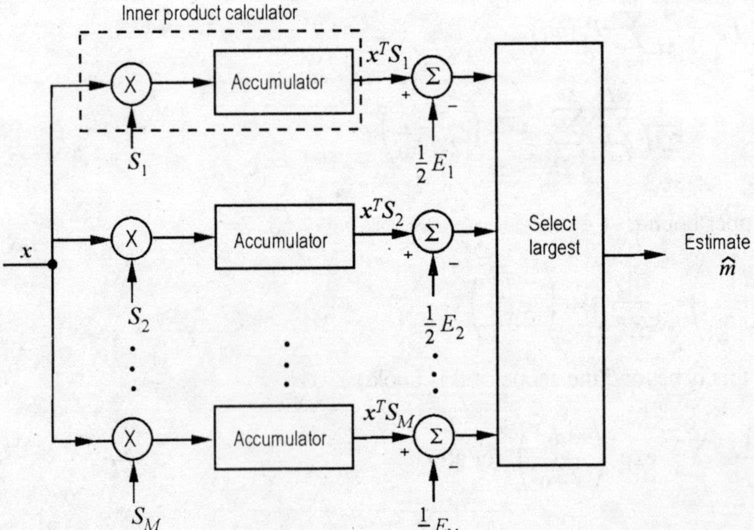

Fig. 4.4(b) *Signal decoder.*

It has two parts:

– The correlator part at the front end with a bank of product integrators which gives the components of the observation vector $\boldsymbol{x}$.
– The maximum likelihood decoder, which selects the largest product among the products of $\boldsymbol{x}$ elements and $\boldsymbol{S}$ elements.

Figure 4.4 illustrates both the above parts, viz., detector and decoder.

4.7 Non-coherent Reception

So far in the discussions, the phase of the basis functions in received signal are assumed to be known. Hence, it was 'coherent reception'. If the phase is unknown, we have to embark on 'non-coherent reception'. A full length discussion on non-coherent reception is beyond the scope of this book. However, three models of non-coherent receivers are presented in Fig. 4.5 namely (a) quadrature

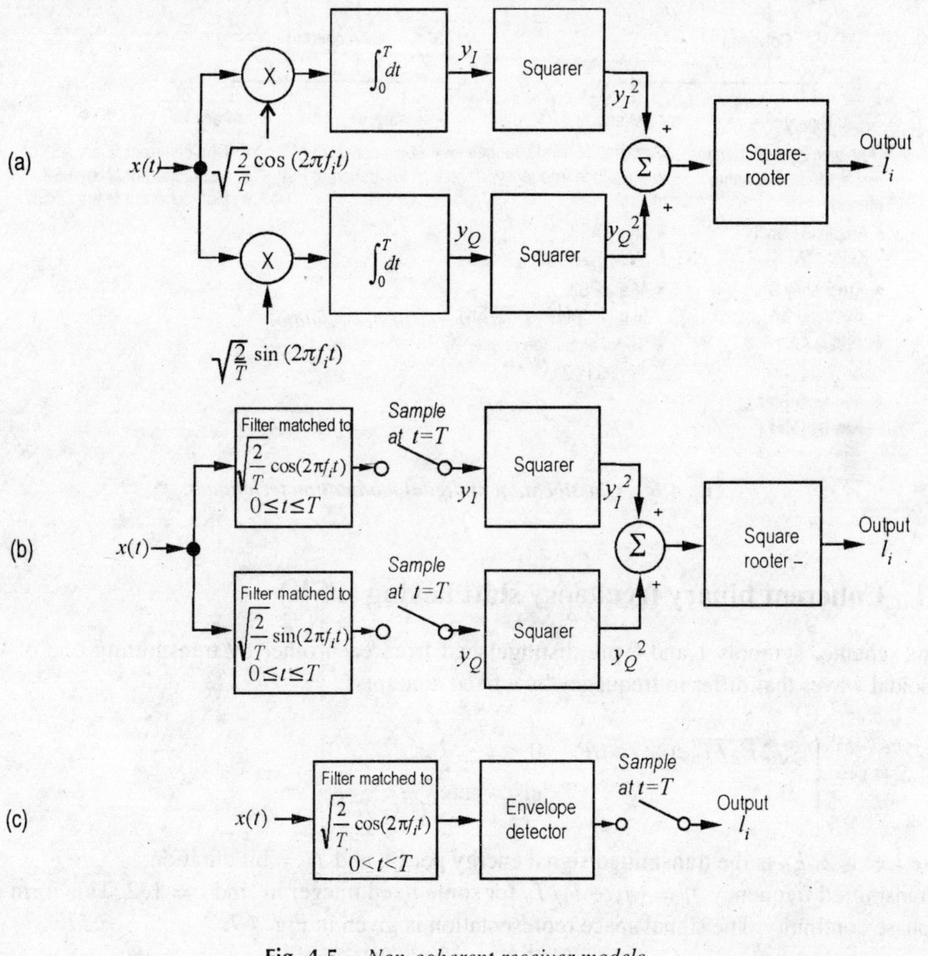

Fig. 4.5 *Non-coherent receiver models.*

receiver using correlators, (b) quadrature receiver using matched filter, and (c) matched filter with envelope ditector.

4.8 Hierarchy of Digital Modulation Techniques

In Sections 4.3 to 4.7, an understanding of optimum receiver for coherent and non-coherent reception was developed. In this section, various modulation techniques for digital transmission will be discussed. With the ideas developed in previous sections, we will be equipped to do noise effect analysis and BER performance analysis of various digital transmission systems.

Digital modulation techniques are classified in Fig. 4.6. Coherent binary FSK, PSK, coherent MSK, coherent QPSK, QAM, DPSK, and M-ary FSK are discussed in this book.

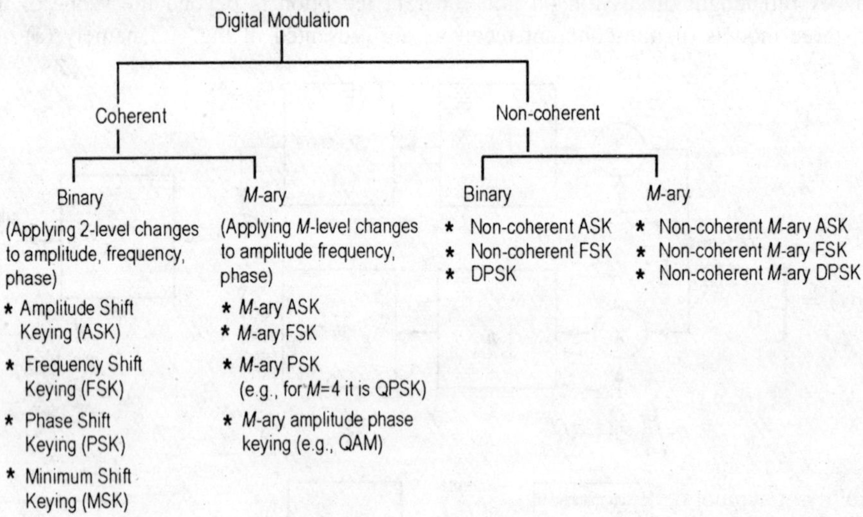

Fig. 4.6 *Classification of digital modulation techniques.*

4.8.1 Coherent binary frequency shift keying (FSK)

In this scheme, symbols 1 and 0 are distinguished from each other by transmitting one of the two sinusoidal waves that differ in frequency by a fixed amount.

$$S_i(t) = \begin{cases} \sqrt{2E_b/T_b}\cos(2\pi f_i t) & 0 \le t \le T_b \\ 0 & \text{elsewhere,} \end{cases}$$

where $i = 1, 2$, E_b is the transmitted signal energy per bit and T_b = bit duration.

Transmitted frequency $f_i = (n_k + i)/T_b$ for some fixed integer n_k and $i = 1, 2$. This form of FSK has phase continuity. The signal space representation is given in Fig. 4.7.

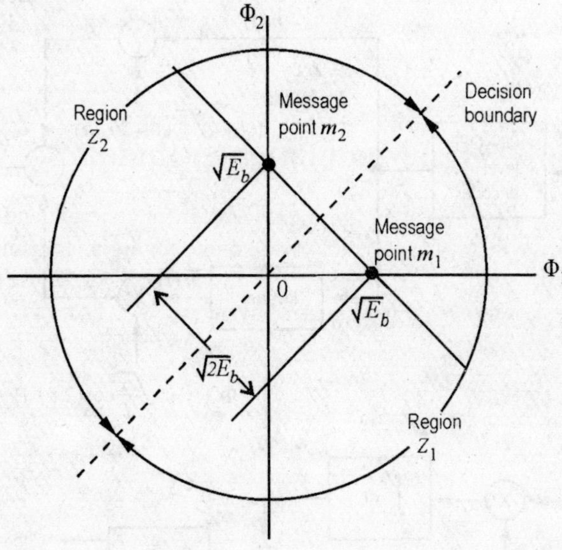

Fig. 4.7 *Signal space representation of coherant binary FSK.*

The orthonormal basis functions are:

$$\Phi_2(t) = \begin{cases} \sqrt{2/T_b}\cos(2\pi f_i t) & 0 \le t \le T_b \\ 0 & \text{elsewhere} \end{cases}$$

$$S_1 = [\Phi_1\Phi_2]\begin{bmatrix} \sqrt{E_b} \\ 0 \end{bmatrix}; \quad S_2 = [\Phi_1\Phi_2]\begin{bmatrix} 0 \\ \sqrt{E_b} \end{bmatrix}.$$

A block diagram for the generation and detection of coherent FSK is given in Fig. 4.8.

Probability of symbol error can be derived as derivation not presented here, can be found in Ref.[1].

$$P_e = \frac{1}{2}\,\text{erfc}\left(\sqrt{\frac{E_b}{2N_0}}\right).$$

($N_0/2$ is power spectral density of white Gaussion noise.)

4.8.2 Coherent binary phase shift keying (PSK)

In this scheme, the signals $S_1(t)$, $S_2(t)$ representing symbols 1 and 0 respectively are

$$S_1(t) = \sqrt{\frac{2E_b}{T_b}}\cos(2\pi f_c t),$$

$$S_2(t) = \sqrt{\frac{2E_b}{T_b}}\cos(2\pi f_c t + \pi) \quad = -\sqrt{\frac{2E_b}{T_b}}\cos(2\pi f_c t),$$

where $0 \le t \le T_b$; E_b is the transmitted energy per bit, and T_b is the bit duration.

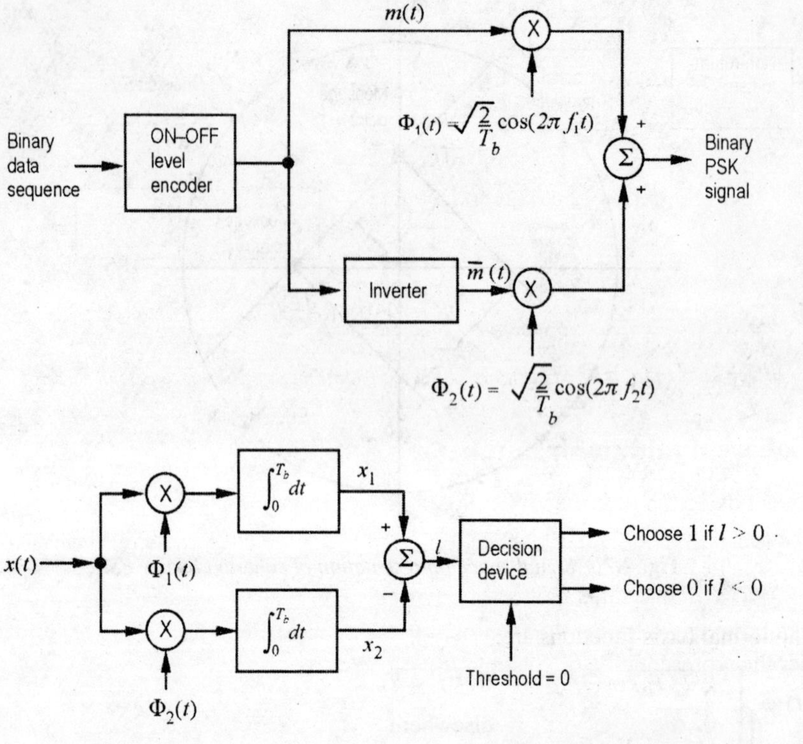

Fig. 4.8 *Block diagram of (a) binary FSK transmitter and (b) coherant binary FSK receiver.*

The orthonormal basis function is only one.

$$\Phi_1(t) = \sqrt{\frac{2}{T_b}} \cos(2\pi f_c t), \quad 0 \le t \le T_b,$$

$$S_1(t) = \sqrt{E_b}\,\Phi_1(t), \quad 0 \le t < T_b,$$

$$S_2(t) = -\sqrt{E_b}\,\Phi_1(t), \quad 0 \le t < T_b.$$

The signal space representation is one-dimensional.

$$-\sqrt{E_b} \qquad \sqrt{E_b}$$

★ ———— ★

The block diagram for generation and detection of coherent binary PSK signals is given in Fig. 4.9.

Probability of error can be derived as

$$P_e = \frac{1}{2}\,\text{erfc}\sqrt{\frac{E_b}{N_0}}.$$

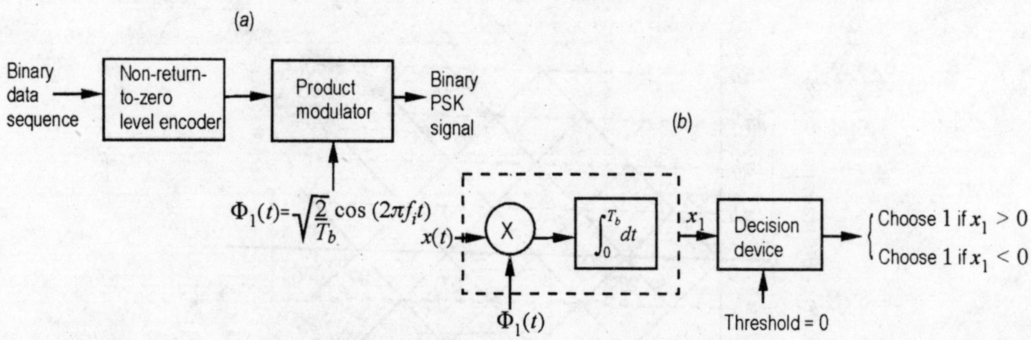

Fig. 4.9 *(a) Binary PSK transmitter, (b) Coherent receiver.*

4.8.3 Coherent minimum shift keying (MSK)

In the coherent detection of binary FSK, the phase information in the received signal was utilized for synchronization of the receiver to the transmitter. By imparting more utility to phase information, noise performance can be improved. It can be observed from the expression of coherent FSK that by sending symbol 1 the phase of the signal increases, and by sending symbol 0 the phase decreases. This property is utilized in framing a different modulation scheme.

Consider the representation of coherent FSK:

$$S(t) = \begin{cases} \sqrt{\dfrac{2E_b}{T_b}}\cos[2\pi f_1 t + \theta(0)] & \text{for symbol } 1 \\ \\ \sqrt{\dfrac{2E_b}{T_b}}\cos[2\pi f_2 t + \theta(0)] & \text{for symbol } 0 \end{cases} \qquad (A)$$

where E_b = transmitted signal energy per bit,

T_b = bit duration, and

$\theta(0)$ = phase at time $t = 0$, depending on the past status of modulation process.

In the conventional FM/PM (frequency modulation/phase modulation) representation form, physically, it means that the nominal carrier frequency f_c is the arithmetic mean of the frequencies f_1 and f_2, and h the deviation rate is the difference between f_1 and f_2 normalized w.r.t. $1/T_b$.

At $T = T_b$

$$\theta(T_b) - \theta(0) = \pi h \text{ for symbol 1,}$$

$$\theta(T_b) - \theta(0) = \pi h \text{ for symbol 0.}$$

i.e., the sending of symbol 1 increases the phase of FSK signal by πh, and the sending of symbol 0 reduces it by an equal amount. Thus, we have mathematically verified the starting assumptions.

The variation of phase $\theta(t)$ with time t follows a path consisting of a sequence of straight lines the slopes of which represent frequency changes. The possible paths starting from $t = 0$ are shown in Fig. 4.10 (called a phase tree).

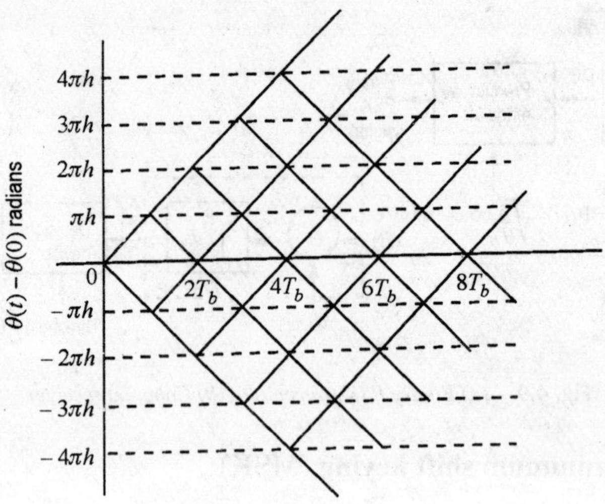

Fig. 4.10 *Phase tree.*

A special situation arises when $h = 1/2$. In this case, phase can take on two values, namely $\pm\pi/2$ at odd multiples of T_b, and only two values 0 and π at even multiples of T_b. The resulting graph called 'phase trellis' for the binary sequence 1101000 is shown in Fig. 4.11. Each path in the trellis from left to right corresponds to a specific binary sequence input. In other words, the sequence of phase states of incoming binary sequence is taken 'modulo 2π', assuming at $t = 0$ the phase $\theta(0)$ is zero.

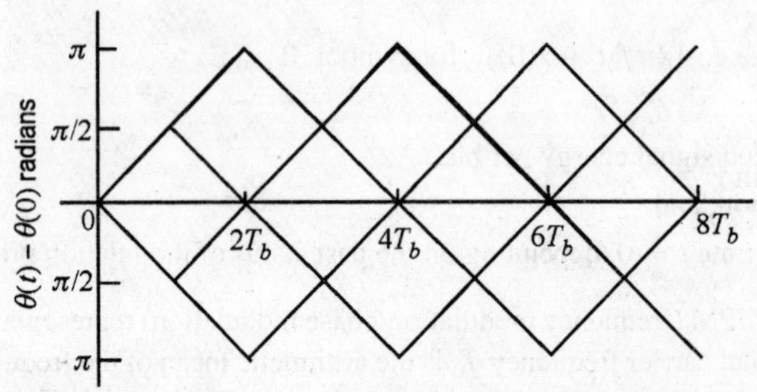

Fig. 4.11 *Phase trellies for the sequence 1101000.*

4.8.4 Why the name MSK?

With $h = 1/2$, the frequency deviation (i.e. normalised frequency difference between f_1 and f_2) equals half the bit rate. This is the minimum frequency spacing that allows two FSK signals representing symbols 1 and 0 to be coherently orthogonal in the sense that they do not interfere with one another in the process of detection.

Hence the name minimum shift keying (MSK).

Representation of MSK in terms of orthonormal basis functions

The expressions for in-phase and quadrature components of MSK signal are

$$S_I(t) = \pm\sqrt{\frac{2E_b}{T_b}}\cos\left(\frac{\pi}{2T_b}\right) \quad -T \le t \le T_b,$$

$$S_Q(t) = \pm\sqrt{\frac{2E_b}{T_b}}\sin\left(\frac{\pi}{2T_b}t\right) \quad 0 \le t \le 2T_b.$$

In $S_I(t)$, the + sign corresponds to $\theta(0) = 0$, and the $-$ sign corresponds to $\theta(0) = \pi$.

In $S_Q(t)$, the + sign corresponds to $\theta(T_b) = \pi/2$, and the $-$ sign corresponds to $\theta(T_b) = -\pi/2$.

Since $\theta(0)$ and $\theta(T_b)$ can each assume one of the two values given above, the following four possibilities arise:

$\theta(0) = 0,\ \ \theta(T_b) = -\pi/2:$ transimission of symbol 1

$\theta(0) = \pi,\ \ \theta(T_b) = -\pi/2:$ transimission of symbol 0

$\theta(0) = 0,\ \ \theta(T_b) = -\pi/2:$ transimission of symbol 1

$\theta(0) = 0,\ \ \theta(T_b) = -\pi/2:$ transimission of symbol 0

This implies that the MSK singal can assume any one of the four possible forms depending on the values of $\theta(0)$ and (T_b).

The orthonormal basis functions $\Phi_1(t)$ and $\Phi_2(t)$ for MSK are a pair of sinusoidally modulated carriers

$$\Phi_1(t) = \sqrt{\frac{2}{T_b}}\cos\left(\frac{\pi}{2T_b}t\right)\cos(2\pi f_c t) \quad 0 \le t \le T_b,$$

$$\Phi_1(t) = \sqrt{\frac{2}{T_b}}\sin\left(\frac{\pi}{2T_b}t\right)\sin(2\pi f_c t) \quad 0 \le t \le T_b,$$

and the signal

$$S(t) = S_1\Phi_1(t) + S_2\Phi_2(t),$$

where

$$S_1 = \int_{-T_b}^{T_b} S(t)\Phi_1(t)dt = \sqrt{E_b}\cos[\Phi(0)] \quad -T_b \le t \le T_b,$$

$$S_2 = \int_{0}^{2T_b} S(t)\Phi_2(t)dt = \sqrt{E_b}\sin[\theta(0)] \quad 0 \le t \le 2T_b.$$

The signal constellation of MSK is 2-dimensional (Fig. 4.12). P_e can be derived as 1/2 erfc $(\sqrt{E_b/N_0})$. The MSK transimitter and reviver block diagrams are shown in Fig. 4.13.

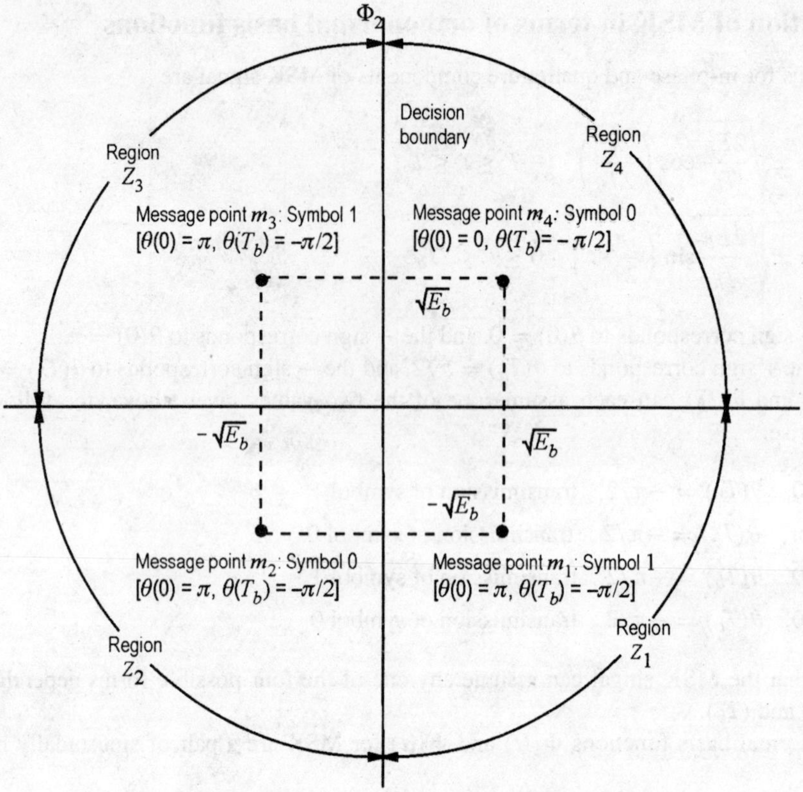

Fig. 4.12 *Signal space diagram for MSK system.*

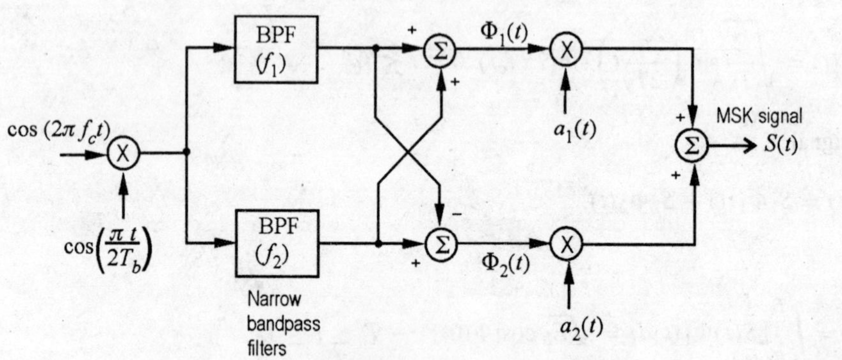

Fig. 4.13(*a*) *Block diagram of MSK transmitter.*

Note: $a_1(t), a_2(t)$ are two binary waves of bit rate $1/2T_b$ extracted from the incoming binary sequence, taking the sequence of phase states modulo 2π.

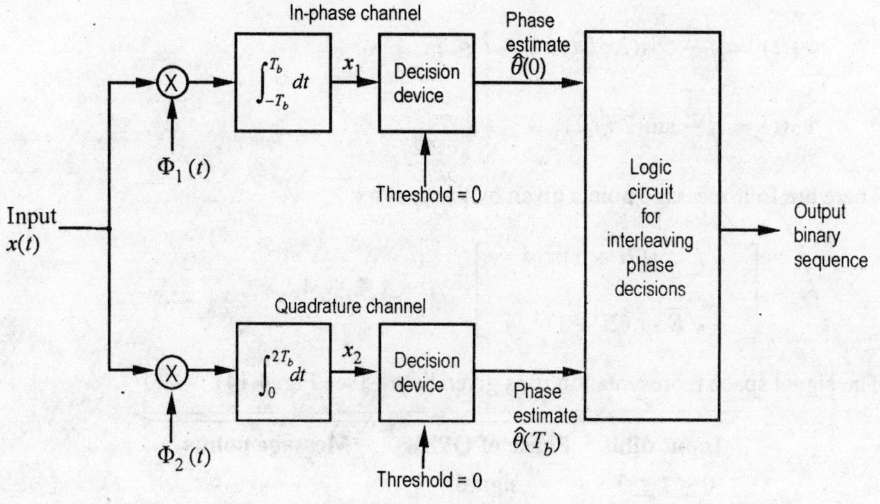

Fig. 4.13(*b*) *Block diagram of MSK receiver.*

4.8.5 Coherent quadrature phase shift keying (QPSK)

QPSK scheme is the coherent M-ary PSK scheme for $M = 4$. Its speciality is low probability of error as well as efficient utilization of bandwidth.

In this scheme, the phase of the carrier takes on one of the four equally spaced values $\pi/4$, $3\pi/4$, $5\pi/4$, $7\pi/4$. The transmitted signal is represented by

$$S_i(t) = \begin{cases} \sqrt{\frac{2E}{T}} \cos[2\pi f_i t + (2i-1)\frac{\pi}{4}] & 0 \le t \le T \\ 0 & \text{elsewhere} \end{cases}$$

where $i = 1, 2, 3, 4,$

E = transmitted signal energy per bit,

and T = symbol duration.

Carrier frequency $f_c = n_k/T$ for some fixed interger n_k. Each possible value of the phase corresponds to a unique pair of bits called 'dibit'. Gray-encoded set of dibits are chosen to correspond to the four phases.

$S_i(t)$ represented above can be expressed as

$$S_i(t) = \sqrt{\frac{2E}{T}} \cos\left[(2i-1)\frac{\pi}{4}\right] \cos(2\pi f_c t) - \sqrt{\frac{2E}{T}} \sin\left[(2i-1)\frac{\pi}{4}\right] \sin(2\pi f_c t),$$

where $i = 1, 2, 3, 4$.

This form of expression reveals that

(i) There are two orthonormal basis function $\Phi_1(t)$, $\Phi_2(t)$ are contained in $S_i(t)$ where

$$\Phi_1(t) = \sqrt{\frac{2}{T}}\cos(2\pi f_c t) \quad 0 \le t \le T,$$

$$\Phi_2(t) = \sqrt{\frac{2}{T}}\sin(2\pi f_c t) \quad 0 \le t \le T.$$

(*ii*) There are four message points given by the vectors

$$S_i = \begin{bmatrix} \sqrt{E}\cos(2i-1)\pi/4 \\ -\sqrt{E}\sin(2i-1)\pi/4 \end{bmatrix} \qquad i = 1, 2, 3, 4.$$

(*iii*) The signal space representation is as given below (see Fig. 4.14).

Input dibit $0 \le t \le T$	**Phase of QPSK signal**	**Message points**
10	$\pi/4$	$(+\sqrt{E/2}, -\sqrt{E/2})$
00	$3\pi/4$	$(-\sqrt{E/2}, -\sqrt{E/2})$
01	$5\pi/4$	$(-\sqrt{E/2}, +\sqrt{E/2})$
11	$7\pi/4$	$(+\sqrt{E/2}, +\sqrt{E/2})$

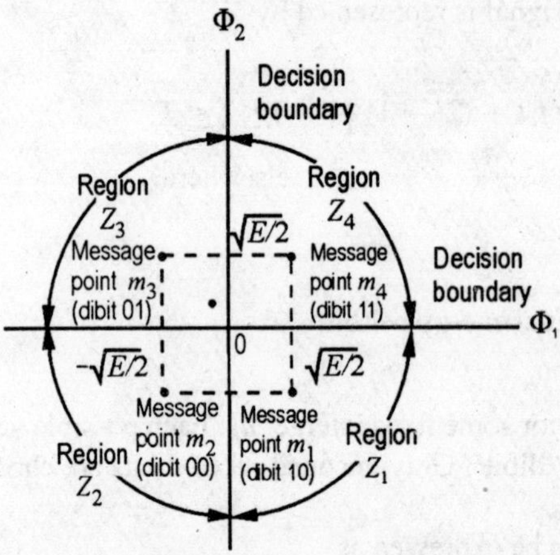

Fig. 4.14 *Signal space diagram of coherant QPSK.*

Probability of error P_e can be derived as

$$P_e = \text{erfc}\sqrt{E_b/N_0}.$$

Because of the use of dibit, BER $= 1/2$ erfc $(\sqrt{E_b/N_0})$. Block diagrams showing generation and detection of QPSK signals are given in Fig. 4.15.

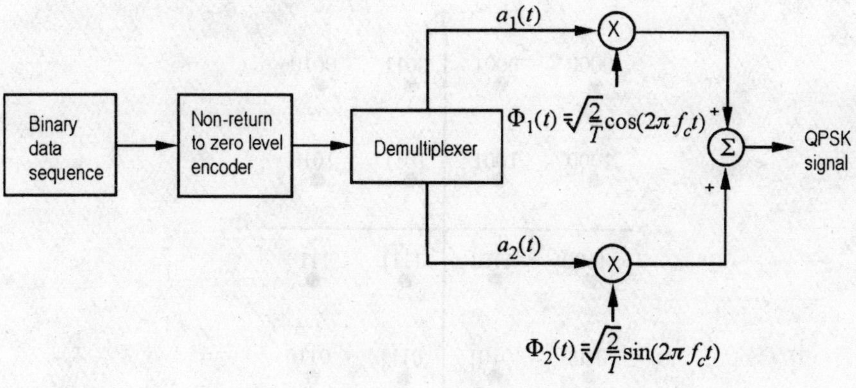

Fig. 4.15(*a*) *Block digram of QPSK transmitter.*

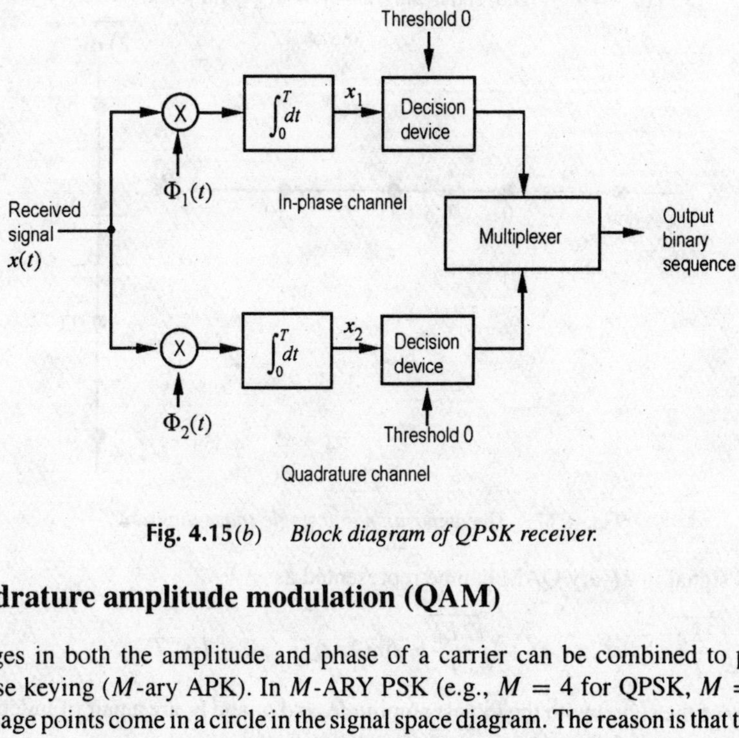

Fig. 4.15(*b*) *Block diagram of QPSK receiver.*

4.8.6 Quadrature amplitude modulation (QAM)

Discrete changes in both the amplitude and phase of a carrier can be combined to produce M-ary amplitude phase keying (M-ary APK). In M-ARY PSK (e.g., $M = 4$ for QPSK, $M = 8$ for OCTA-PSK), the message points come in a circle in the signal space diagram. The reason is that the orthonormal basis functions (in-phase and quadrature components) are interrelated in such a way that the signal envelope is a constant.

If this constraint of constant envelope is removed, i.e. the in-phase and quadrature components are permitted to be independent, a new modulation scheme is available. It is called the M-ray amplitude phase keying (M-APK) or M-ary quadrature amplitude modulation (QAM).

The signal constellation of QAM has square lattice of message points (see Fig. 4.16.) This constellation can further be decomposed into constellation diagram for in-phase and quadrature components (see Fig. 4.17).

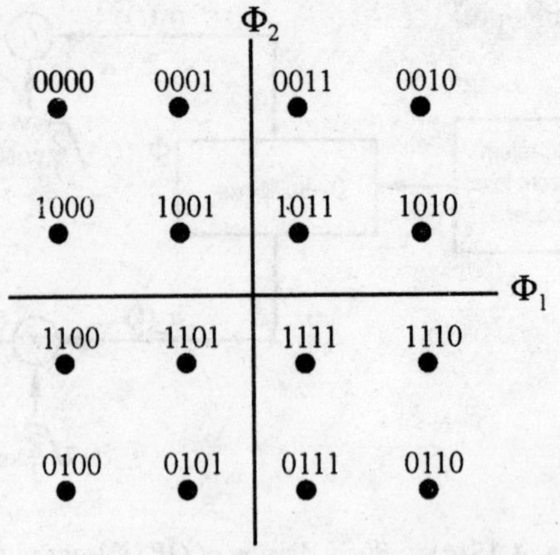

Fig. 4.16 *Signal space diagram of M-ary QAM for M = 16.*

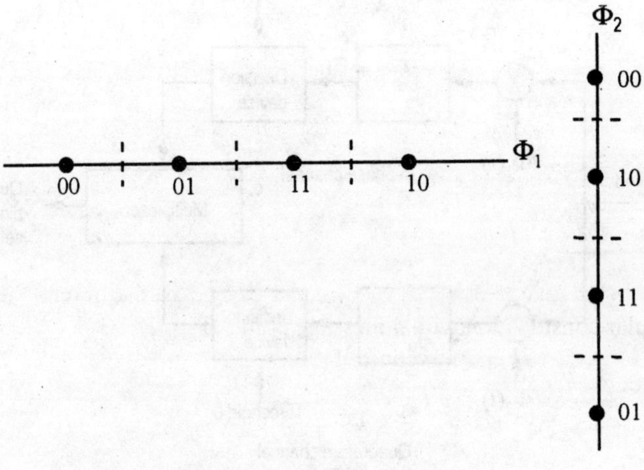

Fig. 4.17 *Decomposition into single space diagram.*

Transmitted signal in M-ary QAM can be represented as

$$S_i(t) = \sqrt{\frac{2E_0}{T}} a_i \cos(2\pi f_c t) + \sqrt{\frac{2E_0}{T}} b_i \sin(2\pi f_c t) \quad 0 \leq t \leq T.$$

E_0 is the energy of the signal with the lowest amplitude, and a_i and b_i are a pair of independent integers chosen as per location of the message point.

$S_i(t)$ consists of two phase-quadrature carriers, each of which is modulated by a set of discrete amplitudes:

$$\Phi_1(t) = \sqrt{\frac{2}{T}} \cos(2\pi f_c t) \quad 0 \leq t \leq T,$$

$$\Phi_2(t) = \sqrt{\frac{2}{T}} \sin(2\pi f_c t) \quad 0 \leq t \leq T.$$

Coordinates of the ith meassage point are $a_i\sqrt{E_0}$ and $b_i\sqrt{E_0}$, where (a_i, b_i) are elements of an $L \times L$ matrix $(L = \sqrt{M})$:

$$\{a_i\ b_i\} = \begin{bmatrix} (-L+1, L-1) & (-L+3, L-1) & \dots & (L-1, L-1) \\ (-L+1, L-3) & (-L+3, L-1) & \dots & (L-1, L-3) \\ \dots & \dots & \dots & \dots \\ (-L+1, -L+1) & (-L+3, -L+1) & \dots & (L-1, -L+1) \end{bmatrix}$$

For 16 QAM, $L = \sqrt{16} = 4$.

$$\{a_i, b_i\} = \begin{bmatrix} (-3, 3 & (-1, 3) & (1, 3) & (3, 3) \\ (-3, 1) & (-1, 1) & (1, 1) & (3, 1) \\ (-3, -1) & (-1, 1) & (1, -1) & (3, -1) \\ (-3, -3) & (-1, -3) & (1, -3) & (3, -3) \end{bmatrix}$$

The probability of error can be evaluated as

$$P_e = 2\left(1 - \frac{1}{\sqrt{M}}\right) \operatorname{erfc}\left(\sqrt{\frac{3E_{\text{av}}}{2(M-1)N_0}}\right),$$

where $E_{\text{av}} = \dfrac{2(M-1)E_0}{3}$.

It may please be noted that M-ary PSK, QPSK and 16 QAM are different. QPSK is M-ary PSK for $M = 4$. It has circular constellation with 4 message points (Fig. 4.18(a)).

M-ary PSK for $M = 16$ has circular constellation with 16 message points (Fig. 4.18(b)).

16 QAM is M-ary APK for $M = 16$ and has square lattice constellation (Fig. 4.18(c)).

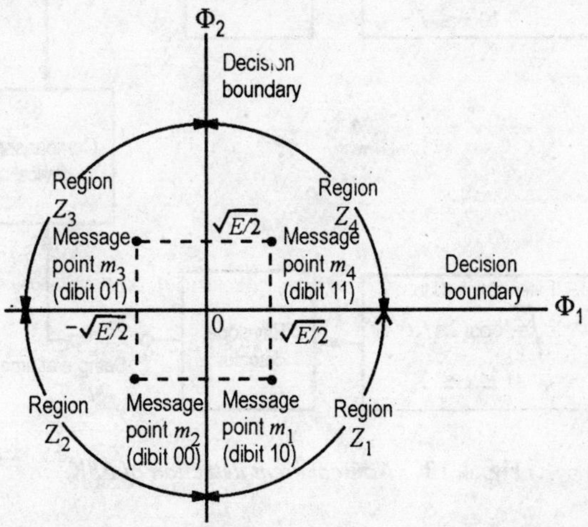

Fig. 4.18(a) *Signal space diagram for QPSK (M-ary PSK with $M = 4$).*

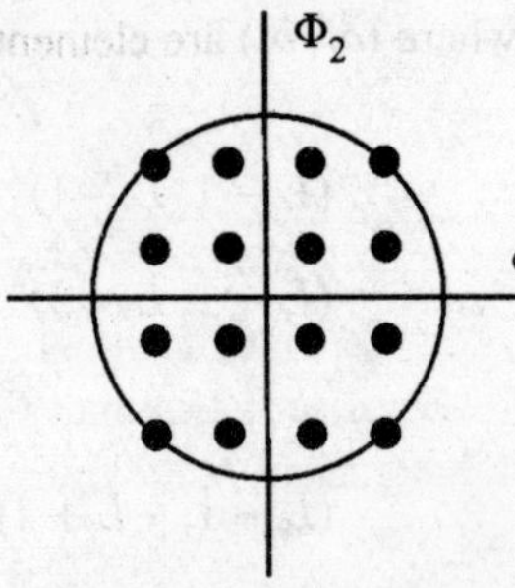

Fig. 4.18(*b*) Signal space diagram for M-ary PSK with $M = 16$.

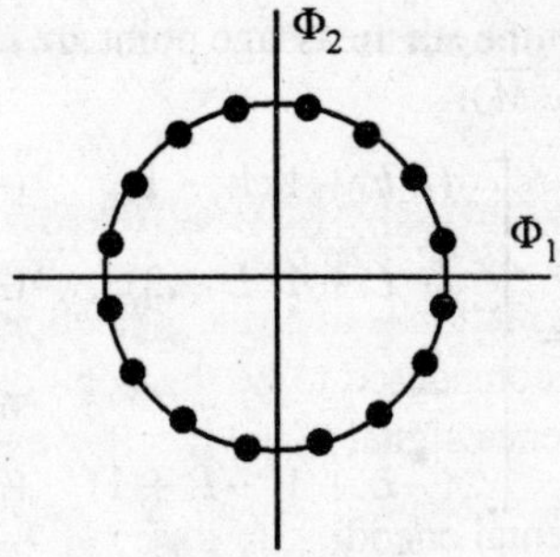

Fig. 4.18(*c*) Signal space diagram for M-ary APK with $M = 16$.

4.8.7 Non-coherent binary FSK

In binary FSK, the transmitted signal is defined by

$$S_i(t) = \begin{cases} \sqrt{\frac{2E_b}{T_0}}\cos(2\pi f_c t) & 0 \le t \le T_b \\ 0 & \text{elsewhere.} \end{cases}$$

$f_i = n_i/T_b$, where n_i is an integer and takes two different values to give f_1, f_2 for symbol 1 and 0. For non-coherent detection of this signal, we use filters matched to $\sqrt{2/T_b}\cos(2\pi f_1 t)$ and $\sqrt{2/T_b}\cos(2\pi f_2 t)$ (over $0 \le t \le T_b$) and the envelope detector. The output of the envelope detectors are sampled at $t = T_b$ (see Fig. 4.19).

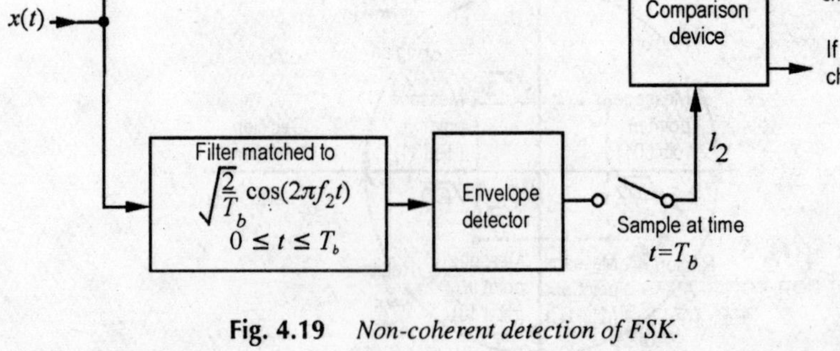

Fig. 4.19 *Non-coherent detection of FSK.*

The envelope samples of both paths are compared and the symbol is decided in favour of the larger sample.

Probability of error is $P_e = \frac{1}{2} \exp\left(\frac{-E_b}{2N_0}\right)$.

4.8.8 Differential phase shift keying (DPSK)

When the information is in phase, detection has to the coherent, i.e., there is no non-coherent PSK. But DPSK can be considered to be the non-coherent version of PSK because it eliminates the need for a coherent reference signal at the receiver by combining two basic operations at the transmitter:

(*i*) differental encoding of the input binary wave,
(*ii*) phase-shift keying.

Hence, it is termed differental phase shift keying (DPSK). (Many people mistake DPSK for dual PSK—we already have binary PSK, there is no need for dual PSK.)

To send symbol 1, the phase of the current signal waveform is advaned by 180°, and to send symbol 0, the phase is unchanged. The receiver has 'memory' to show the relative phase difference between the waveforms received during two successive bit durations.

Suppose the carrier transmitted in DPSK is

$$\sqrt{\frac{E_b}{2T_b}} \cos(2\pi f_c t) \text{ for } 0 \le t \le T_b,$$

where T_b is the bit duration and E_b the signal energy per bit.

Let $S_1(t)$ be the transmitted DPSK signal for $0 \le t \le 2T_b$ for the case where symbol 1 is at the input for the second part of the interval (i.e., $T_b \le t \le 2T_b$). For symbol 1, carrier phase is unchanged. So, $S_1(t)$ will be

$$S_1(t) = \begin{cases} \sqrt{\frac{E_b}{2T_b}} \cos(2\pi f_c t) & 0 \le t \le T_b \\ \sqrt{\frac{E_b}{2T_b}} \cos(2\pi f_c t) & T_b \le t \le 2T_b. \end{cases}$$

Let $S_2(t)$ denote the transmitted DPSK signal for $0 \le t \le T_b$ for the case when symbol 0 is at the transmitter input for $T_b \le t \le 2T_b$. The transmission of a 0 advances the phase by π, so $S_2(t)$ shall be

$$S_2(t) = \begin{cases} \sqrt{\frac{E_b}{2T_b}} \cos(2\pi f_c t) & 0 \le t \le T_b \\ \sqrt{\frac{E_b}{2T_b}} \cos(2\pi f_c t + \pi) & T_b \le t \le 2T_b. \end{cases}$$

It can be seen that $S_1(t)$, $S_2(t)$ are orthogonal over two sccessive bit intervals $0 \le t \le 2T_b$, i.e., DPSK is a special case of non-coherent orthogonal modulation with $T = 2Tb$ and $E = 2E_b$.

Probability of error $P_e = \frac{1}{2} \exp\left(\frac{-E_b}{N_0}\right)$.

Signal space diagram of DPSK is given in Fig. 4.20. Generation and detection of DPSK are illustrated in Fig. 4.21.

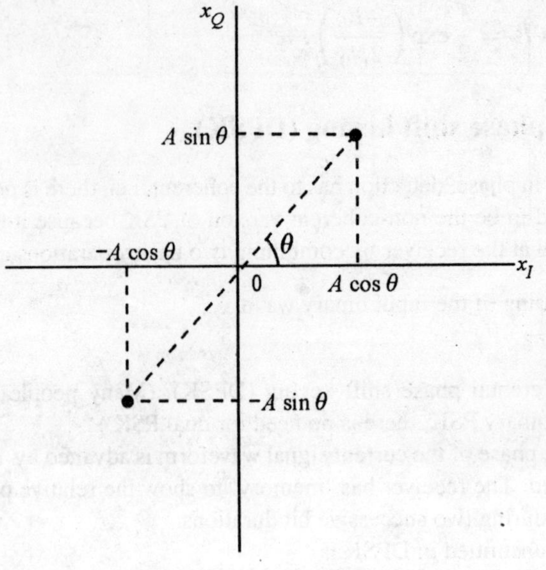

Fig. 4.20 *Signal space diagram of DPSK.*

4.8.9 Non-coherent M-ary FSK

In M-ary PSK, the transmitted signals are defined by

$$S_i(t) = \sqrt{\frac{2E}{T}} \cos\left[\frac{\pi}{T}(n_c + i)\right], \quad 0 \le t \le T.$$

where $i = 1, 2, \ldots, M$ and $f_c = n_k/2T$ for some fixed integer n_k.

The transmitted signals are of equal duration T, and have equal energy E. Since individual frequencies are $(1/2T)\pi$ apart, the signals are orthogonal.

$$\int_0^T S_i(t)S_j(t)dt = 0 \text{ for } i \neq j.$$

In coherent M-ray FSK, we need a bank of correlators with exact phase references. This can be costly and difficult to maintain. In non-coherent M-ary FSK, we use matched filters followed by envelope detectors.

$$P_e = \sum_{k=1}^{M} \frac{(-1)^{k+1}}{k+1} \binom{M-1}{K} \exp\left[\frac{-kE}{(k+1)N_0}\right],$$

where $\binom{M-1}{K}$ is a binomial coefficient given by

$$\binom{M-1}{K} = \frac{(M-1)!}{(M-1-k)!K!}.$$

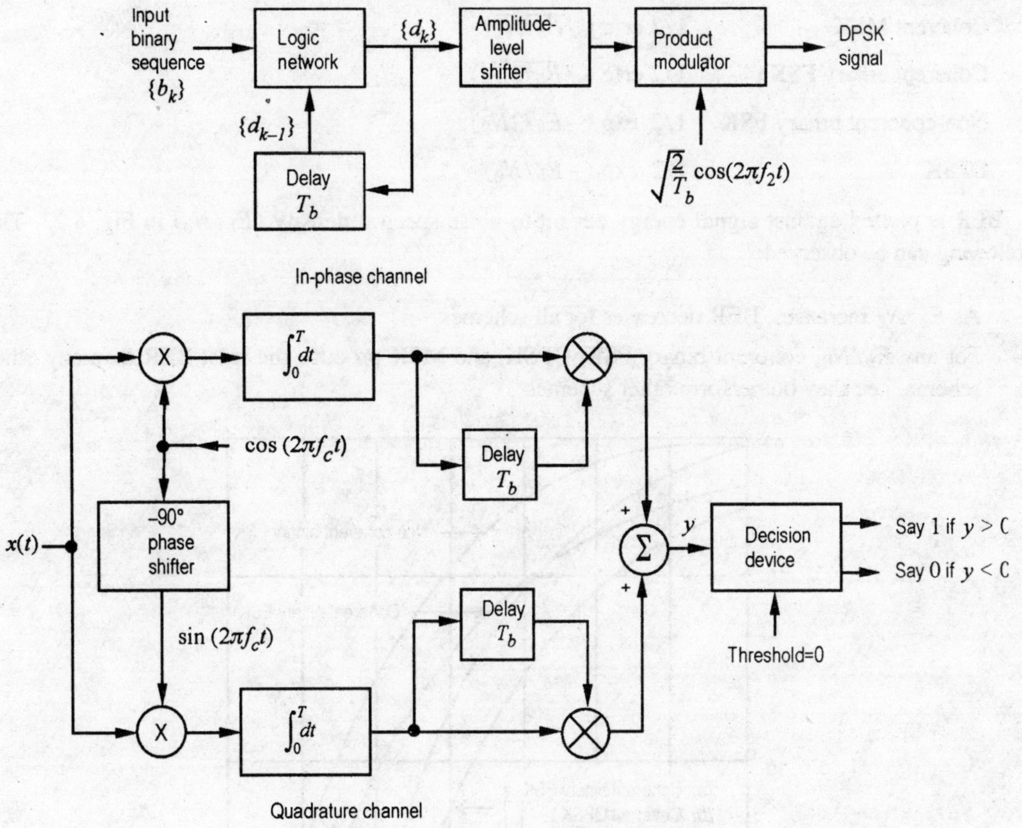

Fig. 4.21 *Generation and detection of DPSK.*

The leading term is the upper bound for the series, i.e.,

$$P_e \leq \frac{M-1}{2} \exp\left(\frac{-E}{2N_0}\right)$$

for non-coherent M-ary FSK.

4.9 BER Performance of Various Modulation Schemes

Expressions for probability of error (or BER) are summarized below.

Coherent binary PSK	$1/2$ erfc $(\sqrt{E_b/N_0})$
Coherent QPSK	$1/2$ erfc $(\sqrt{E_b/N_0})$

Coherent MSK	$1/2 \text{ erfc } (\sqrt{E_b/N_0})$
Coherent binary FSK	$1/2 \text{ erfc } (\sqrt{E_b/2N_0})$
Non-coherent binary FSK	$1/2 \exp(-E_b/2N_0)$
DPSK	$1/2 \exp(-E_b/N_0)$

BER is plotted against signal energy per bit-to-noise spectral density (E_b/N_0) in Fig. 4.22. The following can be observed:

- As E_b/N_0 increases, BER decreases for all schemes.
- For any E_b/N_0, coherent binary PSK, QPSK, and MSK produce the least BER than any other scheme, i.e., they outperform other schemes.

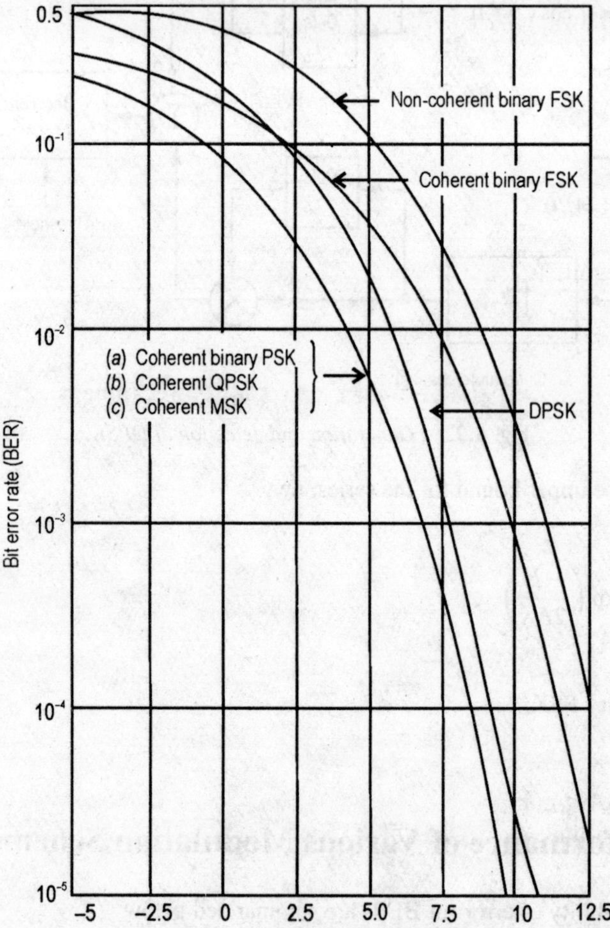

Fig. 4.22 *BER versus signal energy per bit.*

- For the same BER performance, coherent binary PSK/DPSK require 3 dB less power than FSK, coherent/non-coherent binary FSK, respectively
- At high E_b/N_0, the performance of DPSK and non-coherent binary FSK are almost the same
- In coherent QPSK, two orthogonal carriers (cosine, sine) are used and two independent bit streams can be transmitted and detected simultaneously
- In coherent MSK, two orthogonal carriers are modulated by two antipodal symbol shaping pulses over $2T_b$ interval, where T_b is the bit duration. So the receiver applies coherent phase decoding process over two successive bit intervals to recover the original bit stream. Hence, MSK receiver must have a memory (2-dimensional signal space) though the transmitted signal has one-dimensional signal space
- M-ary QAM outperforms M-ary PSK for $M > 4$
- In M-ary FSK, increasing M reduces power requirement for a given BER. But this is at the cost of increased bandwidth

4.10 Why QPSK for Digital MW Radio?

In section 4.5, we have seen that coherent binary PSK, coherent QPSK and coherent MSK outperform other schemes. Further, we have also noted that the transmission and detection of two bit streams simultaneously using sine and cosine carriers give QPSK an added advantage. Now, let us compare power–bandwidth requirements of M-ray PSK and binary PSK for BER $= 10^{-4}$.

M	BW (M-ary) / BW (binary)	Average power (M-ary) / Average power (binary)
4	0.5	0.34 dB
8	0.333	3.91 dB
16	0.25	8.52 dB
32	0.2	13.52 dB

$M = 4$, i.e. QPSK offers the best trade-off between power and bandwidth requirements.

Keeping the power requirement low is very useful for long haul digital radio as the savings on power plant would be tremendous, that too without compromise on BER and with efficient utilisation of bandwidth.

Though M-ray QAM outperforms M-ary PSK for $M > 4$, the benefit can be realised only if the channel is free of non-linearities (which is not so in a radio path).

4.11 Hierarchy of Digital Multiplexing

So far in this chapter the hierarchy of modulation techniques was discussed. Now, the other ingredient of transmission is multiplexing, and its hierarchy is discussed next.

The basic entry in digital transmission is 30 channel PCM (2 MB E1 frame) as per CCITT standards. The multiplicative factor between hierarchies is 4, i.e.,

- 2 Mbps for 30 channels
- Four 2 Mb: 8 Mbps for 120 channels
- Four 8 Mb : 34 Mbps for 480 channel including framing and other information.
- Four 34 Mb : 140 Mbps for 1920 channels
- Four 140 Mb : 565 Mbps for 1920 channels

TDM hierarchry is illustrated in Fig. 4.23. The frame structure details of 2 Mb, 8 Mb, 34 Mb and 140 Mb are described in detail in Chapter 6.

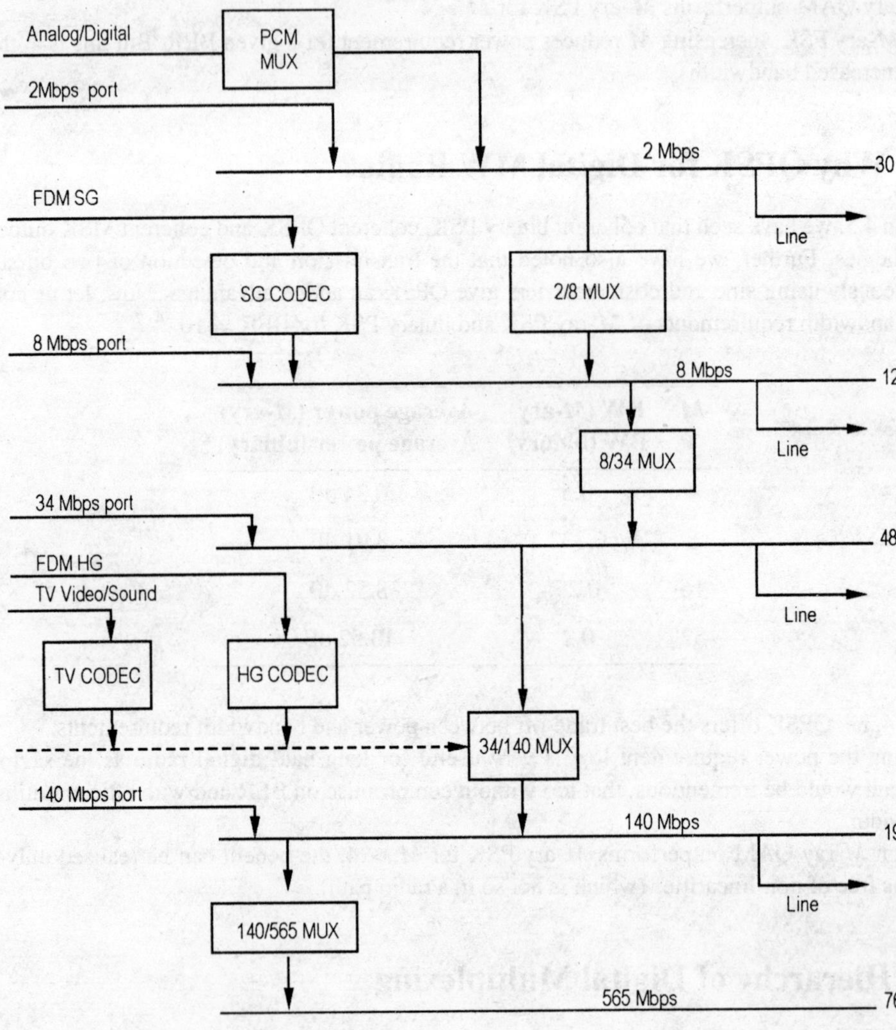

Fig. 4.23 *TDM hierarchy.*

Reference

[1] Simon Haykins, *Communication Systems*, chapter 10, PHI.

Exercises

1. In the hierarchy of digital multiplexing, the multiplication factor is 4. Then how do we get 34 Mb after 8 Mb, and 140 Mb after 34 Mb?
2. Distinguish between probability of symbol error and probability of bit error (or BER).
3. In M-ary FSK, increasing M reduces the power requirement. Then why can't we try a system with higher M?
4. Though M-ary QAM outperforms M-ary PSK (for $M > 4$), why do we still use QPSK in some of the digital MW radio systems?

5

Digital Microwave Communications System

The general form of a digital communications system has been described briefly in Section 2.8. A digital microwave (MW) communications system has the same constituents, with digital transmission hierarchies duly incorporated. Information in the form of audio/video/picture/data is converted to digital format through pulse code modulation (PCM has been dealt with in detail in Section 3.4). Thus, source coding is taken care of by PCM. However, instead of simple primary multiplexing of 30 channels as 2 Mbps data frame, the hierarchy of multiplexing up to III-order (i.e., 34 Mbps data frame with 480 channels), or IV-order (i.e., 140 Mbps data frame with 1920 channels), is supported by a digital MW communications system. The channel capacity of a digital MW link is planned as per traffic requirement and budgetary resources. Higher the channel capacity, better the utilization of the transmission bandwidth available with radio carrier in the microwave band.

Channel encoding/decoding and modulation/demodulation are taken care of in a digital microwave radio equipment. Channel encoding provides some immunity against degradation that sets in due to noise/distortion over transmission medium (in this case, RF band in MW range over the hop length). The modulation scheme is chosen as per design considerations presented in Section 4.8.

The radio channel consists of (*i*) the waveguide from the radio equipment to the antenna of the transmitting station, (*ii*) the hop or the line-of-sight path between the two adjacent microwave stations, and (*iii*) and the waveguide from the antenna to the radio equipment at the receiving station. Separate carriers are used for transmission and reception over a hop, which propagate through the same radio channel described above. Hence, it can be easily understood that a digital MW communications system provides for full duplex communication for all channels of the hierarchy.

A digital MW communications system thus comprises digital MW radios with antennas, waveguides and accessories, placed over a series of hops supporting III-order/IV-order digital multiplexing hierarchy. A 34 Mbps (480 channel) digital MW network of South Central Railways and 140 Mbps (1920 channel) digital MW network of DOT (now BSNL) are presented in this chapter.

Refer to Fig. 6.1 in the next chapter showing a 34 Mbps digital MW network on Vijayawada–Madras–Renigunta route of South Central Railway. As can be seen from the figure, a digital MW station can be one of the following:

- Terminal station
- Repeater station without channel dropping
- Repeater station with channel dropping

In terminal station we have the 34 Mbps/140 Mbps digital radio and associated higher order multiplexing equipment with the required number of primary multiplexing equipment as per the channel capacity planned for the link. The example presented in Fig. 5.1 has 34 Mbps digital radio and III-order MUX equipment.

At the repeater station without channel dropping, there are only digital MW radio equipment connected back-to-back.

At repeater station with channel dropping, we again have the digital MW radio, higher order multiplexing equipment, and primary multiplexing equipment as per the requirement of channel dropping.

At every station, the radio equipment has a hot standby. Further, the digital MW radio can be with/without space diversity . In a radio equipment without space diversity, the main and the standby transmitter, as well as main and standby receiver are connected to one antenna only. However, in a radio equipment with space diversity, while the main and standby transmitter and the main receiver are connected to the main antenna, the standby receiver alone is connected to diversity antenna. (Diversity techniques are discussed in detail in Chapter 15.)

The interconnection between primary multiplexing equipment, higher order multiplexing equipment and digital MW radio at all the three types of digital MW stations are shown in Fig. 5.1. It may please be noted that the output of each primary MUX, i.e. 2Mbps data frame, is considered a tributary. To drop channels at a repeater, the tributary containing those channels is tapped from the 34 Mbps/140Mbps data frame through hierarchical demultiplexing, by suitably deploying higher order MUX equipment.

Let us now turn our attention to the 140 Mbps digital MW network of the DOT(now called the BSNL). A 6 GHz, 140 Mbps network is used by the DOT for trunk route applications. In this system, each RF channel (carrier) carries 1920 channels. The capacity of the system is 7+1 radio channels (carriers), i.e. seven carriers each with a load of 1920 channels, transmitted simultaneously by the same antenna. (Such massive channel capacity is required by the DOT network to support its voluminous traffic.) The '+1' indicates one protection channel (RF carrier) common to all seven carriers. Though equipped for 7+1, at present 5+1 is used, with one carrier reserved for satellite communications. On the trunk routes where the traffic volume is yet to pick up, a 3+1 system is used, which can be upgraded to 7+1. The 6 GHz system of the DOT operates in the 6.430–7.110 GHz band with carrier spacing of 40 MHz. Like the 34 Mbps network, three types of stations exist:

- Terminal station
- Repeater station without channel dropping
- Repeater station with channel dropping

Each radio station has space diversity reception, i.e., a standby receiver connected to diversity antenna. Further, the protection carrier (i.e. the +1) provides standby for the needy transmitter (on first-come-first-served basis, or with priority for a particular main carrier that has suffered a failure). Once the failed main transmitter is restored, the protection carrier is made free for any other needy transmitter. Thus the protection carrier is shared by the main seven carriers (or three carriers) in a 7+1 or 3+1 system, as the case may be.

Refer to Fig. 5.2 which shows the trunk route digital MW network of the DOT. The configuration is:

- Bombay–Madras: 3+1 configuration, 33 hops with back-to-back terminal stations at Panjim, Mangalore, Mysore, Bangalore
- Dhulia–Nagpur: 3+1 configuration, 12 hops with back-to-back terminal station at Akola

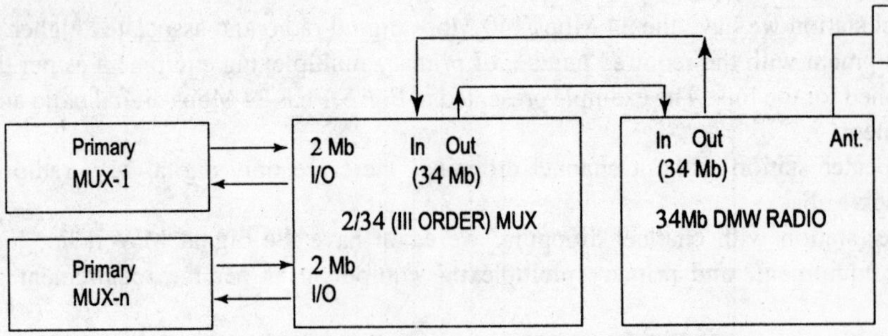

Fig. 5.1(*a*) *Interconnections of equipment at terminal station.*

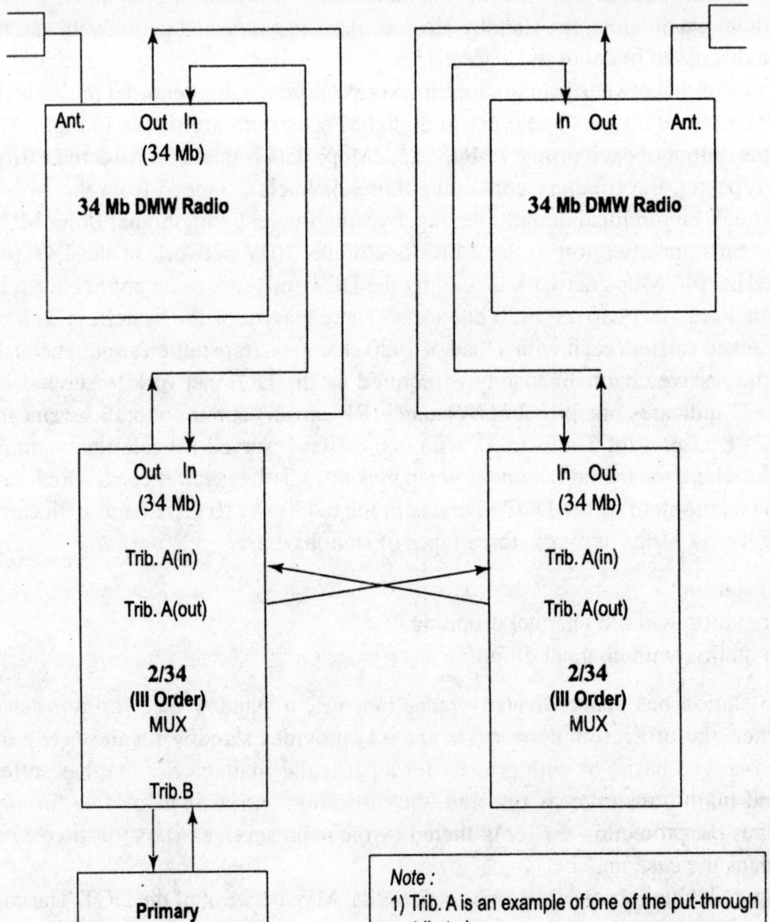

Fig. 5.1(*b*) *Interconnection of equipment at repeater station with channel dropping.*

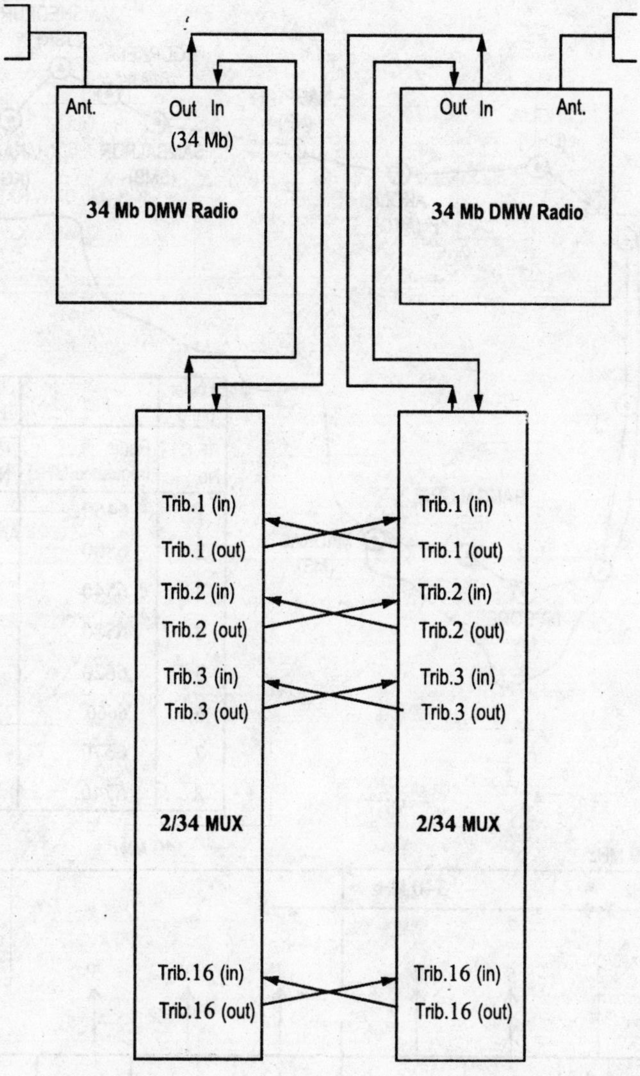

Fig. 5.1(*c*). *Interconnection of equipment at repeater station without channel dropping (tributary in/out connections cross-patched for all tributaries).*

- Dhulia–Nagpur: 3+1 configuration, 12 hops with back-to-back terminal station at Akola
- Sambalpur–Jamshedpur: 3+1 configuration, 8 hops with back-to-back terminal station at Rourkela
- Jamshedpur–Calcutta: 3+1 configuration, 8 hops with back-to-back terminal station at Kharagpur
- Madras–Calcutta, Madras–Nagpur: 7+1 configuration networks have been partly commissioned. The DOT uses 34 Mbps network also, but only for junction working (13 GHz band, 34 Mbps network) and thin-route applications (7 GHz band, 34 Mbps network).

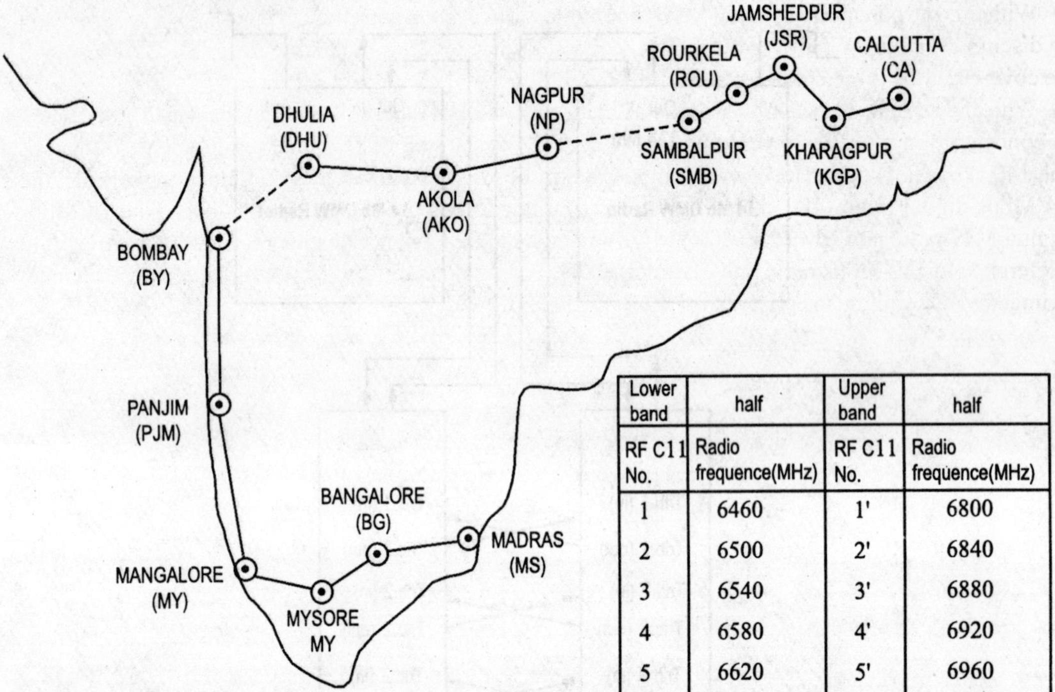

Lower band	half	Upper band	half
RF C11 No.	Radio frequence(MHz)	RF C11 No.	Radio frequence(MHz)
1	6460	1'	6800
2	6500	2'	6840
3	6540	3'	6880
4	6580	4'	6920
5	6620	5'	6960
6	6660	6'	7000
7	6670	7'	7040
8	6740	8'	7080

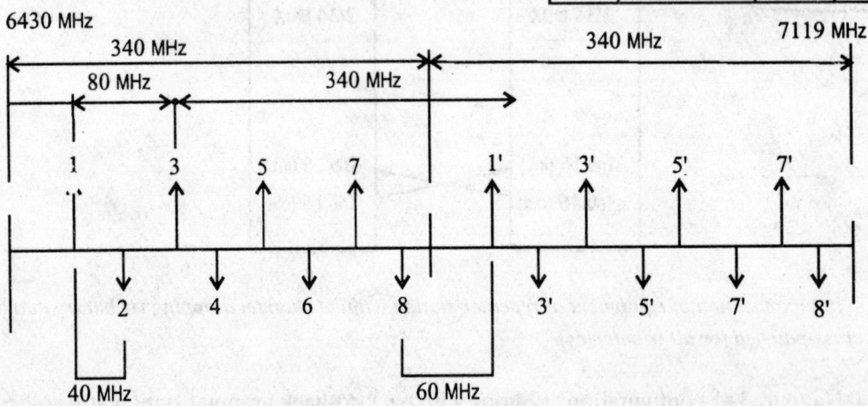

f_0 : 6770 MHz

lower half of band: $f_n = f_0 - 350 + 40\,n$

upper half of band: $f_n = f_0 - 10 = 10 + 40\,n$

n = 1, 2, 3, 4, 5, 6, 7 or 8

Fig. 5.2 *6GHz (140 Mb) digital MW network of DOT and radio frequency (channel) arrangement.*

With this introduction on digital MW communications system presented in this chapter, we are ready to discuss digital MW radio (34Mbps/140 Mbps), primary MUX, and higher order MUX. These will be covered in the three chapters that follow.

With the advent of synchronous digital hierarchy or SDH (more of this in Chapter 29), the DOT is concentrating more on optical fiber networks for very large traffic volume requirements instead of the 140 Mbps networks. However, for moderate traffic volumes, thin routes and junction working, the 34 Mbps digital radio (480 channel) systems continue to be used. Hence, in this book, the 34 Mbps digital MW radio and III-order (2/34) MUX are covered in detail (in Chapters 6 and 8 respectively) in preference to 140 Mbps radio and IV-order MUX. For any hierarchy, the basic block is the 30 channel primary MUX, and this is covered in detail in Chapter 7.

6

34+2 Mb Digital Microwave Radio Equipment

6.1 34 Mb and 2 Mb Traffic

In III-order multiplexing a 34 Mbps data frame for 480 channels is generated. If so, the radio has to be a 34 Mb radio. Why is it called 34 + 2 Mb radio? The answer involves the concept of wayside traffic. Normally a 34 Mbps data frame of 480 channels caters to mainstream traffic where the channels are taken to relatively important destinations without dropping/inserting tributaries at intermediate stations. The 2 Mbps data frame of 30 channels caters to relatively less important destinations with channel requirement less than 30. The requirement of all such destinations can be dealt with by a separate 2 Mbps traffic termed Wayside (WS) traffic, without disturbing the mainstream 34 Mbps traffic. The same radio equipment carries the 34 Mbps as well as the 2 Mbps traffic via separate ports, hence it is called 34+2 Mb radio.

The advantages of having WS traffic are as follows:

- Unnecessary dropping/insertion operations on the 34 Mbps mainstream are avoided. This reduces jitter accumulation as can be seen later in Chapter 16
- Channel dropping/insertion on WS is dealt with at primary MUX level. We can thus save on the cost of higher order MUXes

For example, in the Vijayawada–Madras (Chennai)–Renigunta (BZA–MAS–RU) digital MW link of the South Central Railway, tributaries are dropped from the 34 Mbps mainstream only at Undavalli (UDL), Guntur (GNT), Bitragunta (BTTR), Gudur (GDR), Nayudupeta (NYP), Renigunta (RU) and Madras (MAS). At these stations III-order or 2/34 MUX and primary MUXes are used. At (Bapatla (BPP), Ongole (OGL) Ulavapadu (UPD), Narsimhakonda (NKD), and Srikalahasti (SKHT), channels are dropped from the 2 Mbps WS stream and hence only primary MUXes (drop/insert MUXes) are used at these stations. Drop/insert MUX is a primary MUX that facilitates insertion of new channels in the time-slots of dropped channels without resorting to voice frequency (VF) level patching. Please refer to Fig. 6.1 showing the deployment of equipment on the BZA–MAS–RU digital MW link.

Having understood the necessity of WS traffic, we now turn our attention to the integration of 2 Mbps traffic with mainstream traffic of 34 Mbps so that it is carried on the same RF carrier.

WS traffic is integrated with the mainstream traffic in Tx-DPU (transmitter data processing unit) on the 'transmit' path. It is tapped out after phase demodulation on the receive path. In addition to WS traffic, we have the analog service channel (ASC) traffic and the digital service channel (DSC) traffic.

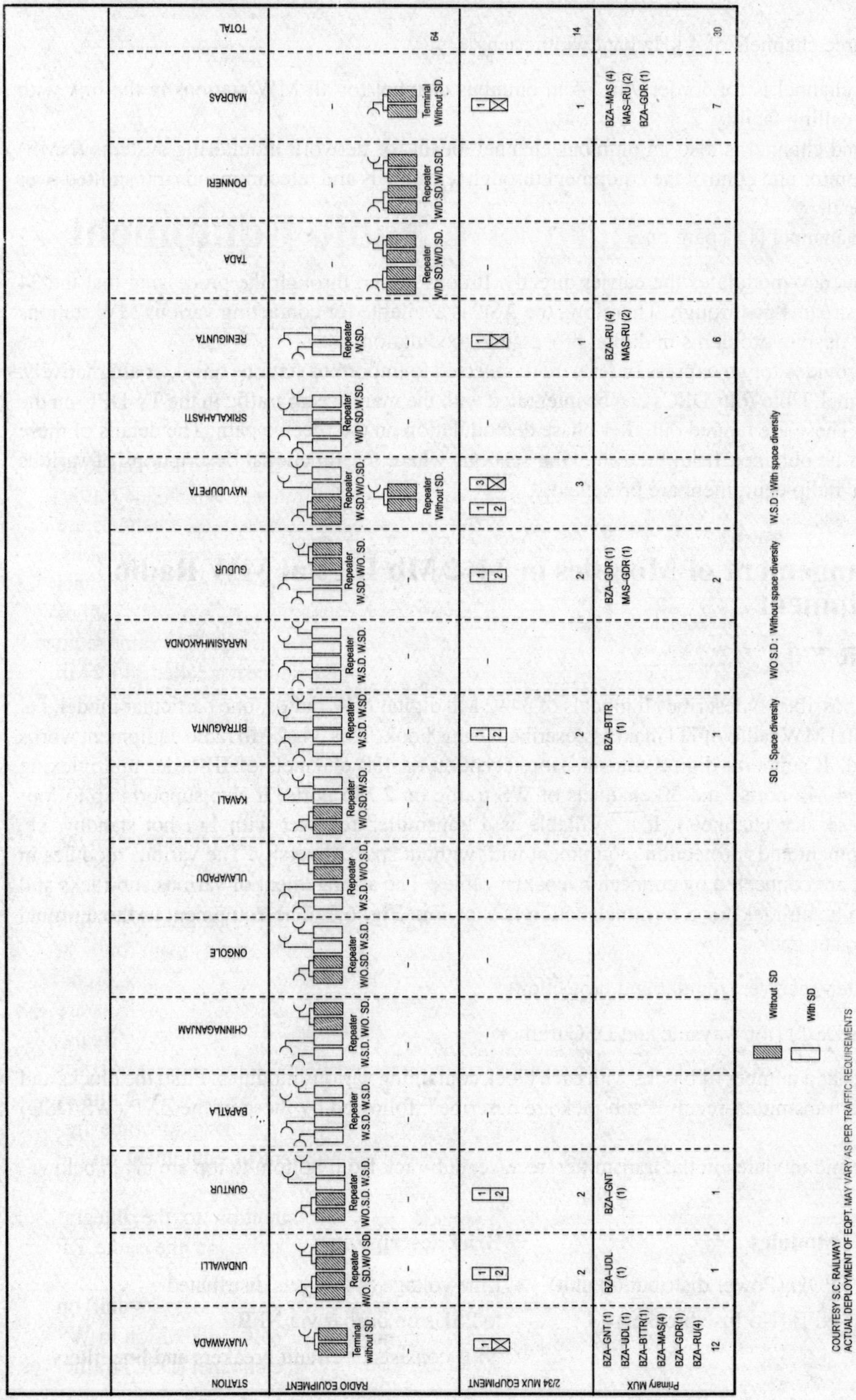

Fig. 6.1 *South Central Railway's Vijaywada—Renigunta—Madras digital microwave system (34Mbps, 480 channels)*

ASC has three channels of 4 kHz bandwidth each.

- The first channel is for 'order wire'—an omnibus channel for all MW stations in the link with selective calling facility.
- The second channel is also an omnibus channel meant for network monitoring systems (NMS) which monitor and control the equipment through telesignals and telecommands transmitted over this channel.
- The third channel is a spare one.

The ASC frequency-modulates the carrier directly. It does not go through the processing that the 34 Mb and 2 Mb streams go through. Therefore, the ASC is available for contacting various MW stations in the network despite problems in digital processing, modulation, etc.

The DSC provides for an express order wire (connected to important stations only), or alternatively, a 64 kbps channel. Upto four DSCs can be integrated with the mainstream traffic in the Tx-DPU on the transmit path. These are tapped out after phase demodulation on the receive path. The details of these operations can be obtained from the following sections where the functional descriptions of various modules of the radio equipment are presented.

6.2 Arrangement of Modules in 34+2 Mb Digital MW Radio Equipment

6.2.1 DMR 770

Since it is not possible to describe all models of 34+2 Mb digital MW radios, one particular model, i.e. DMR 770 digital MW radio of ITI make is described in this book. This 34+2 Mb radio equipment works in 7 GHz band. It supports digital transmission/reception of 480 channels of III-order multiplexing hierarchy on 34 Mb ports, and 30 channels of WS traffic on 2 Mb ports. It also supports up to four DSCs (digital service channels). It is available as a transmitter–receiver with 1+1 hot standby, i.e, 'regular' equipment and 'protection' equipment with/without space diversity. The various modules in the equipment are connected by connectors/coaxial cables. The arrangement of various sub-racks and module within a sub-rack for a terminal station is shown in Fig. 6.2. The equipment at the terminal station has two sub-racks:

- Transmitter–receiver (regular and protection)
- ZXP(WS/DSC) (for wayside and DSC traffic)

Each sub-rack has a number of blocks with each block containing various modules. First, the blocks and modules in the transmitter–receiver sub-rack are described, followed by those of the ZXP (WS/DSC) sub-rack.

The blocks and modules in the transmitter–receiver sub-rack from bottom to top are given below:

Block	Modules	Brief description
PDU-NFB	PDU (Power distribution unit) NFB (No fuse breaker)	Line voltage (48V) gets distributed to PSUs on busbar via NFB. NFB consists of circuit breakers and line filters.

Block	Modules	Brief description
DPU	Tx-DPU (1+1) Rx-DPU (1+1)	DPU has four modules: -two Tx-DPUs (REG/PROT) and -two Rx-DPUs (REG/PROT)
DEM	PH-DEM (1+1) BIT-COMB (1+1)	DEM has four modules: -two PH-DEMs (REG/PEOT) and -two BIT-COMBs (REG/PROT)
AUX	SWO SWO-CONT	SWO (switch-over) and SWO-CONT (switch-over control) modules are present here (only one set)
ALM	Alarms unit	This is an optional module
SC-TEL-DISP	SC-TEL DISPLAY	Service channel telephone, display unit (optional) for display of alarms
Transceiver (Tx–Rx)	Tx (1+1) Rx (1+1)	Transmitter and receiver modules (REG & PROT) are housed here
BR-CKT	Circulators, duplexers and associated terminations	Circulators and duplexers segregate Tx and Rx paths from waveguide ports
IDB	Data in/out connectors for 34 Mb and 2 Mb; waveguide ports	34 Mb and 2 Mb data connections with MUX are extended from here; also waveguide ports are available here for termination of waveguide coming from the antenna.

The blocks and modules in the ZXP (WS/DSC) sub-rack from bottom to top are given below.

Block	Modules	Brief description
PS	Power supply (REG) Power supply (PROT)	DC–DC converters for power supply to REG and PROT equipment (48 V input, +5 V, +12 V and −12 V output)
AUX	AUX board	It has indicators showing normal operation of DC–DC converters (REG/PROT)
SC-INTFC-D (2 blocks)	SC-REP-D2 (two nos.) and SC-OW-REP-D1 A/B/C (two nos.), i.e., a total of four modules can be housed in each SC-INTFC-D block as per system configuration	Each SC-INTFC-D block can provide interface for two DSCs; since two such blocks are available, four DSCs can be interfaced (more details in Section 6.6)
ALM	Alarm circuit	Optional

Block	Modules	Brief description
DSC	Four DSC-INTFC modules can be housed, one for each DSC channel interface on 34 Mb Tx-Rx side. If only one SC-INTFC-D block is equipped, two DSC-INTFC modules are present in this block.	It accepts DSC data on 72 kbps interface and presents it to the main 34 Mb transmitter–receiver on a 89.5 kbps interface
WS	WS-INTFC (two nos., one each for REG and PROT), WS-SWO	It accepts WS traffic on 2.048 Mb interface and presents it to main 34 Mb transmitter–receiver on 2.4615 Mb interface. It has 1 + 1 hot standby with switch-over

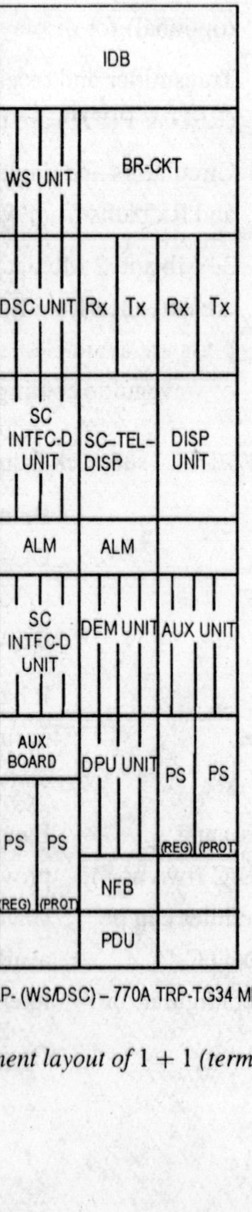

Fig. 6.2 *Equipment layout of* 1 + 1 *(terminal) DMR 770 radio.*

6.3 Signal Flow in 34+2 Mb Digital MW Radio Equipment

A quick description of signal flow (refer to Fig. 6.3 (*a&b*)) is presented in this section to give an overall idea of the functioning of DMR 770 model. The levels of signals at various points and also the impedance and data rate/signal frequency (as may be the case) are presented in the tables which are part of Fig. 6.3(*a*) and Fig. 6.3(*b*). While Fig. 6.3(*a*) is for the main 34 Mb transmitter–receiver, Fig. 6.3(*b*) is for ZXP (WS/DSC). Let us first examine Fig. 6.3(*a*).

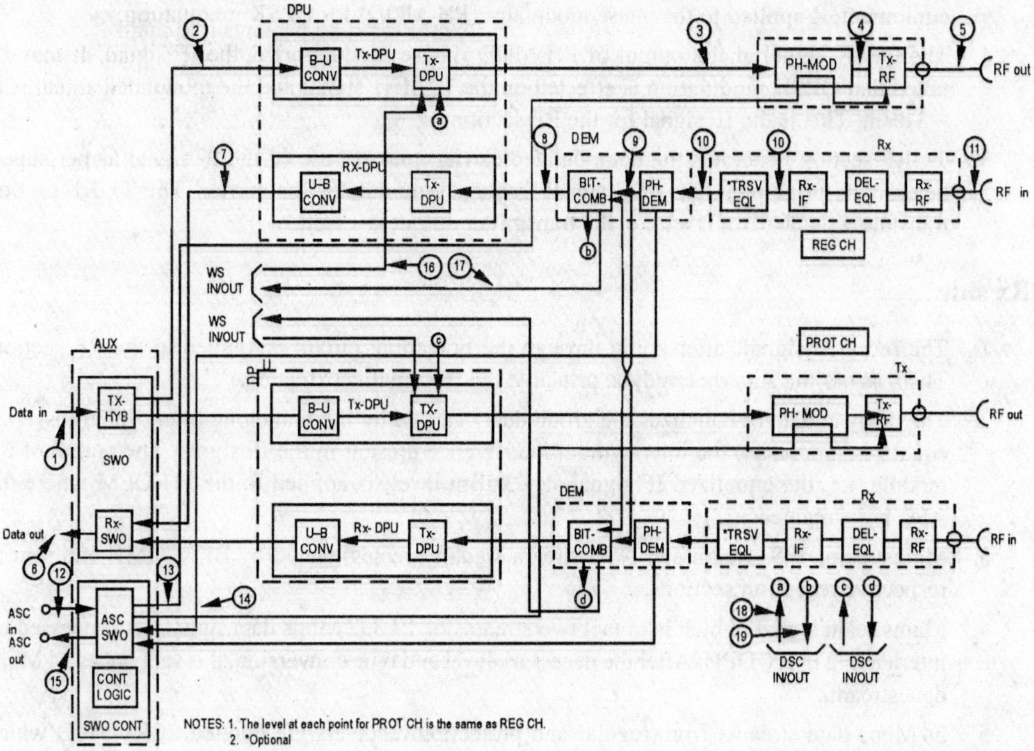

Point	Normal level	Data speed/Frequency	Impedance	Point	Normal level	Data speed/Frequency	Impedance
1	$1.0\,V_0-p \pm 0.1$ V	34.368 Mbps	75 Ω, unbalanced	11	–40 dBm	Assigned Tx RF frequency	50 Ω, unbalanced
2	$0.5\,V_0-p \pm 0.05$ V	34.386 Mbps	75 Ω, unbalanced	12	–30 dBm	0.3 to 12 kHz	600 Ω, balanced
3	0 to 2V	19.332 Mbps	75 Ω, unbalanced	13	–34.5 dBm	0.3 to 12 kHz	600 Ω, balanced
4	–3 dBm	70 MHz	75 Ω, unbalanced	14	–20 dBm	0.3 to 12 kHz	600 Ω, balanced
5	+ 30 dBm	Assigned Tx RF frequency	75 Ω, unbalanced	15	–20 dBm	0.3 to 12 kHz	600 Ω, balanced
6	$1.0\,V_0-p \pm 0.1$ V	34.368 Mbps	75 Ω, unbalanced	16	TTL	2.4165 Mbps	–
7	$1.0\,V_0-p \pm 0.1$ V	34.368 Mbps	75 Ω, unbalanced	17	TTL	2.4165 Mbps	–
8	0 to 2V	19.332 Mbps	75 Ω, unbalanced	18	TTL	89.5 kbps	–
9	0 to 2V	19.332 MHz	75 Ω, unbalanced	19	TTL	89.5 kbps	–
10	– 3 dBm	70 MHz	75 Ω, unbalanced	20	–	–	

Fig. 6.3(*a*) *Block and level diagram of Tx–Rx.*

Tx side

1. 34 Mbps data signal from MUX equipment given to the data-in port is applied to the SWO unit where it is duplicated by Tx-HYB.
2. The duplicated 34 Mbps signals are applied to the Tx-DPUs of regular and protection equipment. DSC data and WS data are also applied to the Tx-DPU for integration with the mainstream.
3. The output of Tx-DPU, which is two streams of 19.332 Mbps in both regular and protection equipment, is applied to the phase modulator (PH-MOD) for QPSK modulation.
4. The QPSK signal at the output of PH-MOD (phase modulator) is the IF signal. It may be noted that QPSK modulation is effected on the 70 MHz signal and the modulated signal is at −3 dBm. This is the IF signal for the RF section.
5. Tx-RF section generates the demodulated carrier, making use of the IF signal as per superheterodyne principle. The ASC signal frequency-modulates the carrier. The Tx-RF section transmits at +30 dBm level, i.e. the transmitter output is 1 watt.

Rx side

1. The received signal, after going through the branching circuit is applied to the RF section. Then, as per the superheterodyne principle the IF signal is extracted.
2. The delay equalizer equalizes the group delay caused by the branching filter. The transversal equalizer suppresses the inter-symbol interference present in the IF signal. The output of Rx module, i.e. the equalized IF signal at −3 dBm level, is applied to the PH-DEM where the ASC is segregated.
3. Mainstream, WS stream and DSC stream signals are segregated by BIT-COMB and sent to respective reception sections.
4. Mainstream signal, which is in fact two streams of 19.332 Mbps data signals, is processed by interleaving in Rx-DPU. After the necessary level and type conversion, it is sent out as 34 Mbps data stream.
5. 34 Mbps data streams from regular and protection receivers are applied to Rx-SWO which selects the better signal under the control of SWO-CONT.
6. The 34 Mbps data output from Rx-SWO is given to 2/34 MUX.

Let us now turn our attention to Fig. 6.3(*b*).

DSC block

The 72 kbps data from terminal equipment interface (SC-INTFC-D block) is converted to 89.5 kbps (one DSC-INTFC module each for one channel; four such modules for four DSC channels; one SC-INTFC-D for two DSC channels, hence two such blocks for four DSC channels).

If the system configuration is only for two DSC channels (in most of the cases it is so—one for express order wire and the other for G.703, i.e. 64 kbps interface), only one SC-INTFC-D block is equipped and only two DSC-INTFC modules are present in the DSC block. Further, modules in SC-INTFC-D are chosen, suitable for G.703 interface and express order wire.

- SV-REP-D1 with SC-REP-D2 for G.703
- SV-OW-REP-D1C with SC-REP-D2 for express order wire

WS block

The 2.048 Mbps data applied to the WS-SWO is duplicated and given to the WS-INTFC (REG and PROT). WS-INTFCs converts the input 2.048 Mbps stream to 2.4615 Mbps stream suitable for the main transmitter interface in Tx-DPU.

On the receive side, the WS-INTFCs (REG/PROT) gets the 2.4615 Mbps data from the main receiver (Rx-DPU), and converts it to 2.048 Mbps data. The WS-SWO module selects the REG/PROT 2.048 Mbps stream from WS-INTFCs under the control of SWO-CONT of the main transmitter-receiver.

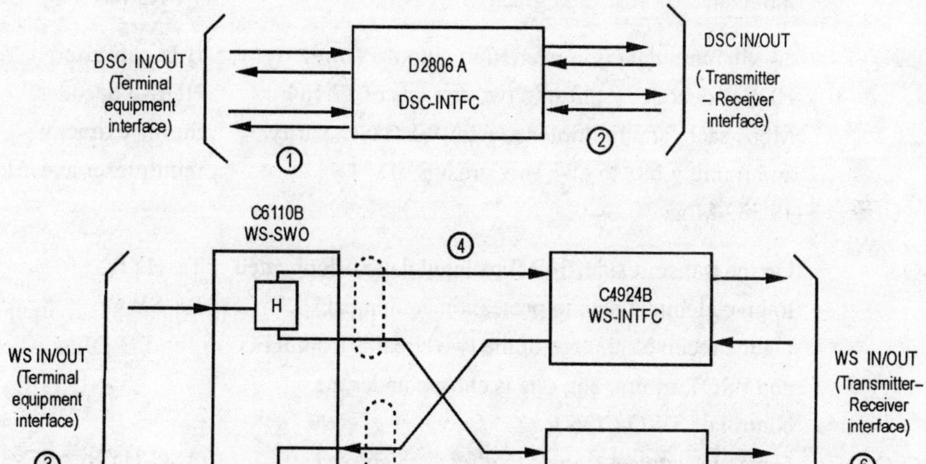

Position	①	②	Main data transmission capacity	DSC-INTFC Code No
Bit rate	72 kbps	89.5 kbps	34 Mbps × 1	D2806A
Level	TTL	TTL	–	–
Impedance	130 Ω	130 Ω	–	–

Position	③ ⑤	④	⑥	Main Data Transmission Capacity	DSC INTFC Code No
Bit rate	2.048 Mbps	2.048 Mbps	2.4165 Mbps	34 Mbps × 1	C49913
Level	2.37 V_{o-p}	0.82 V_{o-p}	TTL	TTL	–
Impedance	75 Ω	75 Ω	130 Ω	–	–

Fig. 6.3(*b*) *Block and level diagram of ZXP-WS/DSC.*

6.4 Modules and Sub-modules in 34+2 Mb Digital MW Radio

Figures 6.3(*a*) and 6.3(*b*) are shown with more detail as Figs. 6.3(*c*) and 6.3(*d*) respectively, i.e., showing sub-modules in each module. Table 6.1 shows the modules and sub-modules in the transmitter-receiver sub-rack, and Table 6.2 the modules and sub-modules in ZXP (WS/DSC) sub-rack.

Table 6. 1 *Modules and sub-modules in transmitter–receiver sub-rack*

Module	Function in brief	Sub-modules
Tx	In the PH-MOD part, the carrier is QPSK modulated by data signals. In the Tx-RF part, it is frequency-modulated by ASC and given to Tx port	PH-MOD has SUM logic, QPSK modulator, limiter; Tx-RF has VCO and mixer.
Tx-DPU	34 Mb input data is converted to unipolar form, HDB-3 decoded, split into two streams of 17.184 Mbps each, and multiplexed with WS, DSC, parity and framing bits to give two streams of 19.332 Mbps	B-U converter, HDB-3 decoder, timing extractor, multiplexer and ALM
SWO	On the transmit side, 34 Mbps input data is duplicated to give identical data to protection equipment. On the receive side, out of the two streams from REG and PROT equipment, one is chosen under the control of SWO-CONT	Tx-HYB, Rx-SWO
SWO-CONT	Generates control signals to effect switch-over	ASC-HYB, ASC-SWO and CONT logic
Rx	In the Rx-RF part, IF signal is generated from the received RF signal.	Rx-RF has pre-amplifier, mixer, LO and BPF
	In the DEL-EQL part, group delay caused by the branching circuit filters is equalised	DEL-EQL has tapped delay line filter
	The Rx-IF section gives steady IF signal at −3 dBm level to PH-DEM	Rx-IF part has IF-BPF, IF-AMP, AGC amplifier and alarm circuit
TRSV-EQL	Suppresses intersymbol interference in IF signal	Optional
PH-DEM	Two streams of 19.332 Mb are regenerated after demodulation along with a clock	4-PH demodulator, Rx-Logic

Table 6. 1 *Continued*

BIT-COMB	Hitless switching between data streams of regular and protection channels; two streams of 19.332 Mbps are applied to Rx-DPU	Hitless switch, delay adjustor-1, delay adjustor-2, switching logic, frame sync and alarm control circuit
Rx-DPU	34 Mbps main data stream is generated from the two 17.184 Mbps streams and applied to Rx-SWO	Demultiplexer, delay adjustor, unipolar-to-bipolar converter, HDB-3 encoder, alarm control circuit

Table 6. 2 *Modules, sub-modules in ZXP (WS/DSC) sub-rack*

Module	Function in brief	Sub-modules
WS-SWO	Switch-over of wayside traffic from REG to PROT and vice versa	Hybrid, selector
WS-INTFC	In the transmit path of WS traffic, accepts 2.048 Mbps data on WS data-in port and presents it as 2.4156 Mbps to the main Tx	Tx path has B-U CONV and WS-Tx-DPU
	In the receive path of WS traffic, accepts 2.4165 Mbps data from the main Rx and presents it as 2.048 Mbps to WS data-out port	Rx path has U-B CONV, WS-Rx-DPU and U-B CONV
DSC-INTFC	Each DSC-INTFC module accepts the two DSC channels' data at 72 kbps from SC-INTFC-D block and presents them at 89.5 kbps rate to the main Tx (Tx-DPU)	Tx path has DSC-T, input PLS DET and TIM GEN
	In receive path, it accepts two DSC streams at 89.5 kbps from the main Rx and presents them at 72 kbps to SC-INTFC-D block	Rx path has input PLS DET, F-SYNC, DSC-R, TIM GEN and output PLS DET

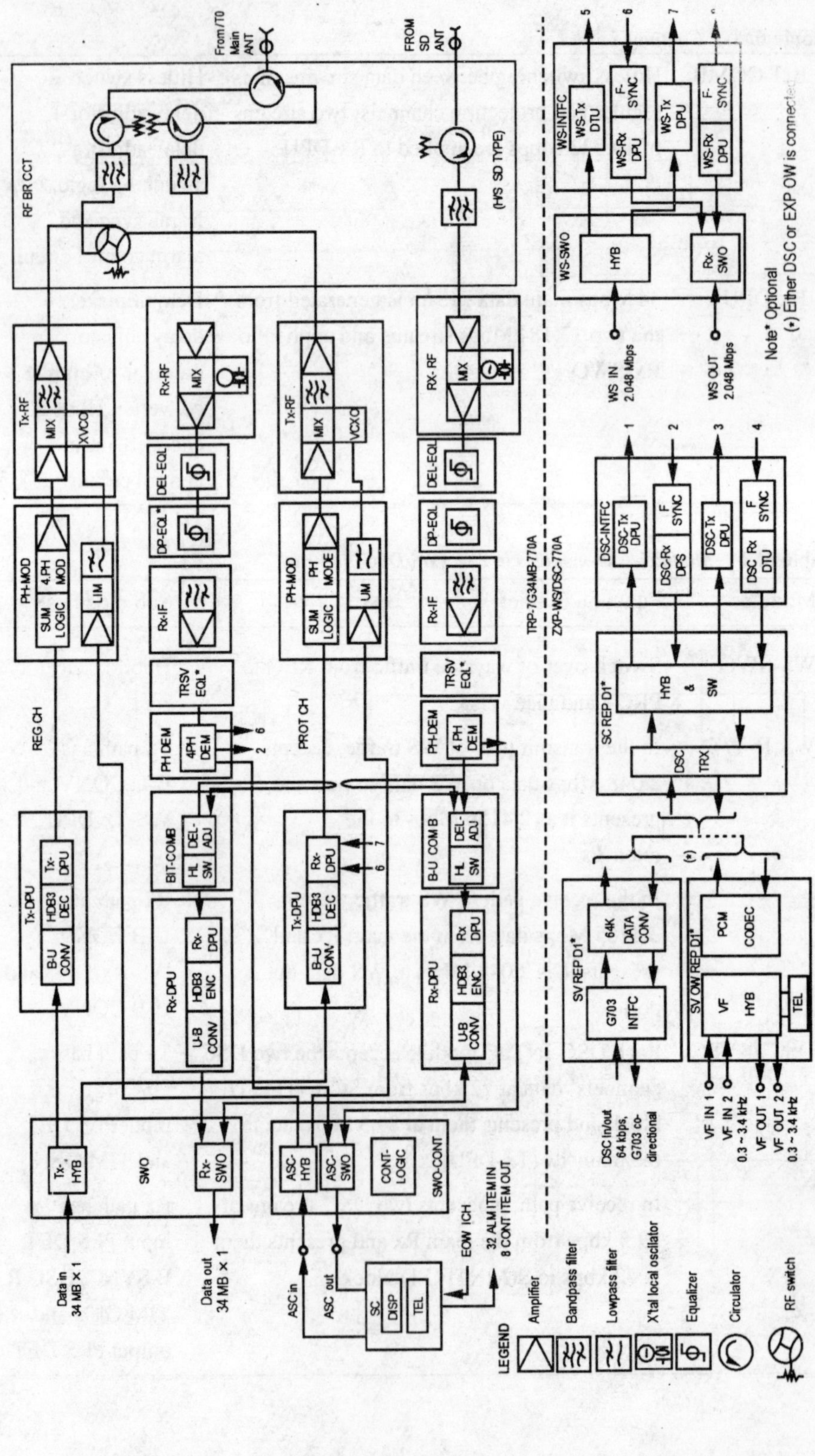

Fig. 6.3(c) *Modules and sub-modules in $T_x - R_x$.*

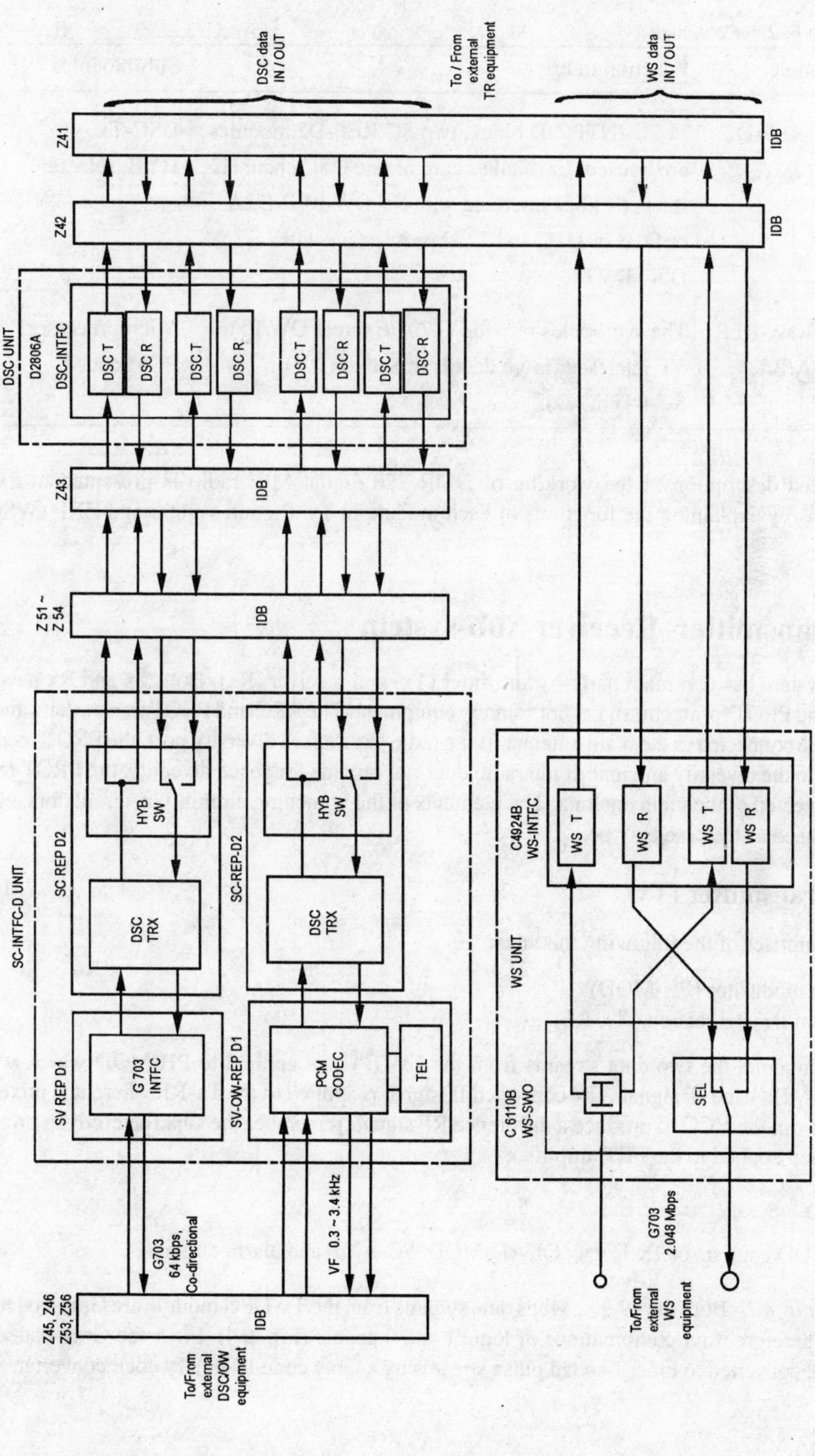

Fig. 6.3(*d*) *Modules and sub-modules in ZXP (WS/DSC).*

Table 6. 2 *Continued*

Module	Function in brief	Sub-modules
SC-REP-D2	In SC-INTFC-D block, two SC-REP-D2 modules are housed. Each takes care of one DSC channel. It has 64 kbps interface with SV-OW-REP-D1A or D1B or D1C and 72 kbps interface with DSC-INTFC	DSC-Tx, HYB, selector
SV-OW-REP D1A/B/C	These modules provide G.703/Express OW/Ext VF interface (more details in Section 6.8 on SC-INTFC-D).	Microprocessor, PCM codec

A detailed description of the working of DMR 770 digital MW radio is presented in the next two sections by explaining the functions of each module in Tx–Rx sub-system and ZXP (WS/DSC) sub-system.

6.5 Transmitter–Receiver Sub-system

This sub-system has two main parts—transmitter (Tx) and receiver (Rx). Both Tx and Rx have REG (regular) and PROT (protection), i.e. hot standby equipment. The REG and PROT transmitters and REG receivers are connected to the main antenna. If the radio has a space diversity port, the PROT receiver is connected to the diversity antenna. If the radio does not provide for space diversity, the PROT receiver is also connected to the main antenna. The elements of the branching circuit, viz. circulators, etc., are suitably placed as the case may be.

6.5.1 Transmitter (Tx)

The Tx comprises of the following modules:

- Phase modulator (PH-MOD)
- Transmitter RF section (Tx-RF)

In the Tx module, the two data streams from the Tx-DPU are applied to PH-MOD which converts them to 70 MHz band IF signal. The converted IF signal is applied to the Tx-RF where it is mixed with RF signal from the VCO to produce a difference RF signal (remember the superheterodyne principle), which is then applied to the FET amplifier.

PH-MOD

The PH-MOD consists of Tx-Logic, QUAD-MOD, SC-AMP and alarm circuits.

Tx-Logic circuit: Both the 19.332 Mbps data streams from the Tx-DPU module are Gray-coded pulse streams. Therefore, four combinations of logic 1 and 0 appear (Fig. 6.4). First, the Gray-coded pulse streams are converted to binary-coded pulse streams by a Gray code-to-binary code converter.

Quaternary	Gray code	Binary code
0	0 0	0 0
1	0 1	0 1
2	1 1	1 0
3	1 0	1 1

Fig. 6.4 *Gray code to binary code relationship.*

Data streams are differentially encoded and in the binary-coded pulse stream, phase is accumulated in quaternary numeration at every time-slot (see Fig. 6.5). The resultant output data streams are then converted to Gray-coded pulse streams by a binary code-to-Gray code converter.

When a non-modulated 70 MHz IF carrier is required for the purpose of measurement—for example Tx-RF measurement—the MOD-ON/OFF switch is set to OFF.

QUAD MOD circuit: Two 19.332 Mbps data streams from the Tx-Logic circuit are amplified by a pulse amplifier and sent to a lowpass filter circuit for shaping the spectrum of the pulse streams in such

Time-slot				1	2	3	4	5	6	7	8	9	10	11
PH–MOD	Tx–LOGIC input	Data 1		0	1	1	0	1	0	0	1	0	1	1
		Data 2		0	0	1	1	1	0	1	0	0	1	1
		Quaternary		0	3	2	1	2	0	1	3	0	2	2
	SUM logic output	Quaternary		0	3	1	2	0	0	1	0	0	2	0
		Stream 1		0	1	0	1	0	0	0	0	0	1	0
		Stream 2		0	1	1	0	0	0	1	0	0	0	0
	Tx-LOGIC output	MOD PLS 1		0	1	0	1	0	0	0	0	0	1	0
		MOD PLS 2		0	0	1	1	0	0	1	0	0	1	0
	Modulation phase			0	$3\pi/2$	$\pi/2$	π	0	0	$\pi/2$	0	0	π	0
PH–DEM	Pulse regeneration circuit output (quaternary)	Demodulation local carrier (Lol), phase	$\pi/4$	2	1	3	0	2	2	3	2	2	0	2
			$3\pi/4$	1	0	2	3	1	1	2	1	1	3	1
			$5\pi/4$	0	3	1	2	0	0	1	0	0	2	0
			$7\pi/4$	3	2	0	1	3	3	0	3	3	1	3
	DIFF logic output	Quaternary		(0)	3	2	1	2	0	1	3	0	2	2
		Stream 1		0	1	1	0	1	0	0	1	0	1	1
		Stream 2		0	1	0	1	0	0	1	1	0	0	0
	Rx-LOGIC output	Data 1		0	1	1	0	1	0	0	1	0	1	1
		Data 2		0	0	1	1	1	0	1	0	0	1	1

Fig. 6.5 *Logic differential operation.*

a way as to limit the transmitter output power spectrum. The pulse streams thus shaped are applied to two $0 - \pi$ modulators.

A 70 MHz IF carrier generated at the 70 MHz oscillator is split into two carriers (LO1 and LO2), one for each of the $0 - \pi$ modulators. The phase difference between the two 70 MHz carriers is $\pi/2$ radians. In each of the $0 - \pi$ modulators, the phase of the 70 MHz carrier is modulated to a 0-phase or π-phase by the input pulse stream. The outputs of the two $0 - \pi$ modulators are then combined for a four-phase assignment onto a 70 MHz IF carrier (see Fig. 6.6). The four-phase modulated IF signal thus obtained is amplified by the IF amplifier and sent to the Tx-RF module at an output level of −3 dBm.

To obtain a stable output level, the IF output is regulated to −3 dBm by changing the impedance of the PIN diode under the application of ALC (automatic level control) signal from the ALC circuit. The ALC circuit is a part of the IF-AMP (IF amplifier).

SC-AMP circuit: The SC-AMP circuit amplifies the ASC signal and applies it to the Tx local oscillator in the Tx-RF module. The RF carrier is thus frequency-modulated by the ASC signal.

Alarm circuit: The PH-MOD has an alarm detection circuit for monitoring the output data streams of the (*i*) Tx-logic circuit; (*ii*) the IF output signals level of the QUAD-MOD circuit; and (*iii*) the alarm information from the Tx-RF module. In normal condition, a '1' (about 0 V) appears, and in an alarm condition, a '0' appears (about −8 V), lighting up the Tx-ALM indicator red.

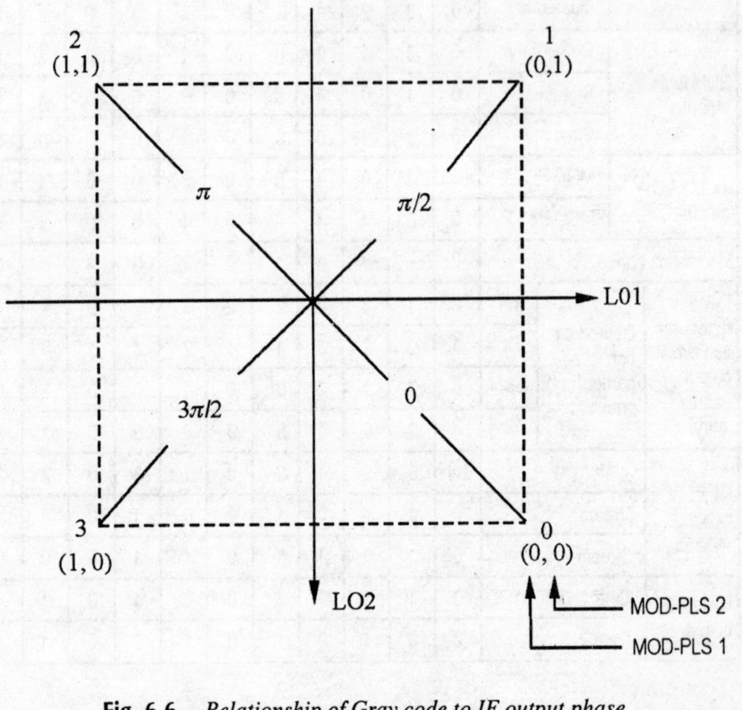

Fig. 6.6 *Relationship of Gray code to IF output phase.*

Tx-RF

Tx-RF consists of a mixer (MIX), a 5-stage bandpass filter (5-BPF), a FET-AMP and the Tx local oscillator (Tx-LO) circuit.

MIX circuit: The IF signal from PH-MOD is mixed with the local RF signal from the VCO to produce the transmitting frequency.

5-BPF circuit: In 5-BPF, only the desired RF signal is allowed to pass and all the undesired frequencies are eliminated.

FET-AMP circuit: The FET-AMP circuit has a series of GaAs FET amplifiers with microstrip circuits, an RF output monitoring circuit, an ALC circuit and a protection circuit for the FET amplifier.

In the FET-AMP circuit, RF signal from the mixer circuit is amplified up to the specified level by the GaAs FET and the signal thus obtained is sent through an RF-OUT jack. Isolators are connected to the input and output terminals to improve impedance. Other functions of the FET-AMP circuit are described below in brief.

Transmit output stabilizing function: The ALC circuit compensates for the variation of other conditions like temperature, and ensures stable level for the Tx output.

Transmit power monitoring function: The transmit output power is monitored and a feedback given to PH-MOD in the form of a DC voltage.

Tx-LO circuit: The Tx-LO circuit contains an APC-OSC (automatic phase controlled oscillator) and a HYB-IC (hybrid amplifier IC). The APC-OSC consists of a VCO and an APC circuit with VCXO (reference oscillator), PH-DET, LOOP-AMP and SEARCH- OSC. The RF output frequency of Tx-LO circuit is phase-locked with the reference frequency from the VCXO by means of APC operation.

VCO function: The VCO has a tuning circuit with a varactor diode. It generates a 7 GHz frequency corresponding to the RF frequency band used. A reference voltage from the loop amplifier circuit controls this VCO.

APC function: The VCO's fundamental RF frequency enters the automatic phase control (APC) circuit that keeps it exactly within the specified value. A part of the VCO output enters the phase detector, which samples it with a reference frequency from the VCXO (voltage controlled crystal oscillator) to produce a reference dc voltage. When the phase of the output voltage either leads or lags, the phase detector produces a dc voltage to drive the VCO in such a way that the deviated frequency is brought back to the exact desired value. To frequency-modulate the RF carrier with the service channel signal, a 0.3 to 12 kHz service channel signal is fed to the VCXO (voltage controlled crystal oscillator) through the SC-AMP on the PH-MOD module.

6.5.2 Receiver (Rx)

The Rx comprises the following modules:

- Receiver RF section (Rx-RF)
- Delay equalizer (DEL-EQL)

- Receiver IF section (Rx-IF)
- Transversal equalizer (TRSV-EQL)

The RF input signal, after passing through the branching circuit, is mixed with the RF signal from the local oscillator and converted to IF signal in the 70 MHz band. The converted IF signal is then applied to the Rx-IF through the DEL-EQL. The IF signal is filtered and amplified in the Rx-RF and sent to the TRSV-EQL for equalization.

Rx-RF

The Rx-RF consists of a pre-amplifier (pre-RF AMP), a receive-frequency converter (Rx-FREQ-CONV) and the Rx local oscillator (Rx-LO) circuits.

Pre-RF AMP: When the received signal from the antenna port passes through branching filter, the unwanted signals (frequencies) are eliminated. The 'cleansed' signal is applied to the RF pre-amplifier whose noise figure should be 4.0 dB or less.

Rx-FREQ-CONV: The Rx-FREQ-CONV forms a balanced mixer with Schottky barrier silicon diode, integrated isolator and pre-IF amplifier. The mixer is of an image-cancelling type, hence an image frequency rejection filter is not necessary. In this mixer circuit, the RF signal is passed through an isolator and mixed with a local signal from the Rx-LO circuit to produce the IF signal. The IF signal is amplified by a pre-IF amplifier to the required level.

Rx-LO circuit: The Rx-LO contains a crystal controlled oscillator with an APC circuit. The operation is similar to Tx-LO crystal controlled oscillator, except that there is no ASC signal here.

DEL-EQL

The DEL-EQL equalizes group delay caused by branching filters.

Rx-IF

The Rx-IF consists of IF bandpass filter (IF-BPF), main IF amplifier (MAIN IF-AMP), automatic gain control circuit amplifier (AGC-AMP) and AGC alarm (AGC-ALM) circuits.

The IF-BPF passes the desired IF signal while eliminating unwanted signals generated in the mixer.

The MAIN-IF-AMP has a five-stage hybrid IC. Maximum gain of this amplifier is about 50 dB and is controlled by an AGC circuit to keep the IF output level at −3 dBm.

The Rx-IF has an AGC detection circuit for monitoring the operating status. In alarm conditions, the IF output decreases to −4 dBm at the MAIN-IF-AMP output, prompting the appearance of an alarm signal.

TRSV-EQL

The TRSV-EQL module consists of a three-tapped transversal type filter (variable band-pass filter) and control signal generator circuits. In the control signal generator, signals for controlling the taps of the three-tapped transversal filter are produced according to error signals applied by the associated PH-DEM module. In the three-tapped transversal filter, intersymbol interference in the IF signal caused by selective fading is automatically suppressed with tap control signals. The desired IF signal at the output of TRSV-EQL module is applied to PH-DEM module.

6.5.3 Demodulator (DEM)

The DEM comprises of the following modules:

- Phase demodulator (PH-DEM)
- Bit combiner (BIT-COMB)

The received IF signal is applied to the 4-PH-DEM and converted to two binary-coded pulse streams. The converted pulse streams are then sent to Rx-DPU through BIT-COMB. BIT-COMB selects two data streams with hitless switching between protection (PROT) and regular (REG) channels in response to any degradation detected in the bit error rate (BER) of the received signal.

PH-DEM

The PH-DEM consists of a four-phase demodulator circuit (4-PH-DEM) and a receive-logic circuit (Rx-Logic) (see Fig. 6.7).

4-PH DEM circuit: The 4-PH-DEM circuit consists of two phase detectors (PH-DET), a carrier recovery circuit (VCO and PH-ERROR-DET), a sampler, a clock synchronizer (CLK-SYNC) and an SC amplifier (SC-AMP).

The input IF signal (phase-modulated) is split into two and applied to the phase detector (PH-DET) circuit where the phase information is detected.

The voltage controlled oscillator (VCO) generating the carriers (LO1, LO2) for coherent detection is controlled by the output of the phase error detector (PH-ERROR-DET) circuit. The phase error of the output signal of the PH-DEM is detected at the PH-ERROR-DET to synchronize the carrier signal with the input IF signal.

The four-phase vector-to-carrier relationship is shown in Fig. 6.8. Subsequently, two data signals (D1, D2) and a clock pulse regenerated at the clock synchronizer (CLK-SYNC) circuit enter the sampler to discriminate the polarity of each bit, and then regenerate the two NRZ signals (R1, R2) at standard logic circuit level.

Rx-Logic circuit: Two 19.332 Mbps data streams of regenerated pulses and the clock pulses undergo differential decoding. The Rx-Logic circuit ogperates reverse of the differential encoding operation of the Tx-Logic circuit. The Gray code-to-binary code converter circuit converts the Gray-coded pulse streams to binary-coded pulse streams. In binary-coded pulse streams, the phases of the signals one time-slot before the current one are subtracted from the phases of the signals of the current time-slot in the form of quaternary numeration, and are thus decoded to Gray-coded data streams. These streams are then given as input to the frame-synchronizing unit (an LSI) along with the 19.332 MHz recovered clock pulse. The functions of the LSI circuit are frame synchronization establishment, demultiplexing (or segregating) the inserted information bits and the DSC signal bits, and BER detection.

Channel synchronization is obtained by selecting the frame synchronization bits for frame synchronization function (ID-CODE function) at both Tx and Rx ends.

The PH-DEM has an alarm circuit (DEM-ALM) for monitoring the operating status. Alarm signals appear under the following conditions:

1. Loss of output of 19.332 Mbps data stream
2. BER degradation exceeding 3×10^{-4} (known through parity check after frame synchronization)
3. Loss of frame synchronization

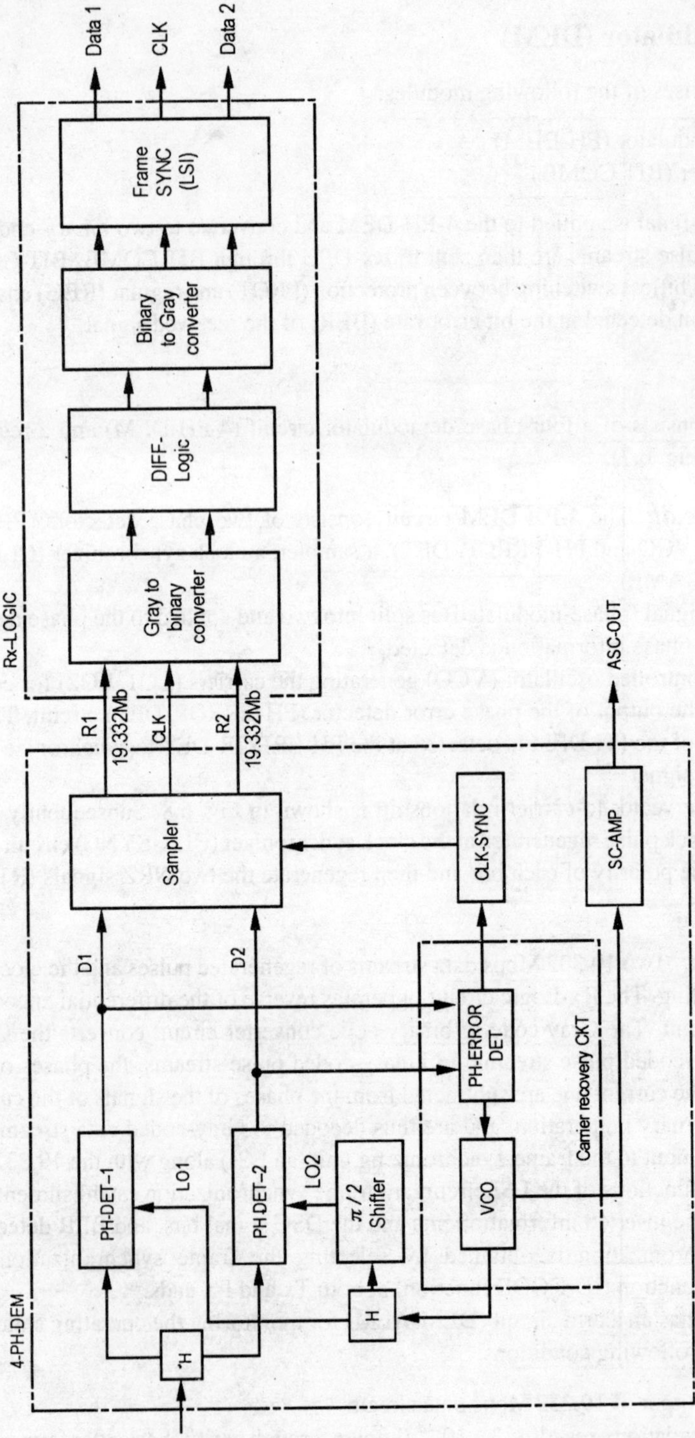

Fig. 6.7 *Simplified block diagram of PH-DEM.*

Phase of local carrier (LOI)	ψ = π/4					ψ = 3π/4					ψ = 5π/4					ψ = 7π/4				
Phase of IF carrier	PH-DET output		PLS regenerator output		QUATE	PH-DET output		PLS regenerator output		QUATE	PH-DET output		PLS regenerator output		QUATE	PH-DET output		PLS regenerator output		QUATE
	D1	D2	R1	R2		D1	D2	R1	R2		D1	D2	R1	R2		D1	D2	R1	R2	
0	+	+	1	1	2	–	+	0	1	1	–	–	0	0	0	+	–	1	0	3
$\pi/2$	+	–	1	0	3	+	+	1	1	–2	–	+	0	1	1	–	–	0	0	0
π	–	–	0	0	0	+	-	1	0	3	+	+	1	1	2	–	+	0	1	1
$3\pi/2$	–	+	0	1	1	-	-	0	0	0	+	–	1	0	3	+	+	1	1	2

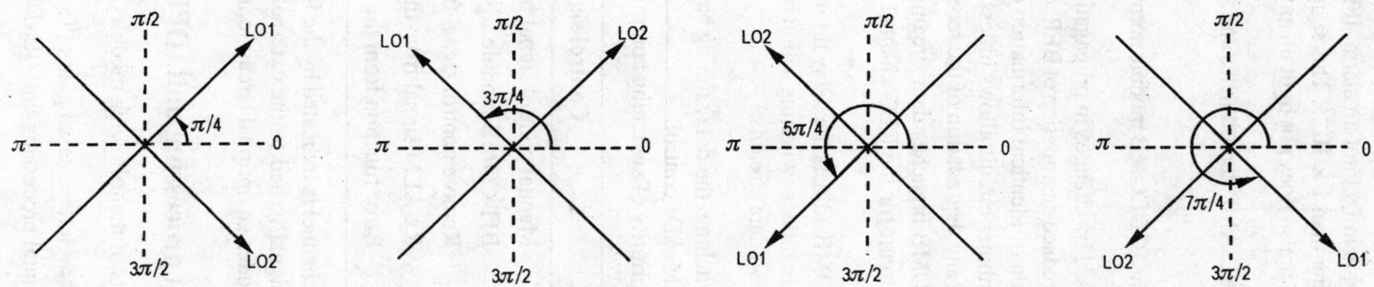

Fig. 6.8 *Four-phase vector-to-carrier relationship.*

Normally, a +5 V or 'high' signal appears when an alarm is created. When there is no alarm, a 'GND' or a 'low' signal appears.

The PH-DEM facilitates insertion/dropping of DSC and WS signals into the main data streams. The LOW-BER-ALM is in fact not an alarm. It is an assurance signal for transmission quality monitoring when BER is better than 1×10^{-7}. This signal does not actuate any switching for changeover between REG and PROT; it also does not light up any LED.

SC-AMP: The SC-AMP demodulates and amplifies ASC signals from the APC at the carrier recovery circuit.

BIT-COMB: The BIT-COMB module normally employed for the diversity system operates as follows:

- Fading caused by changes in propagation conditions can degrade the bit error rate (BER) of data signals. To reduce degradation of BER, diversity reception is employed. This includes transmission and reception of identical information over two channels identified as REG and PROT channels. The bit combiner circuit allows hitless or interruption-free switching between these channels in response to any degradation of the received signal detected in the BER.
- The BIT-COMB branches data streams from the PH-DEM into two—one enters the built-in bit combiner circuit for the REG channel and the other enters the other BIT-COMB for the PROT channel.
- The BIT-COMB adjusts the delay timing of each data stream in the REG and PROT channels and accomplishes hitless switching action by triggering the four switching mode signals so as to select one of the two data streams.

For hitless switching, the delay timing between the REG channel and the PROT channel is adjusted to zero by the TIM-ADJ control.

The order of priority of switching controls is as follows:

Priority	Control signal
1	Manual control signal from the REG-AUTO-PROT switch on the BIT-COMB module
2	Remote control signal from the external control equipment
3	Rx-ALM signal from the SWO-CONT module (automatic control)
4	Error function from the PH-DEM module (automatic control)

The selected channel is indicated by the REG or PROT indicator (LED) of the BIT-COMB module, and an answer signal issued to the external circuit. For the DSC and/or WS data transmission, the BIT-COMB includes an optional circuit to insert/drop DSC or WS signals into the main data streams.

6.5.4 Digital processing unit (DPU)

The DPU consists of the following modules:

- Transmitter digital processing unit (Tx-DPU)
- Receiver digital processing unit (Rx-DPU)

Tx-DPU

The Tx-DPU facilitates bipolar-to-unipolar conversion; serial-to-parallel conversion; insertion of frame pattern; parity check information; switch (SW) signals; DSC and WS signals. The assembled data is made out as two data streams, each at 19.332 Mbps. The data stream is scrambled and sent to the PH-MOD module.

The Tx-DPU consists of a line equalizer, HDB-3-to-unipolar converter, timing extractor, multiplexer and alarm control circuits.

Line equalizer circuit: The HDB-3 coded signal, which suffers distortion in the interconnecting cable, is applied to the fixed equalization circuit to pre-compensate distortion. This signal is then applied to the equalization circuit which is under automatic control. The frequency characteristics of the equalization process are controlled by the current from the DET circuit, and the output level of the equalization section is maintained a constant.

HDB-3 to unipolar converter and timing extractor: High density bipolar-3 (HDB-3) coding is used at the interface of any MUX and radio equipment (refer to Chapter 12 on Line Codes). HDB-3-to-unipolar conversion with simultaneous extraction of timing (clock) is done in this sub-module.

Serial-to-parallel converter circuit: Figure 6.9 shows the block diagram of serial-to-parallel converter. The 34.368 Mbps data signal is applied to two memories—M1 and M2. The data signals are read out from these memories by a 1/2 divided input clock. The signal from M1 is re-timed at M3 and sent out together with the signal from M2.

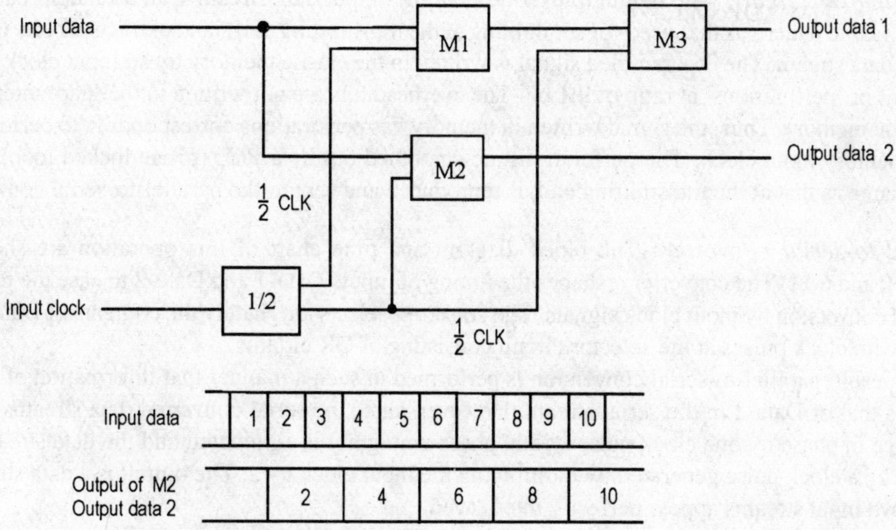

Fig. 6.9 *Functional block diagram of serial-parallel converter.*

Multiplexer (MUX): In this circuit, overhead bits are inserted into the two data streams at intervals of eight bits. For Data-1 stream, FS (frame synchronazation) bit and the parity bit are inserted, while for Data-2 stream, FS bit and DSC bit are inserted. The FS bit is used for establishing frame synchronization and the parity check bit is used for monitoring the quality of the channel.

Scrambler (SCRB): SCRB scrambles each data stream with a 432-bit pattern generated by control/time base generator. In such scrambling, the transmission signal mark ratio between radio stations is averaged and the transmit output power spectrum is kept constant, disregarding the mark ratio of the output multiplex equipment in the terminal station.

Alarm control circuit: The Tx-DPU has an alarm circuit (Tx-DPU-ALM) for monitoring the operating status. Alarm signaling conditions include loss of input 34.368 Mbps data streams, output 19.332 Mbps data streams and loss of read-out clock signal (19.332 MHz) at the buffer memory circuit. Under normal conditions, a '1' signal appears and in alarm condition, the Tx-DPU-ALM indicator lights red and a '0' signal appears.

Rx-DPU

The Rx-DPU descrambles, removes overhead bits, corrects delay timing difference between the input data stream and the regular and protection channels, and converts a unipolar NRZ signal to an HDB-3 signal. It functions converse to that of the B-U converter operation at the Tx end, and feeds the original HDB-3 signal to the SWO module or the PCM multiplex equipment.

The Rx-DPU comprises demultiplexer, delay adjuster, unipolar-to-HDB-3 converter and alarm control circuit.

Demultiplexer circuit: The demultiplexer descrambles input data stream with a random pattern of 432 bits (same pattern as that used for scrambling at the transmitting end), and extracts all bits inserted into the data stream. The descrambled signal is written in the elastic memory by an input clock having time gaps or 'perforations' at ratio of 9:1 bit. The overhead bits are not written in the 'gap' intervals in the elastic memory. Thus, the signal written in memory has perforations corresponding to perforations in the memory input clock. The perforations are smoothed out by a PLL (phase-locked loop), and a signal, same as that at the transmitting end, is reproduced and sent to the parallel-to-serial converter.

Parallel-to-serial converter: The block diagram and time chart of this operation are shown in Figs 6.10 and 6.11. The converter reshapes the timing of inputs Data-1 and Data-2 to ease the parallel-to-serial conversion by input clock signals. The converter selects alternately the contents of Data-1 and Data-2 with clock pulses at the selector circuit consisting of OR circuits.

As a result, parallel-to-serial conversion is performed in such a manner that information of Data-1 precedes that of Data-2 in the serial stream. However, since the serial converted data streams have a difference in phase by one clock pulse-width, phase ambiguity is eliminated and the data streams are retimed by a clock pulse generated by multiplying the input clock by 2. The output is a data stream in which two input streams appear perfectly interleaved.

Unipolar to HDB-3 converter: The input NRZ data stream is fed to the data splitter circuit which divides the input data into the '0' and '1' position signal respectively. The '0' position signal is applied to a '0000' or a '0001' or '1001' encoder which converts the 4-bit continuous input 0 sequence to '0001' or '10001' sequence.

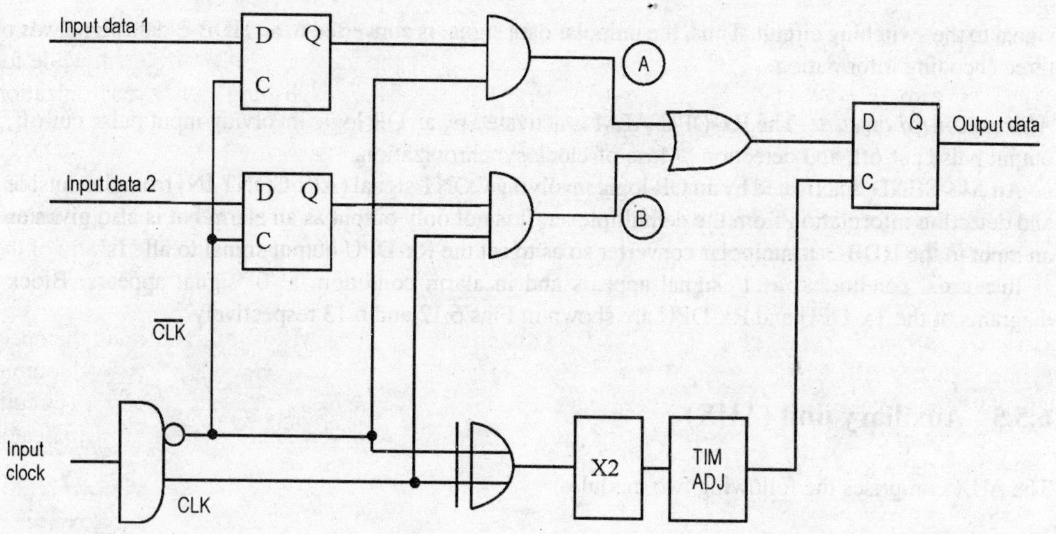

Fig. 6.10 *Functional block diagram of a parallel-serial converter.*

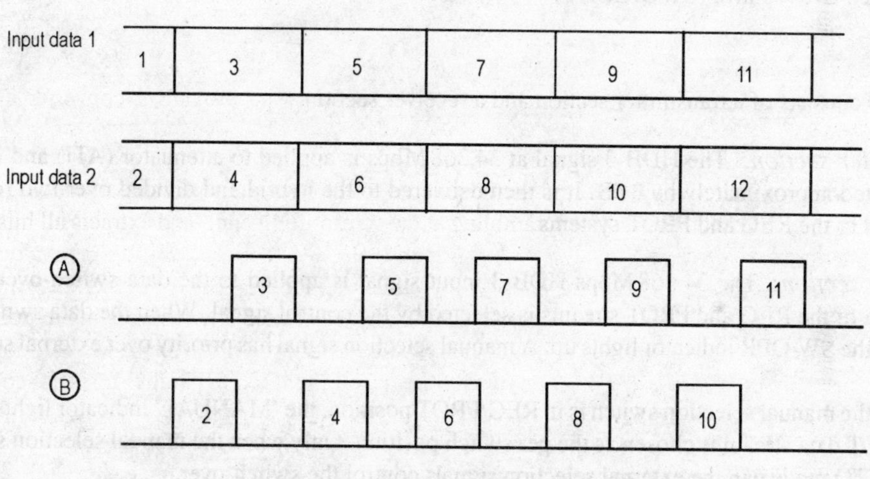

Fig. 6.11 *Time chart of a parallel-serial converter.*

Part of the output encoded data is then applied as reset input to the Modulo-2 logic circuit where the split '1' position signal from the data-splitter circuit is processed sequentially. The Modulo-2 logic circuit detects whether the number of '1s' included in the violation pulse interval is odd or even. If an even number is detected, a bipolar violation pulse is inserted.

On the other hand, the data stream is also applied to a 5-bit shifter where an insertion coincident with the timing of the bipolar violation pulse is effected. This is inserted into the main data signal.

The output data from the '0000' to '0001' encoder is applied to the switching circuit through a 4-bit shifter and is used as a trigger for the violation pulses.

The data-stream of the main signal is also applied to a polar discriminator circuit as well as a switching circuit. The polar discriminator circuit supplies the decoded polar information of the main signal to the switching circuit. Thus, the unipolar data signal is converted to an HDB-3 data signal with three encoding information.

Alarm control circuit: The Rx-DPU-ALM is activated by an OR logic involving input pulse cut-off, output pulse cut-off, and detection of loss-of-clock synchronization.

An AIS-SEND is activated by an OR logic involving CONT signal (AIS-CONT-IN) from the outside and detection information from the demultiplexer. It is not only output as an alarm but is also given as an input to the HDB-3-to-unipolar converter so as to set the Rx-DPU output signal to all '1s'.

In normal conditions, a '1' signal appears and in alarm condition, a '0' signal appears. Block diagrams of the Tx-DPU and Rx-DPU are shown in Figs 6.12 and 6.13 respectively.

6.5.5 Auxiliary unit (AUX)

The AUX comprises the following two modules:

- Switch-over (SWO)
- Switch-over control (SWO-CONT)

SWO

The SWO consists of a transmitter section and a receiver section.

Transmitter section: The HDB-3 signal at 34.368 Mbps is applied to attenuator (ATI) and its level is attenuated approximately by 3 dB. It is then delivered to the hybrid and divided over two routes to be applied to the REG and PROT systems.

Receiver section: The 34.368 Mbps HDB-3 input signal is applied to the data switch-over circuit where one of the REG and PROT streams is selected by the control signal. When the data switch-over operates, the SW-OPR indicator lights up. A manual selection signal has priority over external selection signal.

When the manual selection switch is in REG/PROT position, the 'MANUAL' indicator lights up and REG/PROT data stream is chosen as the per switch position. Only when the manual selection signal is in the AUTO mode can the external selection signals control the switch-over.

SWO-CONT

The SWO-CONT consists of the ASC circuit, and the alarm and control circuits.

Alarm and control circuits: The SWO-CONT indicates an alarm condition and also outputs alarm and control signals to the external equipment. Alarm signals are monitored in each of the respective modules and the control signals are gathered at the SWO-CONT module for processing according to the need of the situation.

ASC circuit: The ASC signal applied to the hybrid is divided over two routes to be applied to the REG and PROT equipment through an attenuator. The received ASC signal is switched over by a manual switch, an alarm or a remote control signal.

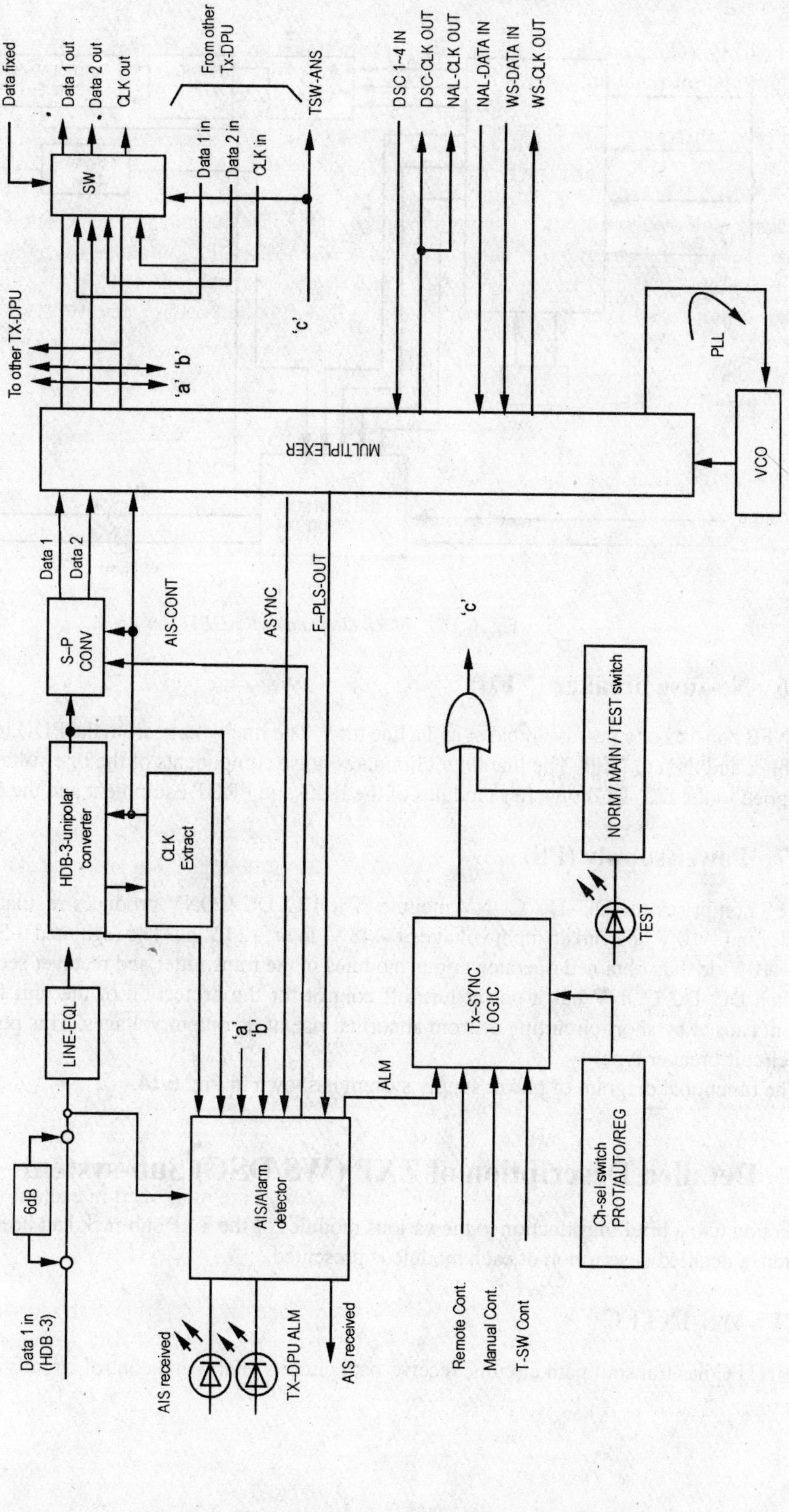

Fig. 6.12 *Functional block diagram of Tx-DPU.*

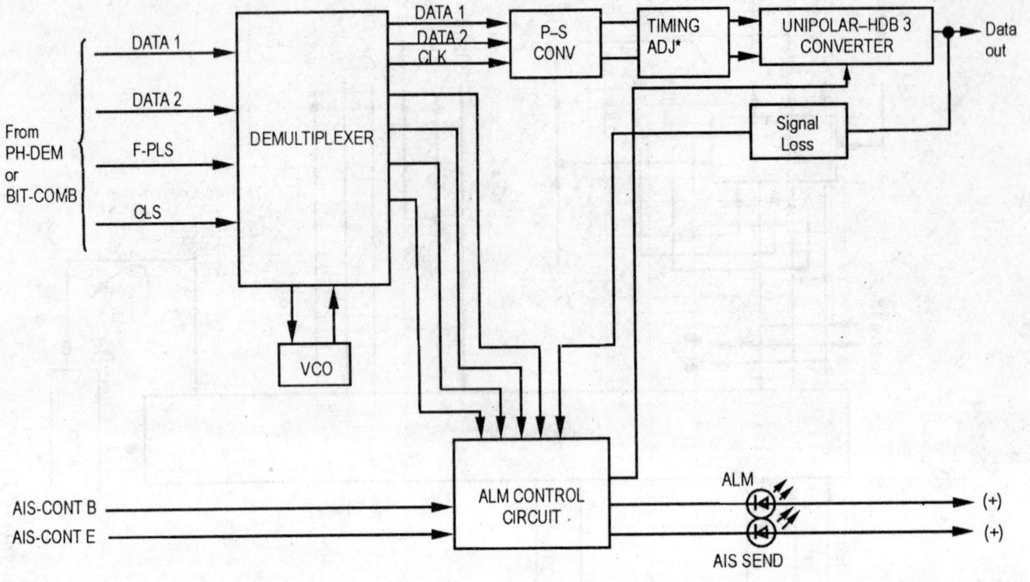

Fig. 6.13 *Block diagram of Rx-DPU.*

6.5.6 No-fuse breaker (NFB)

The NFB consists of a no-fuse-breaker and a line filter. The line voltage from the PDU is applied to the line filter and then to NFB. The line filter eliminates noise components of the line voltage. The output is applied to the DC–DC converter modules of the REG and PROT equipment and the SC modules.

6.5.7 Power supply (PS)

The PS comprises the DC–DC CONV module. The DC–DC CONV produces regulated +5V dc, + 9 V dc, and −10 V dc from an input voltage of −48 V dc or −24 V dc. The regulated +5V dc, +9 V dc and −10 V dc thus obtained operate various modules in the transmitter and receiver sections.

Each DC–DC CONV has a power turn-off control for the protection of the unit from excessive current caused by short-circuiting or from abnormal rise of dc output voltages. This power turn-off is of a circuit-breaker type.

The functional diagram of power supply system is shown in Fig. 6.14.

6.6 Detailed Description of ZXP (WS/DSC) Sub-system

In Section 6.4, a brief introduction to the various modules of the ZXP sub-rack had been given. In this section, a detailed description of each module is presented.

6.6.1 WS-INTFC

WS-INTFC has transmit path circuits, receive-path circuits, alarm and control circuits.

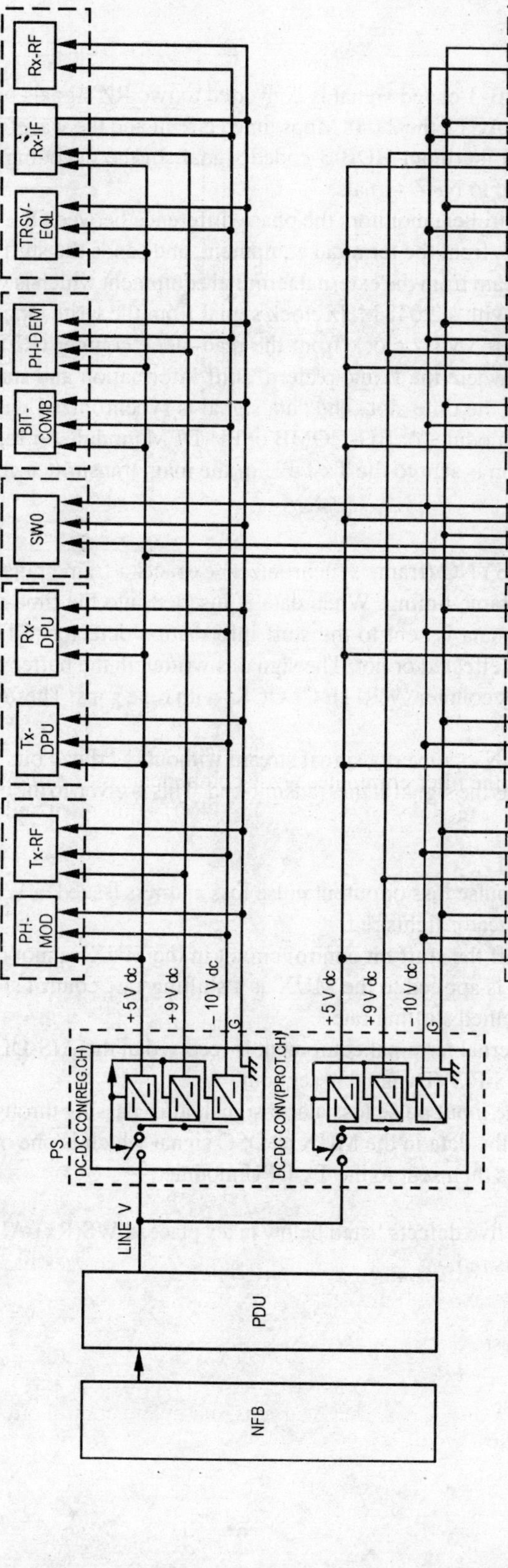

Fig. 6.14 *Block diagram of power supply for the terminal equipment.*

Transmit path

Refer to Fig. 6.15. The input HDB-3-coded signal is converted to two RZ signals by a B–U converter. The clock extractor (CLK-EXTRACT), the 2.048 Mbps tuned circuit and the waveform shaping circuit regenerate the clock signal from the input HDB-3-coded signal. In the decoding process, violation pulses are removed and converted to NRZ signals.

The STUFF-CONT (stuff controller) monitors the phase difference between the input master clock from Tx-DPU and the input clock from the terminal equipment, and sends the stuff information to the multiplexer (MUX). The data stream from the external terminal equipment which is written in the buffer memory (BUFFER MEMORY) with a 2.048 MHz clock signal from the write clock counter (WRITE CLOCK), is read out with a 2.4165 MHz clock from the read-clock counter (READ CLOCK). The data stream is sent to the MUX where the frame pattern, stuff information and stuff bits are inserted at the position of overhead bit of the time-slot. The data signal is synchronized with the input master clock from the Tx-DPU or DPU module (or BIT-COMB or PH-DEM module for regenerative repeater station). Then the WS information is sent to the Tx-DPU of the main transmitter–receiver equipment.

Receive path

Refer to Fig. 6.15 again. The F-SYNC (frame synchronizer) extracts a frame pulse (F-PLS) by 1-bit shift-hunting method to obtain frame timing. When data is inserted into the time-slot of the stuff-bit slot at the transmitting end, the data is sent to the stuff information detector (STUFF-INF-DET) to determine whether the stuffing is effected or not. The signal is written in the buffer memory by a clock signal generated at the write clock counter (WRITE CLOCK) with time gaps. These gaps are smoothed by the PLL.

The ENCODER and U–B CONV convert a signal stream without overhead bits to a HDB-3-coded bipolar stream which is the same as the signal at the transmit-end. This is given to the external equipment.

Alarm circuits

Transmit side: When an input pulse loss or output pulse loss alarm is issued, a WS(Tx)-ALM signal is sent out and the WS-ALM indicator lights red.

When the input pulse is cut-off the stuff bit control circuit in the MUX cannot operate. Therefore, the input pulse loss alarm signal is applied to the MUX as a stuffing-rate control signal to set the data signal from the MUX to the specified stuffing rate.

When an AIS signal from external terminal equipment is received at the AIS-DET, the AIS-RCVD indicator lights amber and an AIS-RCVD signal is sent out.

The AIS-RCVD signal and the input pulse loss alarm signal, after passing through the DSC as AIS information, are multiplexed on the data in the MUX as DSC signal which is one of the additional bit for the radio link. The resultant is then sent to the Tx-DPU module.

Receive side: When one of the five defects listed below takes place, a WS(Rx)-ALM signal is issued and the WS-ALM indicator lights red.

- Input pulse is lost
- Frame synchronization is lost
- Bit error rate is degraded
- VCO synchronization is lost
- Output pulse is lost

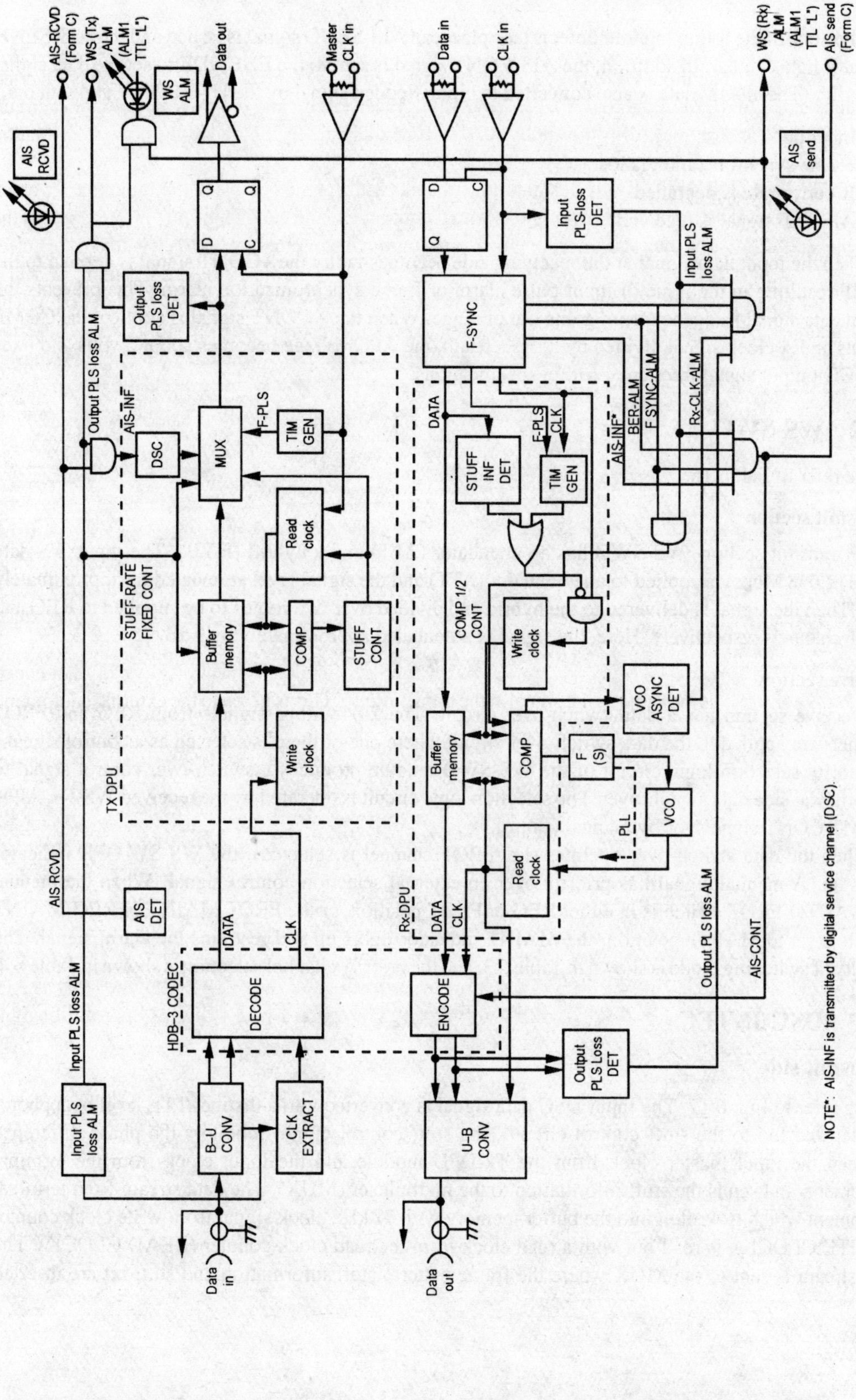

NOTE*: AIS-INF is transmitted by digital service channel (DSC).

Fig. 6.15 *Functional block diagram of WS-INTFC.*

When one of the following four defects take place, an AIS-SEND signal is issued and the AIS-SEND indicator lights amber. In addition, the AIS-CONT signal is sent to the ENCODE to set the data signal to all '1s'. (The all '1s' data is also converted to bipolar-coded signal by the U–B CONV and sent out.)

- Input pulse is lost
- Frame synchronization is lost
- Bit error rate is degraded
- AIS-INF signal is received

When the input data signal at the receiving side becomes faulty the +CONT signal is applied to the COMP resulting in the issue of input pulse alarm or frame synchronization alarm. This prevents the output data signal frequency from going out of range. When the +CONT signal is applied, the COMP outputs an Rx clock signal divided by 2. As a result, the APC voltage becomes about 2 volts and fixes the VCO output signal frequency within specifications.

6.6.2 WS-SWO

Please refer to Fig. 6.16.

Transmit section

In the transmit section, WS-SWO has an attenuator (ATT) and a hybrid (HYB). The input WS data signal (2.048 Mbps) is applied to an attenuator (ATT) and the signal level attenuated by approximately 6 dB. Then the signal is delivered to the hybrid and divided over two routes to be supplied to REG and PROT channels respectively. Here, the signal is atttenuated approximately by 9 dB.

Receive section

The receive section has a data switch-over circuit. The 2.048 Mbps signals from REG and PROT channels are applied to the data switch-over circuit where one of them is selected as a control signal.

Priority selection logic circuit of the WS-SWO module produces a switch-over control signal to perform wayside data switch-over. The selection logic circuit is operated by the received WS-Rx-ALM or RMT-CONT signals, or by manual control.

When the data switch-over operates (i.e. PROT channel is selected), the WS-SW-OPR indicator lights up. A manual signal has priority over an external selection control signal. When the manual REG-AUTO-PROT switch is in either REG or PROT position, or the PROT-MAINT or REG-MAINT switch is set to MAINT position, the MAINT indicator lights up and restrains the alarm signal. The priority of switching mode is shown in Table 6.3, and the priority channel selection is shown in Table 6.4.

6.6.3 DSC-INTFC

Transmit side

Please refer to Fig. 6.17. The input DSC data signal is converted into a distinct TTL level by a photo-coupler and fed to the stuff control circuit. The stuff control circuit monitors the phase difference between the input master clock from the Tx-DPU module and the input clock from the terminal equipment, and sends the stuff information to the multiplexer (MUX). The data stream from terminal equipment which is written into the buffer memory with 72 kHz clock signal from write clock counter (WRITE CLOCK), is read out with a read clock from the read clock counter (READ CLOCK). The data stream is sent to the MUX where the frame pattern, stuff information and stuff bit are inserted

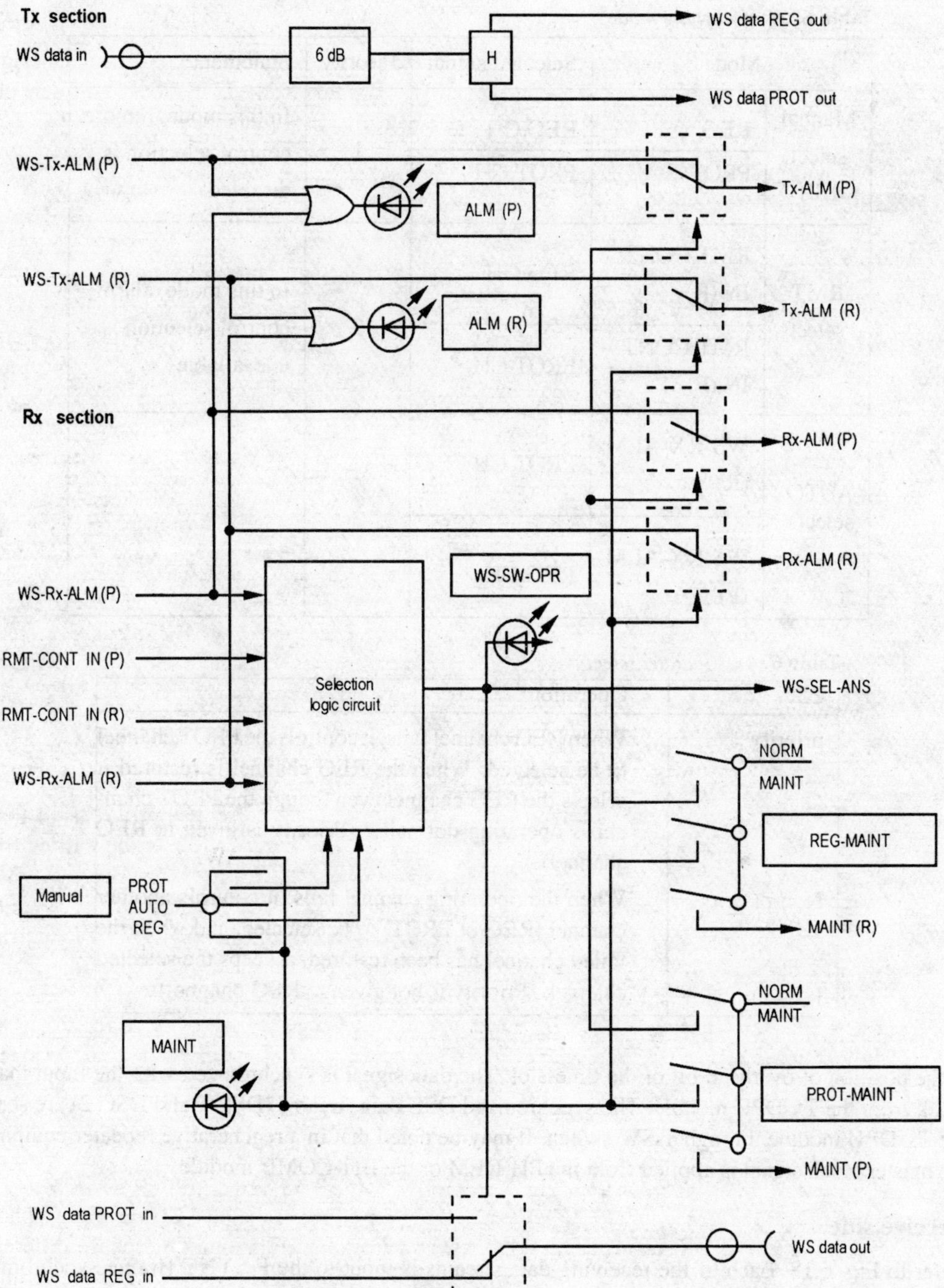

Fig. 6.16 *Functional block diagram of WS-SWO.*

Table 6. 3 *Switching mode*

Mode		Selected signal	Priority	Statement
Manual select (SI)	REG	REG CH	1	In this mode, remote control selection is disabled
	PROT	PROT CH		
RMT select	RMT-CONT IN (R)	REG CH	2	In this mode, alarm control selection is unavailable
	RMT-CONT IN (P)	PROT CH		
AUTO select	WS-RX-ALM (R)	PROT CH	3	
	WS-RX-ALM (P)	REG CH		

Table 6. 4 *Channel selection*

Z5	Operation
priority	When REG channel fails, it controls the PROT channel to be selected. When the REG channel is restored it selects the REG channel even though the PROT channel is operating normally. (Priority is given to REG channel)
Non-priority	When the operating channel fails, it controls another channel (REG or PROT) to be selected, and when the failed channel has been restored, it keeps the selected channel. (Priority is not given to REG channel)

at the position of overhead bit of the time-slot. The data signal is synchronized with the input master clock from the Tx-DPU module. The synchronized DSC data streams (DSC 1 and DSC 2) are sent to the Tx-DPU module through T-SW switch. It may be noted that in a regenerative repeater equipment the master clock signal is applied from the PH-DEM or the BIT-COMB module.

Receive side

Refer to Fig. 6.18. Each of the receiving data streams is input to the F-SYNC (frame synchronizer) circuit. F-SYNC initiates a frame pulse (F-PLS) by 1-bit-shift hunting method to obtain frame synchronization. The stuff information detector detects a stuff pulse and the timing generator detects a frame pulse. When the detectors detect a stuff pulse or a frame pulse, the one clock pulse applied to the write counter is rejected.

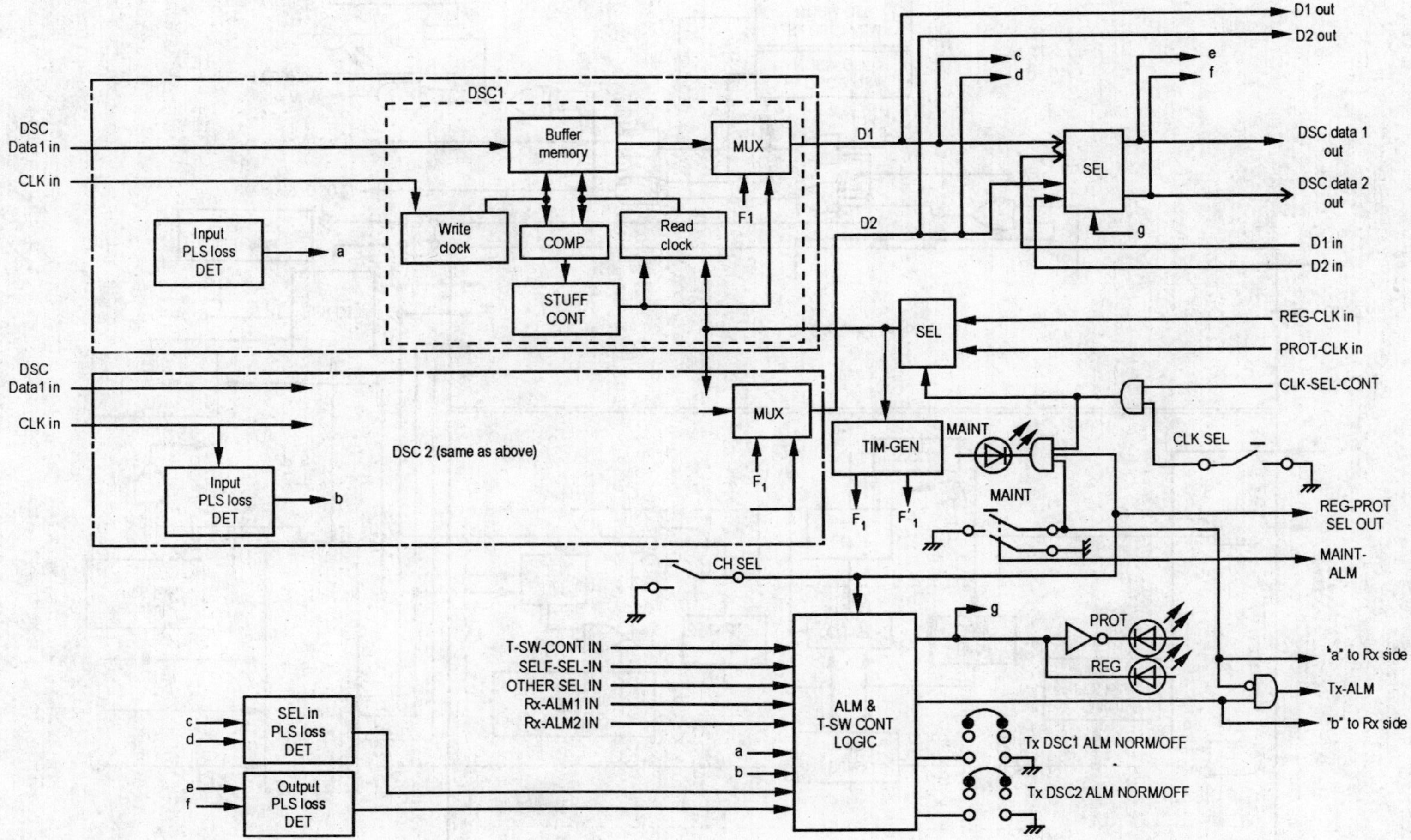

Fig. 6.17 *Block diagram of the transmitting section of D2806A DSC-INTFC module.*

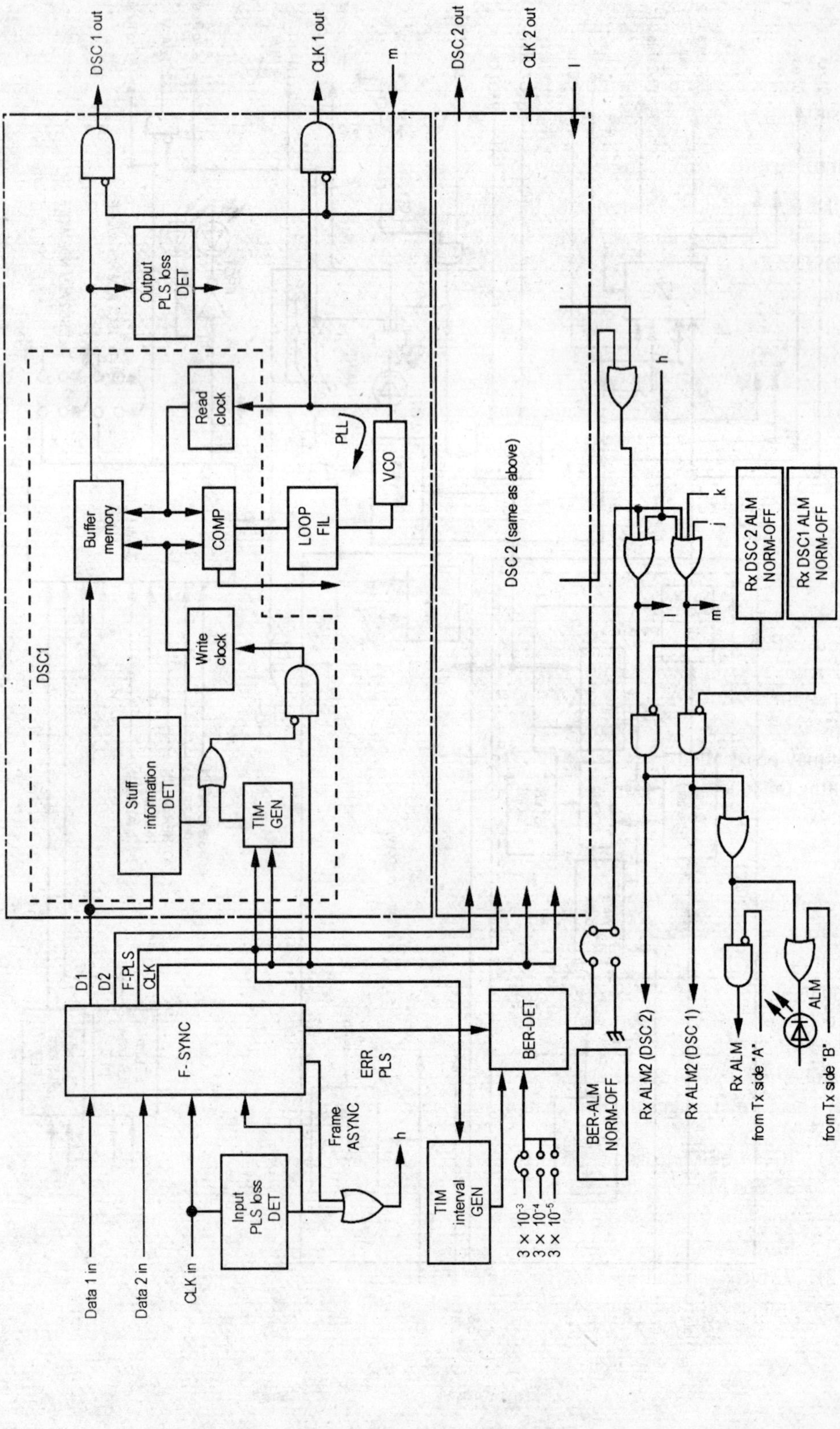

Fig. 6.18 *Block diagram of the receiving section of D2806A DSC-INTFC module.*

The signal is written into memory by a write clock having time gaps. At this time the overhead bits (F-PLS and STUFF-PLS) are not written in the buffer memory, thus the signal written in memory has gaps. The gaps are smoothed by a PLL (phase-locked loop). The original signal—same as that of the transmitting end—is reproduced and sent to the external terminal equipment.

Alarm circuit

On the transmit side, when one of the input pulse streams is lost or the output pulse stream is lost, or the selector input pulse stream is lost, an alarm is issued at DSC 1 or DSC 2 as the case may be. Tx-DSC-ALM (1 or 2) signal is generated and the ALM indication lights red. The DSC channel with alarm can be inhibited by setting the plug in such a way as to allow the other traffic to flow.

On the receive side, Rx-ALM (1 or 2) is generated whenever there is VCO-asynchronization, poor BER, signal loss, or output pulse loss. The bad DSC channel can be inhibited by setting the plug in such a way that it permits other traffic to flow. The BER alarm threshold (1×10^{-3} or 1×10^{-4} or 1×10^{-5}) can be decided by strap setting. Section 6.8 treats SC-INTFC-D in detail.

6.7 Various Data Rates Encountered in 34+2 Mb Digital MW Radio

We come across data streams of various rates in a 34 + 2 Mb digital MW radio. We turn our attention to these data rates and understand the calculations involved Data rates of 34 Mbps and 2 Mbps (34.368 Mbps and 2.048 Mbps to be precise) have been explained earlier. We have various other data rates like 72 kbps and 89.5 kbps associated with DSC; 2.4615 Mbps associated with WS interface in Tx DPU; 19.332 Mbps associated with the two-streams at the output of Tx-DPU or the input of Rx-DPU. We will first look at the calculation pertaining to 19.332 Mbps.

It may be recalled from Section 6.5 that two 19.332 Mbps streams are output by the Tx-DPU after splitting the incoming 34.368 Mbps into two streams and multiplexing them with WS and DSC traffic, parity bits, and framing bits. While multiplexing, an elastic memory of eight bits is used for *each* of the 17.184 Mbps stream. For every 8 bits of 17.184 Mbps, one bit of FS/parity, WS and DSC bits are added in a pre-determined fashion to *each* of the 17.184 Mbps stream. Thus, with each eight bits getting one more bit makes the data rate equal to 17.184 × (9/8), which equals 19.332 Mbps. This is the 'Rule of 9 to 8' in the multiplexing (demultiplexing) operation of the Tx-DPU(Rx-DPU). Obviously, the output of Tx-DPU has two streams of 19.332 Mbps corresponding to the two streams of 17.184 Mbps at the input.

In other words, we can consider the process as follows: Two 19.332 Mbps put together make 38.664 Mbps, which is in fact the result of multiplexing (*i*) 34.368 Mbps mainstream, (*ii*) 2.4165 WS stream, (*iii*) four DSC streams each at 89.5 kbps, (*iv*) parity and framing patterns, and (*v*) switch-over data. This format makes the explanation simpler.

1) It can be seen from Fig. 6.19 that the master clock (derived from the received stream, i.e. 38.664 Mbps) is divided by 72 to get the sub-frame pulses. All the 72 bits together constitute one sub-frame. From the 'Rule of 9 to 8', we have 64 bits of mainstream (34.368 Mbps) traffic.
2) 72 bits constituting one sub-frame implies that there are (38.664/72) Mbps, i.e. 537 k sub-frames. For each sub-frame we have one parity bit (the odd/even parity bits pertaining to the

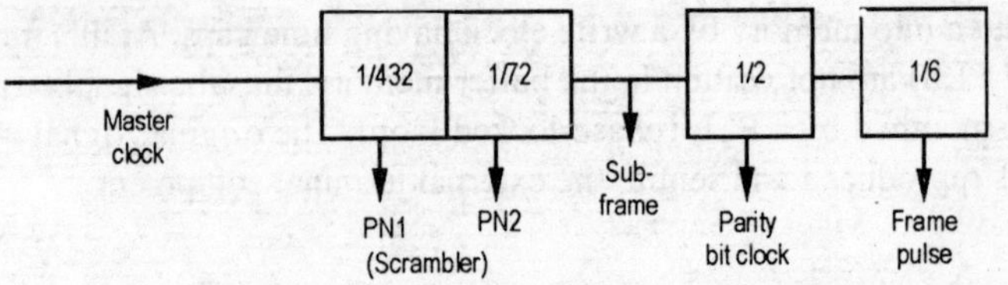

Fig. 6.19 *Sub-frame and parity clocks.*

two 17.184 Mbps streams are transmitted with every alternate sub-frame). Hence, we have 537 k parity bits.

3) Framing pattern is chosen via jumper-setting two bytes—FS1 and FS2. These two bytes make 16 bits which are repeated in sequence for every 16 sub-frames, i.e., in a set of two successive sub-frames, we have bits from FS1 and FS2, alternately. For 537 k sub-frames we have 537 k FS bits.

4) WS stream pumps in data at 2.4165 Mbps. Four DSC streams put together pump in 358 kbps (4 × 89.5 kbps). These are accomodated as follows:

   – WS data at the rate of 9 bits for two sub-frames
   – DSC data at the rate of 2 bits for three sub-frames

   We also have switch-over (NAL) data. This is accomodated at the rate of 5 bits for 6 sub-frames, i.e. 537 k $\times 5/6 = 447.5$ kbits

5) Now the grand total can be worked out as the sum of the following:

   34.368 M or 34368 kb mainstream
   2.4165 M or 2416.5 kb WS stream
   358 kb DSCs
   537 kb parity
   537 kb FS data
   447.5 kb NAL data
   38664 kb or 38.664 Mb

   Since this is output as two streams, each stream has a data rate of 19.332 Mbps.

6) The above data rates can be cross-checked as follows:

   A set of six consecutive sub-frames form a mini frame. The bit map for a mini frame is given in Table 6.5.

Table 6. 5 *Bit map for a mini-frame*

Sub-frame No.	Main stream bits	Parity bits	FS bits	WS bits	DSC bits	NAL bits	Total bits
1	64	1	1	4	1	1	72
2	64	1	1	5	-	1	72
3	64	1	1	4	1	1	72
4	64	1	1	5	1	-	72
5	64	1	1	4	1	1	72
6	64	1	1	5	-	1	72

7) Fractions of bit assignments in a 72-bit sub-frame are given in Table 6.6.

Table 6. 6

Data-stream	Rate of occurrence	Fraction
Main stream	64 in 72	8/9
WS	9 in 72 × 2	1/16
DSC	2 in 72 × 3	1/108
Parity	1 in 72	1/72
FS	1 in 72	1/72
NAL	5 in 72 × 6	5/432
		Total 1

8) All the clocks correspond to the fraction arrived at in Table 6.6.

For WS, $38.664 \times 1/16 = 2.4165$ Mbps

For one DSC, $38.664 \times 1/108 \times (1/4) = 38.664 \times 1/432 = 89.5$ kbps

Thus, clocks of 2.4165 Mbps and 89.5 kbps can be easily derived from 38.664 Mbps incoming clock (recovered). This is why 2.4165 Mbps and 89.5 kbps are chosen for WS and DSC interface on the main transmitter–receiver side.

9) It may please be noted that a mini frame which has six sub-frames has 6 × 72 bit, i.e., 432 bits. This is the same as the scramble pattern length chosen.

6.8 SC-INTFC-D

6.8.1 Why 72 kbps intermediate data rate in processing DSC data?

In the previous section we have seen that each of the DSC data stream is given as an input to the main transmitter (Tx-DPU) at 89.5 kbps to facilitate integration. 89.5 kbps being a sub-multiple of the master clock (38.664 Mbps), integration was simple. In section 6.4 it was mentioned that DSC-INTFC has 72 kbps interface with DSC channels and 89.5 kbps interface with the main transmitter–receiver. DSC being the digital channel for express order wire/G.703 /Ext VF interface, the data rate should be 64 kbps. Then why again this 72 kbps data rate, i.e., one more step of data rate enhancement? Can we not go directly to 89.5 kbps from 64 kbps?

The answer involves the concept of 'SV' or 'supervisory system'. This calls for more discussion. We already have NMS or the network monitoring system that makes use of ASC. NMS monitors radio and other equipment parameters through telesignals and facilitates control/switch-over between REGand PROT equipment via telecommands.

Then why this SV?

Digital MW radio is designed with the idea of a nation-wide network rather than a single route solution. A nation-wide network provides for route diversity with provision for switching the traffic onto different routes at branching stations through a nation-wide supervisory system (SV). While NMS is confined to a route, or at most to a small network, SV has a nation-wide network of several routes under its purview. SV only uses DSC and not ASC.

Thus, DSC has to provide a 64 kbps channel for exchanging information about the status of various routes, and enable a 'digital path' (even low data rate paths would suffice) for control commands of SV. To take care of this aspect, SC-INTFC-D is designed to accept up to four 1.33 kbps channels. SC-INTFC-D also facilitates a drop/insert type configuration for the DSC channel instead of the simple omnibus-type arrangement of the ASC.

Four channels with 1.33 kbps data rate serve as AUX-data channels. The 64 kbps channel and four AUX-data channels are integrated in the SC-INTFC-D to give a data rate of 72 kbps, which is treated as one DSC. The frame structure for 72 kbps DSC line data is given in Fig. 6.20.

6.8.2 Frame structure of DSC data

The features of the DSC data frame are as given below.

1. Nine bits form one sub-frame and three sub-frames form one mini frame (i.e., 27 bits in one mini frame)
2. In every mini frame, one bit is for frame synchronization (F1, F2, F3 and F4 bits of the entire pattern are placed in four consequent mini frames)
3. AUX-CH data occupies two bits. In two consequent mini frames, four bits—one pertaining to each AUX channel data—gets accomodated.
4. The remaining 24 bits are for PCM data (Express OW/G.703/Ext.VF) which pertain to the 64 kbps stream.
5. As per the above description, the makeup of various data in one mini frame of 27 bits are:

   PCM data : 24/27

   AUX data: 2/27 (all channels put-together or 1/54 for each channel)

   PN data: 1/27

   Multiplying each by the DSC line rate of 72 kbps, we get:

   PCM data: $24/27 \times 72$ kbps $= 64$ kbps
   AUX data: $1/54 \times 72$ kbps $= 1.33$ kbps
   PN data: $1/27 \times 72$ kbps $= 2.66$ kbps
   Thus, all data rates stand tallied.

6.8.3 SC-INTFC-D configuration

SC-INTFC-D (service channel interface unit for DSC) has the following configurations:

Type 1 For use at the terminal station with order-wire telephone (Fig. 6.21).
It handles two channels of 72 kbps DSC signals, one for SV and OW, and the other for external VF channel.

Type 2 For use at repeater station without order-wire telephone (Fig. 6.22). It handles two channels of 72 kbps DSC signal, both for external VF channel.

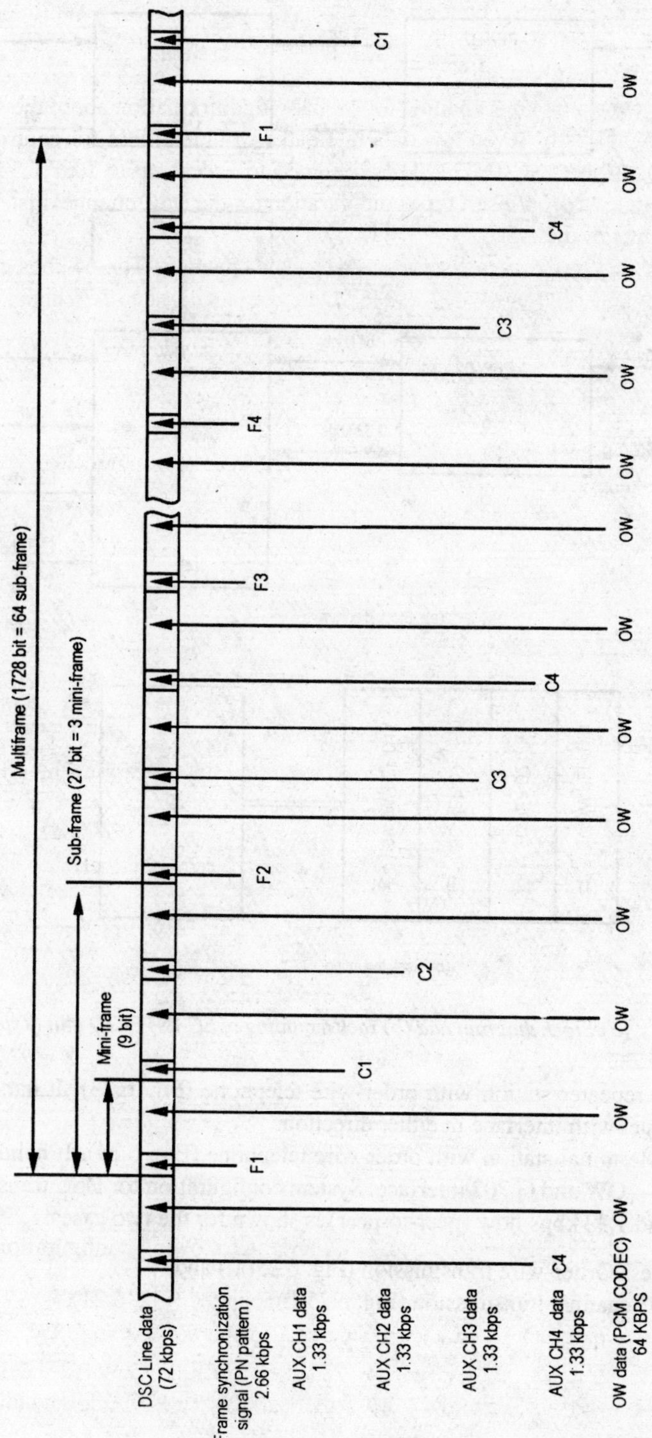

Fig. 6.20 *Frame format of DSC signal.*

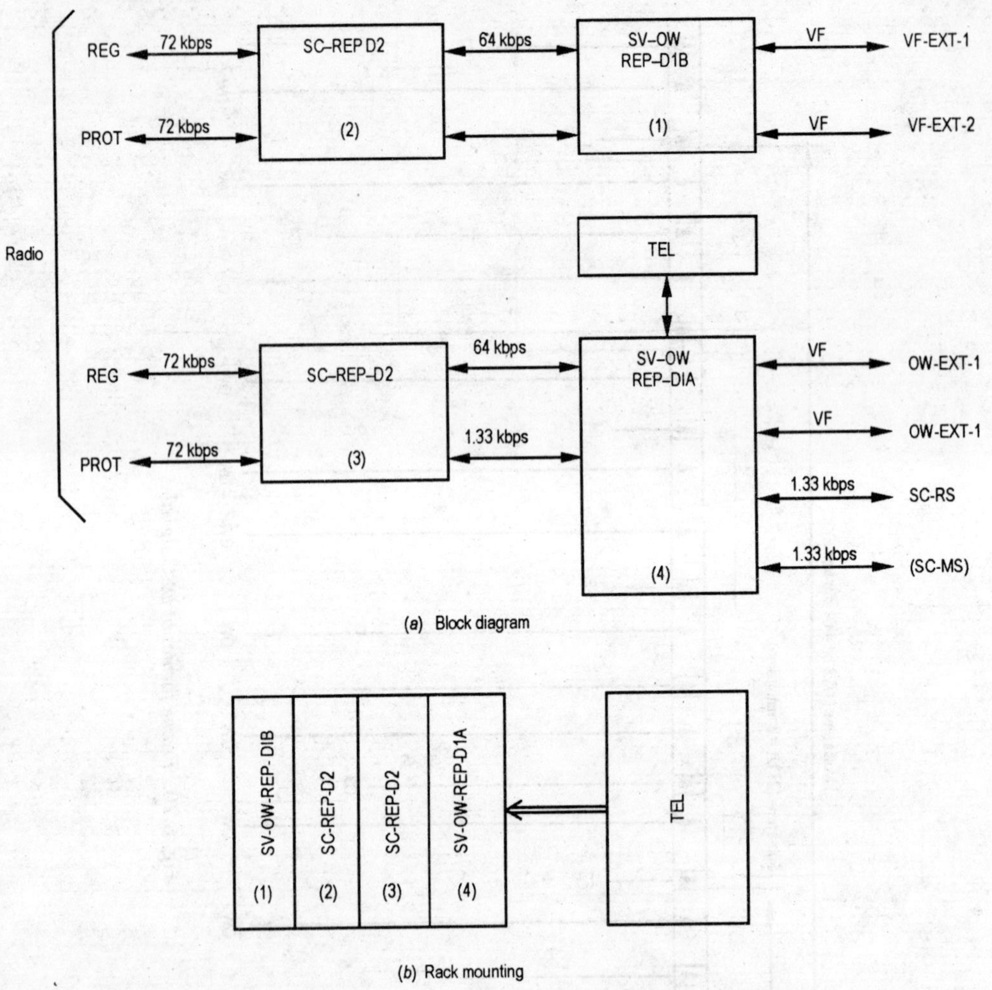

(a) Block diagram

(b) Rack mounting

Fig. 6.21 *(a) Block diagram and (b) rack mounting of SC-INTFC-D unit (Type 1).*

Type 3 For use at repeater station with order-wire telephone (Fig. 6.23). It handles one 72 kbps DSC channel with interface in either direction.

Type 4 For use at terminal station with order-wire telephone (Fig. 6.24). It handles two channels of 72 kbps—OW and G.703 interface. System configuration for DSC transmission showing 64 kbps and 1.33 kbps flow (peer-to-peer) is shown for the two cases:

(*i*) Express Order-wire transmission (Fig. 6.25(a)) and

(*ii*) G.703 channel transmission (Fig. 6.25(b)).

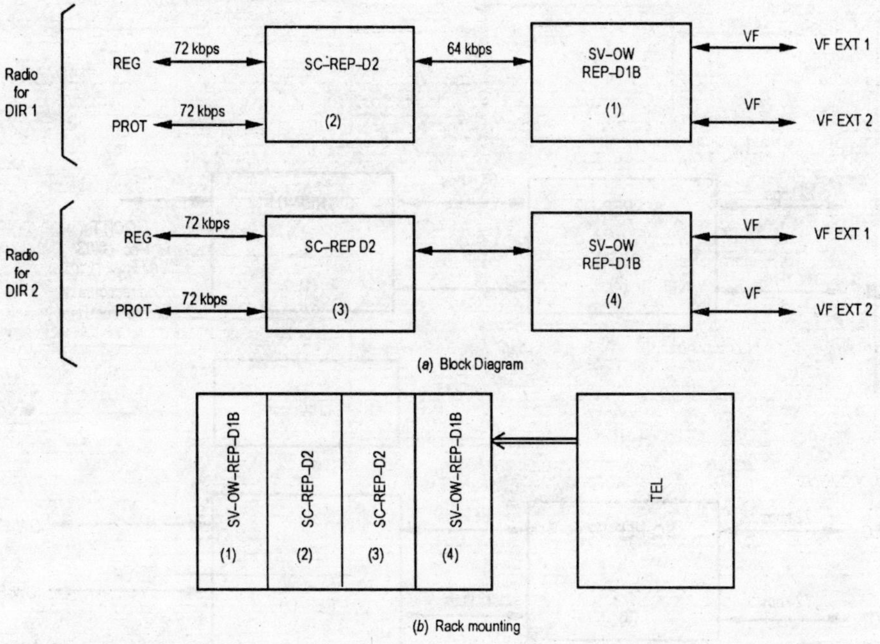

Fig. 6.22 *(a) Block diagram and (b) rack mounting of SC-INFC-D unit (Type 2).*

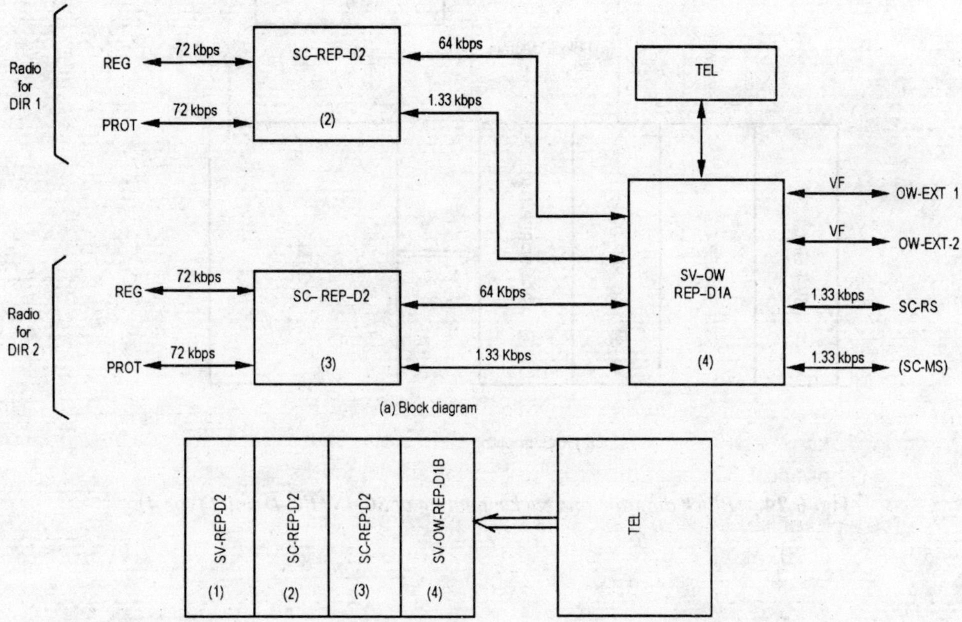

Fig. 6.23 *(a) Block diagram and (b) rack mounting of SC-INTFC-D unit (Type-3).*

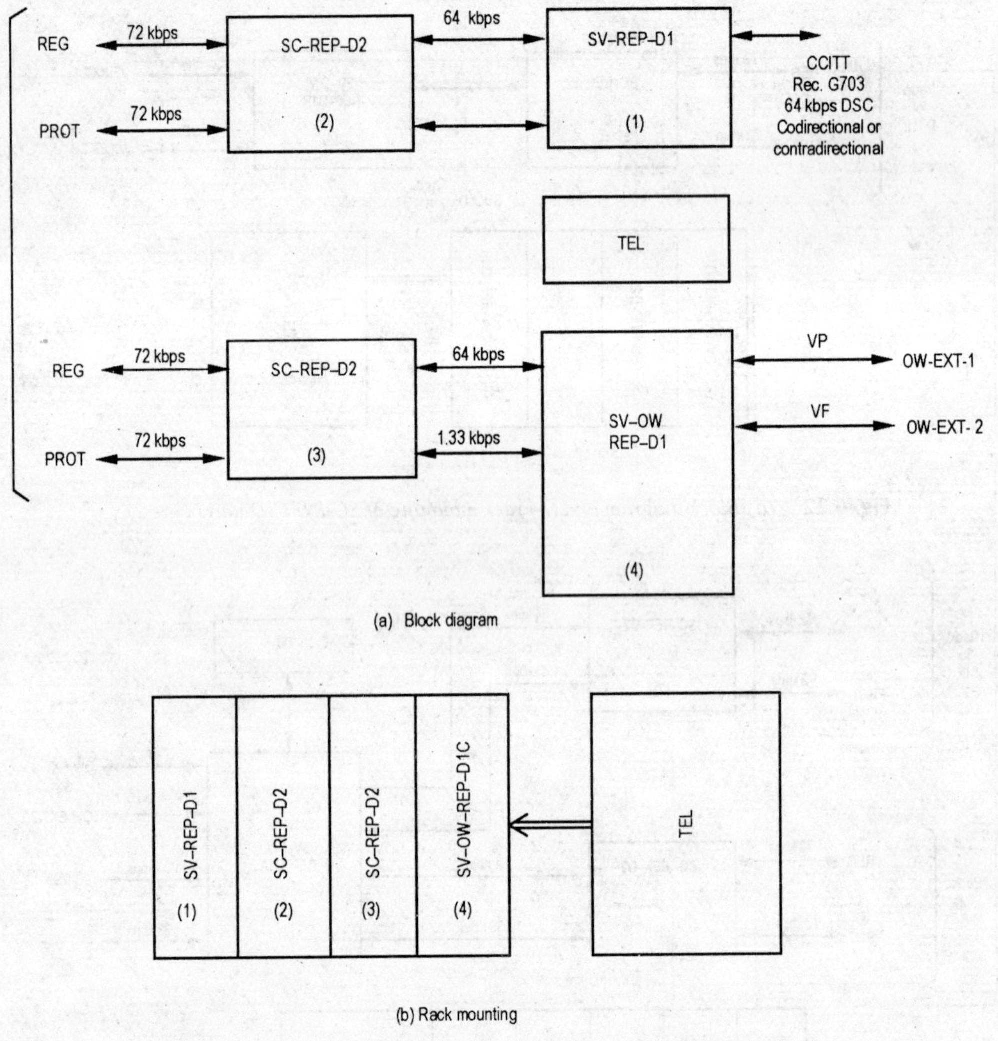

Fig. 6.24 *Block diagram and rack mounting of SC-INTFC-D unit (Type 4).*

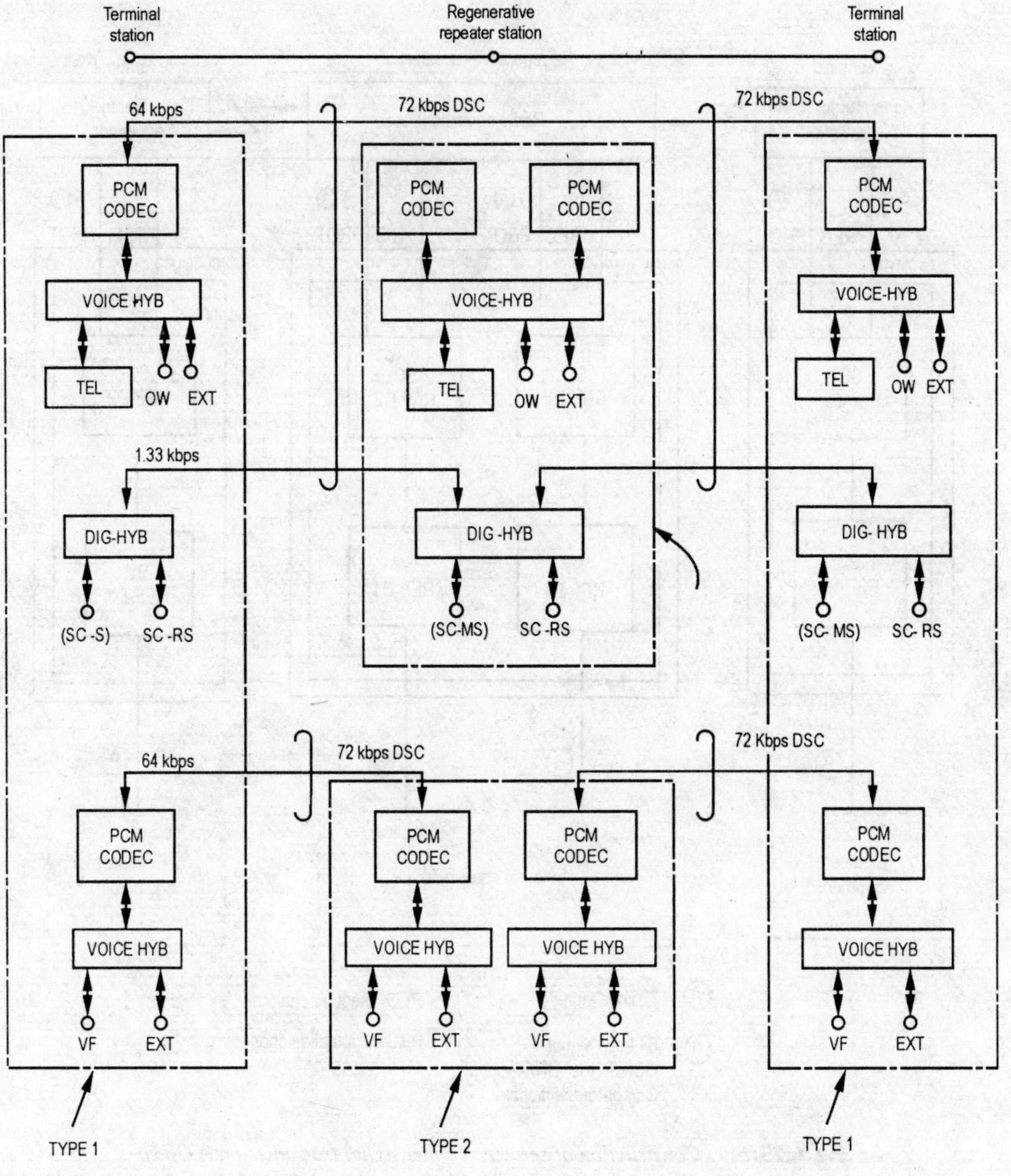

Fig. 6.25(*a*) *System configuration of DSC signal transmission (OW).*

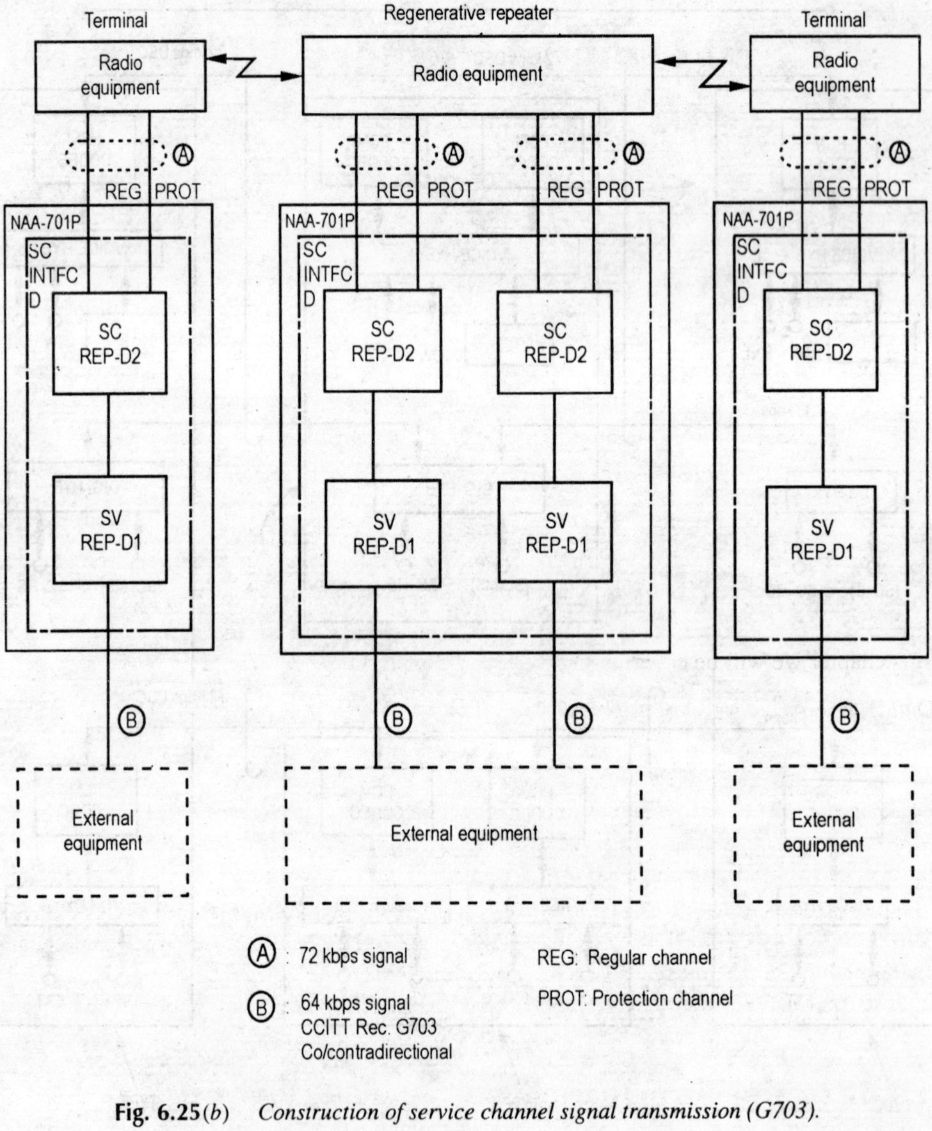

Fig. 6.25(*b*) *Construction of service channel signal transmission (G703).*

7

30 Channel Primary MUX

7.1 Introduction

In Chapter 3, we have seen how by time division multiplexing the primary MUX combines PCM data of thirty channels to form a 2.048 Mbps data frame. This frame is referred to as E1 frame. The E1 frame carries PCM data of thirty channels in thirty time-slots; this along with signalling data of all the thirty channels in two time-slots make it a total of 32 time-slots. Since in each time-slot the data rate is 64 kbps, we have a total data rate of (64 × 32) kbps, i.e., 2048 kbps or 2.048 Mbps.

The 30-channel primary MUX generating this 2.048 Mbps frame (E1 frame) is the basis for the following digital hierarchy:

- II-order (120 channels or 8 Mbps)
- III-order (480 channels or 34 Mbps)
- IV-order (1920 channels or 140 Mbps)

In this chapter we will be covering in detail the following:

- Data frame structure of 30-channel primary MUX and provisions thereof to carry signalling information
- Types of signalling in telecommunication, with emphasis on exchange-to-exchange signalling
- Bit assignments for various signalling states in the relevant time-slots of data frame of the primary MUX so as to provide suitable signalling interface with exchanges
- System description of 30-channel primary MUX
- Functional description of various modules of a primary MUX equipment (model PDM 30B of M/s Precision Electronics, New Delhi is chosen for explanation); various voice channel and data channel interfaces of a primary MUX
- Interface cabling requirements for connecting primary MUX to the exchange

7.2 Data Frame Structure of 30 Channel Primary MUX

As already mentioned, the 30-channel primary MUX operates at a data rate of 2.048 Mbps. It combines thirty channels using TDM to form the E1 frame, with two more channels (time-slots) utilized for carrying signalling and synchronization information. In each time-slot or channel the information is encoded as 8 bits, and hence an E1 frame has 32 × 8, i.e. 256 bits. The structure of an E1 frame generated/accepted by a 30-channel PCM is shown in Fig. 7.1.

TS-0	TS-1	TS-2	. . .	TS-16	TS-17	. .	TS-30	TS-31

Fig. 7.1. *Structure of E1 frame. TS-1 carries Ch # 1 data, TS-2 carries Ch # 2 data, .. and TS-15 carries Ch # 15 data; TS-17 carries Ch # 16 data, TS-18 carries Ch # 17 data, ...and TS-31 carries Ch # 30 data.*

In an E1 frame, time-slot 1 (TS-1) carries channel 1 data; TS-2 carries channel 2 data and so on till TS-15 carrying channel 15 data; thereafter, TS-17 carries channel 16 data, TS-18 carries channel 17 data and so on till TS-31 carrying data of channel 30. If in each E1 frame the encoded data of thirty channels are carried, one in each time-slot, how to accommodate the signalling information of all the thirty channels as well as the frame synchronization information in the remaining two time-slots?

The answer lies in 'multi-frame'. Sixteen E1 frames (F0 to F15) form a multi-frame. Refer to Fig. 7.2 showing the structure of a multi-frame.

F0	F1	. . . .	F15

Fig. 7.2 *Structure of a multi-frame.*

1) In TS-16 of each frame, signalling information of two channels is carried (4 bits per channel). Thus, fifteen frames accommodate signalling information of thirty channels. In other words, TS-16 of F1 carries signalling information of channel 1 and channel 16; TS-16 of F2 carries signalling information of channel 2 and channel 17, and so on.
2) TS-16 of frame F0 carries multi-frame alignment pattern and multi-frame alarm information.
3) TS-0 of every alternate frame carries the 'frame alignment word (FAW)'. The eight bits in TS-0 are 'X0001101'. The bits 0001101 constitute the FAW. The first bit 'X' can be 1 or 0. If the PCM is to be interlinked to international network, the first bit is set to '1'.
4) TS-0 of frames which do not carry FAW are used to carry supervisory and alarm signals.
5) To distinguish TS-0 of frames carrying the supervisory/alarm signals from those carrying FAW, bit-2 in TS-0 of these frames is set to '1'. The bit map of TS-0 for FAW and supervisory/alarm transmissions appear as shown in Table 7.1.

Table 7. 1 *Bit map of TS-0 for FAW/alarm transmission*

Frame no.	**TS-0 bit numbers**								**Transmitted pattern**
	B1	**B2**	**B3**	**B4**	**B5**	**B6**	**B7**	**B8**	
F0	X	0	0	0	1	1	0	1	FAW
F1	X	1	Y	Y	Y	1	1	1	Alarm
F2	X	0	0	0	1	1	0	1	FAW
F3	X	1	Y	Y	Y	1	1	1	Alarm

6) In frames F1, F3, etc., bits B3, B4, B5 denote the various types of alarms such as:
   - If Y = 1 in B3 position, it indicates frame synchronization alarm.
   - If Y = 1 in B4 position, it indicates high-error density alarm.
   - If Y = 1 in B5 position, it indicates codec alarm (codec means PCM encoder–decoder).
   - If there is no alarm condition, B3, B4 and B5 are set to '0'.
   - An urgent alarm has all '1's from B2 to B8 in TS-0.
7) On the arrival of a digital stream, the time-slots are checked for FAW. The incoming bits are compared with the FAW pattern stored in the receiver. If successive tests fail to confirm the presence of FAW, a search is initialized.
8) During the search period, signalling units at the receiving end are blocked and the speech path is also disconnected to prevent noise from entering the speech circuit.
9) It may be recalled from (5) that TS-0 of every alternate frame has a FAW. If TS-0 of three consecutive frames with FAW (i.e. frames $n, n+2, n+4$) do not have the standard FAW code, a loss of alignment is declared.
10) The frame alignment is considered to be recovered if FAW is recognized for the second time in a duration of three frames, i.e.,

    FAW is recognized in frame n,
    FAW absent in frame $n+1$,
    FAW present in frame $n+2$.

11) Each frame has a duration of 125 μs. And sixteen frames, i.e. one multi-frame, has the duration of 2 ms. Thus, signalling information is transmitted at the rate of 500 bps. This would be sufficient as signalling does not vary as rapidly as the data (speech) signal, and hence need not be sampled at the rate of 8 kHz (as is done for speech signal).

So far we have looked at the provision for transmission of signalling information in the frames/multi-frames of 30-channel PCM. We will now discuss the signalling interface between primary MUX, i.e. the transmission system, and the exchange or the switching system. We start the discussion with a briefing on signalling in telecommunications.

7.3 Signalling in Telecommunication

The transmission of information required for the setting up of, control and supervision of a call between two subscribers in a telecommunication network is called signalling. Signalling can be (*a*) dc or ac signalling (in the form of certain frequency tones); (*b*) carried on the same physical wires carrying the channel or on separate wires; (*c*) in the same band as the speech signal or ouside it; (*d*) in a channel (or time-slot) associated with the main data transmission or on a separate channel setup during the time of signalling.

The signalling information has to traverse a complex path—from the subscriber at the calling end to the exchange; between various control units within the exchange; from the exchange (switching system) to the primary MUX (transmission system) for onward transmission over the digital hierarchy to another exchange through the transmission medium; to the exchange at the far end; and ultimately to the subscriber at the receiving end.

Exchanges have evolved over time from Strowger, to Crossbar to electronic exchanges (with several varieties). Whatever be the type of exchange in use, the PCM terminal equipment is expected to convert the signalling information to a digital code so that it can be transmitted over the PCM system.

In Section 7.1 we have already seen how provision exists in the PCM frames/multi-frames to carry signalling information. The primary MUXes have VF (voice frequency) interface cards which convert the signalling information coming from the exchange into the form of bits so as to fit them in the nominated time-slots of PCM frame/multi-frame. The design of VF interface card depends on the signalling format received from the exchange, which in turn depends on the type of signalling that is followed by the exchange in question. To comprehend the signalling format we need to have an understanding of the classification of signalling.

7.3.1 Classification of signalling

The only signalling information with the first ever telephone system, i.e. the magneto telephone, was the ringing current generated by the magneto. This was sent over a dedicated line between two subscribers. With the advent of manual switching (manual exchanges), additional signals got included. These were 'call-request' and 'call-release'. The number of signals used increased further with the invention of electromechanical automatic exchanges, and even more with the advent of stored program control (SPC) electronic exchanges.

Signalling can be broadly classified into:

- subscriber-line signalling (signalling between the subscriber and the exchange), and
- inter-exchange signalling (signalling between exchanges).

Subscriber-line signalling

In automatic exchanges the power is fed over the subscriber's loop by a centralized battery at the exchange (48 V dc). Power is fed irrespective of the state of the subscriber—idle, busy or talking. The subscriber can be a calling subscriber or a called subscriber. Thus, we have two groups of signals in the subscriber-line signalling, viz.,

- calling subscriber line signals, and
- called subscriber line signals.

Calling subscriber line signals: These are Call report signal, Address signal, End-of-selection signal, Answer-back signal, Release signal, Permanent line or permanent glow (PG) signal.

Call report signal: When the subscriber is idle the line impedance is high. The line impedance falls as soon as the subscriber lifts the handset, resulting in an increase of line current. This is detected as a new call signal by the exchange, which connects an appropriate equipment/module to the line to receive the address information (which the subscriber is going to dial) and notifies its readiness to receive the 'to- be-dialed number information', by extending the dial tone to the subscriber.

Address signal: After the receipt of the dial tone the subscriber dials the number (through decadic dialing or multi-frequency push-button dialing) and sends the address digits (called the subscriber's number) in a sequence.

End of selection signal: The addressed receiver is disconnected after receipt of the complete address. After the connection is established or if the attempt has failed, the exchange sends one of the following signals known as 'End of selection signals':

- Ring-back tone if the called line is free
- Busy tone if the called line is busy or otherwise inaccessible
- Recorded announcement to give reasons for the call failure other than 'line-busy'

Answer back signal: As soon as the called subscriber lifts the handset (on hearing the ringing), a battery reversal signal is transmitted on the line of the calling subscriber. This may be used to operate a special equipment attached to the calling subscriber (e.g., short-circuiting the transmitter of a coin collection box in a public telephone till a coin is inserted).

Release signal: When the calling subscriber goes on hook, i.e. releases the call, the line impedance goes high. This is detected by the exchange which releases all equipment/modules involved in the call.

Permanent line or permanent glow (PG) signal: A PG signal is sent to the calling subscriber if he/she fails to release the call even after the called subscriber has gone on-hook and the call is released after a time delay. The PG signal is also sent if the subscriber takes too long to dial.

Called subscriber line signals: These signals are Ring signal, Answer signal, Release signal, and Register recall signal.

Ring signal: On receipt of a call to a subscriber whose line is free, the terminating exchange sends the ringing current to the called telephone. This is typically 25 Hz or 50 Hz, with suitable interruptions. Ring-back tone is also fed back to the calling subscriber by the termination exchange.

Answer signal: When the called subscriber lifts the handset on the receipt of a ring, the line impedance goes low. This is detected by the exchange which cuts off the ringing current and ring-back tone.

Release signal: If the called subscriber goes on-hook before the calling subscriber (after the conversation is over), the line impedance of the called subscriber goes high which is detected by the exchange. The exchange sends PG to the calling line and releases the call after a time delay if the calling subscriber fails to clear by then.

Register recall signal: Using DTMF (dual tone multi-frequency) telephones, another number can be dialed while holding on to the call in progress in order to set up a call to a third subscriber. The signal to recall the dialing phase during the talking phase is called 'register recall signal'. It is achieved through interruptions on the calling subscriber's loop for a duration less than the release signal.

Inter-exchange signalling

Signalling information between exchanges is transmitted over trunks. Inter-exchange signalling can be classified based on the type of transmission or configuration of the trunk circuit.

Classification based on the type of transmission: Based on the type of transmission, inter-exchange signalling is classified as:

- pulsed signalling,
- continuous signalling,
- compelled signalling.

Pulsed signalling: The signal is transmitted as pulses in pulsed signalling. The signal is characterized by a change from idle condition to an active state for a particular duration.

Continuous signalling: A transition from one condition to another characterizes the presence of a signal, while a steady state implies no signalling information.

Compelled signalling: This is similar to pulsed mode signalling, but the transmission of signalling continues till the an acknowledgement from the receiving unit is received by the sending unit.

Classification based on trunk circuit configuration: Based on the method of connecting trunks between exchanges, inter-exchange signalling is classified as:

- loop signalling,
- E&M signalling,
- single frequency tone signalling.

Loop signalling: In loop signalling or dc signalling, the trunk connection between exchanges comprises a dc loop similar to exchange-subscriber loop. This is known as dc loop since dc conditions such as high ohmic loop, low ohmic loop, normal polarity on a lead, reverse polarity on a lead, and earth/battery on a lead are detected and interpreted by the system. Signalling and speech are carried on the same pair of limbs or wires (known as A and B limbs) which form the dc loop. The signalling conditions on the loop are shown in Table 7.2.

E&M signalling: In E&M signalling, an additional pair of line is utilized per trunk. One wire is dedicated to the forward signals (M-wire for transmit or mouth) which corresponds to receive or R-lead of the destination exchange. The other wire is dedicated to the backward signals (E-wire for receive or ear) which corresponds to the transmit or send wire or S-lead of the destination exchange. The E&M signalling states are shown in Table 7.3.

Single frequency tone signalling: Exchanges separated by long distance cannot use any form of dc line signalling as dc loops or E&M leads cannot be extended over long distances. Suitable interfaces are to be interposed in-between to enable the transmission of signalling states over long distances. A single frequency or a tone can be used to carry the ON/OFF hook information. Also, dial-pulses can be transmitted by pulsing the states. The states involved are shown in Table 7.4.

As can be seen, the set of signals is small. The signals can be transmitted in-band or out-of-band. For in-band signalling, the tone frequency chosen is 2600 Hz or 2400 Hz. As this frequency lies within speech band, to prevent premature disconnection simulation of tone-ON condition indicating the end of the call by the speech signal has to be guarded against.

Table 7. 2 *dc Loop signalling states*

State	Outgoing end	Incoming end
Idle	High resistance loop	−ve potential on A limb +ve potential on B limb (normal potential)
Seizure	Low resistance loop	- do -
Answer	Low resistance loop	+ve potential on A limb −ve potential on B limb (reversed potential)
Clear forward	Disconnection followed	Normal/reverse potential
Clear back	High resistance	Change to normal potential
Release guard	Disconnection followed by high resistance loop	On recognizing forward disconnection signal, open circuit/reversed potential, followed by normal potential when incoming equipment has been released.
Trunk offering and re-ringing	Earth on both A and B limbs	
Blocking	High resistance loop	Disconnection of potential

Table 7. 3 *E&M signalling states*

State	Outgoing exchange		Incoming exchange	
	M-lead	E-lead	M-lead	E-lead
Idle (on hook)	Earth	Open	Earth	Open
Forward				
Seizure (off hook)	Battery	Open	Earth	Earth
Release (on hook)	Earth	Earth/Open	Battery/Earth	Open
Backward				
Answer (off hook)	Battery	Earth	Battery	Earth
Clear back (on hook)	Battery	Open	Earth	Earth
Blocking	Earth	Earth	Battery	Open

Table 7.4 *Single frequency signalling states*

State	Forward	Backward
Idle (on hook)	ON	ON
Forward		
Seizure (off hook)	OFF	OFF
Release (on hook)	ON	OFF/ON
Backward		
Answer (off hook)	OFF	OFF
Clear back (on hook)	OFF	ON
Blocking (off hook)	ON	OFF

Out-of-band signalling uses a tone frequency of 3825 Hz which lies beyond the speech band and thus the problem of simulation of signalling tone by speech tone is avoided. But the additional hardware required for this purpose in the out-of-band signalling costs more.

Line signalling versus register signalling

The three types of signalling presented so far are operative throughout the duration the call and are hence known as 'line signalling' between exchanges.

Trunk service can be managed in a better way by using 'trunk register' which is basically an address digit register to hold address information. In this method signalling is operative only during the relatively short phase of call-setting—essentially for transmitting the address information. This is known as register signalling between exchanges.

Register signals are interchanged between registers during the phase between the receipt of the trunk seizure signal and the switching of the exchange to speech phase. These signals are (*i*) the 'proceed to send (PTS)' signals, (*ii*) address signals and (*iii*) signals indicating the result of a call attempt.

The register signals may be transmitted out-of-band or in-band. In out-of-band transmission, only limited range of signals may be used; a single out-of-band tone may be selected and the information sent as pulses.

In register signalling in-band transmission can be used easily as there is no possibility of interference with speech. To reduce transmission time and to increase reliability, a number of frequencies can be used as sets. Normally, two out of six frequencies are used, and for reliability a compelled sequence is used. Hence this type of signalling is termed 'compelled sequence multi-frequency (CSMF)' signalling. In CCITT terminology, it is called R2 signalling. Section 7.4 is devoted to R2 signalling as PCM systems follow R2 signalling.

End-to-end signalling v/s link-by-link signalling

When more than two exchanges are involved in the setting up of connections, the signalling may be done in either end-to-end signalling, or link-by-link signalling.

In end-to-end signalling, the signalling is always between the ends of a connection as the call progresses. For example, in a 3-exchange (A–B–C) connection, the signalling is between A–B, then

B–C, and finally, A–C to establish connection.. In link-by-link signalling, signalling is confined to individual links, i.e., A–B and B–C, and not A–C.

7.4 R2 Signalling

7.4.1 Signal frequencies in R2 signalling

Compelled sequence multi-frequency (CSMF) signalling has been standardized as R2 signalling by CCITT as already mentioned. In this signalling, two out of six frequencies, viz., 1380, 1500, 1620, 1740, 1860, and 1980 Hz respectively are used for forward signals (i.e., a total of 15 combinations are possible). For backward signals also, two out of six frequencies, viz., 1140, 1020, 900, 780, 660, 540 Hz respectively are used (i.e., a total of 15 combinations are possible).

The DOT (now BSNL) prescribes R2 signalling with slight modification for an exchange–PCM interface. It is termed 'modified R2 signalling'. In modified R2 signalling, 1980 Hz is not used in the forward group and 540 Hz is not used in the backward group. Thus there are ten possible combinations in both the directions.

Since it is difficult to mention the frequencies each time, the frequencies are denoted by an index (f0 to f4) or a weight-code (0 to 4). Thus the two frequencies chosen out of the five available ones are denoted by an index or a weight code. For example, 0 + 1 implies f0 and f1 (1380 Hz and 1500 Hz) signals. Signal frequency indices and weight codes are given in Table 7.5.

The significance of forward and backward signals are given in Table 7.6.

7.4.2 Bit assignments in R2 signalling

It can be recalled from Section 7.1 that PCM provides for four bits for signalling of each channel in TS-16. R2 signalling uses two of the four bits available. The bit assignments for forward and backward directions are given in Table 7.7 (forward direction is along the originating end to terminating end of a call, and backward direction is its converse).

The reader might wonder why R2 signalling under-utilizes the bits available for signalling in TS-16 (only two out of available four bits are used, as already explained). The reason is as follows:

In the existing telecom network of DOT (BSNL) and also Indian Railways, a trunk-offer service is still needed to connect manual exchanges. The additional signalling condition of trunk offering is taken care of by the third bit in the forward direction.

Once this constraint is taken care of, we can utilize all the four bits for two channels, i.e., 8 bits in each TS-16 for four channels. We can even think of using one bit per channel but in doing so the signalling speed would be halved. The point to be grasped here is that PCM is capable of flexible and improvised signalling options, whereas an analog system is capable of only one type of signalling as per the wiring of the signalling circuit.

7.4.3 Examples of bit assignments in R2 signalling

The bit assignments shown in the previous section for various signalling conditions shall be made clearer with an example. In this chapter we will be discussing a thirty channel primary MUX model namely PDM 30B. This equipment has different types of VF (voice frequency) interface cards for the thirty channels (two channels per card) of the MUX:

Table 7. 5 *Signal frequency index and weight code*

	Signal frequency (Hz)				
Forward	1380	1500	1620	1740	1860
Backward	1140	1020	900	780	660
Index	f0	f1	f2	f3	f4
Weight code	0	1	2	3	4

Table 7.6(a) *Forward signals and their significance*

Signal	Weight	Group I	Group II
1	0 + 1	Digit 1	Ordinary subscriber
2	0 + 2	Digit 2	Subscriber with priority (Test/Mntc eqpt.)
3	1 + 2	Digit 3	Spare
4	0 + 4	Digit 4	STD barred
5	1 + 4	Digit 5	Spare
6	2 + 4	Digit 6	CCB
7	0 + 7	Digit 7	Changed number to operator
8	1 + 7	Digit 8	Closed number
9	2 + 7	Digit 9	Closed number
10	4+7	Digit 0	Spare

Table 7.6(b) *Significance of backward signals*

Signal	Weight	Group I	Group II
1	0 + 1	Send next digit	Called line free without metering
2	0 + 2	Restart	Changed number
3	1 + 2	Address complete, change over to reception of Gr.II signals	Called line busy
4	0 + 4	Calling line identification for malicious calls	Local congestion
5	1 + 4	Send calling subscriber's category	Number unobtainable
6	2 + 4	Setup speech connection	Called line free, with metering
7	0 + 7	Send last but 1 digit	Route congestion
8	1 + 7	Send last but 2 digit	Spare
9	2 + 7	Send last but 3 digit	Route breakdown
10	4 + 7	Spare	Malicious call booking

Table 7.7 *Signalling conditions in R2 signalling*

Signalling condition	Forward direction		Backward direction	
	Af	**Bf**	**Ab**	**Bb**
Idle	1	0	1	0
Seizure	0	0	1	0
Seizure Ack.	0	0	1	1
Answer	0	0	0	1
Clear back	0	0	1	1
Clear forward	1	0	0	1
		OR		
	1	0	1	1
Release guard	1	0	1	1
Blocked	1	0	1	1

- E&M card in outgoing/incoming mode with E&M signalling
- Outgoing/subscriber/hotline interface with loop signalling
- Incoming/exchange interface card

These cards are all designed to interface with R2 signalling standard. Functional description of these cards is presented in detail in the following sections. In this section however, only the R2 signalling conditions and bit assignments are discussed.

As we have already seen, R2 signalling uses only two out of four bits available in each channel in TS-16 for signalling information. In PDM 30B, the other two bits are fixed as '01'. This will ensure that the same type of MUX is used at either end. Though all the commercially available MUXes are CCITT compatible and follow R2 signalling, the scope to play around with the balance two bits (per channel) in TS-16 makes them distinct. Different manufacturers assign these two bits differently, hence it is essential to check the 'total compatibility' in signalling before using different types of primary MUXes. VF cards can be tested with signalling simulator and PCM channel analyzer to understand the bit assignments in TS-16 for signalling.

The E&M card of PDM 30B for example, which provides VF interface with E&M signalling, can be tested with a simulator (which simulates exchange connection) and the bits in TS-16 observed on a PCM channel analyzer. Such a test for idle condition in the forward direction (i.e. outgoing send bits or bits observed on PCM channel analyzer) shows '10' in the two bits' position pertaining to the signalling information in the relevant TS-16 . Since PDM 30B fixes the next two spare bits as '01', we get '1001' on the channel analyzer. Tables 7.8 to 7.11 show the bit assignments for E&M signalling observed in the forward/backward directions, or on a simulator/analyzer for both normal and E10B exchanges. It may be observed that in all cases the basic two bits for various signalling conditions tally with the contents of the Table 7.7. The test results can also be inferred through LED indications (red/amber) on the card.

Bit assignments in TS-16 for E&M signalling in PDM 30B are given in Tables 7.8a and 7.9 for normal exchange and E-10B exchanges respectively. A note on E10B exchange follows later.

Table 7.8 *Bit assignments in TS-16 for E&M signalling in PDM 30B (Normal exchange)*

State	Bits observed on analyser (O/G send bits) OR forward direction (card to simulator)	Bits on simulator (I/C send bits) OR backward direction (simulator to card)	LED indication on E&M
1. When E&M card under test is operating as O/G (outgoing) converter			
Idle	1001	1001	LEDs OFF
Seizure (Seizure in forward & seizure ACK in backward)	0001	1101	Red LED ON
Answer	0001	0101	Red & amber LED ON
Jn. failure or blocked	1001	1101	Amber LED ON
State	**Bits observed on analyser (I/C send bits) OR forward direction (simulator to card)**	**Bits on simulator (O/G send bits) OR backward direction (card to simulator)**	**LED indication on E&M**
2. When E&M card under test is operating as I/C (incoming) converter			
Idle	1001	1001	LEDs OFF
Siezure (Seizure in forward & seizure ACK in backward)	0001	1101	Red LED ON
Answer	0001	0101	Red & amber LEDs ON
Jn. failure or blocked	1001	1101	Amber LED ON

Bit assignments in PDM 30B for an Outgoing/Subscriber/Hotline card and an Incoming/Subscriber/Hotline cards are given in Tables 7.10 and 7.11 respectively for normal and a E10B exchanges.

A note on E10B exchange

The E10B exchange has evolved from E10A system of digital switching (exchange) used extensively in France. It runs under SPC (stored program control) and switches signals in the digital form, i.e., analog signals are first converted to TDM digital form before switching. The feature which makes E10B a class apart form other electronic exchanges is the segregation of switching and management functions. Switching functions like reception of dialed digits, their storage, analysis, routing of calls, etc., are performed by control units called XEGs which have dedicated processors. Functions like

Table 7.9 *Bit assignments in TS-16 for E&M signaling in PDM30B (E10B exchange)*

State	Bits observed on analyser (O/G send bits) OR forward direction (card to simulator)	Bits on simulator (I/C send bits) OR backward direction (simulator to card)	LED indication on E&M card
1. When E&M card under test is operating as O/G converter			
Idle	0001	0001	LEDs OFF
Seizure (Seizure in forward & seizure ACK in backward)	1001	0101	RED LED ON
Answer	1001	1101	Red & Amber LEDs ON
Jn. failure or blocked	0001	0101	Amber LED ON
State	**Bits observed on analyser (I/C send bits) OR forward direction (simulator to card)**	**Bits on simulator (O/G send bits) OR backward direction (card to simulator)**	**LED indication on E&M**
2. When E&M card under test is operating as I/C (incoming) converter			
Idle	1001	1001	LEDs OFF
Siezure (Seizure in forward & seizure ACK in backward)	0001	1101	Red LED ON
Answer	0001	0101	Red & amber LEDs ON
Jn. failure or blocked	1001	1101	Amber LED ON

subscriber line and circuit group management, faults and alarm management, testing and diagnostics are performed by a mini-computer at the OMC (operations and maintenance centre). All XGEs are connected to OMC via PCM links.

E10B system can be used as:

- Local exchange.
- Tandem exchange, a local transit exchange meant for handling transit traffic. It does not have subscriber line terminating equipment. It handles junction connections only.
- TAX (trunk automatic exchange) for terminations of long distance circuits.

Table 7.10 *Bit assignments in PDM 30 B for Outgoing/Subscriber/Hotline card (loop signaling) in PDM30B*

State	Bits observed on analyser OR forward direction (card to simulator)	Bits on simulator OR backward direction (simulator to card)	LED indication on card
1. Normal exchange			
Idle (switch pressed)	1001	1001	Green LED OFF
Loop (switch released)	0001	1101	Green LED ON
Answer	0001	0101	Red LED ON
Trunk offer (limb 'A' earthed)	0011	1101	Green LED OFF
Jn. failure or blocked	1001	1101	Amber LED ON
State	**Bits observed on analyser OR forward direction (card to simulator)**	**Bits on simulator OR backward direction (simulator to card)**	**LED indication on card**
2. For E10B exchange			
Idle (switch pressed)	1101	1001	Green LEDs OFF
Loop (switch released)	0101	1101	Green LED ON
Answer	0101	0101	Red LED ON
Trunk offer (limb A earthed)	0001	1101	Green LED OFF
Jn.failure or blocked	1101	1101	Amber LED ON

The signalling format, i.e. the bits in TS-16 for various interfaces, differ from those of the corresponding ones in normal exchanges. The VF interface card of the PCM equipment has to be set for E10B or normal exchange, or the feature can also be software-programmable.

Table 7.11 *Bit assignments for Incoming/Exchange card (loop signalling in PDM 30B*

State	Bits observed on analyser OR forward direction (simulator to card)	Bits on simulator OR backward direction (card to simulator)	LED indication on card
1. Normal exchange			
Idle (Normal −48V feed, i.e., A lead more −ve w.r.t B)	1001	1001	Green LED OFF
Loop (Normal feed)	1101	0001	Green LED ON
Answer (reverse feed, i.e., limb B more −ve w.r.t limb A)	0101	0001	Red LED ON
Trunk offer	1101	0011	Green LED OFF
State	**Bits observed on analyser OR forward direction (simulator to card)**	**Bits on simulator OR backward direction (card to simulator)**	**LED indication on card**
2. For E10B exchange			
Idle Normal −48V feed, i.e., A lead more −ve w.r.to B)	1001	1101	Green LED OFF
Loop (Normal feed)	1101	0101	Green LED ON
Answer (reverse feed, i.e., limb B more −ve w.r.t limb A)	0101	0101	Red LED ON
Trunk offer	1101	0001	Green LED OFF

7.5 System Description

7.5.1 Technical description of the equipment

The block diagram of a 30-channel primary MUX (PDM 30B) is given in Fig. 7.3. The equipment can be viewed as a primary multiplexer, working under the supervision of a powerful microprocessor with a human interface. The E1 frarmer makes use of a versatile programmable framer chip that runs under software control. When the power is turned on, the processor card takes control of the framer IC and initialises all its primary functions. All subsequent operations relating to link operation, maintenance

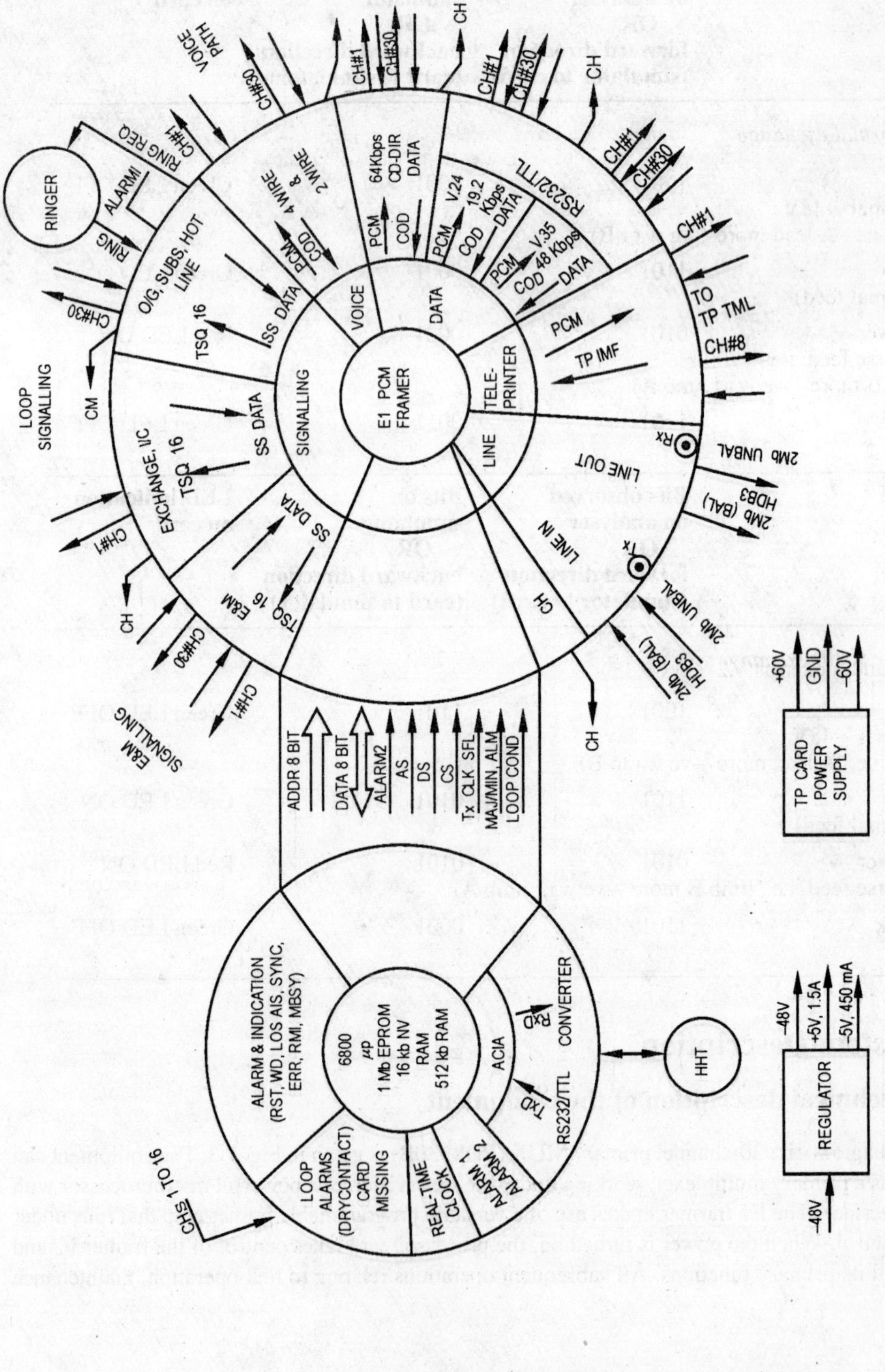

Fig. 7.3 *System block diagram of PDM 30B MUX.*

and error monitoring are controlled under operator commands issued from a hand-held terminal (HHT) hooked up with the processor card.

The chief functions of the framer card are as follows:

- Collection of PCM streams from various interfaces
- 2 Mbps E1 framing as per G.704
- HDB-3 coding of 2 Mbps stream and transmission online
- Timing pulse generation for use by interface cards to strobe transmit/receive operations
- Reception of 2 Mbps stream decoding into TTL level and clock recovery
- Distillation of PCM streams to various interfaces
- Generation of various alarms based on the received HDB-3 stream conditions

PDM 30B provides for (*i*) voice and (*ii*) data interfaces. The voice interface consists of

- 2W voice with loop disconnect signalling, having either of the following field-alterable interface options: (*a*) Outgoing junction interface, Subscriber interface, Hotline interface, Local ringer facility for the subscriber, and Hotline interfaces; (*b*) Incoming junction interface, and Exchange interface,
- 2W/4W voice (field-alterable) with E&M signalling for interchanging E&M junction hook up.

The data interfaces consists of

- 64 kbps codirectional date interface,
- V.24 date interface,
- V.35 date interface,
- teleprinter interface.

Let us understand the terms V.24 and V.35.

V.24: The V.24 standard of the International Telecommunication Union (ITU) is equivalent to the RS-232C standard of EIA (Electronic Industries Alliance) which is the standard for serial transmission interface between data terminal equipment (DTEs) such as computers and terminals, and data communications equipment (DCEs) such as modems, packet assemblers/disassemblers (PADs) and printers. The maximum data rate supported by RS-232C is 115.2 kbps over a distance of 15 meters. The cables are typically 25-wire unshielded twisted-pair (UTP) cables with DB25 type connectors, or 9-wire cables with DB9 connectors.

V.35: V.35 is a serial transmission standard of ITU that defines a physical interface for synchronous communication between DTE and DCE. It supports transmissions upto 1.544 Mbps at distances of up to 1200 metres using balanced lines for signalling and unbalanced lines for control. The V.35 interface is typically used to connect high-speed routers to channel service unit/data service units (CSU/DSUs) for communication over leased lines. V.35 uses UTP cables with block-shaped 34-pin block connector.

7.5.2 Power supply

Power to the MUX terminal is provided by a power supply unit which works on a standard exchange battery (−48 V). It provides well regulated, short-circuit protected output DC voltages of +5 V and −5 V. In case of any over-voltage, the over-voltage crowbar acts, thus shutting off the entire power supply.

Alarm indications are made available on the unit for short-circuit, over-voltage and under-voltage conditions.

7.6 PDM 30B System Specification

The system specifications for PDM 30B are given below.

Pulse code modulation

Speech band	:	300 to 3400 Hz
Modulation	:	PCM
No. of channels per tributary	:	32 (30 for speech, 1 for terminal signalling and for sync. channel)
Sampling rate	:	8000 samples/s
No. of sample bits	:	8 per channel
Total bits per frame	:	256
Bit rate	:	2048 kbps ± 50 ppm
Encoding law	:	A-law as per CCITT G.711
Frames per multiframe	:	16
Synchronisation	:	Frame alignment signal in time-slot 0 in alternate frames, multiframe alignment signal in time-slot 16 in frame 0.

Digital interface at 2048 kbps port

Bit rate	:	2048 kbps ± 50 ppm
Line code	:	HDB-3 with 50% duty cycle
Peak voltage pulses	:	3V ± 10%
Nominal impedance	:	120 ohms balanced

Output port

Peak voltage of mark	:	3.0 ± 0.3V
Peak voltage of space	:	0 ± 0.3V
Pulse width	:	244 ± 25 ns
Pulse mask	:	As per CCITT Fig. 15/G.703
Output jitter	:	As per CCITT G.823

Input port

Permissible attenuation	:	6 dB at 1024 kHz
Return loss	:	> 12 dB (512 kHz–1024 kHz) > 18 dB (1024 kHz–2048 kHz) > 14 dB (2048 kHz–3072 kHz)
Jitter tolerance	:	As per CCITT Table 2/G.823

Alarms and supervisory tools

Alarm indications in control section: These are LOS, AIS, SYNC loss (FAL & MFAL), ERR (BER 10^{-3}), FEF, LOTC, major/minor indication, watchdog time-out failure, card missing alarm, and manual busy indication.

Other Alarms: +5 V power supply fail, loss of 64 kbps signal at codirectional data input port, loss of V.24 signal, loss of V.35 signal.

Supervisory capability provided on HHT: System configuration, current and stored alarms monitoring, link performance monitor (as per CCITT G.821), maintenance tool useage, factory setting reporting, date and time setting.

Monitoring points: are provided for 2 Mbps line signal, trans-clock, voice and data signalling,+5 V and −48 V supply, ringer voltage.

7.6.1 Interface specifications

2W/4W E&M interface

Voice port

Speech band	:	300−3400 Hz
Nominal impedance	:	600 ohm
2W voice	:	As per CCITT G.713
Nominal send level	:	0 dBm (adjustable attenuation 0–15.5 dB in steps of 0.5 dB)
Nominal receive level	:	−2 dBm (adjustable attenuation 0−15.5 dB in steps of 0.5 dB)
Return loss	:	> 12 dB (300−600 Hz)
		> 15 dB (600−3400 Hz)
4 W voice	:	As per CCITT G.712
Nominal send level	:	−3.5 dBm (adjustable attenuation 0−15.5 dB in steps of 0.5 dB)
Nominal receive level	:	−3.5 dBm (adjustable attenuation 0−15.5 dB in steps of 0.5 dB)
Return loss	:	> 20 dB (300−3400 Hz)

E&M signalling: Channel associated signalling in TS-16 as per modified R2 signalling standard is used by the Department of Telecommunications (DOT). Please refer to Tables given in Section 7.3.

2W Outgoing/Subscriber/Hotline interface

Voice port

Speech band	:	300−3400 Hz
Nominal impedance	:	600 ohms
2W voice	:	As per CCITT G.713
Nominal send level	:	0 dBm (adjustable attenuation 0−15.5 dB in steps of 0.5 dB)

Nominal receive level	:	−2 dBm (adjustable attenuation 0−15.5 dB in steps of 0.5 dB)
Return loss	:	> 12 dB (300−600 Hz)
		> 15 dB (600−3400 Hz)

Loop disconnect signalling: Channel associated signalling in TS-16 is as per modified R2 signalling standard used in Indian Posts & Telegraphs. Refer to Tables given in Section 7.3.

64 kpbs codirectional data interface

64 kpbs port

Nominal bit rate	:	64 kbps
Line code	:	Codirectional (as per CCITT Fig 4/G.703)
Nominal Impedance	:	120 ohms balanced
Timing signals	:	64 kHz and 8 kHz octet timing carried along with the signal
Fault indication	:	LOS on 64 kbps input port
Output pulse mask	:	As per CCITT Fig. 5 of G.703
Return loss at input port	:	>12 dB (4 kHz–13 kHz)
		>18 dB (13 kHz–256 kHz)
		>14 dB (256 kHz–354 kHz)

V.24 data interface

Baud rates	:	2400, 4800, 7200, 19200 bps.
Electrical signal characteristics	:	As per CCITT V.28
Fault indication	:	LOS on V.24 input port
Associated handshake signals	:	DCD, DSR, RTS, CTS, DTR, RI
Connector	:	D-9 female.

V.35 data interface

Tx/Rx bit rate	:	Any rate upto 128 kpbs asynchronous/2.1 Mbps synchronous
Electrical signal characteristics	:	As per CCITT V.28
Differential Tx and Rx data	:	As per CITT Rec V.35, Appendix II
Single-ended handshake signals	:	As per CCITT Rec. V.28
Fault indication	:	LOS on V.35 input port
Associated hand-shake signals	:	Any four out of the DCD, DSR, RTS, DTR and CTS
Connector	:	D-25 female.

Teleprinter data interface

Capacity	:	8 teleprinter channels
Sampling rate	:	500 samples/s/channel
Bit used in PCM	:	Bit 4 of NFW
Baud rate	:	50 or 100 baud
Signal	:	Single-ended asyncronous
Peak mark voltage	:	± 60 V (nominal)
Connection	:	Wire wrap termination on power Euro connector

Constant voltage power supply

Input voltage	:	−40 V to −60 V dc (−48 V nominal)
Output voltage	:	+5 V, 1.5 V, −5 V, 450 mA
Protection	:	Diode block to input voltage reversal, short-circuit fold back, over-voltage trip.
Alarm	:	Red LED indication on front panel and extension of loop to bay top in case of power supply failure.

7.6.2 Mechanical dimension

Construction	:	19" CP7 practice
Bay dimensions (H×W×D)	:	2750 × 605 × 225 mm ; optional 1400 × 605 × 225 mm
Subrack dimensions (H×W× D)	:	300×510× 210 mm

7.6.3 Environmental condition

Meets QM 333 Cat. B2.

Ambient operating temperature range	:	0 to 50°C
Maximum relative humidity	:	95% at 35°C

7.7 Module-wise Description of PDM 30B Primary MUX

7.7.1 Processor card

Figure 7.4 gives the functional block schematic of the processor module. The central element in this module is the 32-bit microprocessor MC 68000. It performs two principal tasks:

- Supervisory control on the framer
- Conversation with user hand-held terminal (HHT)

At power-on, the processor card boots from the EPROM and programs the framer for basic input/output operations with default settings. Subsequently, it pools the HHT port to collect user

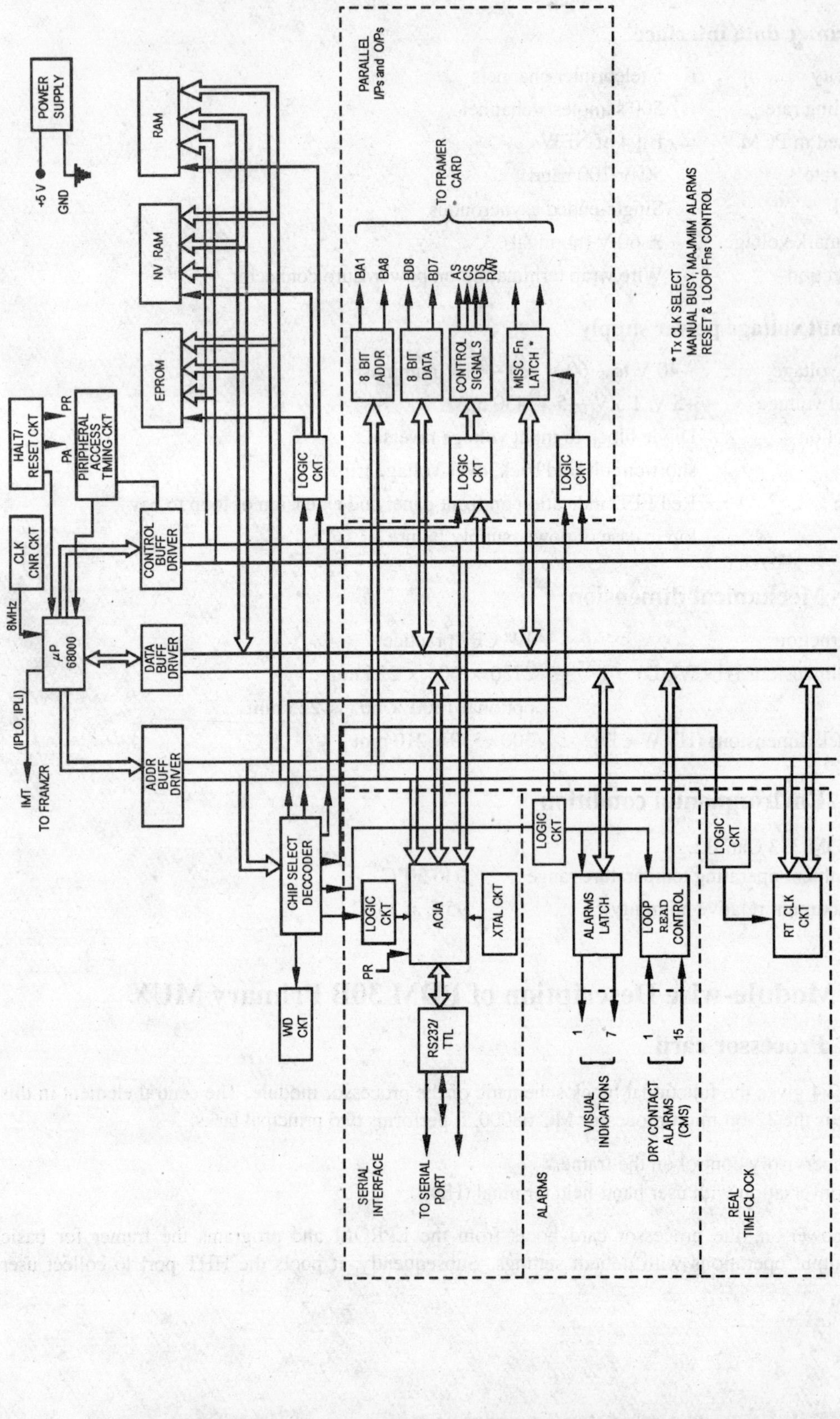

Fig. 7.4 *Block diagram of processor card.*

commands, and branches to the relevant command handler. The menu provided to the user on HHT is as follows:

a) Configuration
b) Alarms
c) Performance
d) Maintenance
e) Factory settings
f) Date and time

Reporting of alarms, link performance, factory settings, etc. against user request is communicated through HHT at the end of command handling procedure.

An asynchronous serial communication block in the processor card provides an RS 232-based serial link with the user HHT. Read/write scratch pad area is provided in the RAM while a 16 kb NVRAM provides for a ten year fail-safe storage of data such as system configuration table and set-in field.

System timekeeping is taken care of by a realtime clock circuit on the processor card, settable and readable by the HHT.

The framer interface consisting of 8-bit multiplexed data, address bus and read/write control signals, is also part of the processor module.

An LED array driven by latches on the front panel of the processor module provides visual indication of different framer alarms. The processor detects alarms by reading the framer status registers and activates the relevant alarm latches.

Software stability of the processor module is supervised by a watchdog timer. In the case of program control going haywire, the watchdog time-out sets off the WDTF alarm display on front-panel LED DS2.

7.7.2 Framer card

Figure 7.5 gives a functional block schematic of the framer module. The central element in this module is a versatile programmable 2 Mbps E1 framer DS2143. It works as a slave under software control of the processor module. An exclusive interface bus exists between the processor and the framer module for this purpose. At power-on, the processor programs the basic transmit/receive functions of the framer by read/write access to various control/status registers in the framer chip. Subsequently, the processor responds to the user commands via HHT by accessing the framer for either control or montoring purpose. The intelligence built-in in the process provides for detailed supervision and maintenance of the equipment and link.

Once programmed, the framer generates the Trans and Receive timing pulses in the form of digit, time-slot, qualified TS-16 and frame pulses. The system backplane distributes these timing signals to different interface slots for strobing the PCM inputs of the respective interface modules jacked-in.

The respective time-slot contribution of each interface card is collected by combiner logic and fed to the framer. The framed output derived as dual TTI streams from the framer is coded into HDB-3 by a transceiver and emitted via a line transformer in either 120 ohms balanced or 75 ohms unbalanced from.

The incoming 2 Mbps E1 stream via the line transformer is used to drive the receive clock by a PLL circuit in the transceiver chip. The received HDB-3 stream is decoded into TTL and fed to the framer for distribution to various interfaces in step with the extracted receive clock. The framer analyses

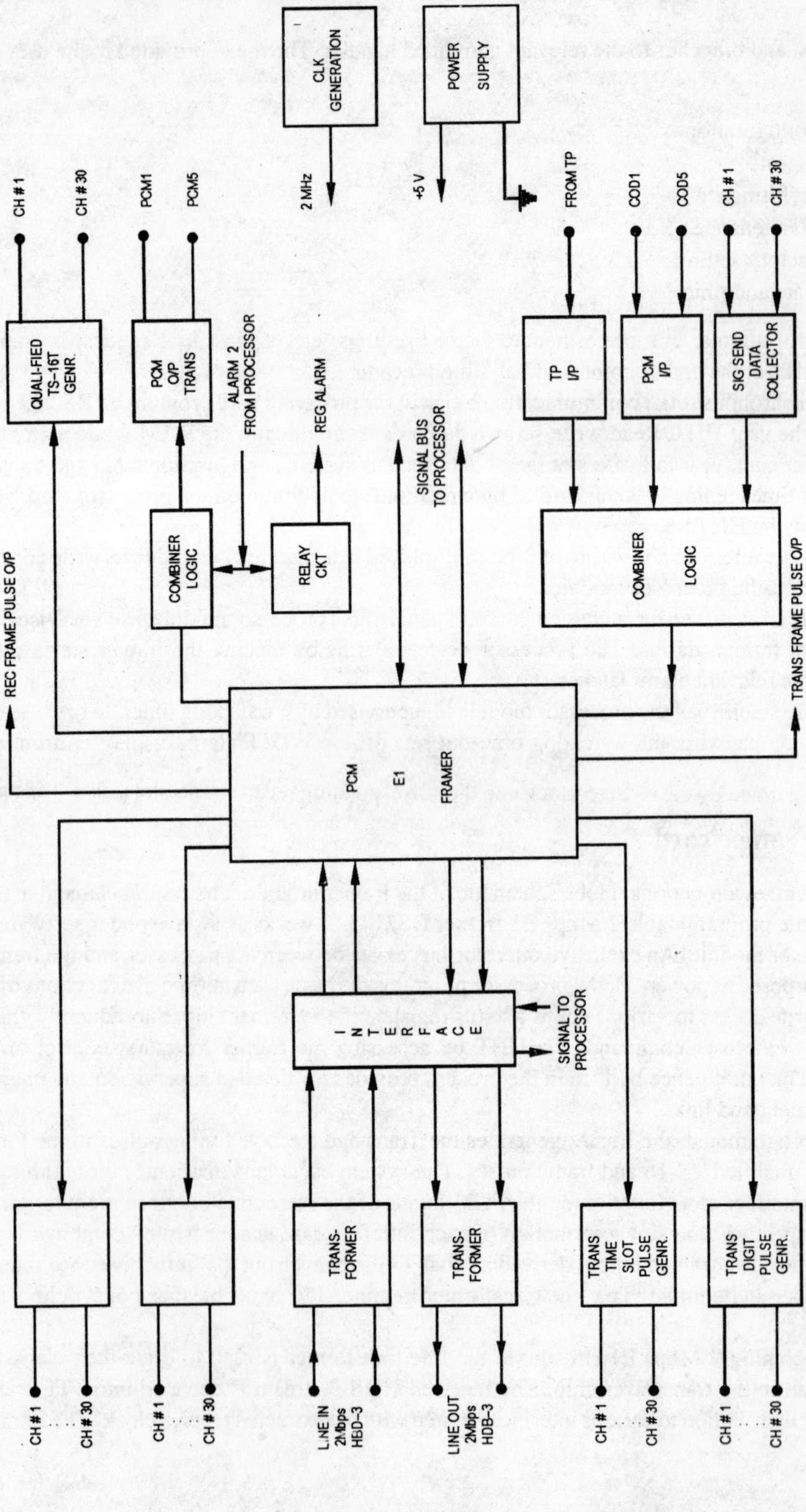

Fig. 7.5 *Block diagram of framer card.*

the received 2 Mbps stream to generate various alarm conditions reflected in its status registers. The processor reads the status registers and generates relevant alarams for display and relay activation. In case of any major alarm, a gating logic disables the PCM distributed by the framer to the interfaces.

Trans-timing of the framer may be based on one of the following options:

1. Internal clock derived from a clock generator oscillator
2. Clock extracted from received signal
3. External clock

The choice can be made from the processor card via HHT. The default select at power-on is the internal clock.

7.7.3 4W/2W E&M interface

The 4W/2W E&M interface is a composite module which can be configured as either a 2W card or a 4W card by altering the bergstick mini jumber straps (see Fig. 7.6).

Trans-direction The 600 ohms voice port receives analog signals via the 2W/4W transformer and a variable attenuator on a codec. The codec performs A-law companding of voice before feeding the 8-bit sample output to the PCM stream directed at the framer. Trans-clock, alongwith the corresponding trans time-slot, is fed to the codec for this purpose.

Signalling status of the M lead sampled during relevant TS-16T is strobed out on SS data by an appropriate digit pulse after combining with SR data received from the far-end E&M card via 2 Mbps stream.

The PCM stream received from the framer is decoded following the A-law companding by the codec, and timed by the receive-clock and the corresponding receive time-slot pulse. The analog output from the codec passes through a variable attenuator pad and an impedance matching transformer, to be available on the 600 ohms voice receive-port.

The signalling information from the remote E&M card is retrieved from the PCM stream by latching the SR1 bit from appropriate TS-16R, and strobing with D1 receive digit pulse. The 'receive' signalling data SR1, after appropriate monoshot delays, is accepted as a reliable signal changeover. It is used to drive the transistor circuit to extend earth on E lead under answer-back condition/JF reversal. It is also used (in combination with M lead bit status) to generate SS data bits in the PCM stream directed at the framer.

Two 2W/4W E&M circuits are accommodated in each interface module.

7.7.4 Outgoing/Suscriber/Hotline interface

Figure 7.7 gives the functional block schematic of the composite interface catering to 2 W outgoing (with trunk offering) subscriber and hotline (H/L) functions. Any of the three functions can be chosen in the field by altering the bergstick mini jumper strapping on the composite module.

The O/G(outgoing)/subscriber/hotline signalling converter card in the O/G mode is capable of working with A/A relay in the case of step-by-step exchange (i.e., auto-to-auto relay of Strowger exchange) or outgoing junction of crossbar exchange. In this mode it is expected to interact with an incoming signalling converter in the remote PCM equipment. In subscriber mode, the card is able to hook up with subscriber telephone instrument on the analog port and interact with a remote-end

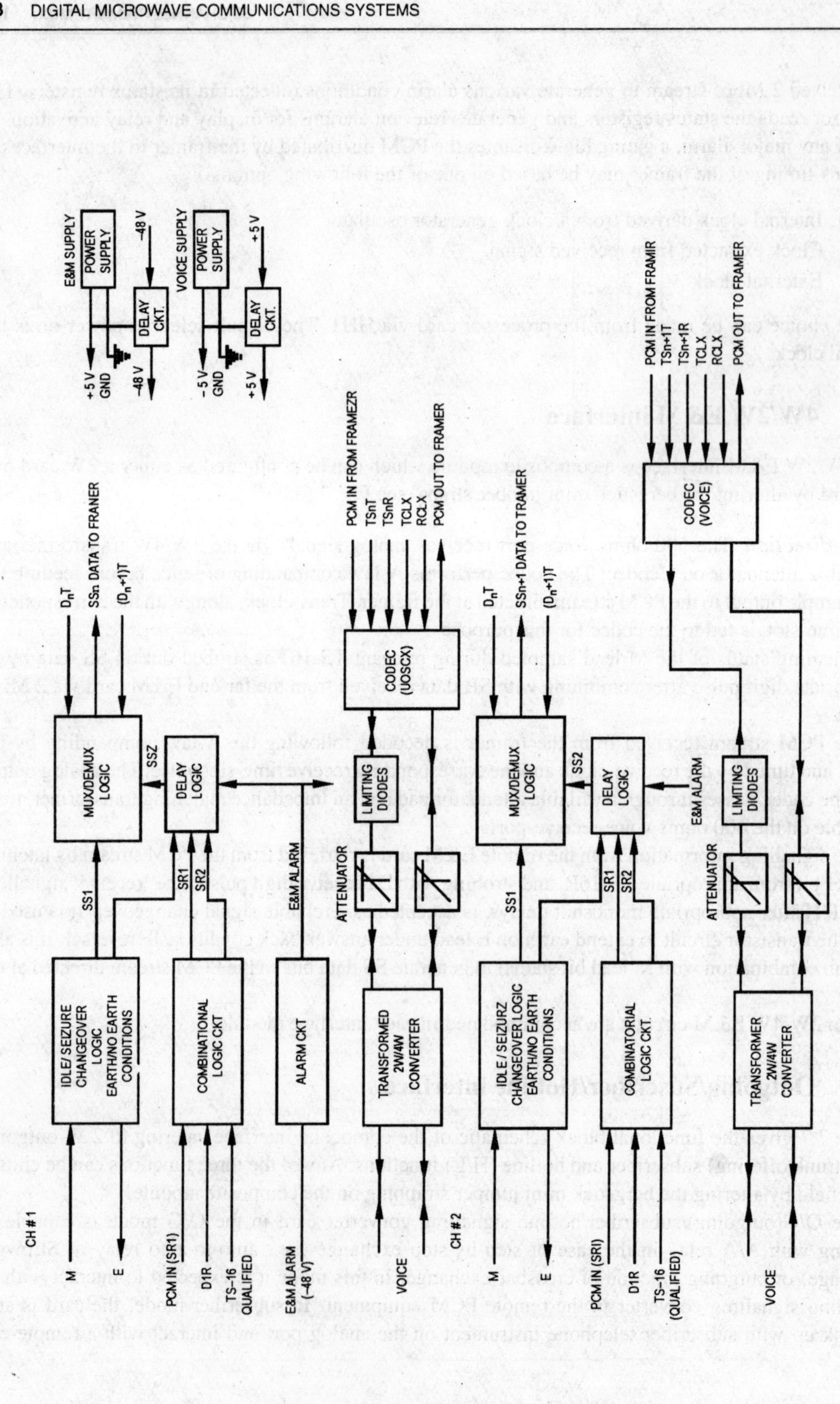

Fig. 7.6 *Block diagram of 2W/4W E&M card.*

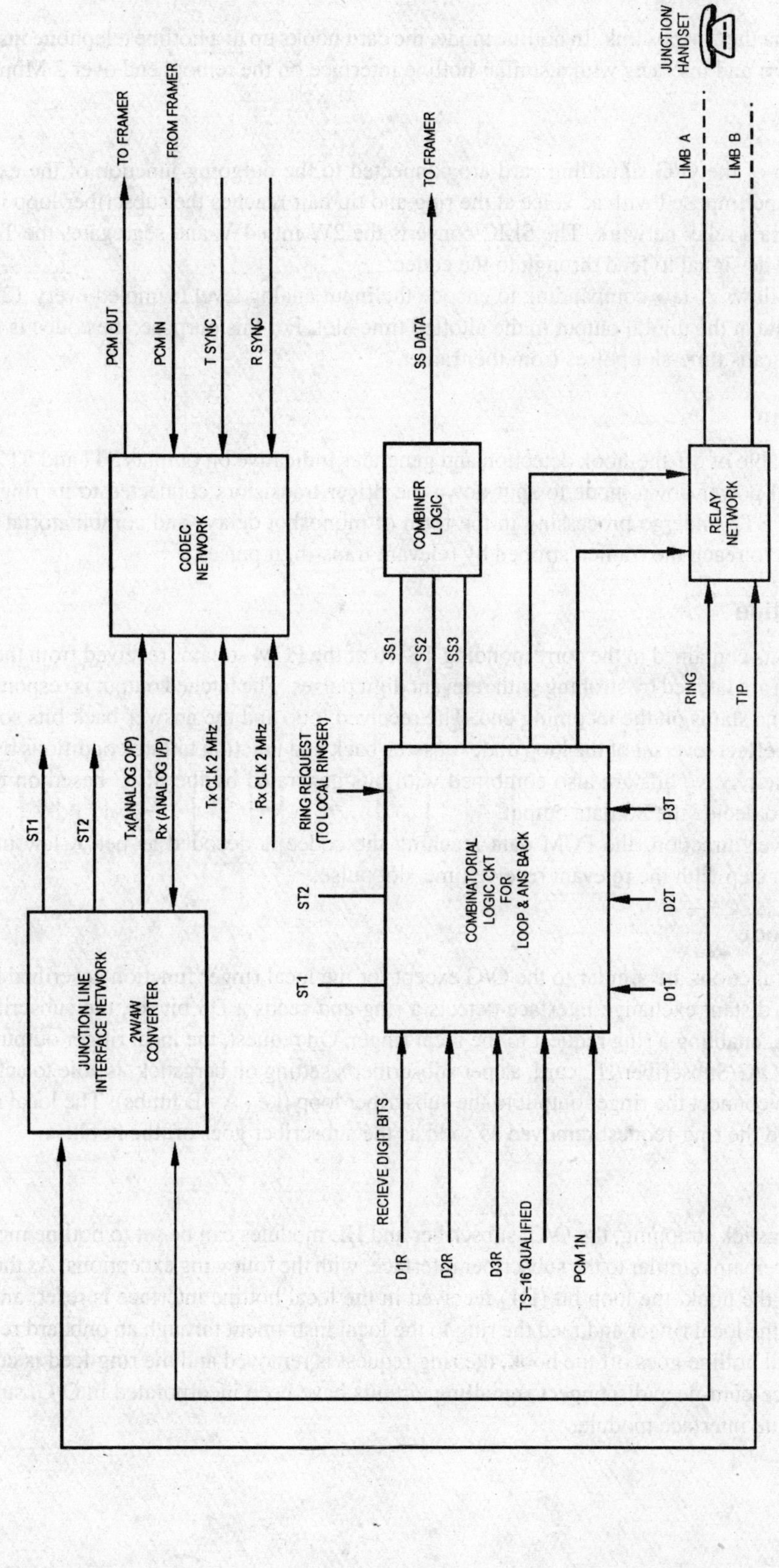

NOTE: (1) Each card supports two channels, block diagram being similar for each.
(2) The card is used in hotline/outgoing/subscriber mode by appropriate setting of jumpers.

Fig. 7.7 *Block diagram of outgoing/subscriber/hot line card.*

exchange card via the 2 Mbps link. In hotline mode, the card hooks up to a hotline telephone instrument on the analog port and interacts with a similar hotline interface on the remote end over 2 Mbps link.

O/G mode

The ring and tip of the O/G signalling card are connected to the outgoing junction of the exchange. Dc signalling superimposed with ac voice at the ring and tip pair reaches the subscriber loop interface circuit (SLIC) via a relay network. The SLIC converts the 2W into 4W, and segregates the Tx-signal (voice) from the dc signal to feed through to the codec.

The codec follows A-law companding to encode the input analog level (sampled every 125 μs) to an 8 bit PCM data at the digital output in the allotted time-slot. For this purpose, the codec is fed with trans-clock and trans time-slot pulses from the framer.

Trans-direction

The SLIC is capable of off-the-hook detection and generates indicative bit outputs STI and ST2. It also employs internal power-down mode to shut down the driver transistors connected to its ring and tip points. ST1 and ST2 undergo processing in the form of monoshot delays and combinatorial logic to produce SS data to reach the framer, strobed by relevant trans-digit pulses.

Receive-direction

The signalling bits contained in the corresponding TS-16 of the PCM stream (received from the framer by the O/G card) are latched by strobing with relevant digit pulses. The latched output is responsible for conveying the line status on the incoming end. The received loop and the answer-back bits so latched are processed to effect reversal of the loop under answer-back and junction failure conditions by a relay change-over. The receive bits are also combined with bits generated by the SLIC based on the local line conditions to decide the SS data output.

In the 'receive' direction, the PCM data reaching the codec is decoded as per A-law using the receive-clock, in step with the relevant receive time-slot pulse.

Subscriber mode

The subscriber functions are similar to the O/G except for the local ringer function described below.

As soon as a distant exchange interface detects a ring and sends a D3 bit set, the subscriber card latches the same, enabling a ring request to the local ringer. On request, the local ringer outputs a ring over a bus. The O/G/Subscriber/HL card, as per subscriber's setting on bergstick, is able to activate an onboard relay to connect the ringer output to the subscriber loop (i.e., A–B limbs). The local ringer is disconnected and the ring-request removed as soon as the subscriber goes off the hook.

Hotline mode

With proper bergstick strapping, the O/G, subscriber and HL modules can be set to hotline mode. The basic operation remains similar to the subscriber interface, with the following exceptions: As the distant hotline goes off the hook, the loop bit (D1) received in the local hotline interface is reset, and this is used to activate the local ringer and feed the ring to the local instrument through an onboard relay. The moment the local hotline goes off the hook, the ring request is removed and the ring feed is cut.

Two 2W voice-cum-deep disconnect signalling circuits have been incorporated in O/G, subscriber and HL composite interface module.

7.7.5 Incoming and exchange interface

Figure 7.8 gives the functional block schematic of the composite interface catering to the 2W 'incoming' and 'exchange' functions. By altering the bergstick mini jumper strapping on the composite module, any of the three different functions can be chosen. The I/C exchange signalling converter card in I/C mode is capable of working with the incoming junction of any electromechanical exchange. In this mode it is expected to interact with an O/G signalling converter in the remote PCM equipment. In the exchange mode the card is able to hook up to the subscriber port of the exchange and interact with a remote-end subscriber card via the 2 Mbps link.

I/C mode

The ring and tip of the I/C signalling card are connected to the incoming junction of an electromechanical exchange. A dc loop signalling superimposed with an ac voice reaches the ring and tip pair from the exchange at the backplane of the module.

Trans-direction

The A–B limb connection at the backplane reaches the trunk interface IC (TIC) and associated circuits, which detect various line conditions such as limb break, normal/reverse battery, etc., to give indicative digital output for driving an opto-isolator. Under idle and JF condition (limb-break /limb- short), this output remains high. With normal/reverse battery applied on ring and tip under loop/answer back, this output goes low to turn the opto ON. The output level of the opto on the 5 V side, indicative of the A–B limb condition, is combined with signal bit SR1 from the far-end O/G interface (duly retrieved after processing receive-stream) . This is used to 'place' send signal bit SS on the transmit data stream by strobing with the help of the corresponding digit pulse and TS-16T.

The ac voice on the line is passed by the TIC to a codec via an impedance matching transformer and variable attenuator pads. The codec follows A-law companding to encode analog signals sampled every 125 μs to form an 8-bit packet that joins the PCM signal directed at the framer in step with the relevant trans time-slot pulse. For this purpose, the codec is fed with trans clock and trans time-slot pulses.

Receive direction

The PCM input to the I/C card from the framer supplies signalling information and voice data from the far end O/G module. The signalling bits (SR1, SR2/SR3) picked up from the PCM input, are latched in the I/C module with the corresponding 'receive' digit pulses which are qualified by the corresponding TS-16R. This is made available for the data generation of suplementary services (SS). In case the trunk offer (TO) bit is found set in the SR bits, the I/C card actuates the TO relay to extend the earth on ring/tip effecting a TO condition. The received PCM stream from the framer is decoded by the codec block following A-law, and timed by the corresponding receive time-slot pulse and receive clock.

The analog output from the codec goes to the TIC via a variable attenuator pad and transformer, and subsequently reaches the A–B limbs superimposed on the dc line condition.

Exchange mode

With appropriate field settings on the bergstick mini jumper straps, the I/C or exchange module can work in exchange mode. The functions under this mode are the same as those in the I/C mode, except that the exchange ring is fed via a mini jumper strap (appropriately set) to a detector circuit. The output

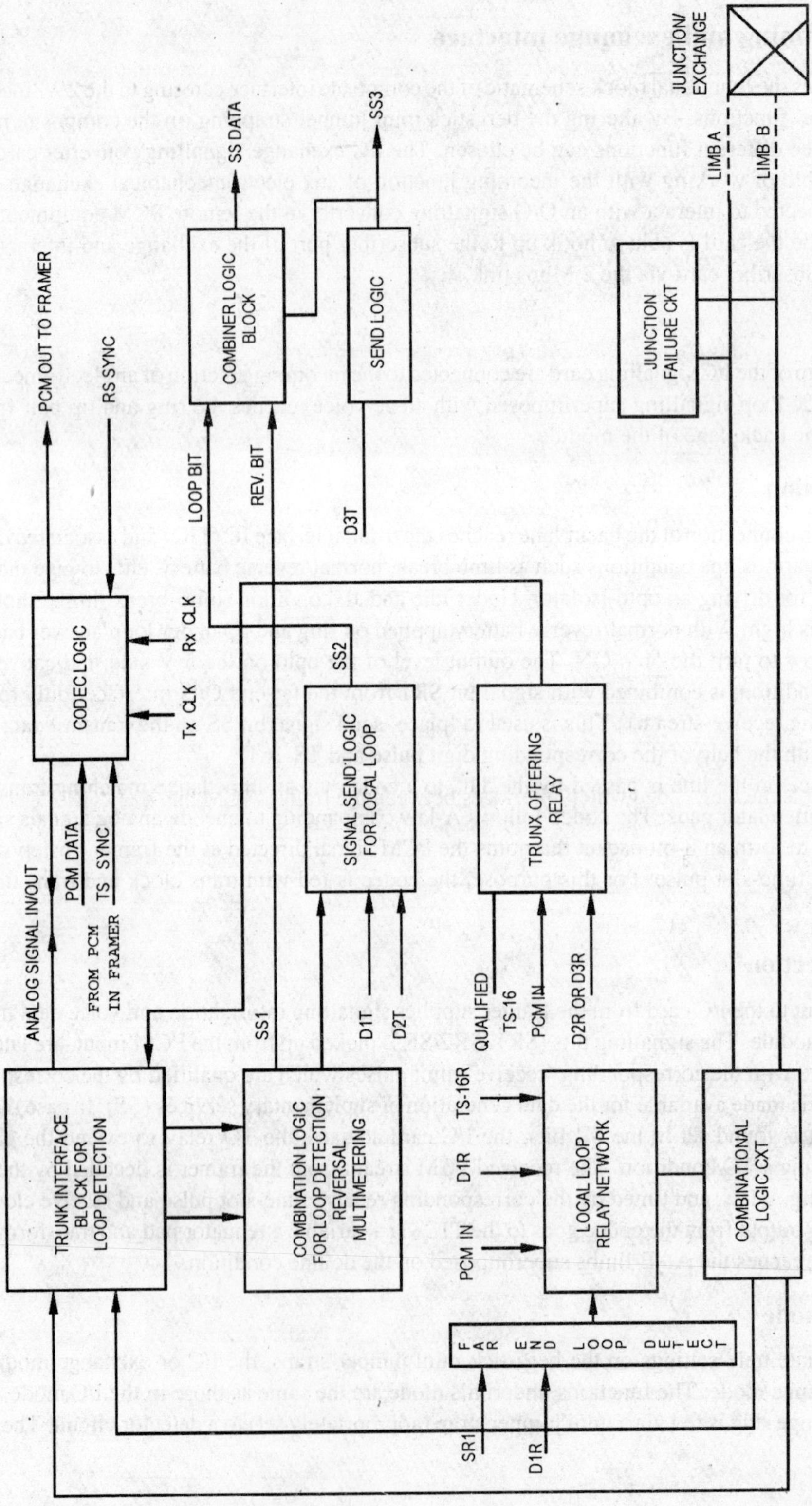

Fig. 7.8 *Block diagram of incoming/exchange card.*

of the detector circuit sets the D3 bit in the SS data to convey the presence of a ring via 2 Mbps stream to the far-end subscriber interface. Two circuits are accommodated in each incoming/exchange interface card.

7.7.6 64 kbps codirectional data card

Figure 7.9 gives the functional block schematic of a G.703 64 kpbs codirectional data interface. The basic elements in the module are the G.703 codirectional line transceiver and the codirectional digital data processor. Codirectional data received over 120 ohms balanced lines via a transformer reaches the transceiver to get converted into dual TTL streams.

The dual TTL streams reach the digital data processor (timed by a 256 kHz clock derived from 2 MHz receive clock) to get framed into the allocated 64 kb time-slot of the PCM. A combining logic on the card collects the outputs of two such circuits timed by their respective trans time-slot pulses, and emits a common PCM output at the backplane.

In the receive direction, the PCM-in from the backplane reaches the digital data processor of each channel, timed by their respective receive time-slot pulses. The digital data processor timed by 256 kHz clock (derived from 2 MHz receive clock) extracts the codirectional data from the PCM sample and makes it available on dual TTL outputs.

The G.703 64 kb codirectional transceiver receives the TTL level signal and translates it into bipolar output to drive the 120 ohms balanced line via a transformer.

7.7.7 Functional description of the V.24 data interface card

Figure 7.10 gives the functional block schematic of the V.24 data interface module.The central element to this module is the data codec which interfaces between the framed PCM bus and the TTL converted RS232 data lines. It uses transition-encoded modulation technique to encode/decode low speed data to and from the 64 kb channel of the PCM bus. The coding/decoding scheme is transparent and can accept either synchronous data at 8 kbps or asynchronous data at 9.6 kbps and 19.2 kbps.

The codec is timed for the trans and receive operation by the relevant (*i*) time-slot pulse, (*ii*) frame pulse and (*iii*) trans/receive clock from the framer by a timing circuit. Bipolar single-ended V.24 data received/transmitted via front panel D9 connector is converted to/from the TTL level by suitable level converters.

The V.24 port caters to the basic trans and receive data at 19.2 kbps along with the following (relatively slow changing) handshake signals: DCD, DTR, DSR, RTS, OTS, RI.

The trans/receive data is taken care of by the data codec as described above. The six handshake lines, after TTL level translation, connects with a parallel-to-serial/serial-to-parallel converter in the trans/receive direction. The serial handshake data is sampled every 2 ms, i.e., once in every multi frame, and timed in/out of the PCM stream over the V.24 time-slot reserved for the purpose of transporting the handshake status on PCM. In the receive direction, the output of the serial-to-parallel converter output is latched only at the end of the V.24 time-slot when the all the handshake outputs have become stable. The combination logic in the trans-direction picks up PCM data from the data codec in the allocated V.24 time-slot in every frame except Frame # 0 . In Frame # 0, instead of the data codec the serial-to-parallel converter is enabled to receive the PCM sample carrying the HS (hand shake) line status. This is subsequently latched to reproduce the HS line conditions on connector after TTL–RS232 level translation.

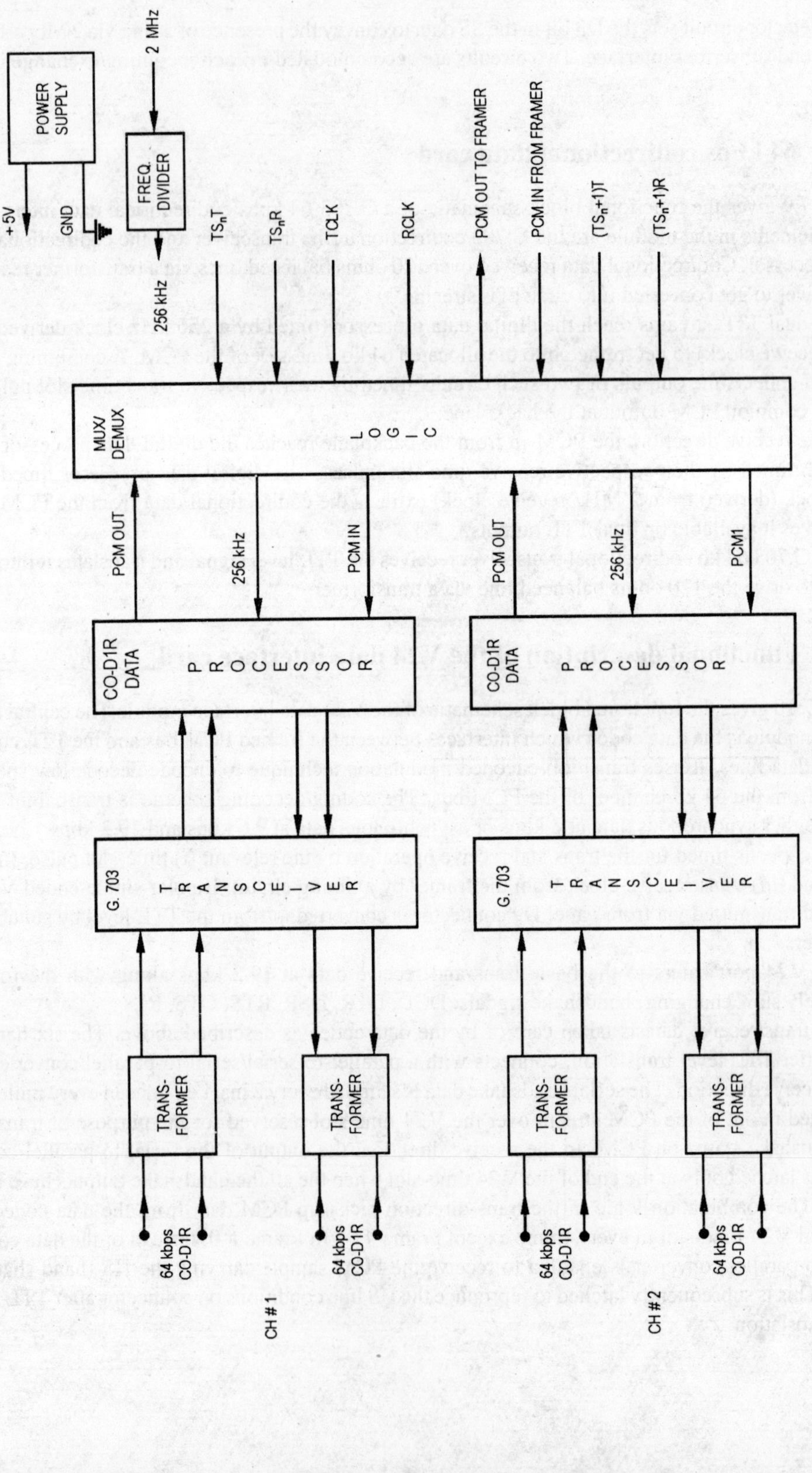

Fig. 7.9 *Block diagram of 64 kbps data card.*

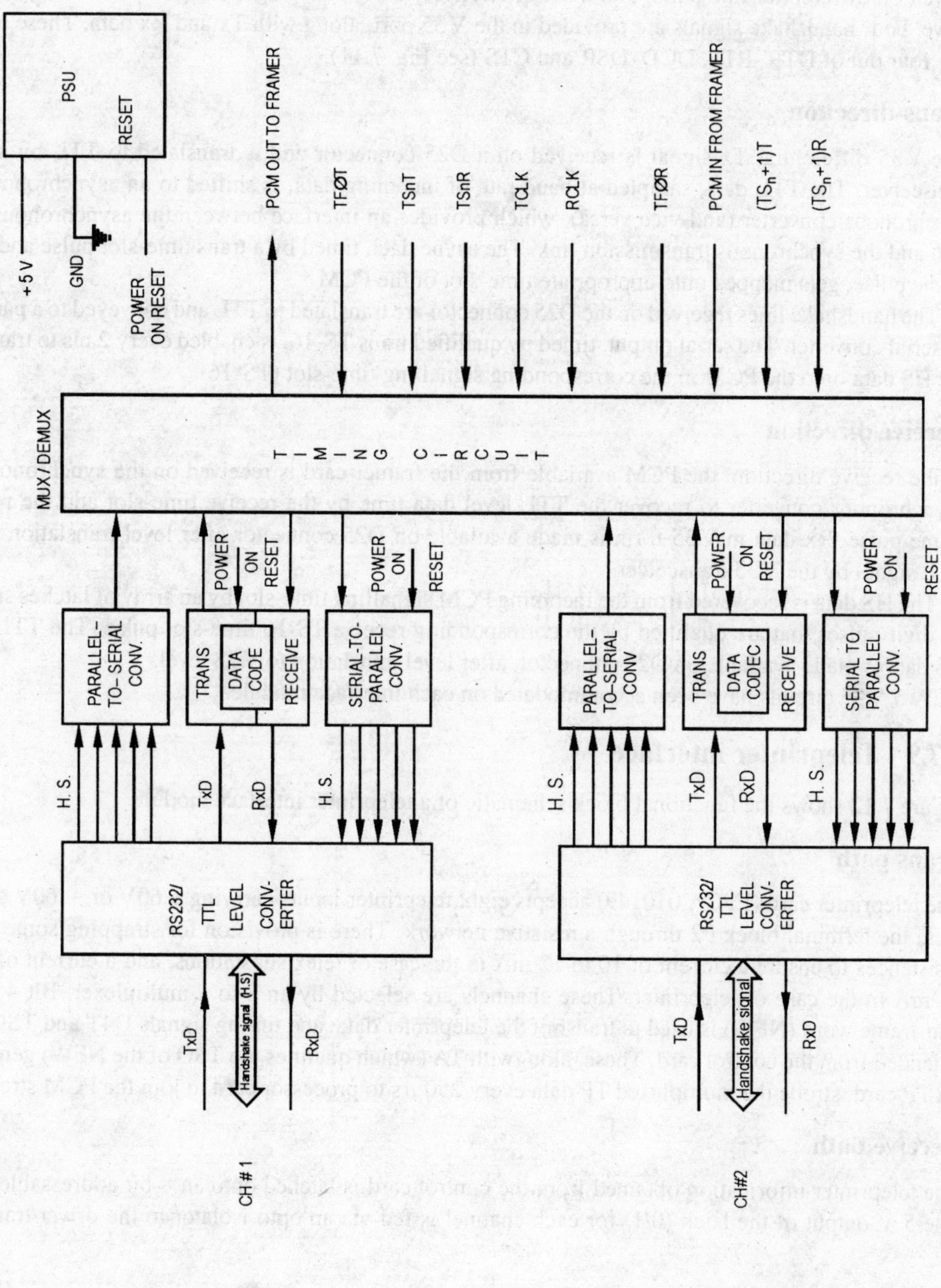

Fig. 7.10 *Block diagram of V.24 data card.*

The two V.24 circuits are accommodated in one card with common combined PCM-out and PCM-in at the backplane, connected to the framer via the motherboard.

7.7.8 Functional description of V.35 interface

The V.35 interface from the remote data terminals is composed of high bit rate trans and receive data driven on differential line pairs, and a set of relatively slow-changing HS signals with single-ended drive. Four handshake signals are provided in the V.35 port, along with Tx and Rx data. These can be any four out of DTR, RTS, DCD, DSR and CTS (see Fig. 7.11).

Trans-direction

The V.35 differential Tx signal is received on a D25 connector and is translated to TTL by a V.35 transceiver. The TTL data, sampled at baud rate of incoming data, is shifted to an asynchronous-to-synchronous converter (and vice versa), which provides an interface between the asynchronous data port and the synchronous transmission link. The async data, timed by a trans time-slot pulse and trans frame pulse, gets mapped onto appropriate time-slot of the PCM.

The handshake lines received on the D25 connector are translated to TTL and conveyed to a parallel-to-serial converter. The serial output, timed by qualified trans TS-16, is enabled every 2 ms to transport the HS data onto the PCM in the corresponding signalling time-slot (TS-16).

Receive direction

In the receive direction, the PCM available from the framer card is received on the synchronous-to-asynchronous converter to recover the TTL level data time by the receive time-slot and the receive frame pulse. Rx data in V.35 form is made available on D25 connector after level translation of the TTL signal by the V.35 transceiver.

The HS data is recovered from the incoming PCM signalling time-slot by an array of latches strobed by digit pulses that are qualified by the corresponding receive TS-16 time-slot pulse. The TTL level HS data is made available on D25 connector, after level translation to V.28 level.

Two V.35 circuits have been accommodated on each interface module.

7.7.9 Teleprinter interface

Figure 7.12 shows the functional block schematic of a teleprinter interface module.

Trans path

The teleprinter card (EPCA 010149) accepts eight teleprinter inputs carrying + 60V or − 60V signals from the terminal block P2 through a resistive network. There is provision for strapping some of the resistances to ensure a current of 10 to 12 mA in the case of telex substations, and a current of 20 to 25 mA in the case of teleprinter. These channels are selected by an 8 to 1 multiplexer. Bit 4 of the non-frame word (NFW) is used to transmit the teleprinter data, and timing signals D4T and TSOT are extended from the control card. These, alongwith TA (which qualifies the TSO of the NFW) generated on TP card, strobe the multiplexed TP data every 250 μs to processor card to join the PCM stream.

Receive path

The teleprinter information obtained from the control card is latched onto an 8-bit addressable latch. The 5 V output of the latch (0/1) for each channel is fed via an opto-isolator to the driver/translator

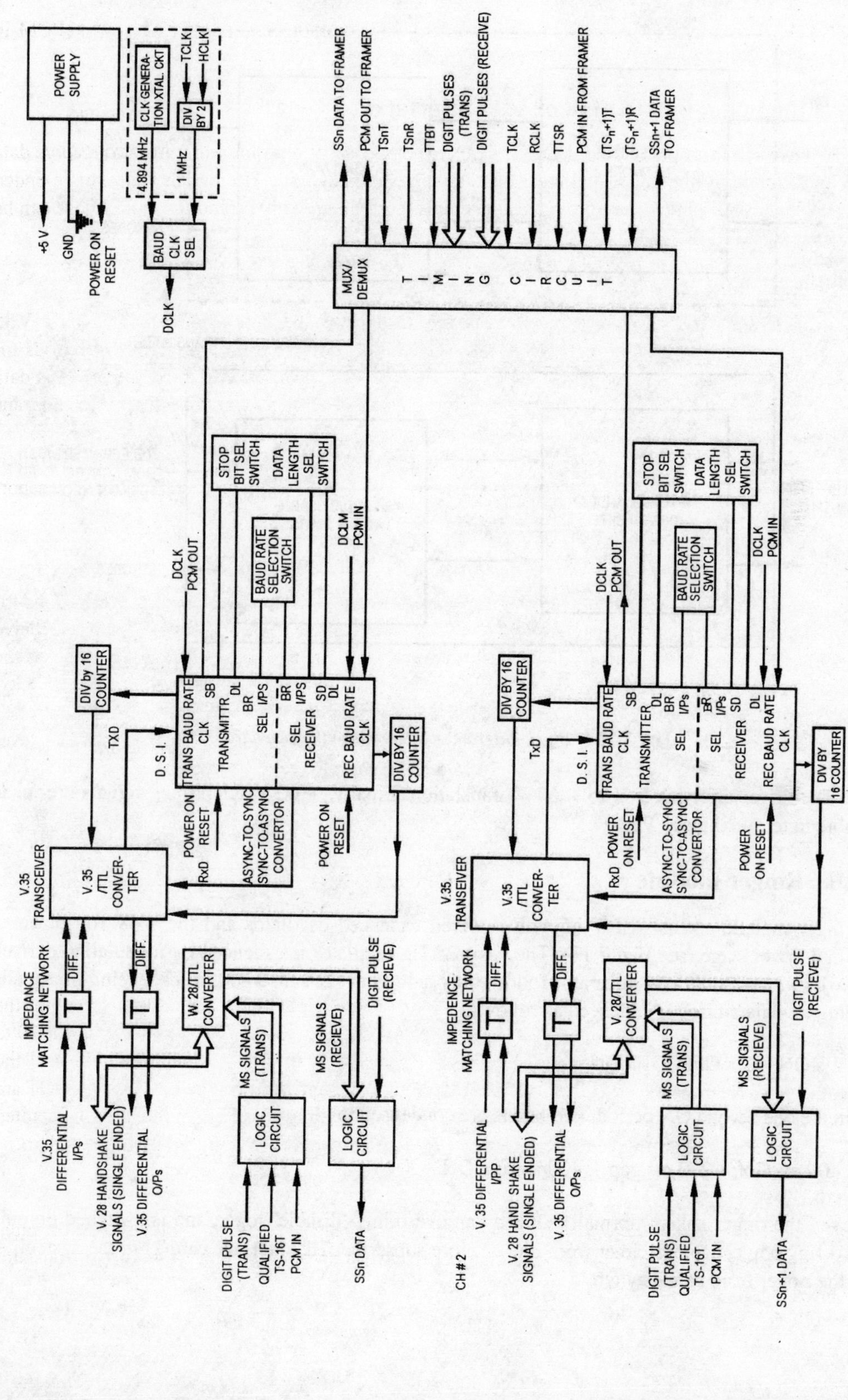

Fig. 7.11 *Block diagram of V.35 data card.*

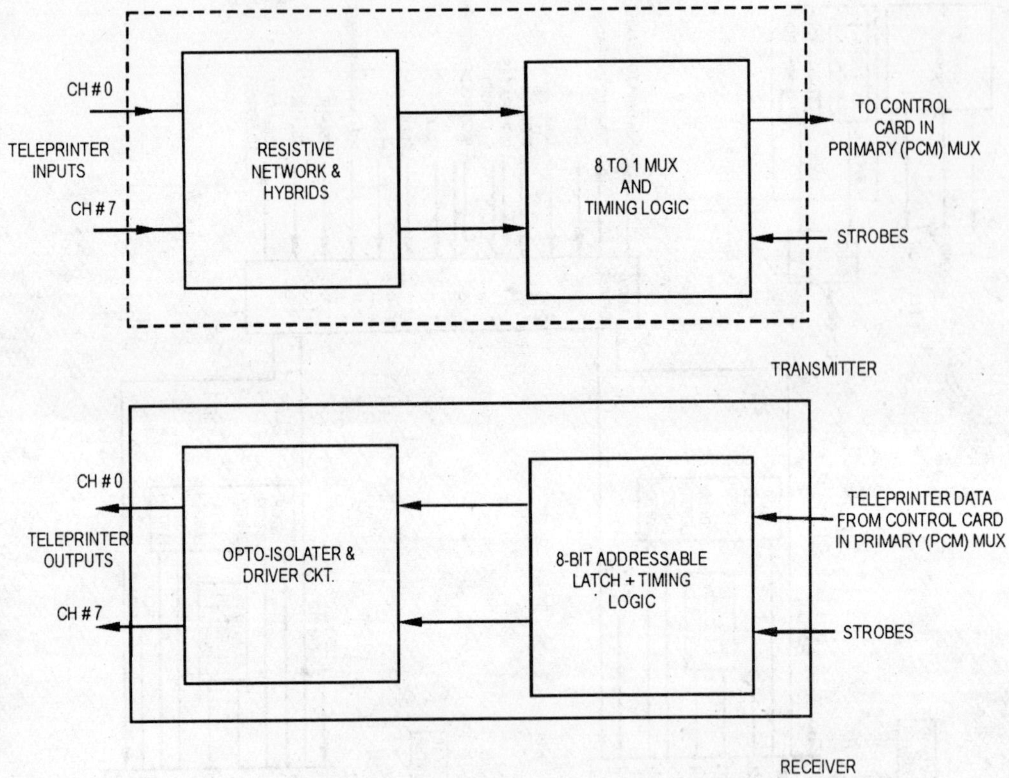

Fig. 7.12 *Block diagram of teleprinter interface card.*

stage. The driver effects a 5 V to + 60 V translation. Finally, + 60 V teleprinter signals are made available in terminal block P2.

7.7.10 Ringer module

The ringer module consists of a +5 volt-powered cadenced oscillator and the −48 volt-powered amplified driver stage (see Fig. 7.13). The basis 25 Hz oscillator is cadenced by modulation derived from a 1 s−2 s−1 s interval generator followed by a 400 ms−200 ms−400 ms cadencing timer. The resulting modulated ring is cadenced as follows:

1 s ON – 2 s OFF – 1 s ON......

Within the one second ON period, two bursts are condensed to

400 ms ON – 200 ms gap – 400ms ring ON...

However, the ringer is kept normally OFF by an overriding 'disable' logic, and is switched on only when a ring request comes either from one or more subscriber/HL interface card(s) in the system, or from the ringer front panel switch.

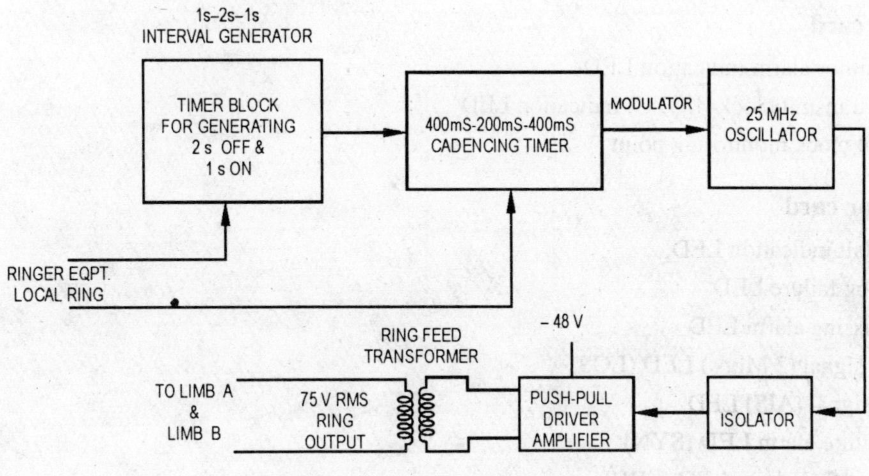

Fig. 7.13 *Block diagram of ringer card.*

The +5 V level cadenced ring is isolated and used for driving an output amplifier powered by the −48 V supply. A step-up ring feed transformer connected as collector load to the power amplifier produces the required 75 V RMS ring output on the secondary. The ringer output is finally fed to the system backplane through ac coupling capacitors to reach all the interface slots.

Ring monitoring points are provided on the front bracket if the module and the field staff is required to press the ring-enable switch while monitoring the ring voltage referred to above.

7.7.11 Functional description of power supply card

The power supply for MUX terminal is provided from the power supply unit. This unit works on standard exchange battery (−48 V). It provides well regulated, short-circuit protected, output dc voltages of +5 V and −5 V. In case of any over-voltage, the over-voltage crobar acts, shutting off the entire power supply. Alarms indications are made available on the unit for short circuit, over-voltage and under-voltage conditions (see Fig. 7.14).

7.8 Alarms/Supervisory Facility in PDM 30B

There exists provision for various types of alarms for supervising the functioning of the MUX, as listed below.

Power supply module

ON/FAIL indication LED's

Monitoring points for + 5 V, −5 V and −48 V

Framer card

Major/minor alarm indication LEDs
Loss of transmit clock (LOTC) indication LED
Transmit clock monitoring point

Processor card

Reset/Halt indication LED.
Watchdog failure LED
Card missing alarm LED
Loss of signal (2 Mbps) LED (LOS)
All 1's signal (AIS) LED
Sync failure alarm LED (SYN)
Bit error 10^{-3} alarm LED (ERR)
Remote terminal failure LED (RMT)
Manual busy indication LED (MBS)
Supervisory hand-held terminal (HHT) interface port.

2W/4W E&M card

M lead ground indication LED
E lead ground indication LED

Subs/OG/HL card

Loop/answer-back condition indication LED
Junction failure indication LED
VF monitoring port.

Exchange incoming (I/C) card

Loop/answer back condition indication LED
Junction failure indication LED
VF monitoring port.

64kb data card

Loss of signal (64kbps) indication LED (LOS)

V.24 data card

Front LED indication—not assigned.

V.35 data card

Front LED indication—not assigned.

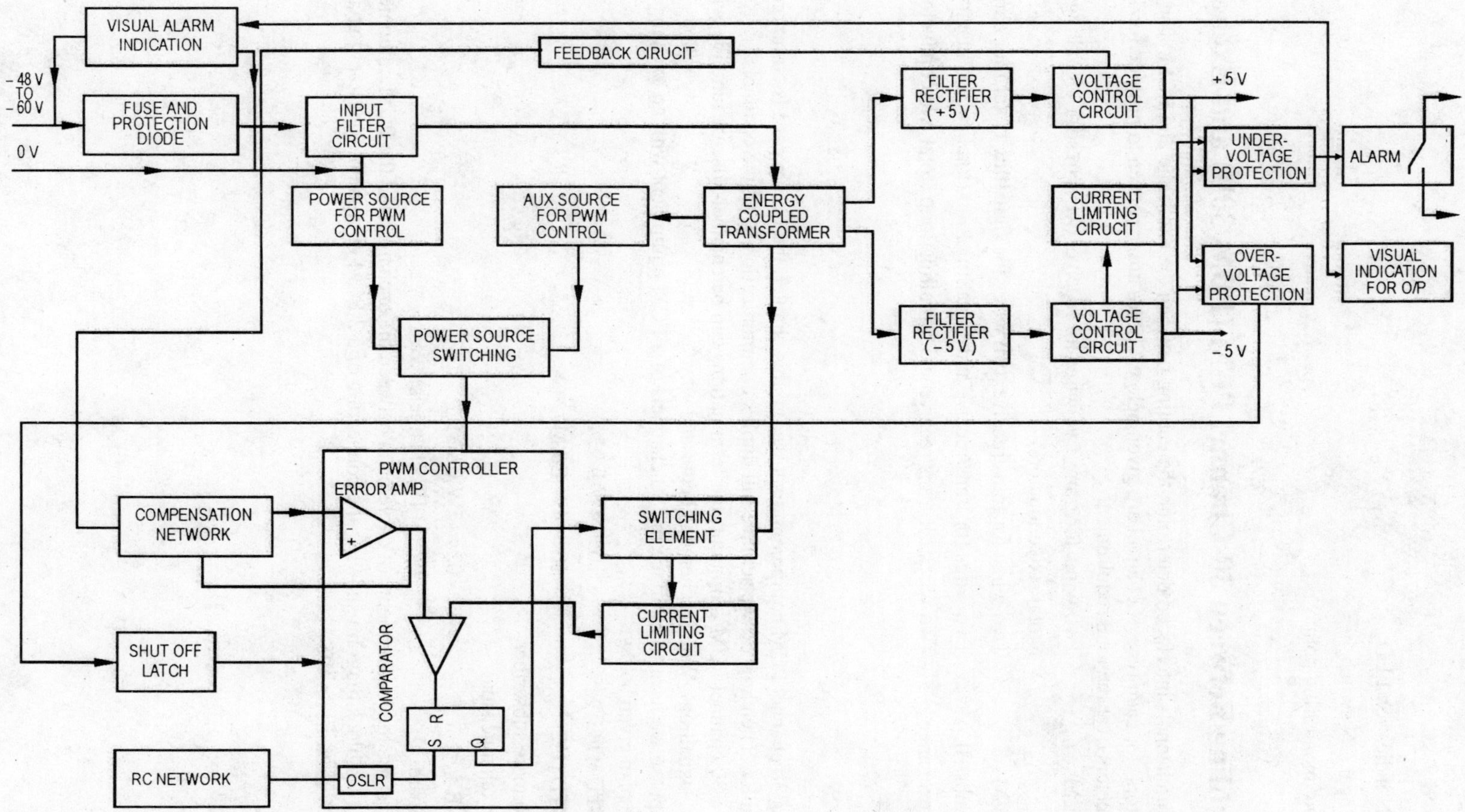

Fig. 7.14 *Block diagram of power supply card.*

Ringer card

Ringer ON indication LED
Manual ringer ON switch
Ring output monitoring point

7.9 Cabling Between 30-Channel PCM (PDM 30B) and Exchange

The types of channels interfaced determine the cabling between the exchange and PCM. The 'wires' or 'conductors' for each type of channel are given below. Cables are is chosen to provide for 10–20% spare conductors over the total requirement.

4W E&M	:	Two wires for trans, two wires for receive, two wires for signalling (total six conductors)
2W E&M	:	Two wires for trans/receive, two wires for signalling (total four conductors)
O/G /Subs/HL	:	Two wires for trans/receive and signalling loop (total two conductors)
I/C or exchange	:	Two wires for trans/receive and signalling loop (total two conductors)

Exercises

1. In a 30 channel PCM equipment discussed in this chapter, the framer card is common for Tx and Rx. Then how do we achieve full duplex communication in all the channels?
2. In a 30 channel PCM quipment, the terminations can be channel-wise or function-wise (Tx, Rx, signalling). Which scheme is better and why?
3. Suppose we have to interface the 30 channels of a PCM equipment with an exchange as per the plan given below:

STD (O/G)	:	10	4W E&M
STD (I/C)	:	10	4W E&M
Remote subscriber (exchange side)	:	04	
CB	:	02	2W E&M
Spare	:	04	(4W channels)

   What are the requirements of cable pairs between the exchange and the PCM equipment ?
4. Draw a block diagram showing the connection of a 2 PCM equipment (one each) at adjacent MW stations on WS.

8

III-Order Multiplexing Equipment

8.1 Introduction

In Section 4.11 the hierarchy of digital multiplexing has been explained. The basic entry in digital transmission being 2 Mbps or 30 channel or E1 data frame, the output data rate for III-order MUX is 34 Mbps. Generally, the III-order MUX equipment available are the 2/34 digital MUX equipment or the skip MUX equipment. This multiplexer multiplexes sixteen plesiochronous 2 Mbps bit streams into one 34 Mbps bit stream on the transmit side, and demultiplexes one 34 Mbps bit stream into sixteen 2 Mbps data streams on the receive side. Internally, an 8 Mbps frame is first generated from four 2 Mbps frames and four such 8 Mbps frames make the 34 Mbps frame as per the CCITT recommended hierarchy (discussed in Section 4.11). Since there is no access to the 8 Mbps externally, the MUX is termed 2/34 skip MUX. The multiplexing principle is cyclic bit interleaving, and positive justification is employed. The alarms and consequent action, and the input and output specifications are as per the relevant CCITT specifications.

Since it is beyond the scope of the book to explain all types of III-order MUX equipment, one model, viz., 2/34 MUX of ITI make is chosen for demonstration. The arrangement of various sub-racks and modules is described first. This is followed by functional description of the equipment. Then, module-wise description with reference to the associated circuits is presented. Finally, the technical specifications of a III-order MUX are presented.

8.2 Description of 2/34 MUX Equipment (ITI Make)

8.2.1 Equipment Rack and Sub-racks

The 2/34 MUX of ITI make comes as a vertical, slim rack with four sub-racks (Fig. 8.1). The four sub-racks from bottom to top are:

- Two sub-racks of 2/34 MUX
- Two sub-racks, one each for either direction of the digital network, house the 2/34 MUX cards or modules. In case of terminal station equipment, only one sub-rack of 2/34 MUX is available.
- Power supply and supervisory sub-rack
- Two power supply modules, one for each MUX sub-rack are housed here. One supervisory module (optional) is also housed here. It caters to both MUX sub-racks. Power consumption of each 2/34 MUX sub-rack is approximately 50 watts.

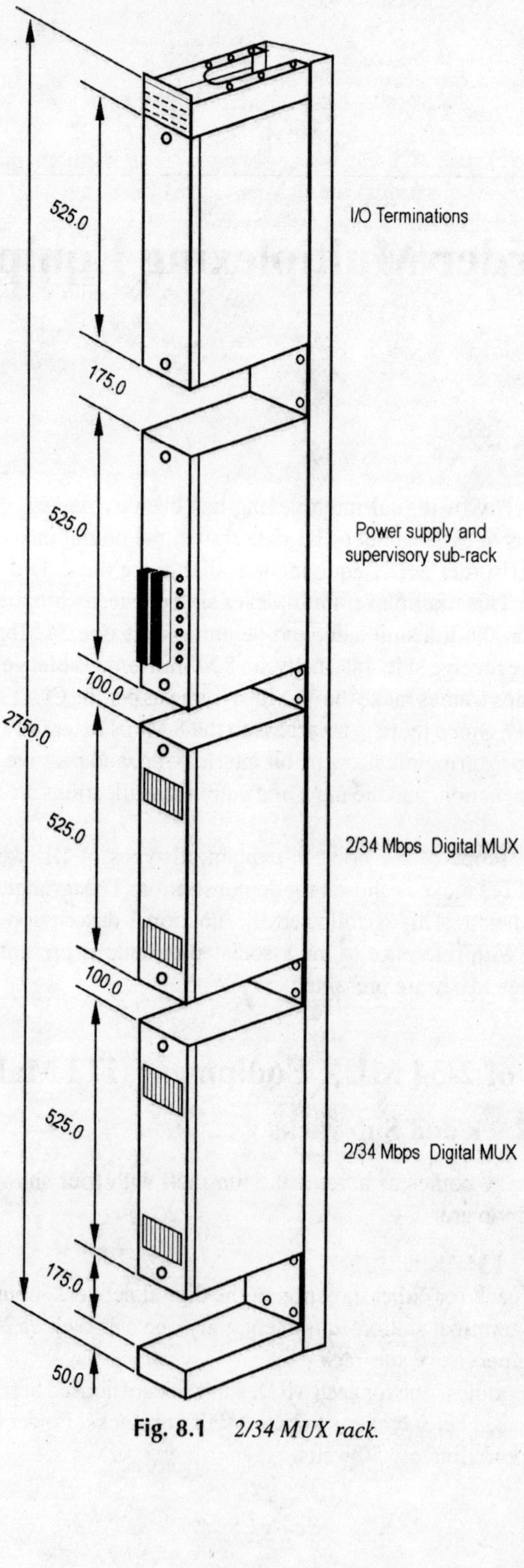

Fig. 8.1 *2/34 MUX rack.*

- I/O termination sub-rack
- The data-in and data-out connections of the 2 Mbps streams—incoming and outgoing—are terminated here.

Each of the 2/34 sub-rack has five cards or modules; four of them are 2/8 modules and the fifth one is an 8/34 module. The various inter-connecting signals between the PC boards are connected through a printed circuit back panel. Custom-built devices, programmable logic devices (EPLD) and clock modules are employed to realize various functions. All the essential alarms are depicted on their respective PC boards for easy identification of fault, and appropriate corrective action.

The power supply module provided in the power supply sub-rack provides the necessary output voltages required for the equipment. The power supply module operates from a normal input supply of −48 V with a supply variation of −40 to −60 V. Control/protection switches are provided in the power supply sub-rack

Supervisory module (optional) equipped in the power supply sub-rack will depict the various MUX alarms on a hand-held terminal. There is a provision for transmitting these alarms on an RS 232 serial port. One supervisory module can cater to two 2/34 MUX equipment.

At the bay top, there is a provision for wire wrapping the 2 Mbps I/O terminations. However, the 34 Mbps I/O terminations are provided on the MUX sub-rack itself. The necessary visual alarms are provided at the bay top for system alarm indication. Terminations are also provided for audible alarm.

8.2.2 Functional Description of Equipment

The 2/34 multiplexing and demultiplexing functions are carried out in two stages. On the transmit path, in the first stage, sixteen 2 Mbps tributaries are multiplexed into four 8 Mbps streams (in each 8 Mb stream, four 2 Mb streams are multiplexed). In the second stage, the four 8 Mbps streams are multiplexed into 34 Mbps stream.

On the receive path, in the first stage, the 34 Mbps stream is demultiplexed into four 8 Mbps streams. In the second stage, each 8 Mbps stream is demultiplexed into four 2 Mbps tributaries.

Refer to Fig. 8.2 showing the multiplexing in the transmit path.

(1) In the first stage, the incoming sixteen 2 Mbps plesiochronous digital streams (tributaries) after equalization/clock extraction, HDB-3 decoding and synchronization (positive justification principle), are multiplexed (by cyclic bit interleaving) to obtain four tributaries of 8 Mbps digital stream. The necessary frame alignment word and service bits are added.

(2) In the second stage, these four 8 Mbps digital streams are again multiplexed in the same way as that of the 2/8 Mbps MUX principle, to obtain one 34 Mbps HDB-3 output.

Refer to Fig. 8.3 showing the demultiplexing in the receive path.

(1) In the first stage, the incoming 34 Mbps HDB-3 input, after station cable equalization/clock extraction, regeneration, HDB-3 decoding and framing, is demultiplexed to realize four tributaries of 8 Mbps digital stream.

(2) In the second stage, these four 8 Mbps digital streams are again demultiplexed to obtain sixteen tributaries of the 2 Mbps (HDB-3 output).

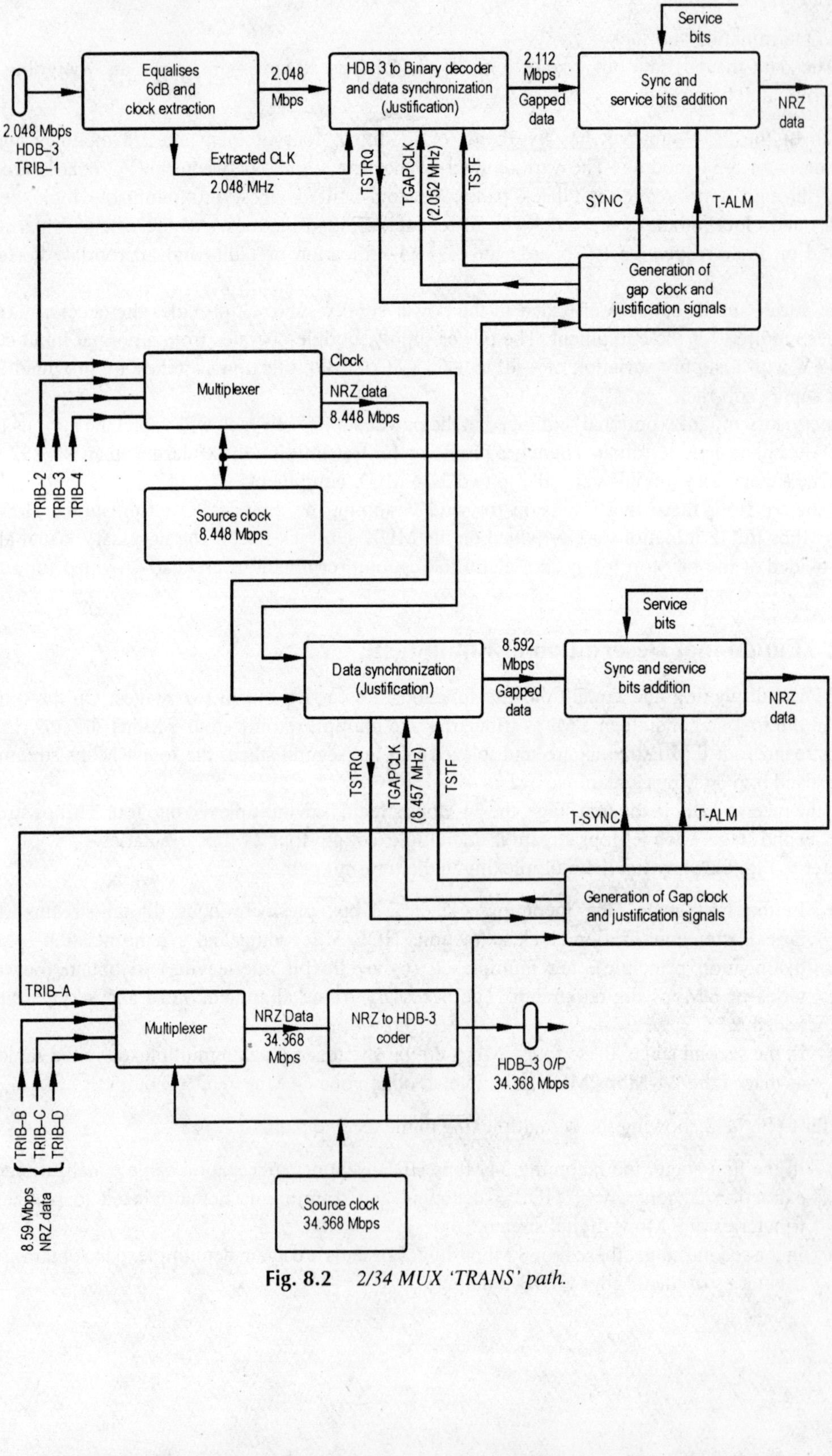

Fig. 8.2 *2/34 MUX 'TRANS' path.*

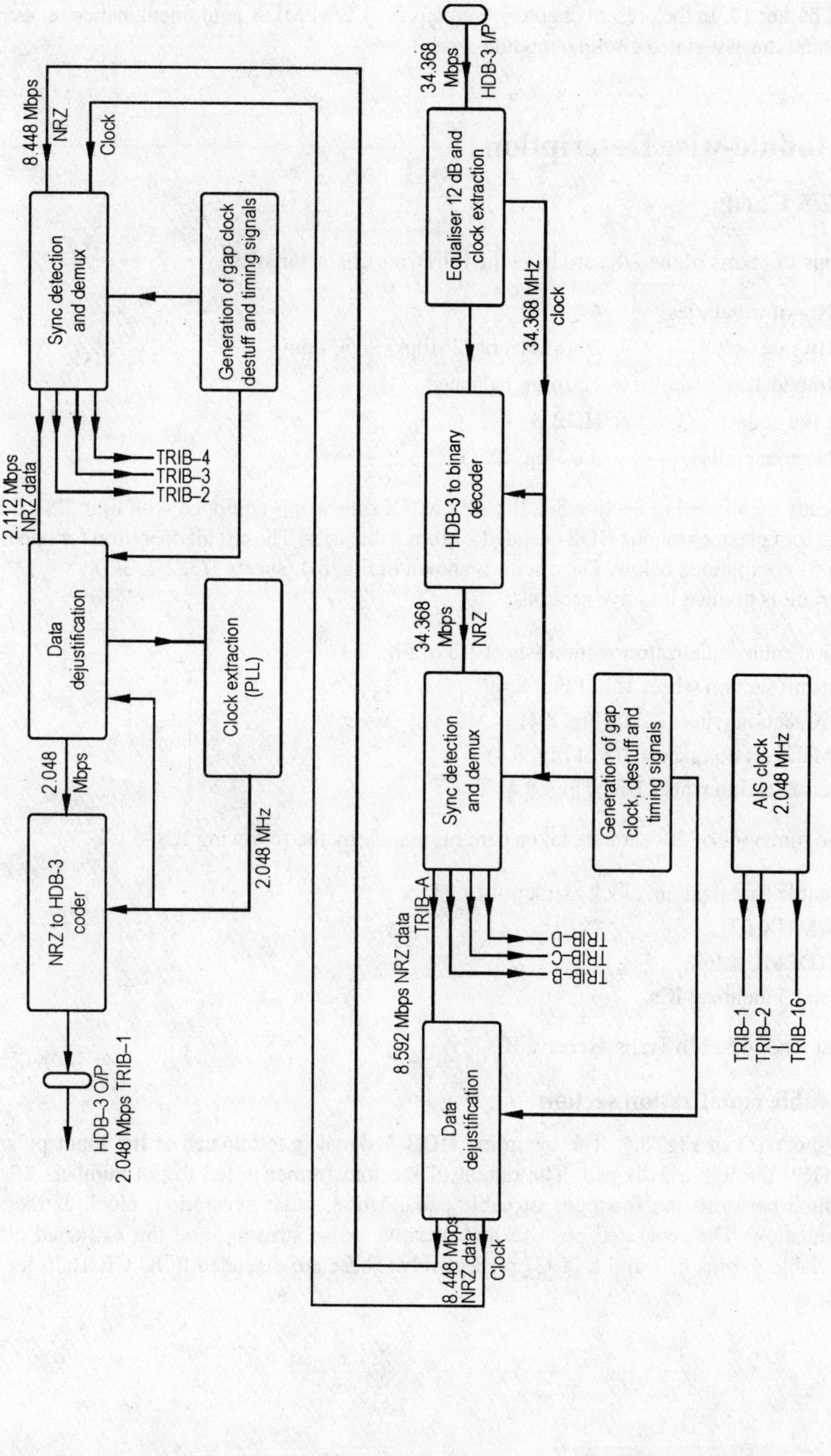

Fig. 8.3 *2/34 MUX 'receive' path.*

Integration of bit streams (four 2 Mbps into one 8 Mbps, four 8 Mbps into one 34 Mbps) and the converse; the concept of justification; the purpose of gap clock or perforated clock are discussed in detail in Chapter 12. In the present chapter, the focus is on 2/34 MUX equipment, hence let us proceed to understand the associated circuits, module-wise.

8.3 Module-wise Description

8.3.1 2/8 Card

The 2 Mbps I/O ports of the 2/8 card have the following characteristics:

- No. of tributaries : 4
- Bit rate : 2048 kbps or (2Mbps) ± 50 ppm
- Impedance : 120 ohm balanced
- Line code : HDB-3
- Nominal pulse : 3.0 Vpp

As already mentioned in Section 8.2, the 2/34 MUX sub-rack is equipped with four 2/8 cards, each catering to four plesiochronous HDB-3 coded 2 Mbps tributaries. The circuit operation for one tributary (tributary 3) is explained below. The circuit is shown in Fig. 8.4 (sheets 1/3, 2/3, 3/3).

The circuit is divided into five sections:

- Station cable equalizaton section (sheet 1/3 of Fig. 8.4)
- Transmit section (sheet 1/3 of Fig. 8.4)
- MUX section (sheet 2/3 of Fig. 8.4)
- DEMUX section (sheet 3/3 of Fig. 8.4)
- Receive section (sheet 1/3 of Fig. 8.4)

The above functions of 2/8 card are taken care of, mainly by the following ICs:

- U7(cable equalization, clock extraction),
- U34(MUX IC),
- U11(DEMUX IC),
- U9 and 3 identical ICs,

i.e., a total 4 nos of 2 Mb Trans–Receive ICs.

Station cable equalization section

Refer to sheet 1/3 of Fig. 8.4. The incoming HDB-3 signal is terminated at the input pulse transformer (TR9) through a 3 dB pad. The output of the transformer is fed to pin numbers 19 and 20 of U7 which performs the functions of cable equalization, pulse separation, clock extraction and jitter attenuation. The separated positive and negative pulse streams, and the extracted clock are made available at pins 6, 7 and 8 of U7 respectively. These are extended to IC CB 1626 for further processing.

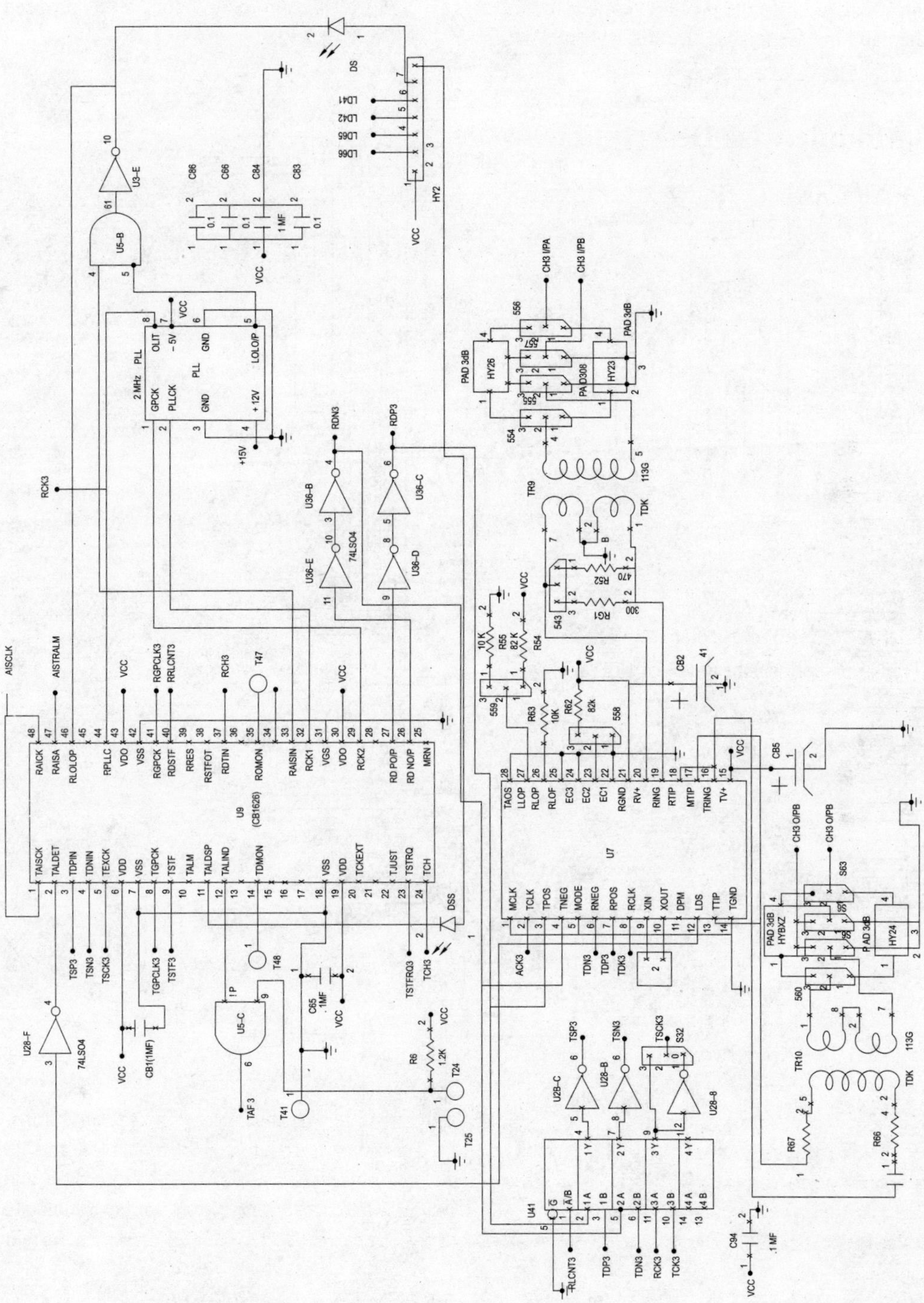

Fig. 8.4 *2/8 card (Sheet 1/3)*

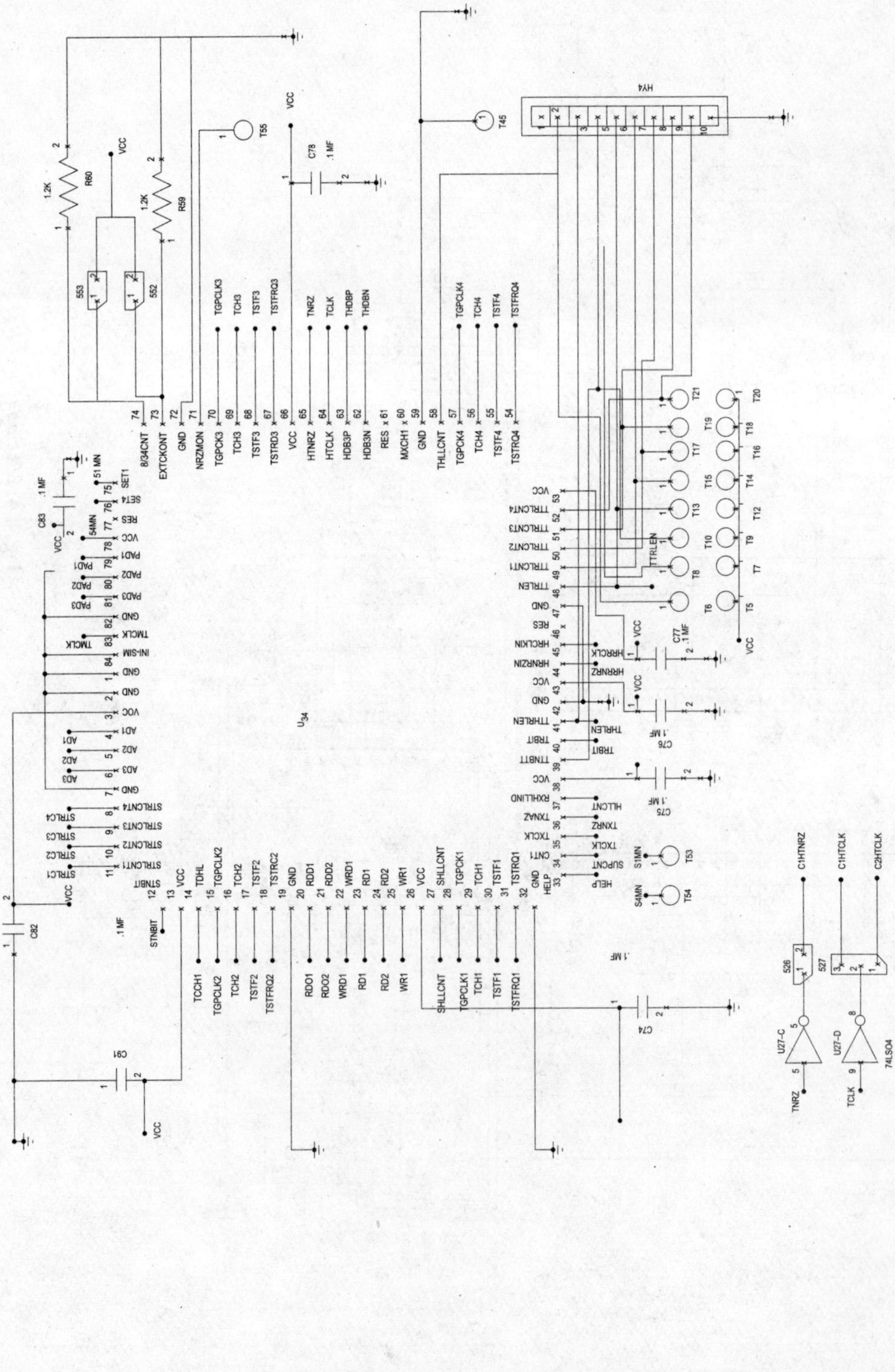

Fig. 8.4 *2/8 card (Sheet 2/3)*

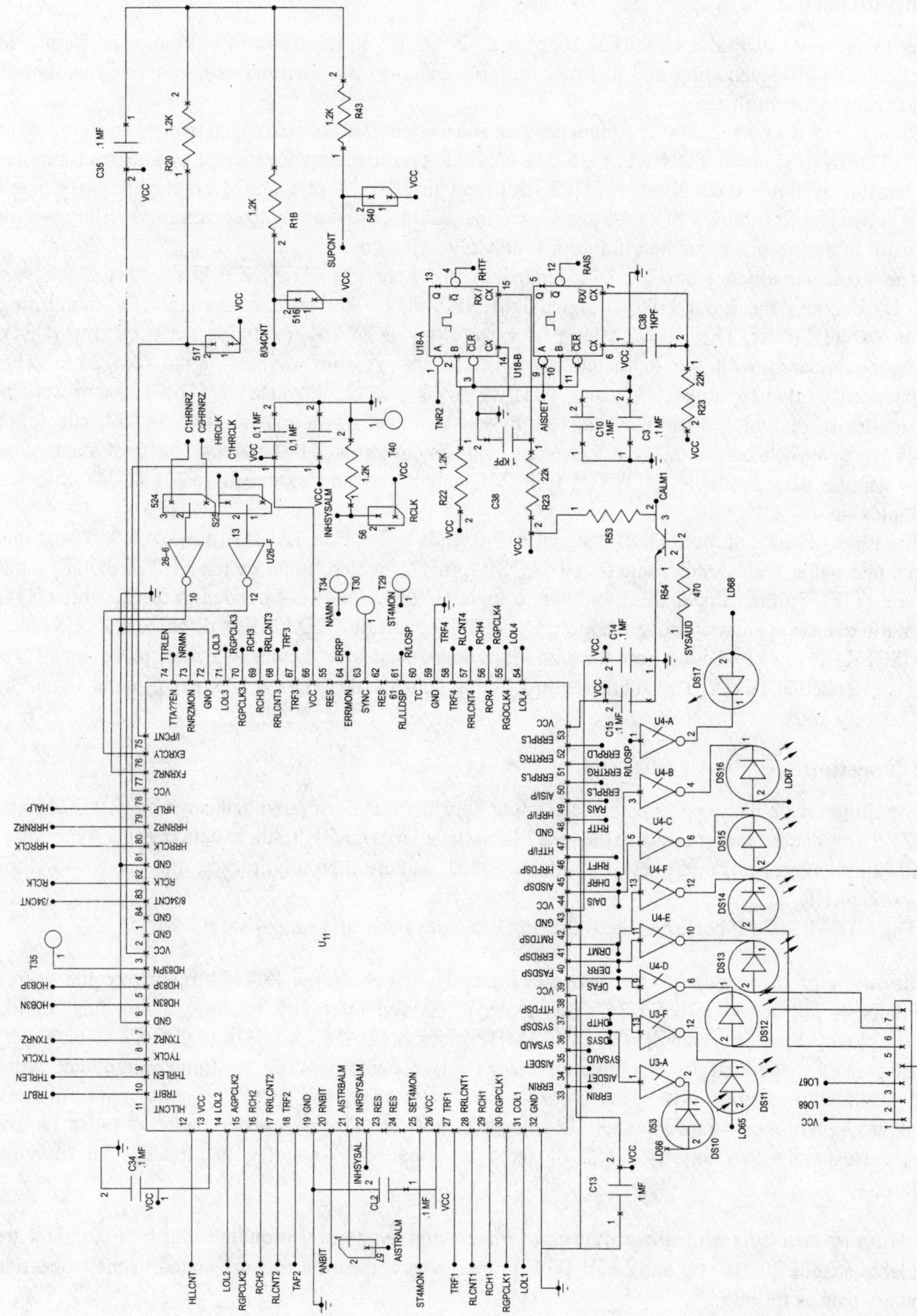

Fig. 8.4 *2/8 card (Sheet 3/3)*

Transmit section

Refer to sheet 1/3 of Fig. 8.4. U9(CB 1626) is a CMOS IC. In the transmit section, it performs the function of HDB-3 decoding and justification (conversion of plesiochronous data into synchronous data stream by bit stuffing).

Refer to Fig. 8.4 (sheet 1/3). The input positive and negative streams extended from U7 are applied at pin 3 (TDPIN) and pin 4 (TDNIN) of U9. The combined positive and negative pulse streams (unipolar RZ data) are available at pin 20 of U9 (TICKEXT) and pin 10 of U9 (TALARM). It may please be noted that U9 pertains to tributary 3. There are 3 more identical ICs pertaining to remaining 3 tributaries on 2/8 card. In the circuits presented in Fig. 8.4, only U9 is shown.

The extracted clock from U7 is extended to pin 5 of U9 (TEXCK). Using this clock signal, U9 decodes the input HDB-3 signal into NRZ data. This data is available for monitoring at pin 14 (TDMON). The 2.0522 MHz perforated clock TGPCLK, extended from Tx chip (U34), is phase-compared with the extracted clock inside U9. A stuff request signal (pin 23 of U9) is extended to the Tx chip (U34) and TSTF removes one clock pulse in TGPCLK corresponding to the second bit position in the fourth set. Using a 10-bit elastic store in U9, the input 2.048 Mbps asynchronous data is converted into synchronous 2.0522 Mbps 'gapped' data. The NRZ gapped data available at TCHX (pin 24) is buffered and extended to Tx (U34) chip for multiplexing.

In case of absence of input HDB-3 signal, U7 extends logic '1' at TALDT (pin 2 of U9). Using this signal internally, U9 extends logic 0 at TALDSP (pin 11), which lights up the LED 'tributary input failure' (TIF). Further, under this condition, a logic '1' (TALIND) is extended to the Tx chip (U34) for alarm detection and consequent action. Under TIF condition, TEXCLK is disabled and AIS clock (TAISCLK) on pin 1 extended from the 8/34 Mbps board is enabled. Under AIS clock mode, an all '1's data is extended at TCHX. The abbreviation AIS thus suits the description 'absence of input signal' or 'all 1s' very aptly.

MUX section

The multiplexer on its transmit path accepts four synchronous perforated data streams (2.0522 Mbps, NRZ). These data streams pertain to the four 2 Mbps tributary signals. To these data streams, sync word, justification control bits and overhead bits are added, and are then multiplexed into one 8.448 Mbps data stream(NRZ).

The MUX IC (U34) performs the following functions (refer to Fig. 8.4 sheet 2/3):

Generation of trans gap or perforated clock: The clock output (8.448 Mbps) from the source oscillator is fed at pin 83 of U34. This clock is divided internally by four. Using this 'divide by 4' clock, the Tx generates a 2.0522 MHz clock (TGAPCLK). The TGAPCLK is a perforated clock generated by suppressing clock pulses corresponding to frame alignment word and justification control bits. This clock is used to convert the plesiochronous input tributary data into synchronous data streams. Whenever a stuff request is received, a stuff pulse (active low) corresponding to time-slot TS2 of set 4 is extended to insert a justification bit (dummy bit).

Addition of sync bits and other overhead bits: After synchronization, the four tributary data are available at pins 29, 16, 69 and 56 of U34. Frame synchronization bits are added to the respective tributary data as follows:

Tributary 1 → 100 Tributary 3 → 10A*
Tributary 2 → 110 Tributary 4 → 10B*
(*A and B stand for overhead bits.)

Multiplexing four tributary data streams (2.112 Mbps) into one 8.448 Mbps data stream: The transmit chip (U34) inserts logic '1' (justification control bits in time-slots TS1 of sets 2 and 4) whenever the tributary data is stuffed. The four synchronous tributary data (2.112 Mbps) which contain the data bits, sync bits, justification control bits and stuff bits (if present), are internally multiplexed in the Tx chip (U34) using the source clock (8.448 MHz).

This multiplexed data (NRZ) is available at pin 65 of U34. This data along with 8.448 MHz clock is extended to 8/34 board for the next stage multiplexing.

Let us now look at the circuit for demultiplexing. Refer to sheet 3/3 of Fig. 8.4. The DEMUX IC (U11) performs the following functions:

Sync detection: The sync detection circuit of U11 searches for the sync word (1111010000). When three consecutive sync words are received correctly, it declares an 'in sync' mode. Similarly, when four consecutive sync words are received correctly, then an 'out of sync' condition is declared.

Demultiplexing: Under sync condition, the demultiplexing circuit of U11 demultiplexes the incoming (8.448 Mbps) NRZ data (extended from the 8/34 card) into four streams of 2.112 Mbps NRZ tributary data, which are available at pins 29, 16, 69, 56 of U11 as RCH1, RCH2, RCH3, RCH4 respectively.

Generation of receive gap or 'perforated' clock: The 8.448 MHz clock received from the 8/34 MUX card at U11 is divided internally by four, and the following timing signals are generated:

RGAPCLK	Receive gap clock
RSTFD	Receive stuffing detect
RDSTF	Receive destuff

The receive gap clock or 'perforated' clock is generated by suppressing clock pulses corresponding to the time-slots provided for frame alignment word and justification control bits. The 2.112 MHz clock is used to recover the tributary data.

RDSTF is an active high signal, corresponding to TS2 of set 4. It destuffs the received tributary data whenever justification control bits 11 are detected.

Receive section

Demultiplexed NRZ data (2.112 Mbps) is extended from the Rx chip (pin 69 of U11) as RCH3 to pin 37(RDTIN) of U9. Further, RRLCNT3 (pin 68 of U11) and RGPCLK3 (pin 70 of U11) are extended to pin 40 (RDSTF) and pin 41 (RGPCK) of U9 respectively. These are internally divided by 12 in U9 and made available as RCK2 and RCK1 at pins 29 and 32 of U9 respectively. Using these two clocks, PLL module gives 2.048 MHz clock (RCK3), which is in phase with destuffed RGPCLK3.

The input NRZ data (2.112 Mbps) is written into 12-bit memory of U9 using destuffed gap clock RGPCLK3 (2.0522 MHz). Using PLL clock, the destuffed data available in the 12-bit memory is fed to the HDB-3 encoder.

The receive NRZ data available at pin 37 is split into positive and negative streams of data (available at pin 27 and 26 respectively), and extended to pin 3 and 4 of U7. The HDB-3 output signal is available at pins 16 and 17 of U7. From here, the streams are extended through TR10 and hybrids as channel 3 O/P A and B (CH3 O/PA, O/PB).

Alarms generated in 2/8 card

The following alarms are generated in a 2/8 card:

- Tributary input failure
- PLL loss of lock

These alarms are extended to the 8/34 card for alarm detection and consequent actions.

In case of AIS (i.e. absence of input data) at 34 Mbps DEMUX port, it is extended on the tributary side (the 8 Mbps side), which in turn reaches the 2/8 cards. Under this condition the PLL clock and loss of lock alarms are inhibited.

On the 2/8 card, the following alarms are provided with visual indications:

- TRF Absence of input 2 Mbps data
- LOL Loss of lock of the PLL

Further, these alarms will also light up the SYS-LED in the 8/34 MUX board, indicating system failure, and consequently, bay top visual indication and audible alarms are generated.

8.3.2 8/34 card

The 34 Mbps I/O port of this card has the following characteristics:

Bit rate	:	34.368 Mbps ± 15 ppm
Impedance	:	75 ohms unbalanced
Line code	:	HDB-3
Normal pulse	:	± 1 Vpp

The circuit is divided into the following four sections:

- Transmit section (sheet 1/3 of Fig. 8.5)
- MUX section (sheet 2/3 of Fig. 8.5)
- DEMUX section (sheet 3/3 of Fig. 8.5)
- Receive section (sheet 1/3 of Fig. 8.5)

The above functions of 8/34 card are taken care of mainly by the following ICs: 8 Mb Trans–Receive ICs (U10 and 3 more identical ICs, i.e., a total 4 ICs for the four 8 Mb streams (in the circuit only U10 is shown); 34 Mb MUX IC (U36); 34 Mb DEMUX IC (U12).

Transmit section

IC CB 1626 (U10) in the transmit section performs HDB-3 coding and conversion of plesiochronous data streams into synchronous data streams. This is explained with reference to one 8 Mbps stream (refer to sheet 1-5 of Fig. 8.5).

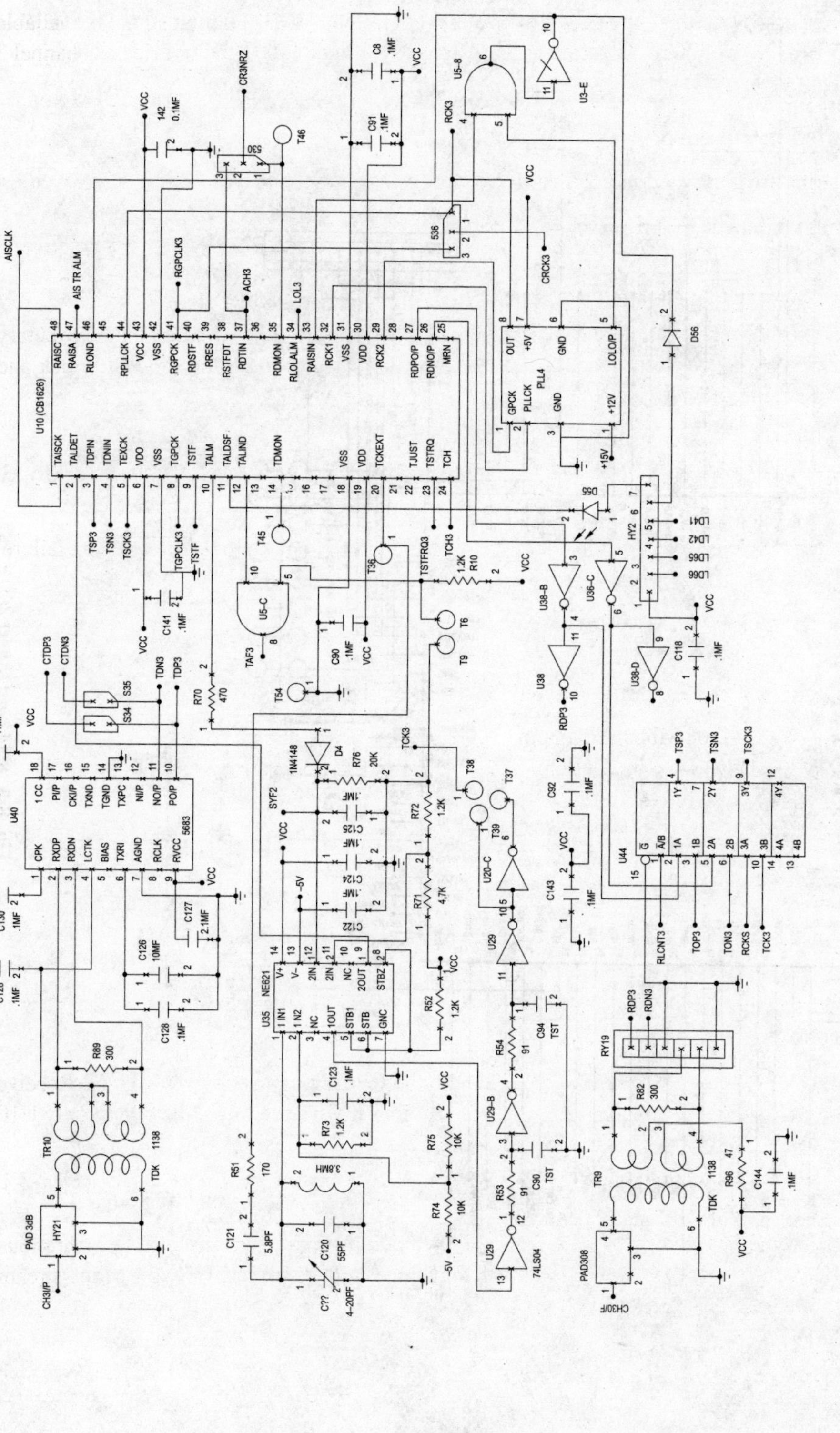

Fig. 8.5 *8/34 card (Sheet 1/3)*

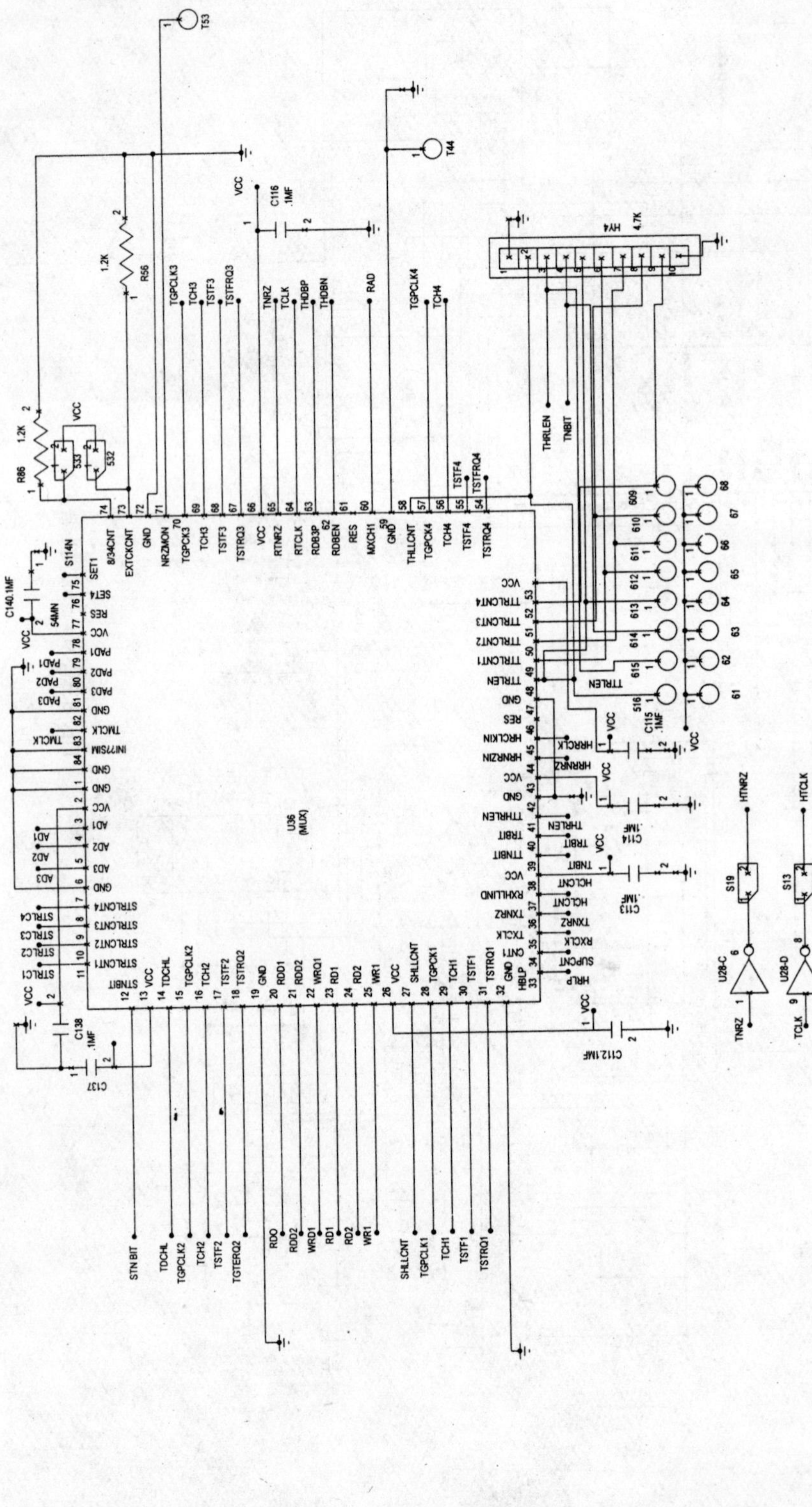

Fig. 8.5 *8/34 card (Sheet 2/3)*

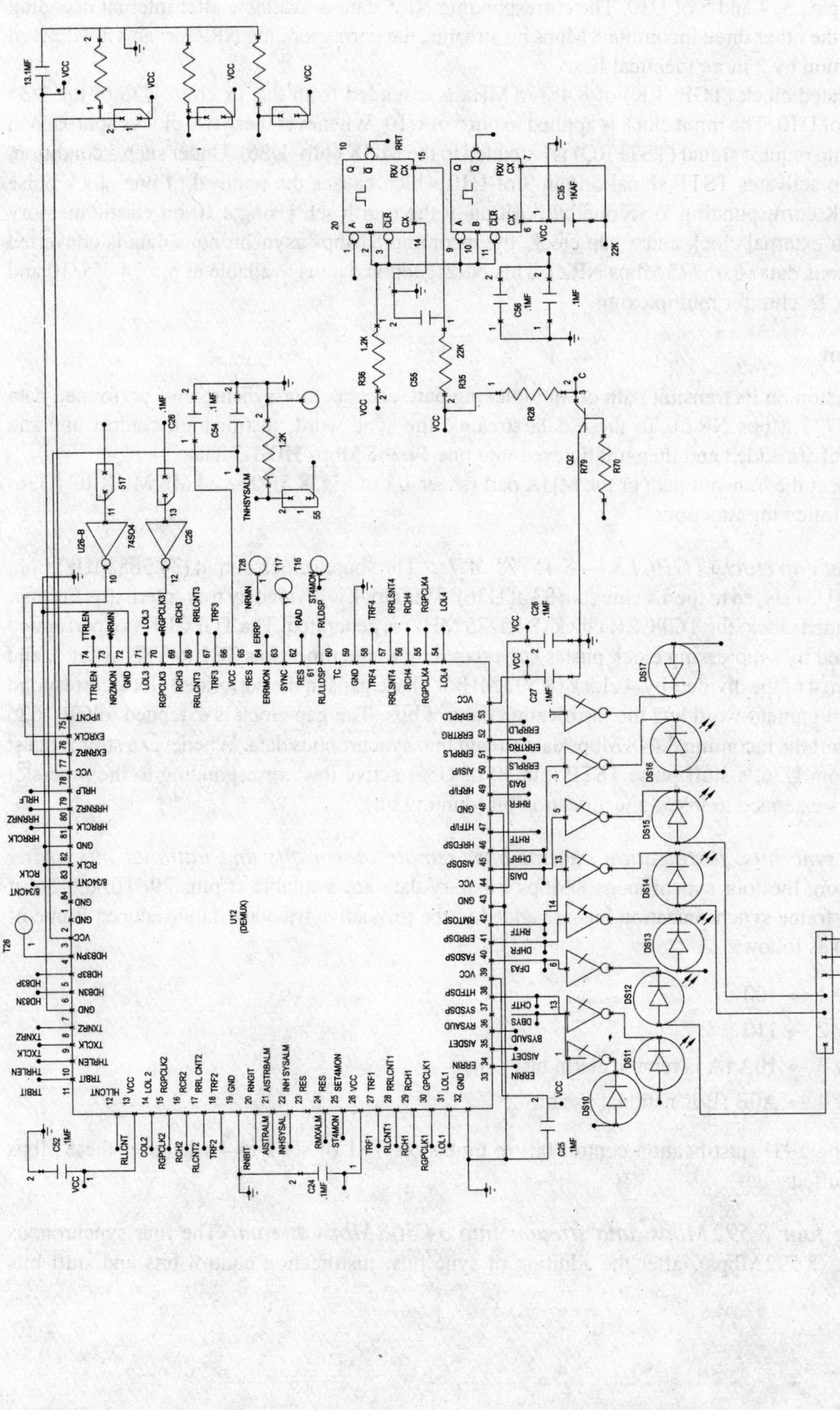

Fig. 8.5 *8/34 card (Sheet 3/3)*

The 8 Mbps bit stream is extended from the 2/8 card in HDB-3 format (positive and negative) along with clock to pins 3, 4 and 5 of U10. The corresponding NRZ data is available after internal decoding at pin 14. For the other three incoming 8 Mbps bit streams, the corresponding NRZ streams are derived in similar fashion by 3 more identical ICs.

The perforated clock (TGPCLK) of 8.45775 MHz is extended from the Tx chip (U36 of the 8/34 card) to pin 8 of U10. The input clock is applied to pin 5 of U10. Whenever these two clocks approach in phase, a stuffing request signal (TSTFRQ) is extended to the MUX chip (U36). Under such a condition, the MUX chip activates TSTF signal on pin 9 of U10 which causes the removal of one clock pulse from TGPCLK corresponding to second bit position of the fourth set. Using a 10-bit elastic memory in U10 and an external clock and a gap clock, the incoming 8 Mbps asynchronous data is converted into synchronous data (8.45775 Mbps NRZ). This NRZ gapped data is available at pin 24 of U10 and is extended to Tx chip for multiplexing.

MUX section

The MUX section on its transmit path or multiplexing part, accepts four synchronous perforated data streams (8.45775 Mbps NRZ). To these data streams, the sync word, justification control bits and national use bit are added and then multiplexed into one 34.368 Mbps HDB-3 data.

Let us look at the transmit path or the MUX part (sheet 4/5 of Fig. 8.5). The 34 Mb MUX IC (U36) performs the following functions:

Generation of gap clock (TGPCLK)—8.45775 MHz: The source clock output (34.368 MHz) from the oscillator (U31) is fed to the Tx chip (pin 83 of U36). This clock is divided by four internally. Further, using this divided clock, the TGPCLK clock (8.45775 MHz) is generated. The TGPCLK is a perforated clock generated by suppressing clock pusles corresponding to the time-slots TS1 to TS3 of set 1, and TS1 of sets 2 to 4 of the divided-by-4 clock (8.592 MHz). These gaps in the perforated clock correspond to the frame alignment word and the justification control bits. The gap clock is extended to CB 1626 (U10) to convert the incoming 8.448 Mbps data stream into synchronous data. Whenever a stuff request is received from U36, a stuff pulse TSTF (pin 30 of U36) active low corresponding to the time-slot TS2 of set 4 is extended to insert a justification bit (dummy bit).

Addition of sync bits, justification control bits, remote alarm bits and national bits: After synchronization, the four synchronous 8 Mbps tributary data are available at pins 29, 16, 69, 56 of U36. To this, frame synchronization bits are added to the respective tributary data (reduced frame of 34.368 Mbps) as follows:

Tributary 1 → 100

Tributary 2 → 110

Tributary 3 → 10A (A is remote alarm bit)

Tributary 4 → A0B (B is national use bit)

U36 inserts logic '1' (justification control bit) in time-slots TS1 of set 2 to 4 whenever the 8 Mbps tributary is stuffed.

Multiplexing four 8.592 Mbps data streams into 34.368 Mbps stream: The four synchronous tributary data (8.592 Mbps), after the addition of sync bits, justification control bits and stuff bits (if present), are internally multiplexed in the Tx chip (U36) using a source clock (34.368 MHz) and internally generated derived clocks.

HDB-3 coding: The multiplexed NRZ data (pin 65) of U36, available for monitoring, is converted into HDB-3 positive and negative streams and are available at pin 62 and 63 of U36. These signals are fed to HDB-3 pulse combiner (in the form of 'child card' mounted on 8/34 card) and the associated pulse transformers.

Transmit failure indication: Whenever there is a failure in 34 Mbps transmission, a higher rate transmit failure (HTF) signal is extended for visual indication (see sheet 3/5 of Fig. 8.5).

DEMUX section

In the 34 Mb DEMUX or receive path, the incoming HDB-3 data (34.368 Mbps) is subjected to cable-loss equalization, lock extraction and positive and negative stream segregation and applied to DEMUX IC (U12).

The DEMUX IC(U12) performs the following functions:

HDB-3 decoding: The input HDB-3 signal on pins 5, 6 of U12 is decoded into NRZ signal internally, making use of the extracted clock. The recovered NRZ data is available at pin 73 of U12 for monitoring. The decoded NRZ data is validated by synchronizing to the frame word (1111010000).

Frame synchronization and demultiplexing: The sync circuit of U12 searches for the sync word. When three consecutive frame words are detected properly, 'in sync' is declared. If four consecutive frame words are received, 'out of sync' is declared; this is displayed by the FAS LED. Under sync condition, the input data is demultiplexed into four streams of 8.592 Mbps, which are made available at pins 29, 16, 69 and 56 of U12.

Generation of receive timing signals: The Rx chip generates the following timing signals from the received clock.

- Divide by 4 (34.816 MHz) clock
- RGPCLK (receive 'gap' or perforated clock)

The 8.45775 MHz perforated clock (RGPCLK) is generated by suppressing clock pulses corresponding to time-slots TS1 of set 2 to 4 of the 8.592 MHz clock. These gaps pertain to frame alignment word and justification control bits.

8.3.3 Power supply card

The power supply card works on the principle of forward converter using current mode control. The input dc power is switched at around 100 kHz and output regulation obtained by controlling the pulse width. The power supply module can work on an input of 42 to 56 V dc, and provide +5 V (6A), 5.5 V (0.8A), +15 V (0.5A), −5 V (0.8A) as output voltages. All the outputs are fully protected against over-current, short circuit and over-voltage. In case of an over-voltage on one or more outputs, the power supply is tripped off.

All the four output voltages and the input voltage can be checked at the monitoring socket provided on the front panel.

8.3.4 Supervisory card

The supervisory card is designed using an 8-bit microcontroller 80C31 with the following features:

- 128 bytes of RAM
- 32 I/O lines
- Two 16 bit timer/counter
- Two-level interrupt structure which can support upto five interrupts
- A full duplex serial port
- On-chip oscillator and clock circuitry

This supervisory card can be used with an optical line terminating equipment (OLTE), which is a front-end transmission equipment in an optical communication link.

Refer to Fig. 8.6 (sheet 1/2). The CPU (8-bit microcontroller 80C31, designated U5) performs the following functions:

The upper 8 lines of address bus (A8 to A15) are extended through U9 (74LS244). The lower 8 lines of address bus (multiplexed with data bus) are extended through U14 (74LS373). The data bus is extended through U8 (74LS245).

The serial port of the microcontroller works in a multi-processor communication mode. The CPU collects all the alarm status from the MUX terminals through polling and stores them in a RAM. When a query is placed through an HHT (hand-held terminal), the alarm status gets displayed on the HHT. Further, through the HHT any one of the 2 Mbps signals at the far end can be looped back. Altogether a total of five telecommands/five telesignals can be issued/monitored. Essential free contacts and relays to operate them on 48 V DC are provided.

A watchdog timer DS1232 is used to restart the microcontroller if it goes out of control, i.e., its bus cycle has some abnormalities. The system program is fused on EPROM [U11(2764A) & U19(2744)].

Refer to sheet 2/2 of Fig. 8.6. To communicate with the HHT and the network monitoring system (NMS if it exists in the MW network), a 'dual synchronous' receiver/transmitter SCN 2681 (U15) is used. It has two full duplex serial ports and converts serial data to parallel and vice versa. It can be interfaced with CPU in polled mode or interrupt mode. In the present circuit, it is interfaced in polled mode. HC 1488 (U8) and MC1489 (U7) are the line diriver/receiver pair to interface between TTL levels and RS23CC levels.

8.4 Technical Specification

8.4.1 Digital interface at 2048 kbps

Input port

No. of tributaries	16
Bit rate	2048 kpbs ± 50 ppm
Code	HDB-3
Impedance	120 ohms balanced (nominal)
Pulse voltage	3.0 V ± 0.3 V
Pulse width	24.4 ns ± 25 ns

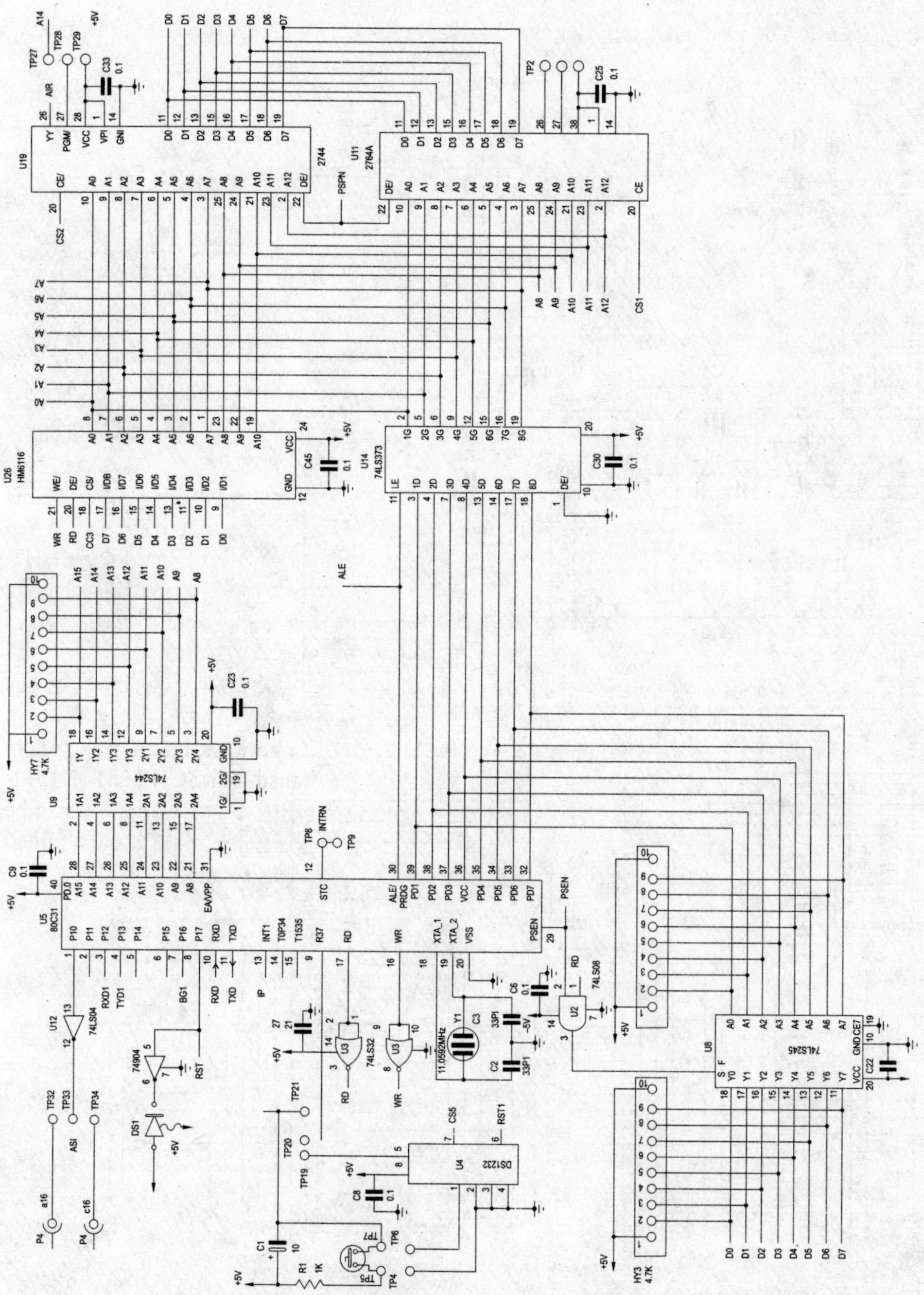

Fig. 8.6 *Supervisory card (Sheet 1/2)*

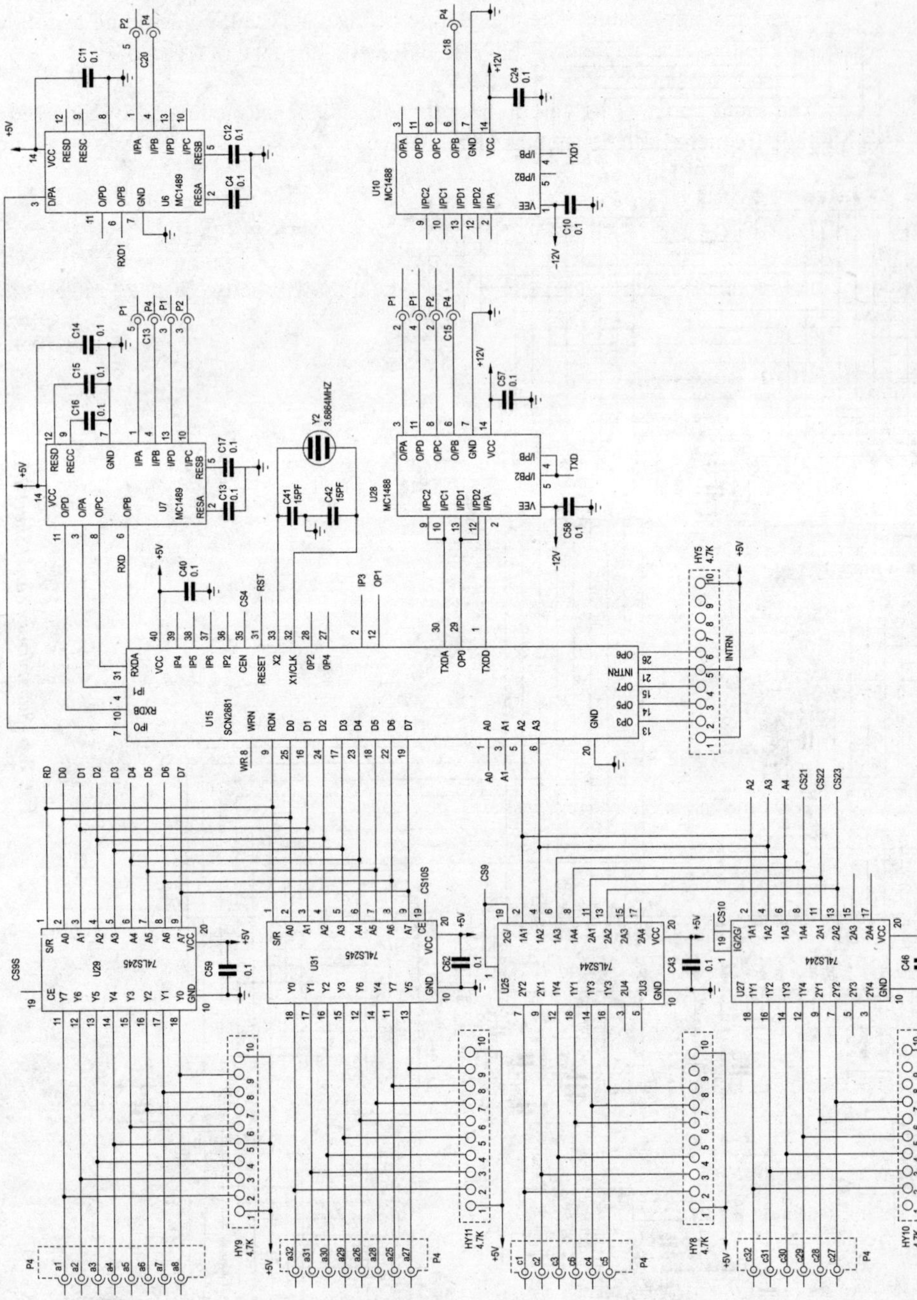

Fig. 8.6 *Supervisory card (Sheet 2/2)*

Station cable equalisation: The digital signal presented at the input port will be modified by the characteristics of the interconnecting cable. The attenuation of this cable will be assumed to follow square root of F law and the loss at a frequency 1024 kHz will be in the range of 0 to 6 dB.

Jitter acceptance: The input port will be able to tolerate a digital signal modulated by sinusoidal jitter having an amplitude frequency relationship as given below.

20 Hz to 2.4 kHz	:	1.5 UI
18 kHz to 100 kHz	:	0.2 UI

The equivalent binary content of the signal with jitter modulation will be two psuedo-random bit sequences.

Return loss at the input port 2048 kbps

Frequencies corresponding to percentage of nominal bit rate	*Return loss*
2.5 to 50% (51.2 kHz–102.4 kHz)	>12 dB
5.0 to 100% (102.4 kHz–2048 kHz)	> 18 dB
100 to 150% (2048 kHz–3072 kHz)	> 14 dB

Output port

No. of tributaries	16
Bit rate	2048 kbps ± 50 ppm
Code	HDB-3
Impedance	120 ohms balanced (nominal)
Pulse voltage	3.0 V ± 0.3V
Pulse width	24.4 ns ± 25 ns

Type of converter

Wire wrapping type connectors are provided in two columns in the I/O sub-rack—each column has five connectors. Each connector caters to four tributaries, and the fifth connector is for OLTE.

8.4.2 Multiplexer section

34 Mb frame

Frame structure (see Fig. 8.7)

Total bits in one frame 1536

Set-1

Frame alignment signal (1111010000)	Bits 1 to 10
Alarm indication to remote unit	Bit 11
Bit reserved for national use	Bit 12
Bits from tributaries	Bit 13 to 384

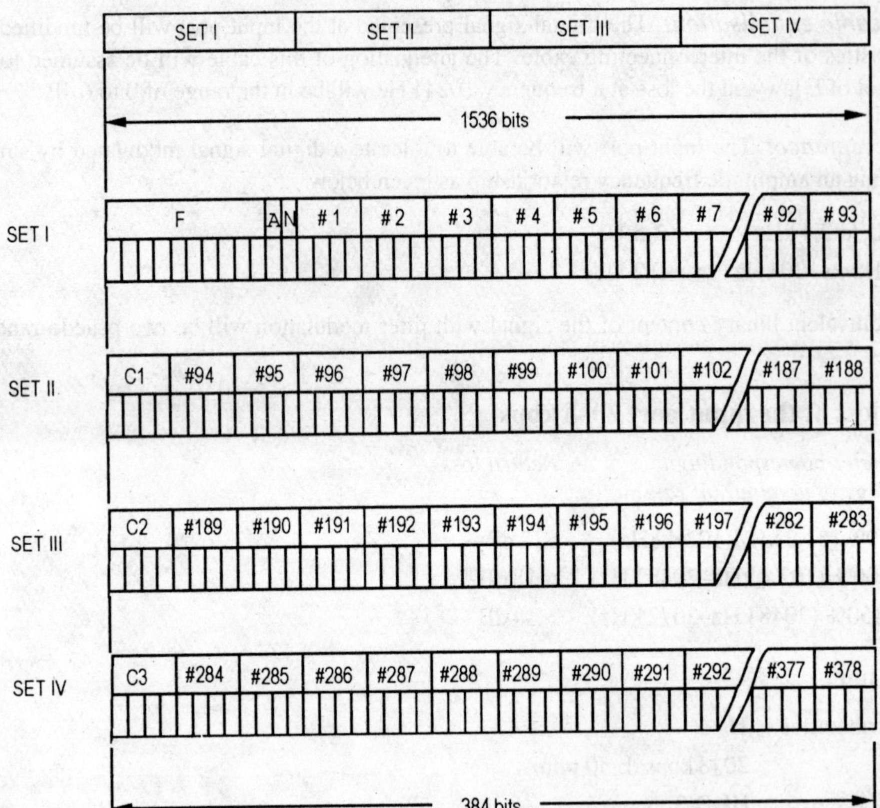

FAS	Frame alignment signal (1111010000)
A	Remote alarm bit (0= Normal, 1= Alarm)
N	This bit is reserved for future use, always set to 1.
# 1 to # 378	In each group, the first bit is assigned to Tributary 1, second bit to Tributary 2, third bit to Tributary 3, and the fourth bit to Tributary 4.
C1 to C3	Groups for carrying the justification control bits of each tributary.
#284	This group carries the justification bit of each tributary. If bits C1 to C3 of a tributary are 000, its bits in this group carries data. If they are 111, its bit in this group carries remote loopback Information; a 0 indicates a request for loopback and a 1 indicates no request.

Fig. 8.7 *34 Mb frame structure.*

Set-2 and Set-3

Justification service bits (C_{jn} ($n = 1$ to 2)	Bits 1 to 4
Bits from tributaries	Bits 5 to 384

Set-4

Justification service bits C_{13}	Bits 1 to 4
Bits from tributaries available for justification	Bits 5 to 8
Bits from tributaries	Bits 9 to 384

Note: C_{jn} indicates the nth justification service bit of the jth tributary.

Loss and recovery of frame alignment: Frame alignment is assumed to have been lost when four consecutive frame alignment signals have been incorrectly received in their predicted positions.

Multiplexing method: The multiplexing method is by cyclic bit interleaving and positive justification.

Service digits: As per CCITT Rec. G-751.

Jitter transfer characteristics (at 2048 Mb/s) 2048 Mb/s): As per CCITT Rec. G-823.

Tributary output jitter: The peak-to-peak jitter at the tributary output in the absence of input jitter should not exceed 0.25 UI when measured in the frequency range upto 100 kHz.

When measured with an instrument incorporating a bandpass filter having a low cut-off frequency of 18 kHz, a roll-off 20 dB/decade, and an upper limit of 100 kHz, the peak-to-peak output jitter should not exceed 0.05 UI with a probability of 99.99% during a measurement period of 10 s.

Multiple signal output jitter: In cases where the transmitting signal is derived from an internal oscillator, the peak-to-peak jitter at the 343768 Kpbs output should not exceed 0.05 UI when measured within frequency range $f1 = 100$ Hz to $f4 = 800$ kHz.

When measured with an instrument incorporating a bandpass filter having a lower cut-off of 3 kHz, a roll-off of 20 dB/decade and an upper limit of 400 kHz, the peak-to-peak output jitter should not exceed 0.05 UI with a probability of 99.99% during a measurement period of 10 s.

8 Mb frame

frame structure (see Fig. 8.8)

Frame structure	Bit number
Frame alignment word (11101000)	1 to 10
Alarm to remote terminal	11
National use	12

Set-1

Bits from tributaries	13 to 212
Justification control bits	1 to 4
Bits from tributaries	5 to 212

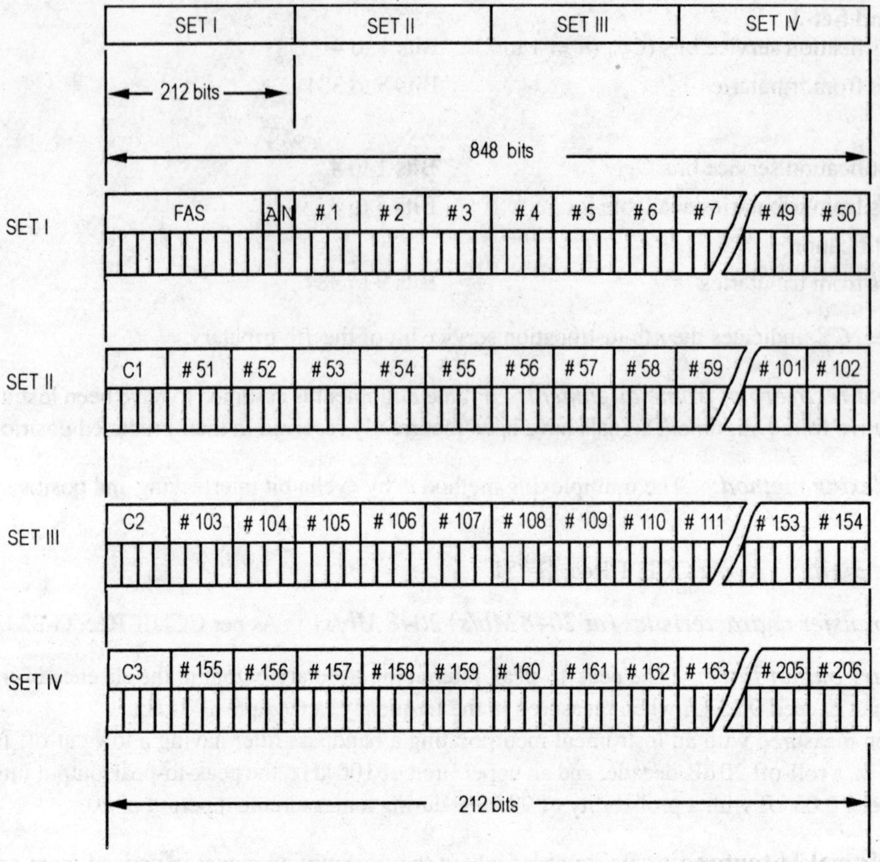

FAS	Frame alignment signal (1111010000)
A	Remote alarm bit (0= Normal, 1= Alarm)
N	This bit is reserved for future use, always set to 1.
# 1 to # 206	In each group, the first bit is assigned to Tributary 1, second bit to Tributary 2, third bit to Tributary 3, and the fourth bit to Tributary 4.
C1 to C3	Groups for carrying the justification control bits of each tributary.
# 155	This group carries the justification bit of each tributary. If bits C1 to C3 of a tributary are 000, its bits in this group carries data. If they are 111, its bit in this group carries remote loopback Information; a 0 indicates a request for loopback and a 1 indicates no request.

Fig. 8.8 *8Mb frame structure.*

Set-2 and Set-3

Justification control bits	1 to 4
Bits from tributaries	5 to 212

Set-4

Justification control bits	1 to 4
Bits for tributaries available for justification	5 to 8
Bits from tributaries	9 to 212
Total length	848 bits
Bits/tributary	206 bits

Note:

1) Loss of frame alignment shall be assumed to have taken place when four consecutive frame alignment signals have been incorrectly received in their predicted positions.
2) When frame alignment is assumed to be lost, the frame alignment device should decide that such alignment has effectively been recovered when it detects the presence of three consecutive frame alignment signals.
3) The frame alignment device, having detected the appearance of a single correct frame alignment signal, should begin a new search for frame alignment signal in one of the two subsequent frames.
4) Cyclic bit interleaving in the tributary numbering order and positive justification is recommended. Positive justification should be indicated by signal 111, and a no signal justification by signal 000. Majority decision is recommended.

8.4.3 Digital Interface at 34368 kbps

Specification at input port

Bit rate	34368 kbit/s ± 20 ppm
Code	HDB-3
Impedance	75 ohms unbalanced (nominal)
Pulse voltage	1 V ± 0.1 V
Pulse width	14.55 ns ± 2.24 ns

Station cable equalisation: The digital signal presented at the input port will be modified by the characteristics of the interconnecting cable. The attenuation of this cable will be assumed to follow square root of F law and the loss at a frequency 17184 kHz will be in the range 0 to 12 dB.

Jitter acceptance: The input port will tolerate a digital signal modulated by sinusoidal jitter having an amplitude–frequency relationship as given below.

100 Hz to 1 kHz	:	15 UI
10 kHz to 800 kHz	:	0.15 UI

Return loss at the input port (34368 kbps)

Frequency corresponding to % of nominal bit rate	Return loss
2.5 to 5% (859.2 kHz–1718.4 khz)	> 12 dB
5 to 100% (1718.4 kHz–34368 kHz)	> 18 dB
100 to 150% (34368 kHz–51552 kHz)	> 14 dB

Specification at output port

Bit rate	34368 kbps ± 20 ppm
Code	HDB-3
Impedance	75 ohms unbalanced (nominal)
Pulse voltage	1 V ± 0.1 V
Pulse width	14.55 ns ± 2.24 ns

Timing signal

The system can synchronise on any one of the following clock sources.

- Internal oscillator at 34368 kbps
- 34368 kbps recovered clock
- External 34368 kHz signal

The selection is a hard-wired connection.

Types of connectors

Spinner connector of 75 ohm coaxial interface is provided for 34 Mbps Tx and Rx.

8.5 Alarms and Consequent Action

8.5.1 Fault conditions

The digital multiplex equipment will detect the following fault conditions:

(*a*) Failure of derived power supply
(*b*) Loss of an incoming signal at any of the sixteen 2048 kbps tributaries at the input of the multiplexer
(*c*) Loss of 34368 kpbs incoming signal at the input of the demultiplexer
(*d*) Loss of the frame alignment in received input signal of 34368 kpbs at the demultiplexer
(*e*) Alarm indication received from the remote multiplexer equipment at the 34368 kpbs input of the demultiplexer
(*f*) AIS received at the input 34368 kpbs demultiplexer
(*g*) Receive tributary failure (loss of lock) at any of the sixteen 2048 kbps tributaries
(*h*) BER alarm of 1×10^{-3} at 34 Mbps input port

8.5.2 Consequent actions

(*a*) Prompt maintenance alarm will be actuated in all the cases except (*e*) and (*f*) above. When an AIS is detected at the input of the demultiplexer, the prompt maintenance alarm indication associated with the loss of frame alignment will be inhibited.

(*b*) AIS will be applied to the output of all tributaries when any of the alarm conditions given in (*c*) and (*d*) is actuated.

(*c*) AIS will be applied to the relevant time-slots of the composite signal (34368 kbps) when alarm (*b*) is actuated.

(*d*) Alarm indication to the remote multiplexer equipment will be generated by changing the bit 11 of set 1 from state 0 to state 1 at the 34368 kbps multiplexer output in situations where alarm condition (*d*) and (*e*) are actuated. The audio and visual alarm will be extended to the exchange for prompt maintenance.

8.5.3 Supervisory facilities

Loop back: It is possible to loop back any one of the tributary output to the corresponding tributary input for remote testing of 2048 kbps tributary path. It is also possible to loop back at the 8 Mb level and 34 Mb level by inserting looping plugs on the top pair on the 2/8 card and 8/34 card respectively.

8.5.4 Alarm indications on 2/8 card

The following alarm indications are available through a tributary wire on the 2/8 card.

TF alarm: Tributary failure alarm, i.e. no input (local)

LL alarm: Loss of link, i.e. no tributary received from the far-end.

8.5.5 Alarm indications on 8/34 card

SYS System alarm

HTF Trans failure (34 Mb)

FAS Receive sync failure

AIS All 1's received at the 34.368 Mbps input port.

RMT Remote alarm

HRF Absence of input 34.368 Mbps signal.

The conditions leading to these alarms are listed.

SYS alarm: The SYS-LED lights up under the following conditions:

- 2 Mbps input tributary failure
- Loss of lock (LOL) of 2 Mbps tributary receive clock
- Sync failure
- Trans failure
- Absence of 34 Mbps input signal (HTF)

HTF alarm: This LED lights up whenever 34.368 Mbps source clock fails.

FAS alarm: This LED lights up whenever frame alignment is not received correctly.

HRF alarm: This LED lights up whenever the 34 Mbps input signal fails. Under all conditions, bay top alarm and audible alarm are energized.

RMT alarm: This LED lights up under failure condition of remote MUX.

AIS alarm: This LED lights up whenever all 1's are received at the 34 Mbps input port.

Exercises

1. Consider an intermediate repeater on a digital MW network where tributaries (2 Mb streams) 1 and 2 are dropped. The remaining 14 streams (out of 16) are put through. How is the wiring done? Draw the sketch.
2. What is the difference between 2/34 MUX and 2/8/34 MUX?
3. In the case of a higher-order MUX, for 2 Mb tributary connections with a primary MUX a paired cable is used, whereas for 34 Mb connections with radio equipment, coaxial card is used. Why?
4. What are the advantages of using vertical and slim rack for higher order MUXes?
5. How do you extend 34 Mb level loop back to the remote-end skip MUX?

Part - II

9

Synchronization in Passband Digital Transmission

9.1 Introduction

In Chapter 4, an analysis of the effects of noise on the performance of a digital transmission systems in passband (bandpass channel with suitable modulation) was presented. First, the Gram–Schmidt orthogonalization procedure was developed for geometric interpretation of transmitted signals, and then channel noise was modelled as additive white Gaussian noise (AWGN). This paved the way for the concept of an optimum receiver, i.e., a receiver that minimizes the average probability of symbol error. Then the concept of coherent receivers or matched filter receivers for the optimum detection of signal in AWGN channel and the corresponding coherent modulation techniques (coherent binary and M-ary) were dicussed. We also developed the non-coherent receiver (BPSK, DPSK) concepts suitable for situations with uncertainities in the carrier phase of the received signal. Finally, we evaluated the P_e performance of each scheme, and the power–bandwidth trade-off involved.

The real challenge in the design of a passband digital transmission equipment is 'synchronization'. This has two aspects—carrier synchronization and symbol synchronization. We deal with these two aspects of synchronization in this chapter.

9.2 Synchronization Concept

The coherent reception of a digitally modulated signal requires that the receiver be synchronous to the transmitter. 'Synchronous' is a relative term; two sequences of events at the transmitter and the receiver are synchronous to each other when the events in one sequence and the corresponding events in the other occur simultaneously. The process of making the sequence of events at the transmitter and the receiver synchronous to each other and maintaining them in that condition is called 'synchronization'.

From the discussion of various digital modulation techniques presented in Chapter 4, the need for two basic modes of synchronization can be identified:

(*i*) When coherent detection is used, knowledge of both frequency and phase of the carrier is necessary. The estimation of carrier phase and frequency is called 'carrier recovery' or 'carrier synchronization'.

(*ii*) In the demodulation process, the receiver has to recognize the beginning and end of individual symbols so that it can determine the sampling time and quench time of the product integrators. The estimation of these times is called 'clock recovery' or 'symbol synchronization'.

These two synchronization processes can be coincident or sequential (in non-coherent systems, carrier synchronization is not relevant). Both these types of synchronization are discussed in the following section.

9.3 Carrier Synchronization

The straightforward approach to carrier synchronization would be to modulate the data-bearing signal onto a carrier in such a way that the power spectrum of the modulated signal contains a discrete component (spectral line) at the carrier frequency. A phase-locked loop (VCO, loop filter and multiplier connected as negative feedback system) can be used to track this component.

But the discrete spectral component representing a carrier does not bear any information and its inclusion is hence a waste of power. Therefore, the receiver employs a suppressed carrier tracking loop, or provides a coherent secondary carrier (subcarrier) reference. Refer to Fig 9.1 where the Mth power loop for M-ary PSK carrier recovery is presented.

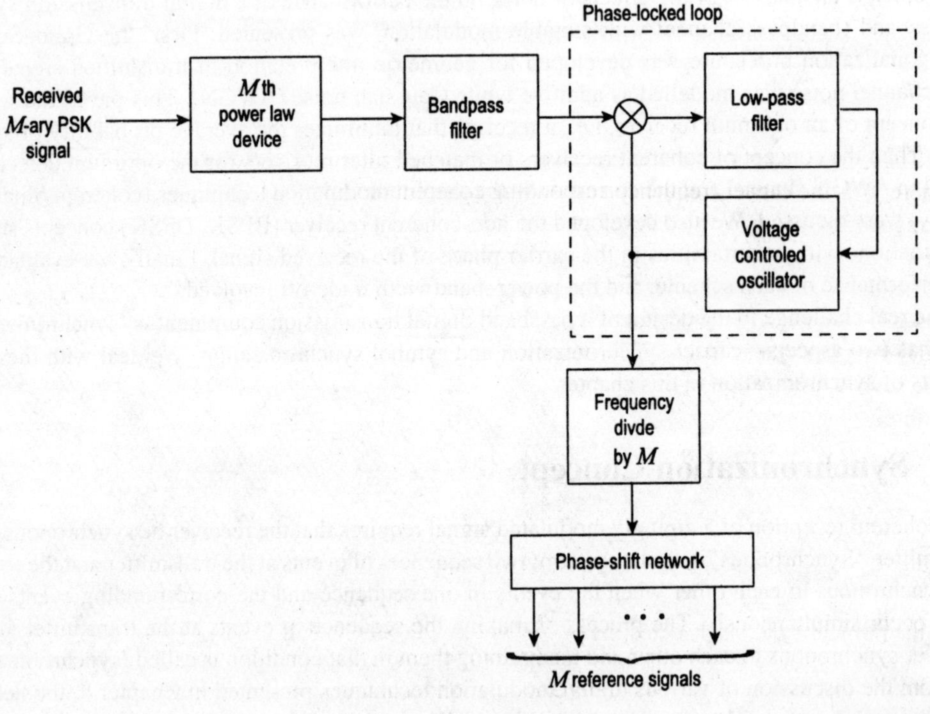

Fig. 9.1 *Mth power loop.*

Consider the case of BPSK, i.e., $M = 2$. The power loop is called 'squaring loop'. It has a squaring device at the input (Mth power law device). In this system, a change in the sign of the input signal leaves the sign of recovered carrier unaltered, i.e., there would be 180° phase ambiguity in the recovered carrier. For $M = 4$, the same problem exists, viz., there is an M-phase ambiguity for any M-ary PSK in the interval $(0, 2\pi)$.

Another method of carrier recovery is the use of 'Costas loop'. This has two paths—in-phase and quadrature. Both have a common VCO (Fig. 9.2).

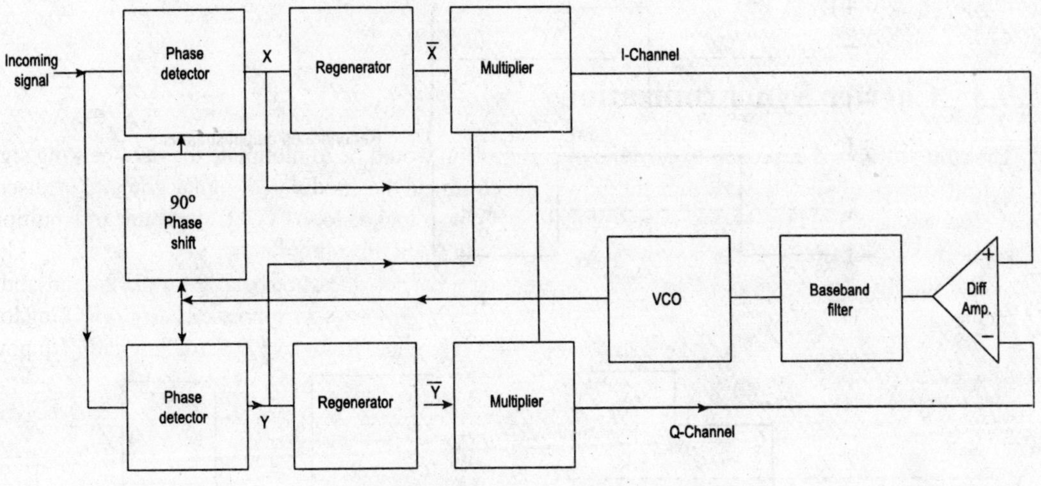

Fig. 9.2 *Costas-type carrier recovery.*

The incoming M-ary PSK signal consists of cos or sin terms of the form $A_c \cos 2\pi f_c t$ or $A_c \sin 2\pi f_c t$ where f_c is the carrier frequency and A_c is the amplitude. For $M = 2$, i.e., BPSK, we have two cosine terms shifted in phase by 180°. Let us take one cosine term and examine how Costas loop works on it.

Let the VCO frequency be f_L. The phase detector comprises product modulator and low-pass filter; hence the received $A_c \cos(2\pi f_c t)$ gets multiplied by $\cos(2\pi f_L t)$ in the I-channel and by $\sin(2\pi f_L t)$ in the Q-channel, giving the 'sum' and 'difference' frequency terms, i.e., $2\pi(f_c + f_L)t$ and $2\pi(f_c - f_L)t$. The sum frequency term is filtered out by the low-pass filter in the phase detector, and we get only the difference frequency terms as

$$X = \frac{1}{2} A_c \cos[2\pi(f_c - f_L)t]$$

and $$Y = -\frac{1}{2} A_c \sin[2\pi(f_c - f_L)t]$$

at the output of the phase detectors.

It may please be noted that there is always a phase error of θ (local, caused by VCO). So, in reality,

$$X = \frac{1}{2} A_c \cos[2\pi(f_c - f_L)t],$$

$$Y = -\frac{1}{2} A_c \sin[2\pi(f_c - f_L)t + \theta].$$

To eliminate θ, we take the products $X\bar{Y}$ and $Y\bar{X}$ ($\bar{X}$, $\bar{Y}$ are the regenerated versions of X, Y respectively) and apply them to a difference amplifier (an example is presented in Fig. 9.3). From the

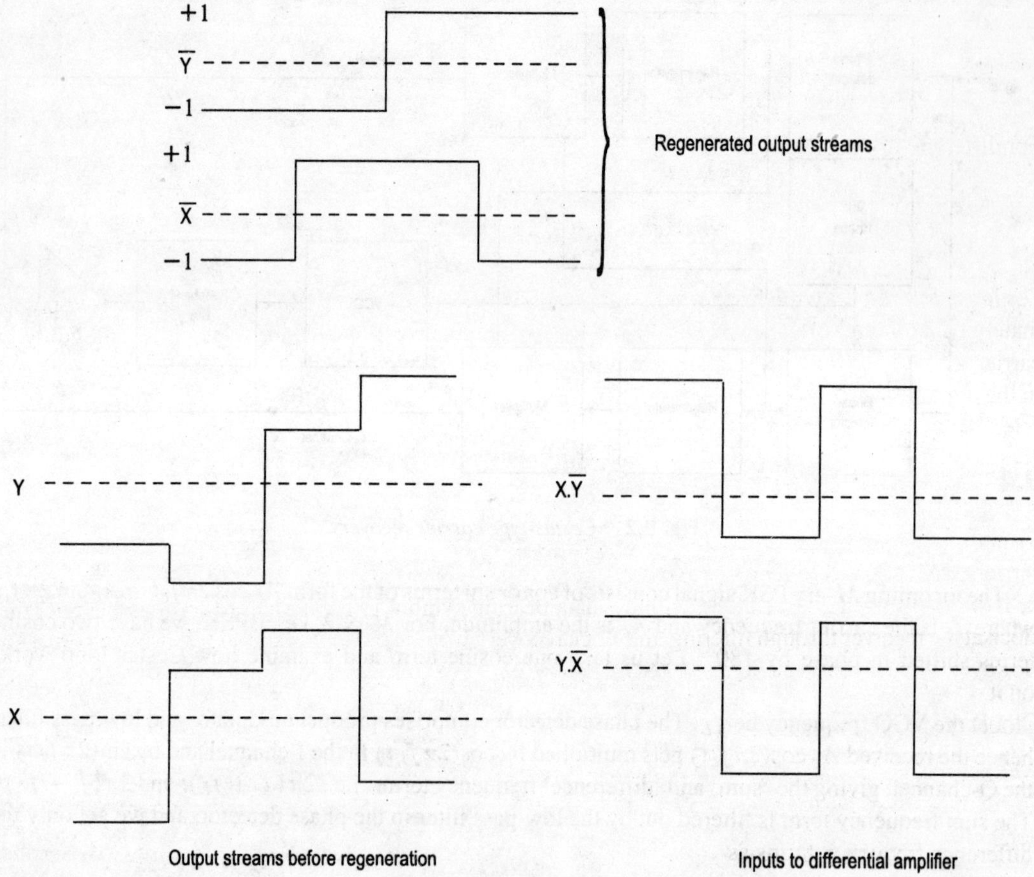

Fig. 9.3 *Costas loop waveforms.*

trignometric identity 'sin A cos B – cos A sin B = sin(A – B)', we get the difference term $1/2A_c \sin\theta$ at the output of baseband filter which corrects the VCO.

The Costas-loop has two problems:

- For $M > 4$, the circuit is prohibitively large
- Phase control ceases with modulation and phase-lock has to be re-established which may cause perceptible distortion in data transmission applications.

9.3.1 Resolution of phase ambiguity in carrier recovery

We have seen that the methods of carrier recovery have phase ambiguity. Differential encoding helps to resolve phase ambiguity. The incoming data stream is differentially encoded before modulation at the transmitter and differentially decoded after detection at the receiver. There is a small degradation in the noise performance.

For differentially encoded PSK,

$$P_e = \operatorname{erfc}\left[\sqrt{\frac{E_b}{N_0}}\right] - \frac{1}{2}\operatorname{erfc}^2\left[\sqrt{\frac{E_b}{N_0}}\right].$$

For differentially encoded QPSK,

$$P_e = 2\operatorname{erfc}\left[\sqrt{\frac{E_b}{N_0}}\right] - 2\operatorname{erfc}^2\left[\sqrt{\frac{E_b}{N_0}}\right] + \operatorname{erfc}^3\left[\sqrt{\frac{E_b}{N_0}}\right] - \frac{1}{4}\operatorname{erfc}^4\left[\sqrt{\frac{E_b}{N_0}}\right].$$

For large E_b/N_0, only the first term is significant, i.e., P_e of differentially encoded QPSK is approximately twice the P_e of coherent QPSK, but it is worth it as we are able to resolve phase ambiguity in carrier recovery. It may be recalled that differentially encoded QPSK is the modulation scheme used in the digital MW radio equipment (refer to Chapter 5).

9.4 Symbol Synchronization

Symbol synchronization or clock recovery can be achieved using two approaches: simultaneous with carriers recovery, or sequential with carrier recovery (precedence can be either way).

If we transmit the clock along with the data bearing symbol in a multiplexed form and extract the clock at the receiver through filtering, the clock recovery would be a simple and straightforward process. However, there is waste of Tx power in this approach.

Clock recovery preceding carrier recovery

We know that clock timing is more stable than carrier phase. Why not then make use of this phenomenon? So, extract the clock through non-coherent detector and then recover the carrier by processing the non-coherent detector output in each clocked interval.

Clock recovery following carrier recovery

In this approach, the clock is extracted by processing the demodulated (not necessarily detected) baseband waveforms, thus avoiding waste of Tx power.

Consider a rectangular pulse and the output of a filter matched to it as shown in Fig 9.4(*a*) & (*b*) respectively. The matched filter output is maximum at $t = T$ and is symmetrical about this point. If it is sampled at $t = T \pm \Delta T$ instead of $t = T$, the absolute values of the two samples obtained are equal but smaller than the value corresponding to the correct sampling at $t = T$.

The symbol synchronizer can be based on this principle (the early–late gate symbol synchronizer of the absolute value type). We can also use correlators to integrate over the entire symbol interval T, with one starting at a time ΔT early relative to transition time estimate, and the other starting ΔT late. The difference between the outputs of correlators at $t = KT$, i.e., the error signal e_{KT}, is low-pass filtered and applied to a VCO. This closed loop is designed to be narrow band relative to symbol rate $1/T$. The instantaneous frequency of the local clock is advanced or retarded in an iterative manner until the equilibrium point is reached. Thus, symbol synchronization is established. (The early–late gate symbol synchronizer is described later with a block diagram in Section 14.1.)

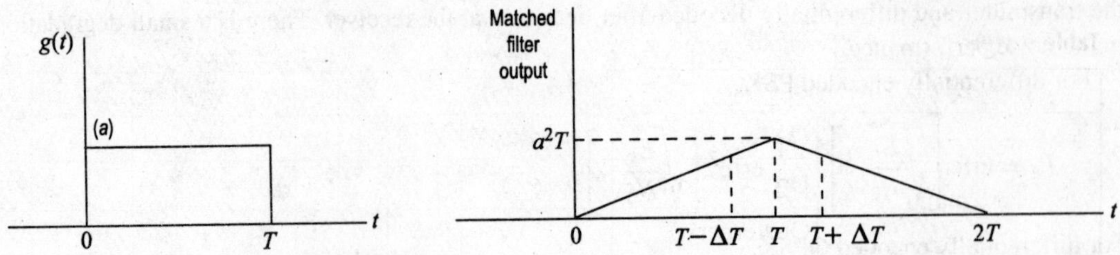

Fig. 9.4 *(a) Rectangular pulse.* **9.4** *(b) Matched filter output.*

9.5 Differentially Encoded QPSK in DMR 770

Please refer to Section 6.2. The Tx logic unit of the phase modulator (QPSK) of DMR 770 digital MW radio equipment (7GHz) takes the two data streams from the Tx-DPU module. They are in Gray code and converted to binary code. The binary-coded pulse streams are differentially encoded in quaternary numeration in each time-slot (Table 9.1 (*a*), (*b*)). The resultant output data streams are then converted to Gray-coded pulse streams by the binary code-to-Gray code converter circuit. The two streams thus obtained are sent to QUAD-MOD circuit (Fig 9.5).

Table 9.1(*a*) *Relationship of Gray code to binary code.*

Quaternary	Gray code	Binary code
0	0 0	0 0
1	0 1	0 1
2	1 1	1 0
3	1 0	1 1

On the receive side, the two 19.332 Mbps data streams of the regenerated pulses and the clock pulses undergo differential decoding. The Gray code-to-binary code converter converts the Gray-coded pulse streams to binary streams. In the binary pulse streams the phases of the time-slot leading one bit before an incoming time-slot are subtracted from that of the incoming time-slot in the form of quaternary numeration, and thus decoded to Gray-coded data streams and sent to the frame synchronizing circuit with the 19.332 MHz recovered clock pulse (Refer to Rx LOGIC portion of Fig. 9.6).

9.6 Carrier Synchronization in DMR770

Refer to Section 6.2. The 4 PH-DEMS consists of a four-phase demodulator circuit and an Rx logic circuit. The Rx logic circuit takes care of the differential decoding of QPSK as explained in the previous section.

Table 9.1(*b*) *Logical differential operation*

		Time–slot	1	2	3	4	5	6	7	8	9	10	11
PH-MOD	Tx logic input	DATA 1	0	1	1	0	1	0	0	1	0	1	1
		DATA 2	0	0	1	1	1	0	1	0	0	1	1
		Quaternary	0	3	2	1	2	0	1	3	0	2	2
	SUM logic output	Quaternary	0	3	1	2	0	0	1	0	0	2	0
		Stream 1	0	1	0	1	0	0	0	0	0	1	0
		Stream 2	0	1	1	0	0	0	1	0	0	0	0
	Tx logic output	MOD-PLS 1	0	1	0	1	0	0	0	0	0	1	0
		MOD-PLS 2	0	0	1	1	0	0	1	0	0	1	0
	Modulation phase		0	$3\pi/2$	$\pi/2$	π	0	0	$\pi/2$	0	0	π	0
PH-DEM	Pulse regeneration ckt output (quaternary) — Demodulation local carrier (LO1) phase	$\pi/4$	2	1	3	0	2	2	3	2	2	0	2
		$3\pi/4$	1	0	2	3	1	1	2	1	1	3	1
		$5\pi/4$	0	3	1	2	0	0	1	0	0	2	0
		$7\pi/4$	3	2	0	1	3	3	0	3	3	1	3
	Differential logic output	Quaternary	(0)	3	2	1	2	0	1	3	0	2	2
		Stream 1	0	1	1	0	1	0	0	1	0	1	1
		Stream 2	0	1	0	1	0	0	1	1	0	0	0
	Rx logic output	DATA 1	0	1	1	0	1	0	0	1	0	1	1
		DATA 2	0	0	1	1	1	0	1	0	0	1	1

A simplified block diagram of a phase demodulator circuit of DMR 770 is presented in Fig. 9.6. The incoming IF signal is applied to the two phase detectors PH-DET1 and PH-DET2 through a hybrid circuit. Both have a common local oscillator from the source VCO with a shift of $\pi/2$. It is a coherent detection process in which the output voltage of the phase-error detector circuit controls the VCO and synchronizes the carrier with the input IF signal.

9.7 Symbol Clock Synchronization in DMR 770

It can be observed from Fig 9.6 that the clock synchronizer is also a part of the PH-DEM, i.e., symbol clock synchronization is being achieved simultaneously. The PH-ERROR-DET controls the VCO as well as the clock synchronization, i.e., the demodulated (not detected) waveform is used in clock recovery as described in Chapter 2. The two data signals D_1 and D_2 from the PH-DET and the recovered clock are given to the sampler. The sampler discriminates the polarity of each bit and generates two 19.332 Mbps signals in NRZ format, which are sent to the Rx logic unit.

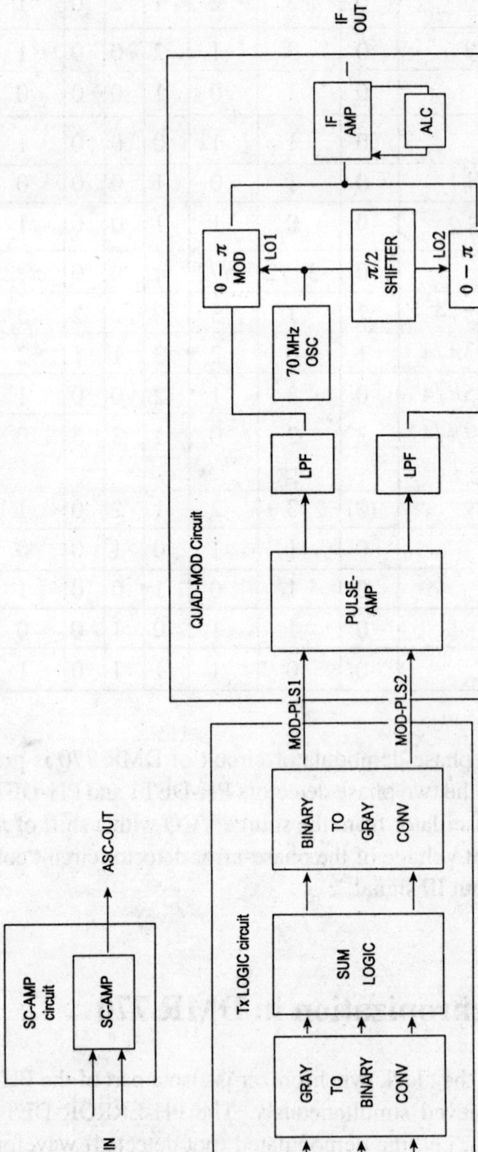

Fig. 9.5 *Tx LOGIC and QUAD-MOD interfaces.*

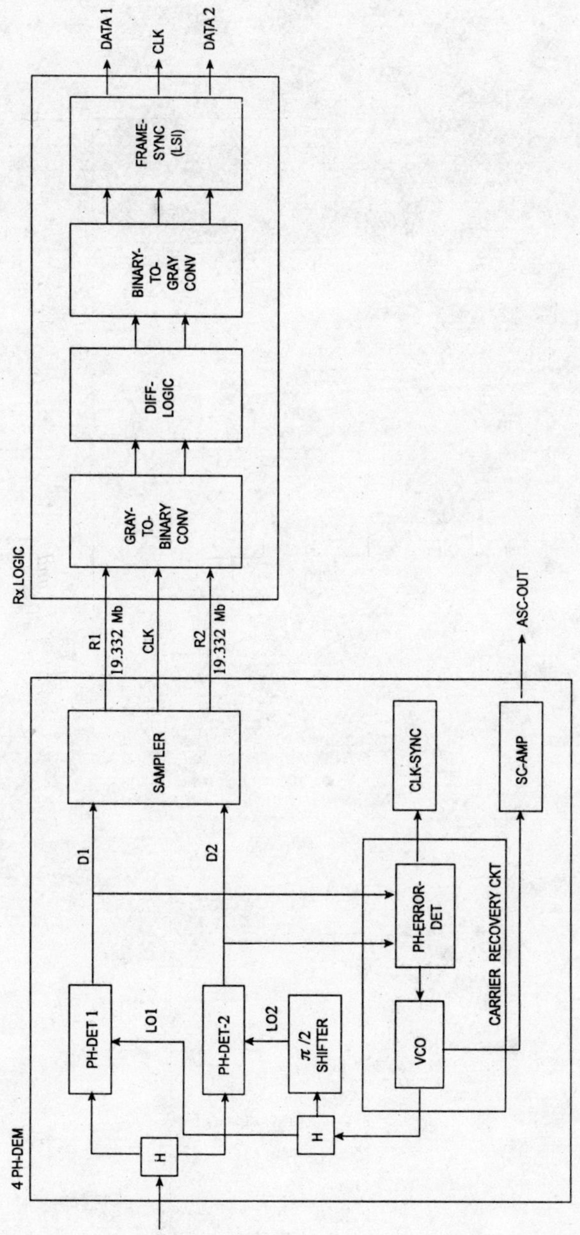

Fig 9.6 *Simplified block diagram of PH-DEM.*

Exercises

1. How is phase ambiguity in carrier recovery resolved?
2. What is the relationship between P_e of differentially encoded QPSK and coherent PSK?
3. How is the locally generated carrier synchronized with input IF (derived) carrier in coherent detection?

10

Control of Intersymbol Interference

10.1 Introduction

The design issues and performance analysis in passband data transmission systems were described in Chapter 4. The problem of synchronization of the carrier and the symbol clock and some of the practical circuits for the same were described in the previous chapter.

Now we are on a comfortable footing as far as AWGN is concerned. But that is not the end of villains in a communications system. We have to face another two—ISI (intersymbol interference) and jitter.

Since bandwidth occupancy is a major factor in the design of digital systems, band-limiting filters are used at the transmitter output, in the transmission medium, and at the receiver input, or combinations thereof. These filters make the channel dispersive and give rise to ISI. It is a major source of bit errors in the reconstructed data stream at the receiver. Unlike noise, it is a signal-dependent form of interference.

In this chapter we will discuss ISI and techniques for combating it. We start the discussion with baseband (low-pass) digital transmission and the propagation effects thereof, and then move on to bandpass systems.

Tackling the third villain—the jitter—is reserved for Chapters 14–16.

10.2 Optimum Receiver and Coding Scheme for Baseband (Digital) Channel

Digital data have a broad spectrum with significant low-frequency content. Hence, for baseband transmission of digital data, a low-pass channel with bandwidth sufficient to accomodate the essential frequency content of the data stream is required. We cannot achieve ideal low-pass filter charteristics in reality, however sharp the cut-off be. This makes the channel dispersive, i.e, each received pulse is affected by the adjacent pulses resulting in intersymbol interference (ISI). This is a source of bit errors in the reconstructed data stream at the receiver. To correct this, pulse shaping techniques are resorted to.

Along with ISI, we also have AWGN. Therefore, the two questions that need to be answered in this context are:

1. What is the optimum receiver to tackle AWGN?
2. What is the coding scheme to tackle ISI?

The answer to the first question is 'Matched Filter'. The answer to the second question is 'Correlative Coding'. We will develop these concepts now.

10.3 Matched Filter

Consider the linear receiver shown in Fig 10.1. It consists of a linear time invariant filter of impulse response $h(t)$ and a sampling circuit at the output of the filter. The input to the receiver is

$$x(t) = g(t) + \omega(t) \ (0 \leq t \leq T),$$

where T is the observation interval and $g(t)$ a pulse signal (symbol 1 or 0) of the digital transmission system. $\omega(t)$ is the AWGN of power spectral density $N_0/2$.

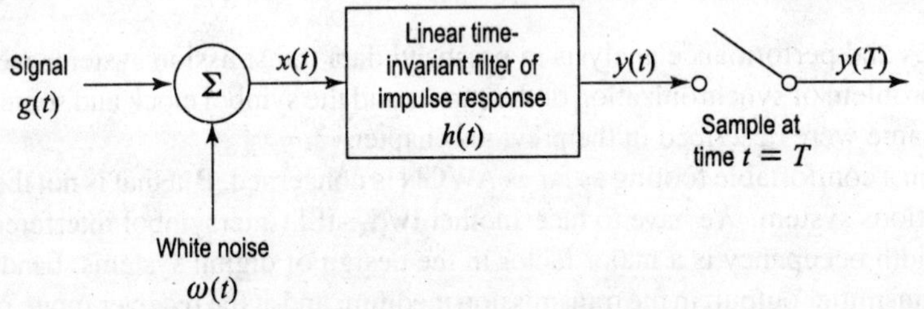

Fig. 10.1 *Linear receiver.*

The receiver has knowledge of the waveform (the coding followed rather) but is uncertain about $\omega(t)$. Its function is to dectect $g(t)$ in an optimum manner, given the received signal $x(t)$.

Since the filter is linear, its output $y(t)$ can be expressed as

$$y(t) = g_0(t) + n(t),$$

where $g_0(t)$ is the response of the filter for $g(t)$ and $n(t)$ the response of the filter for $\omega(t)$.

The output due to the signal component should be considerably greater than the output due to noise component. In actual terms, the peak pulse signal-to-noise ratio η should be maxiamized. η is given by

$$\eta = \frac{|g_0(T)|^2}{E\left[n^2(t)\right]}.$$

If $G(f)$ is the Fourier transform of $g(t)$,

$$g_0(t) = \int_{-\infty}^{\infty} H(f)G(f)\exp(j2\pi ft)df,$$

where $H(f)$ is the Fourier transform of $h(t)$, the impulse response of the filter.

$$|g_0(T)|^2 = \left| \int_{-\infty}^{\infty} H(f)G(f)\exp(j2\pi fT)df \right|^2 \tag{10.1a}$$

$$E\left[n^2(t)\right] = \int_{-\infty}^{\infty} S_N(f)df = \frac{N_0}{2}\int_{-\infty}^{\infty} |H(f)|^2 df \tag{10.1b}$$

where $S_N\ (f)$ is the output noise spectral density.

From 10.1(a) and (b),

$$\eta = \frac{\left| \int_{-\infty}^{\infty} H(f)G(f)\exp(j2\pi ft)df \right|^2}{\frac{N_0}{2}\int_{-\infty}^{\infty} |H(f)|^2 df}$$

The first question of Section 10.2 now gets simplified to:

For a given $G(f)$, what is the particular form of the transfer function $H(f)$ of the filter to make η maximum?

We will make use of Schwarz's inequality to get an answer to the above question.

If $\Phi_1(x)$, $\Phi_2(x)$ are complex functions of the real variable x satisfying the conditions

$$\int_{-\infty}^{\infty} |\Phi_1(x)|^2 dx < \infty \quad \text{and}$$

$$\int_{-\infty}^{\infty} |\Phi_2(x)|^2 dx < \infty,$$

then, according to Schwarz's inequality,

$$\left| \int_{-\infty}^{\infty} \Phi_1(x)\Phi_2(x)dx \right|^2 \leq \int_{-\infty}^{\infty} |\Phi_1(x)|^2 dx \int_{-\infty}^{\infty} |\Phi_2(x)|^2 dx. \tag{10.1c}$$

Back to the question before us—set $\phi_1(x) = H(f)$ and $\phi_2(x) = G(f)\exp(j\pi ft)$ and apply Schwarz's inequality:

$$\left| \int_{-\infty}^{\infty} H(f)G(f)\exp(j2\pi\beta t)df \right|^2 \leq \int_{-\infty}^{\infty} |H(f)|^2 df \int_{-\infty}^{\infty} |G(f)|^2 df. \tag{10.1d}$$

From equations 10.1 (c) & (d), we have

$$\eta \leq \frac{2}{N_0}\int_{-\infty}^{\infty} |G(f)|^2 df.$$

$$\therefore \quad \eta_{max} = \frac{2}{N_0}\int_{-\infty}^{\infty} |G(f)|^2 df. \tag{10.1e}$$

Equality condition applies in Schwarz's inequality if $\phi_1(x) = k\phi_2^*(x)$, where k is a constant and $\phi_2^*(x)$ is the complex conjugate of $\phi_2(x)$.

So, the $H(f)$ corresponding to η_{max} (equality condition) of equation 10.1(e) is

$$H_{opt}(f) = kG(f)\exp(-j2\pi fT). \tag{10.1f}$$

Applying inverse Fourier transform and the property $G(f) = G(-f)$, we get

$$\begin{aligned} H_{opt}(t) &= k\int_{-\infty}^{\infty} G(-f)\exp(-j2\pi fT)\exp(j2\pi ft)df \\ &= k\int_{-\infty}^{\infty} G(-f)\exp(-j2\pi f(T-t))df \\ &= kg(T-t). \end{aligned}$$

This means that the impulse response of the optimum filter is a time-reversed and delayed version of the input signal $g(t)$ except for the scaling factor k, i.e., it is 'matched' to the input signal. Hence the name 'matched filter'.

Suppose we are using a matched filter receiver, i.e., $H(f)$ is $H_{opt}(f)$. Substituting the expression of $H_{opt}(f)$ from equation 10.1(f) in 10.1(b), we get

$$E\left[n^2(t)\right] = \frac{k^2 N_0}{2}\int_{-\infty}^{\infty} |G(f)|^2 df$$

The integral portion in the above expression is the signal energy E of $g(t)$.

$$\therefore \quad E\left[n^2(t)\right] = k^2 N_0 E/2$$

where E is the energy of $g(t)$.

$$\eta_{max} = \frac{(kE)^2}{[k^2 N_0 E/2]} = \frac{2E}{N_0}.$$

This is the most important property of a matched filter.

The peak pulse signal-to-noise ratio of a matched filter depends only on the ratio of the signal energy to the power spectral density of the white noise at the filter input.

As an example, the matched filter output for a rectangular pulse and the output of a integrated dump circuit used for its realization are shown in Fig. 10.2.

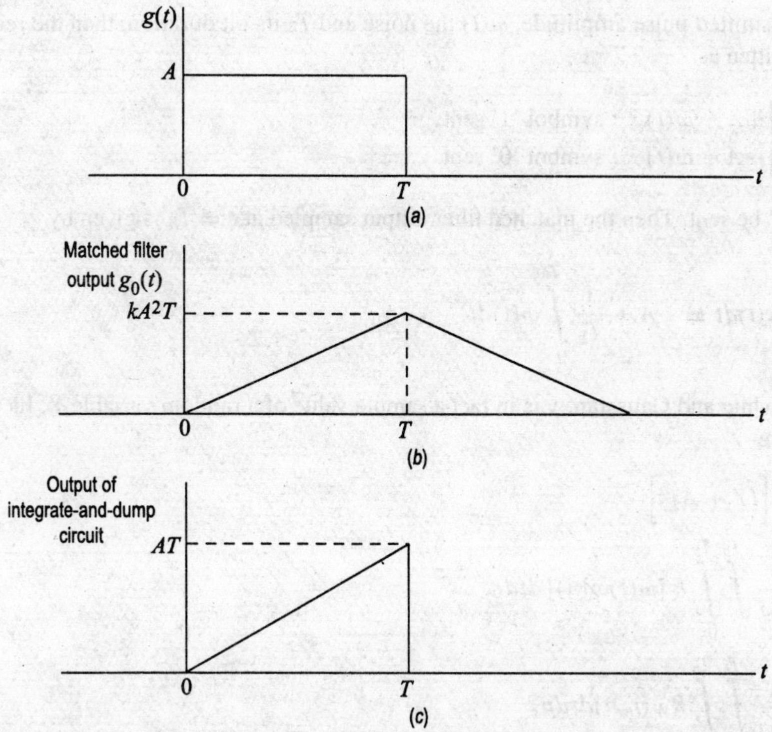

Fig. 10.2 *(a) Rectangular pulse; (b) Matched filter output; (c) Integrator output.*

10.4 Probability of Error in a PCM Receiver using Matched Filter

The receiver for baseband transmission of binary encoded PCM wave using NRZ signalling is shown in Fig. 10.3. There are two kinds of errors:

Error of first kind : Symbol '1' is choosen when '0' was transmitted

Error of second kind : Symbol '0' chosen when '1' was transmitted

The decision device compares output of matched filter (sampled) with decision threshold 'λ' and declares 1 or 0 as the case may be.

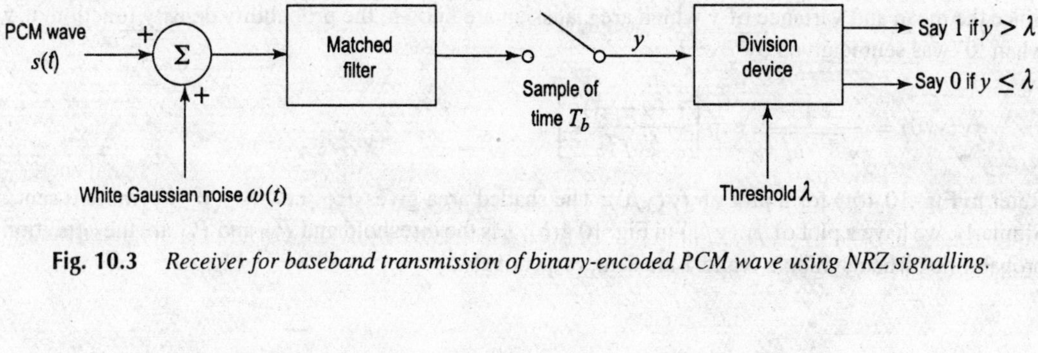

Fig. 10.3 *Receiver for baseband transmission of binary-encoded PCM wave using NRZ signalling.*

If A is the transmitted pulse amplitude, $\omega(t)$ the noise and T_b its bit duration, then the received signal $x(t)$ can be written as

$$x(t) = \begin{cases} +A + \omega(t) & : \text{symbol '1' sent.} \\ -A + \omega(t) & : \text{symbol '0' sent.} \end{cases}$$

Let symbol '0' be sent. Then the matched filter output sampled at $t = T_b$ is given by

$$y = \int_0^{T_b} x(t)dt = -A + \frac{1}{T_b}\int_0^{T_b} \omega(t)dt.$$

Since $\omega(t)$ is white and Gaussian, y is in fact a sample value of a random variable Y. Its mean is $-A$, and its variance

$$\begin{aligned} \sigma_Y^2 &= E\left[(Y + A)^2\right] \\ &= \frac{1}{T_b{}^2}\int_0^{T_b}\int_0^{T_b} E\left[\omega(t)\omega(n)\right] dtdn \\ &= \frac{1}{T_b{}^2}\int_0^{T_b}\int_0^{T_b} R_W(t, n)dtdn, \end{aligned}$$

where $R_W(t, n)$ is the autocorrelation function of $\omega(t)$, i.e., it shows the correlation or dependence of the function values at instants t and n.

Since $\omega(t)$ is the white Gaussian noise with spectral density $N_0/2$,

$$R_W(t, n) = \frac{N_0}{2}\delta(t - n).$$

$$\begin{aligned} \therefore\ \sigma_Y^2 &= \frac{1}{T_b^2}\int_0^{T_b}\int_0^{T_b} \frac{N_0}{2}\delta(t - n)dtdn \\ &= \frac{N_0}{2T_b}. \end{aligned}$$

Since the mean and variance of y which are Gaussian are known, the probability density function of y when '0' was sent is given by

$$f_Y(y/0) = \frac{1}{\sqrt{\pi N_0/T_b}} \exp\left[\frac{-(y + A)^2}{N_0/T_b}\right].$$

Refer to Fig. 10.4(a) for a plot of $f_Y(y/0)$. The shaded area gives P_{e0}, i.e., $P(y > \lambda/\text{symbol 0 sent})$. Similarly, we have a plot of $f_Y(y/1)$ in Fig. 10.4(b); λ is the threshold and P_{e0} and P_{e1} are the condition probabilities when symbols 0 and 1 are sent, respectively.

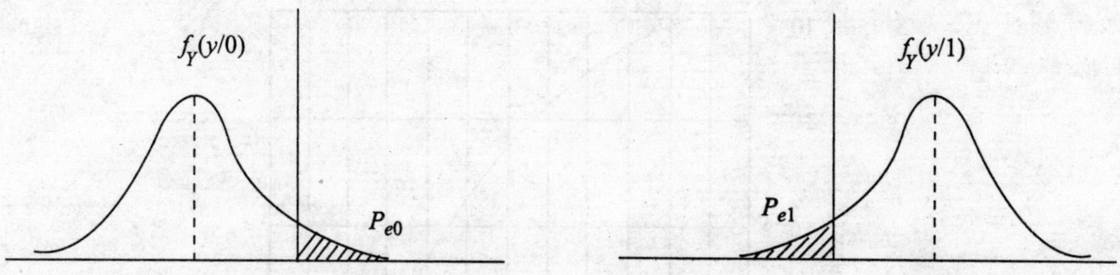

Fig. 10.4(*a*) $f_Y(y/0)$. **Fig. 10.4**(*b*) $f_Y(y/1)$.

Since symbol 0 and 1 occur with equal probability, we have to set the threshold exactly halfway, i.e., $\lambda = 0$.

$$\therefore \quad P_{e0} = \frac{1}{\sqrt{\pi N_0/T_b}} \int_0^{\infty} \exp\left[\frac{-(y+A)^2}{N_0/T_b}\right] dy.$$

Define a variable $z = \dfrac{y+A}{\sqrt{N_0/T_b}}$ for simplicity of expression. Then, P_{e0} becomes

$$P_{e0} = \frac{1}{\pi} \int_{\sqrt{E_b/N_0}}^{\infty} \exp(-z)^2 dz,$$

where E_b is the transmitted signal energy per bit defined by $E_b = A^2 T_b$.

The complementary error function erfc (n) (see Appendix) is

$$\text{erfc}(n) = \frac{2}{\sqrt{\pi}} \int_n^{\infty} \exp(-z)^2 dz.$$

$$\therefore \quad P_{e0} = \frac{1}{2}\text{erfc}\left[\sqrt{\frac{E_b}{N_0}}\right].$$

Similarly, $P_{e1} = \dfrac{1}{2}\text{erfc}\left[\sqrt{\dfrac{E_b}{N_0}}\right].$

$$P_e = p_0 P_{e0} + p_1 P_{e1},$$

where p_0 and p_1 are the probabilities of symbols 0 and 1 being sent, both equal to 1/2.

Hence $P_e = \dfrac{1}{2}\text{erfc}\left[\sqrt{\dfrac{E_b}{N_0}}\right].$

A plot of P_e versus E_b/N_0 is shown in Fig. 10.5. We can see that even a small increase in E_b/N_0 causes great reduction in P_e.

Thus, we have answered the first question of Section 10.2 by developing the concept of matched filter. We have also worked out P_e in PCM system using matched filter receiver. Now, we turn our attention to the second question of Section 10.2.

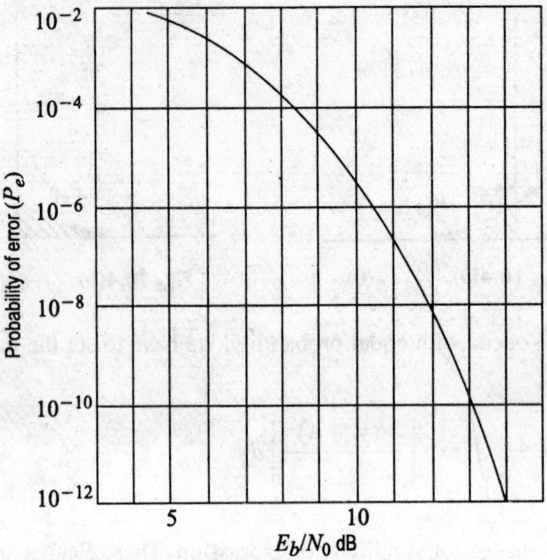

Fig. 10.5 *Plot of P_e versus E_b/N_0.*

10.5 Intersymbol Interference

Consider a baseband digital (binary) transmission system shown in Fig. 10.6. The incoming binary sequence $\{b_k\}$ consists of symbols 1 and 0, each of duration T_b.

The pulse amplitude modulator (since PCM starts as PAM in the baseband, this discussion is worth it) transforms the binary sequence into a new sequence of short pulses with amplitude a_k such that

$$a_k = +1 \text{ if symbol } b_k \text{ is } 1,$$
$$a_k = -1 \text{ if symbol } b_k \text{ is } 0.$$

The transmitted signal (output of the transmit filter of impulse response $g(t)$) is

$$g(t) = \sum_k a_k\, g(t - kT_b).$$

As $s(t)$ goes through the channel, it gets corrupted by noise and becomes $x(t)$ when it reaches the receiver. $x(t)$ is applied to a receive filter of impulse response $c(t)$. The resulting filter output $y(t)$ is sampled synchronously with the transmitter.

$$y(t) = \mu \sum a_k p(t - kT_b) + n(t),$$

where μ is the scaling factor, $p(t)$ the pulse function and $n(t)$ the noise.

$$\mu p(t) = g(t)^* h(t)^* c(t), \quad \text{(the three are in cascade; *implies convolution)}$$
$$\mu P(f) = G(f)\, H(f)\, C(f).$$

($P(f)$ is the Fourier transform of an infinite periodic sequence of delta functions of period T_b whose areas are weighted by sample values $p(t)$).

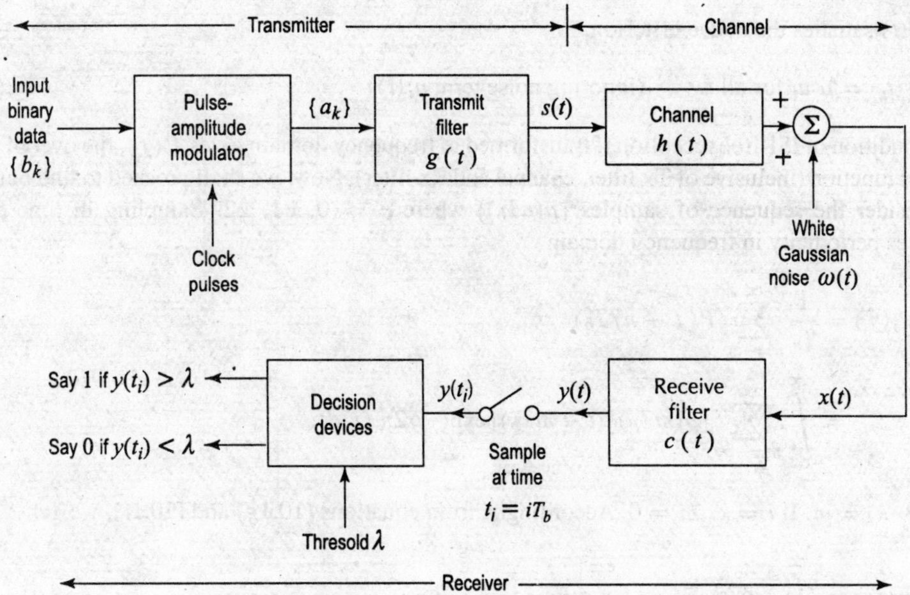

Fig. 10.6 *Baseband binary data transmission system.*

Therefore the output of receive filter sampled at $t_i = iT_b$ is

$$\begin{aligned} y(t_i) &= \mu \sum_{K=-\infty}^{\infty} a_k p\,[(i-k)T_b] + n(t_i) \\ &= \mu a_i + \mu \sum_{\substack{k=-\infty \\ k \neq i}}^{\infty} a_k p\,[(i-k)T_b] + n(t_i). \end{aligned}$$

The first term μa_i is the contribution of the ith transmitted data. The second term represents the residual effect of all other transmitted bits on the decoding of the ith bit (this is due to pulses before and after the sampling instant). The third term is the contribution of noise.

10.6 Criterion for Distortionless Baseband Digital Transmission

Given the transfer function of the channel and the transmitted pulse shape, we have to decide the transfer functions of the transmit and receive filters so as to reconstruct the original binary data sequence.

The receiver reconstructs the binary sequence by sampling $y(t)$ at $t = iT_b$ and decoding the weighted pulse contribution $a_k\ p[(i-k)\ T_b]$. The 'weighted pulse contribution' which represents the residual effect of all other transmitted bits, should be free from the contribution of the pulse at the sampling instant, i.e., for $k = i$, $\Sigma a_k p[(i-k)T_b]$ should be devoid of the term pertaining to $k = i$.

This implies that there should be control on the overall $p(t)$ as shown below.

$$p(iT_b - kT_b) = \begin{cases} 1 & i = k \\ 0 & i \neq k, \end{cases} \tag{10.1g}$$

where $p(0) = 1$ by normalization.

If $p(t)$ satisfies the above criterion,

$$y(t_i) = \mu\, a_i \text{ for all } i. \qquad \text{(ignoring noise term } n_i(t)) \tag{10.1h}$$

This condition of ISI-free reception if transformed to frequency domain gives $P(f)$, the overall system transfer function (inclusive of Tx filter, channel and Rx filter). Now, we shall proceed to find out $P(f)$.

Consider the sequence of samples $\{p(nT_b)\}$ where $n = 0, \pm 1, \pm 2$. Sampling in time domain produces periodicity in frequency domain

$$\begin{aligned} P_\delta(f) &= \frac{1}{T_b}\sum_{n=-\infty}^{\infty} P(f - n/T_b) \\ &= \int_{-\infty}^{\infty} \sum_{m=-\infty}^{\infty} [p(mT_b)\delta(t - mT_b)] \exp(-j2\pi f t)dt. \end{aligned} \tag{10.1i}$$

Let $(i - k) = m$. If $i = k$, $m = 0$. Accordingly, from equations (10.1g) and (10.1i), we get

$$P_\delta(f) = \int_{-\infty}^{\infty} p(0)\delta(t) \exp(j - 2\pi f t)dt = p(0). \tag{10.1j}$$

From (10.1i) and (10.1j) we get the condition of null ISI as

$$\sum_{n=-\infty}^{\infty} P(f - n/T_b) = T_b.$$

This is called the Nyquist criterion for distortionless baseband transmission (in the absence of noise, remember (10.1h)). Thus, Nyquist criterion can be stated as

The overall system transfer function $P(f)$ incorportating Tx filter, channel and Rx filter eliminates ISI for samples taken at intervals T_b if $\sum_{n=-\infty}^{\infty} P(f - n/T_b) = T_b$.

10.6.1 Ideal Nyquist channel

Let us see how we can satisfy Nyquist criterion for distortionless baseband pulse transmission.

If $P(f)$ has the shape of a rectangle (see Fig. 10.7),

$$P(f) = \begin{cases} \frac{1}{2W} & -W < f < W \\ 0 & |f| > W, \end{cases}$$

where $W = 1/2T_b$, the Nyquist criterion is satisfied. The corresponding $p(t)$ is

$$\begin{aligned} p(t) &= \sin(2\pi W t)/2\pi W t \\ &= \text{sinc}(2Wt). \end{aligned}$$

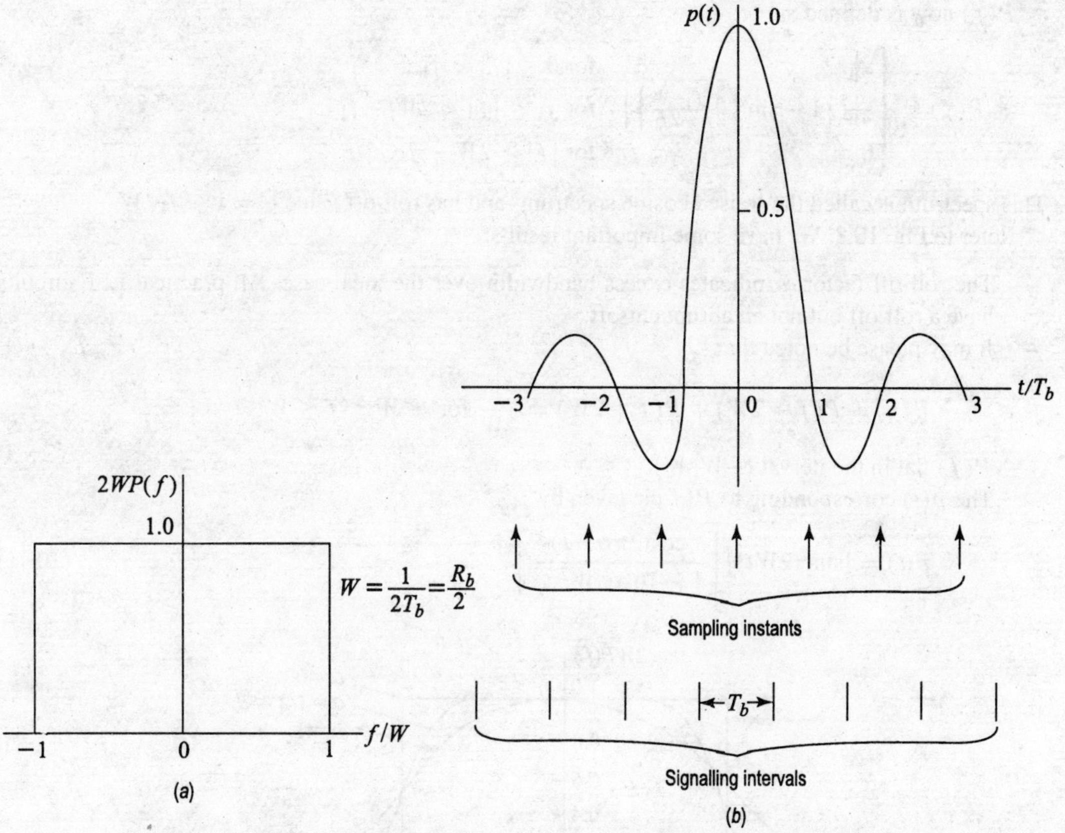

Fig. 10.7 *(a) Ideal amplitude response.* *(b) Ideal basic pulse shape.*

We can arrive at this result in a heuristic manner also. The low-pass characteristic has a rectangular shape. For rectangular function, sinc is the Fourier transform. So, as a 'matched' function, the time domain function should be of sinc form.

The use of ideal Nyquist channel makes ISI null but the difficulties are:

- it requires the amplitude characteristic of $P(f)$ to be flat from $-W$ to W and zero elsewhere. Such abrupt transition is unrealizable,
- $P(t)$ decreases at the rate of $1/|t|$ for large $|t|$, resulting in slower decay (this is also due to abrupt change or discontinuity in $P(f)$ at $f = \pm W$).

This implies that the sampling at the receiver has to be exact and no time margin is allowed.

10.6.2 Raised cosine spectrum

The practical difficulties in the realization of ideal Nyquist channel can be overcome by extending the bandwidth from $1/2T_b$ to a value between W and $2W$. It shall have flat portion in the interval $(-W, W)$, and a roll-off portion in the interval $(-2W, -W)$ and $(W, 2W)$.

$P(f)$ now is defined as

$$P(f) = \begin{cases} \frac{1}{2W} & \text{for } 0 \le |f| < f_1 \\ \frac{1}{4W}\left[1 - \sin\left\{\frac{\pi|f|-W}{2W-2f_1}\right\}\right] & \text{for } f_1 \le |f| < 2W - f_1 \\ 0 & \text{for } |f| \ge 2W - f_1. \end{cases}$$

This spectrum is called the 'raised cosine spectrum' and has roll-off value $\alpha = 1 - f_1/W$.

Refer to Fig. 10.8. We have some important results:

- The roll-off factor α indicates excess bandwidth over the ideal case. All practical LPF circuits have a roll-off but not an abrupt cut-off.
- It may please be noted that

$$P(f) + P(f - 2W) + P(f + 2W) = \frac{1}{2W} \text{ for } -W \le f \le W.$$

$P(f)$ flat in the iterval $(-W, W)$.

- The $p(t)$ corresponding to $P(f)$ is given by

$$p(t) = [\text{sinc}(2Wt)]\left[\frac{\cos(2\pi\alpha Wt)}{1 - 16\alpha^2 W^2 t^2}\right]. \tag{10.1k}$$

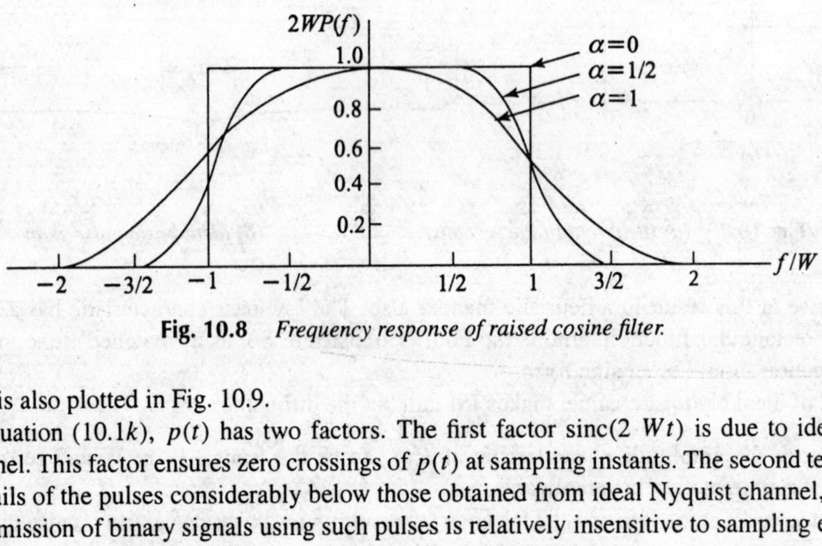

Fig. 10.8 *Frequency response of raised cosine filter.*

$p(t)$ is also plotted in Fig. 10.9.

- In equation (10.1k), $p(t)$ has two factors. The first factor sinc($2\ Wt$) is due to ideal Nyquist channel. This factor ensures zero crossings of $p(t)$ at sampling instants. The second term reduces the tails of the pulses considerably below those obtained from ideal Nyquist channel, so that the transmission of binary signals using such pulses is relatively insensitive to sampling errors.
- ISI decreases as the roll-off factor α increases from 0 to 1.
- When $\alpha = 1,\ f_1 = 0$, it has a full cosine roll-off characteristic.

$P(f)$ and $p(t)$ get simplified as

$$P(t) = \begin{cases} \frac{1}{4W}\,[1 + \cos(\pi f/2W)] & 0 < |f| < 2W \\ 0 & |f| \ge 2W, \end{cases}$$

$$p(t) = \frac{\text{sinc}(4Wt)}{1 - 16W^2 t^2}.$$

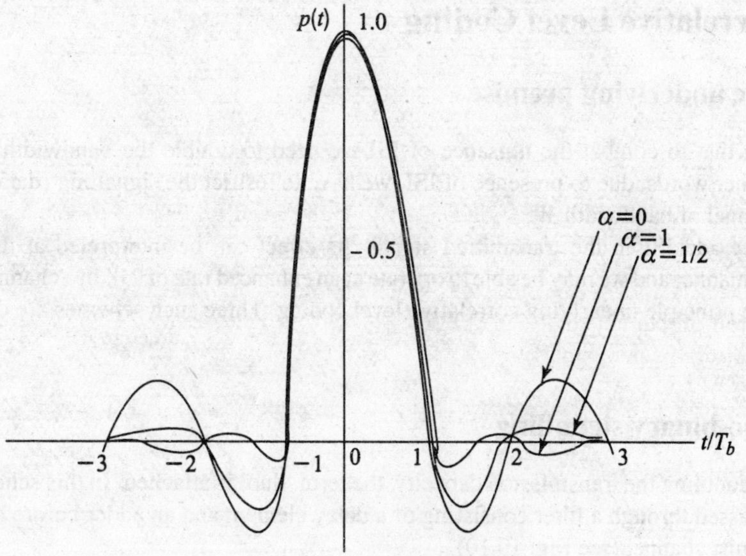

Fig. 10.9 *Time response.*

$p(t)$ has two important properties:

(*a*) At $t = \pm T_b/2 = \pm 1/4W$, $p(t) = 0.5$, i.e., pulse width measured at half amplitude is equal to the bit duration T_b.

(*b*) There are zero crossings at $t = \pm 3\, T_b/2, \ \pm 5\, T_b/2, \ldots$ in addition to the usual zero crossings at $t = \pm T_b \ \pm 2\, T_b$.

These properties help in extraction of timing signal from the received signal for the purpose of synchronization. But we are using twice the bandwidth of the ideal Nyquist channel.

10.6.3 Bandwidth requirement of 30 channel PCM system

A thirty channel PCM has actually 32 slots—one channel for signalling information, one channel for synchronizing information, and thirty for voice/data circuits. Each slot has an 8 bit PCM word.

Sampling rate of 8 kHz implies a 125 μs time frame. So, each bit duration is

$$125\,\mu\text{s}/(8 \times 32) = 0.488\,\mu\text{s}. \quad \text{i.e.,} \quad T_b = 0.488\,\mu\text{s}.$$

$$\text{Bandwidth of ideal Nyquist channel} = \frac{1}{2T_b} = 1024\,\text{kHz}.$$

With $\alpha = 1$ (full cosine roll-off),

$$\text{bandwidth} = (1+\alpha)\frac{1}{2T_b} = 2.048\ \text{MHz}.$$

10.7 Correlative Level Coding

10.7.1 The underlying premise

We have seen that to combat the nuisance of ISI we need to double the bandwidth of the Nyquist channel. In other words, due to presence of ISI, we have to restrict the signalling (data transition) rate to W in a channel of bandwidth W.

Suppose we add ISI in the transmitted signal. Its effect can be interpreted at the receiver in a deterministic manner and we may be able to operate at an enhanced rate of $2W$ in a channel of bandwidth W. This is the principle underlying correlative level coding. Three such schemes are discussed in this section.

10.7.2 Duo-binary signalling

Since we are doubling the transmission capacity, the term 'duo' is attached. In this scheme, a two level sequence is passed through a filter consisting of a delay element and an adder before being applied to the ideal Nyquist channel (see Fig. 10.10).

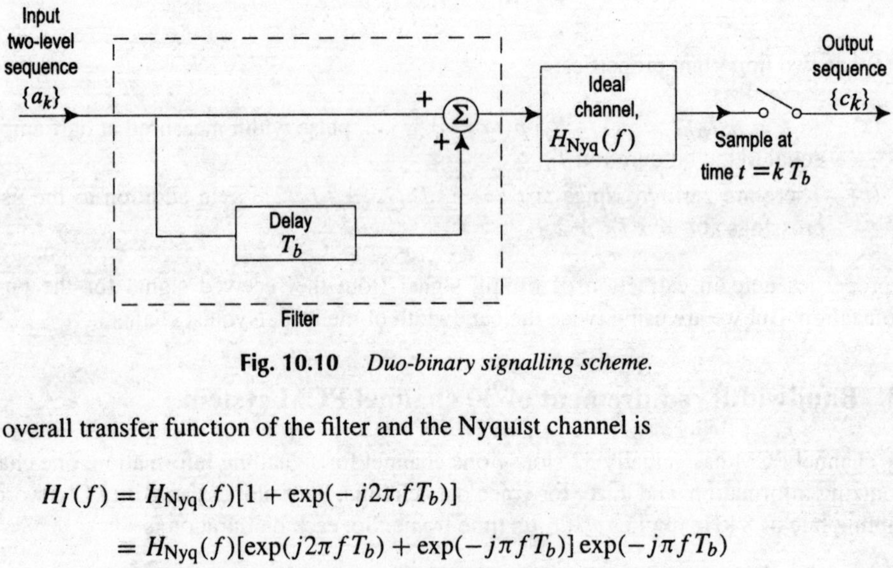

Fig. 10.10 *Duo-binary signalling scheme.*

The overall transfer function of the filter and the Nyquist channel is

$$\begin{aligned} H_I(f) &= H_{\text{Nyq}}(f)[1+\exp(-j2\pi f T_b)] \\ &= H_{\text{Nyq}}(f)[\exp(j2\pi f T_b)+\exp(-j\pi f T_b)]\exp(-j\pi f T_b) \\ &= 2\, H_{\text{Nyq}}(f)\cos(\pi f\, T_b)\exp(-j\pi f T_b). \end{aligned}$$

$$\text{But } H_{\text{Nyq}} = \begin{cases} 1 & |f| \le 1/2T_b \\ 0 & \text{otherwise.} \end{cases}$$

$$\text{So, } H_I\,(f) = \begin{cases} 2\cos(\pi f T_b)\exp(-j\pi f T_b) & |f| \le 1|2T_b \\ 0 & \text{otherwise.} \end{cases}$$

The time domain equivalent is

$$h_I(t) = \frac{\sin(\pi t/T_b)}{(\pi t/T_b)} + \frac{\sin[\pi(t - T_b)/T_b]}{\pi(t - T_b)/T_b} = \frac{T_b^2 \sin(\pi t/T_b)}{\pi t(T_b - t)}.$$

Refer to Fig. 10.11 for $H_I(f)$ and Fig. 10.12 for $h_I(t)$.

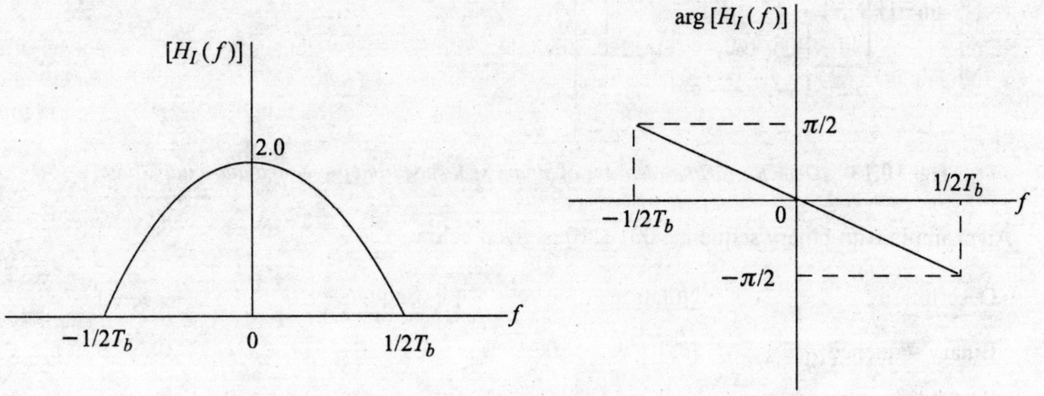

Fig. 10.11 *Frequency response of duo-binary conversion filter.*

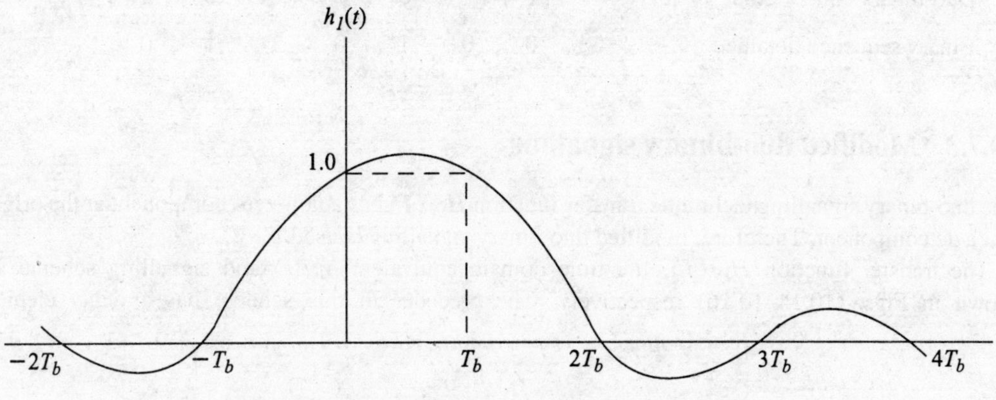

Fig. 10.12 *Time response of duo-binary conversion filter.*

Detection

An inverse of the delay-line operation is used in detection. But in such a scheme, errors once made, tend to propagate. This can be avoided by employing precoding as in Fig. 10.13. The precoding on binary sequence $\{b_k\}$ converts it into another binary sequence d_k defined by

$$d_k = b_k + d_{k-1}.$$

The detector for recovering original binary sequence from the pre-coded duo-binary coder output is shown in Fig. 10.13.

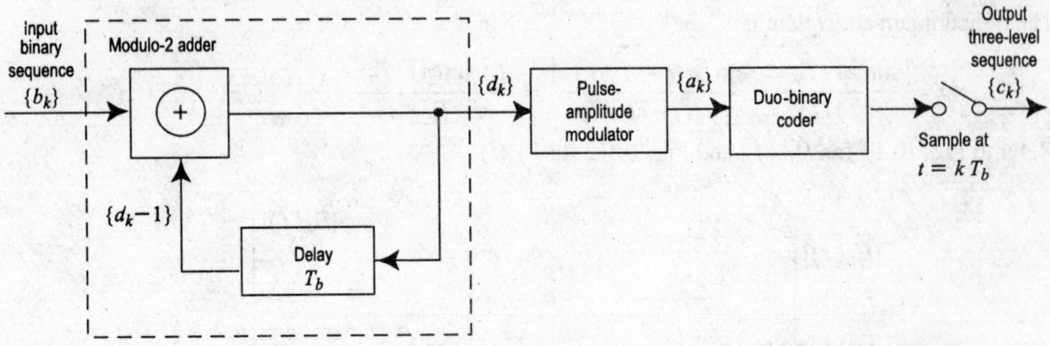

Fig. 10.13 *Detector for the recovery of binary sequence from pre-coded duo-binary data.*

An example with binary sequence 0010110 is given below.

Description	Notation	Sequence						
Binary sequence	$\{b_k\}$	0	0	1	0	1	1	0
Pre-coded sequence	$\{d_k\} - 1$	1	1	0	0	1	0	0
Two-level sequence	$\{a_k\} + 1$	+1	+1	−1	−1	+1	−1	−1
Duo-binary coder output	$\{c_k\}$	+2	+2	0	−2	0	0	−2
Binary sequence obtained		0	0	1	0	1	1	0

10.7.3 Modified duo-binary signalling

The duo-binary signalling technique transfer function $H_{IV}(f)$ has a non-zero component at the origin, i.e., a dc component. Therefore, modified duo-binary signalling is used.

The transfer function $H_{IV}(f)$, the time domain equivalent $h_{IV}(t)$ and signalling scheme are shown in Figs. (10.14–10.16) respectively, The precoder in this scheme has a delay element of $2T_b$.

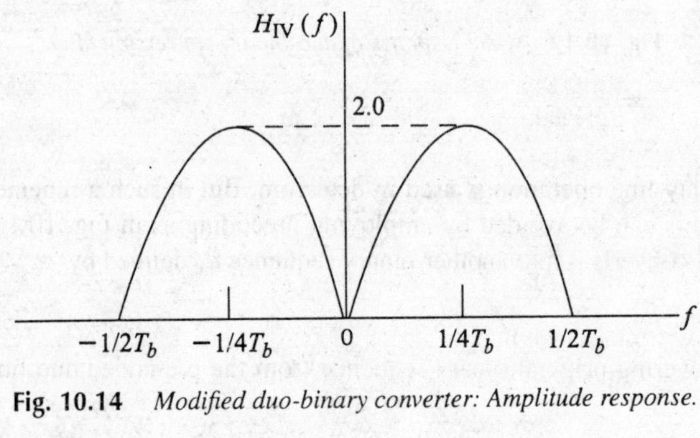

Fig. 10.14 *Modified duo-binary converter: Amplitude response.*

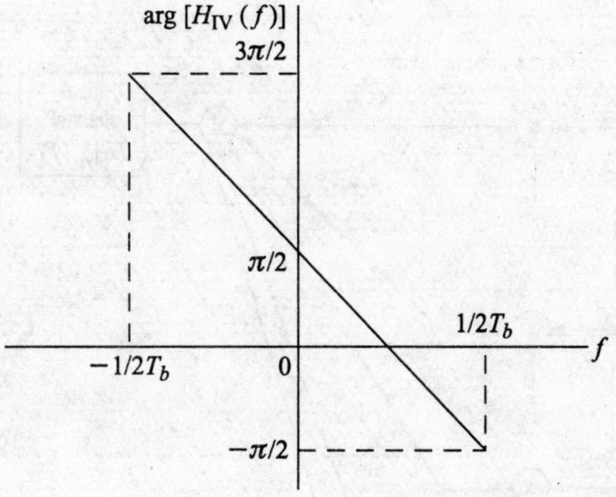

Fig. 10.15 *Modified duo-binary converter: Phase response.*

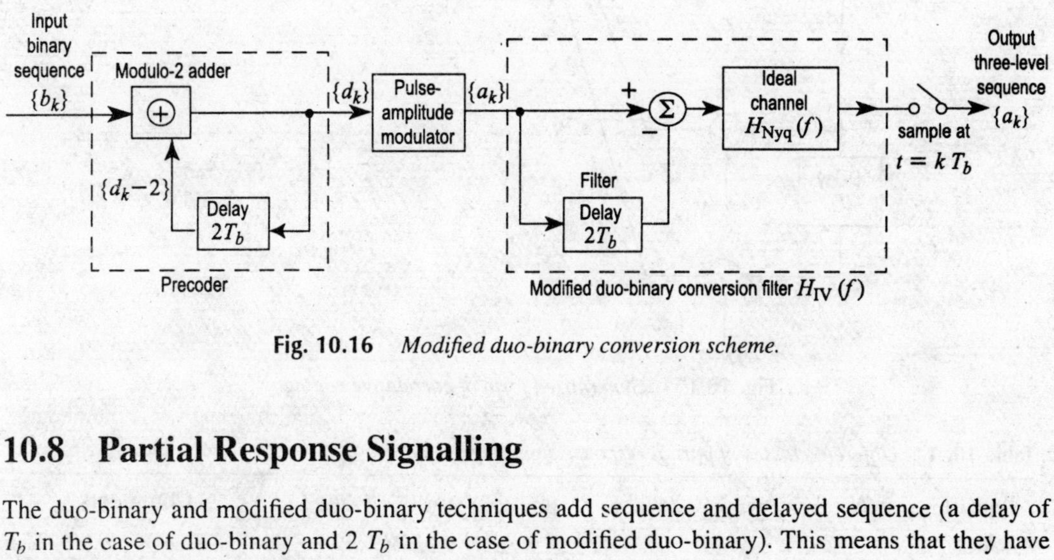

Fig. 10.16 *Modified duo-binary conversion scheme.*

10.8 Partial Response Signalling

The duo-binary and modified duo-binary techniques add sequence and delayed sequence (a delay of T_b in the case of duo-binary and 2 T_b in the case of modified duo-binary). This means that they have correlation spans of T_b, 2 T_b.

The partial response signalling is a generalized form of correlative coding (Fig. 10.17). It involves the use of tapped delay-line filter with tap weights w_0, w_1 ... w_{N-1}. A weighted linear combination of N ideal Nyquist (sinc) pulses $h(t)$, is given by

$$h(t) = \sum_{n=0}^{N-1} w_n \operatorname{sin c}\left[\frac{t}{T_b} - n\right].$$

An appropriate choice of tap weights results in a variety of spectral shapes designed to suit individual applications. Table 10.1 shows the different classes of partial response signalling schemes.

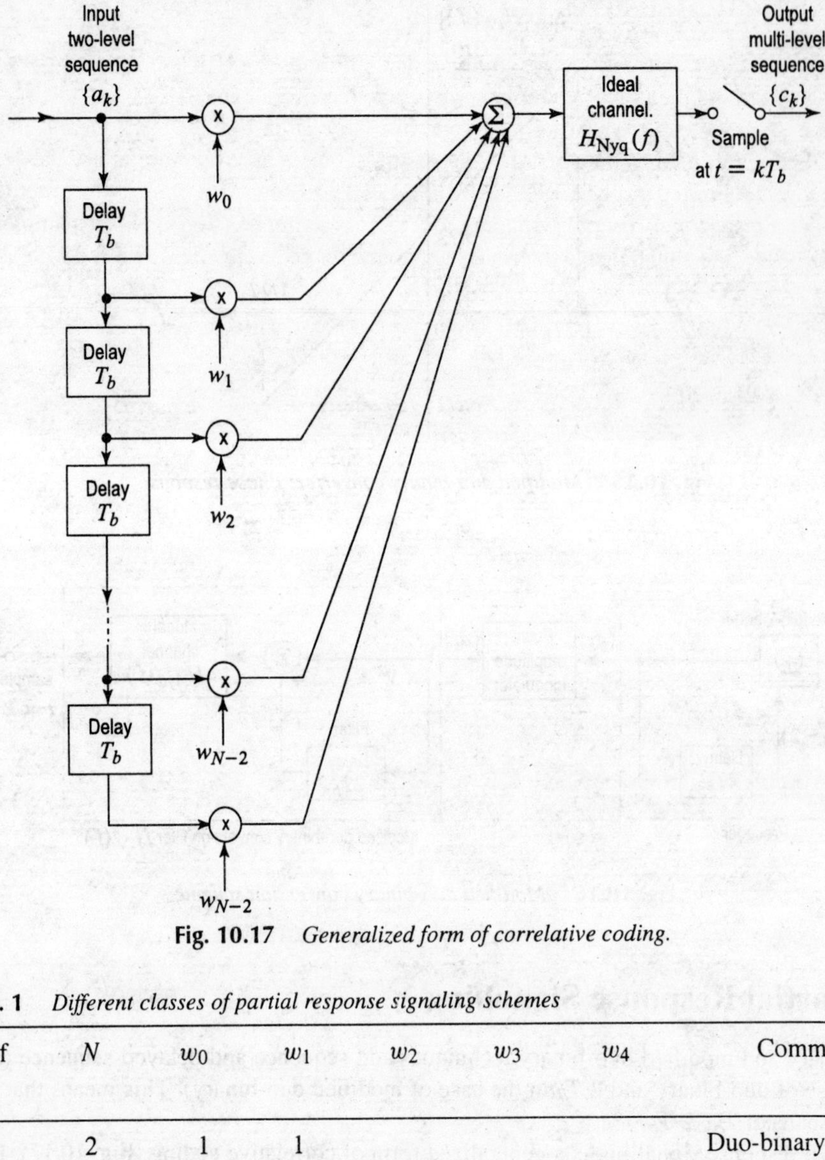

Fig. 10.17 *Generalized form of correlative coding.*

Table 10. 1 *Different classes of partial response signaling schemes*

Type of class	N	w_0	w_1	w_2	w_3	w_4	Comments
I	2	1	1				Duo-binary coding
II	3	1	2	1			
III	3	2	1	−1			
IV	3	1	0	−1			Modified duo-binary coding
V	5	−1	0	2	0	−1	

10.9 Practical Measures to Combat ISI: Equalizers

With the discussions held thus far, we are equipped with the causes of ISI and the strategies to combat it. Now we will examine the practical implementation of these strategies.

Attenuation v/s frequency curve of a telephone line is given in Fig. 10.18. It clearly indicates the dispersive nature of the channel. As a result of dispersion of pulse shape by the channel, the detectability of amplitude levels is limited more by ISI than by noise. We use Tx and Rx filters before modulators and after demodulator respectively for pulse shaping. But there is an imprecision in physical implementation of these filters. Also, the channel characteristics cannot be known exactly. So, we use what is called an 'equalizer'.

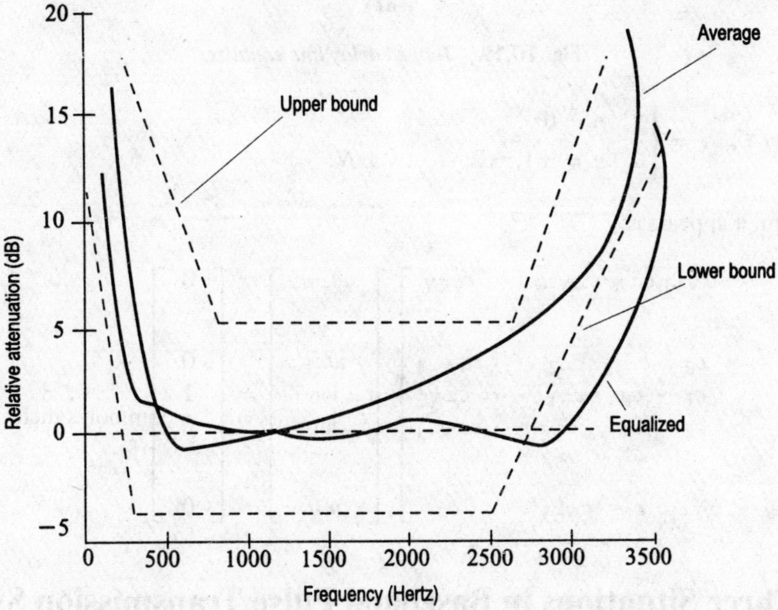

Fig. 10.18 *Attenuation v/s frequency curve of telephone line.*

The equalizer shown in Fig. 10.19 is called a tapped delay-line equalizer. Its impulse response is

$$h(t) \sum_{k=-N}^{N} w_k \delta(t - kT).$$

If $p(t)$ is impulse response of the system, $p(t) = c(t)^{*}h(t)$. It can be shown that at sampling times $t = nT$, if

$$p(nT) = \begin{cases} 1 & n = 0 \\ 0 & n = \pm 1 \pm 2, \ldots \pm N \end{cases}$$

is satisfied, there will be ideal reception. If $C_n = c(nT)$ is the nth sample of $c(t)$, from the discrete convolution we get $(2N + 1)$ simultaneous equations

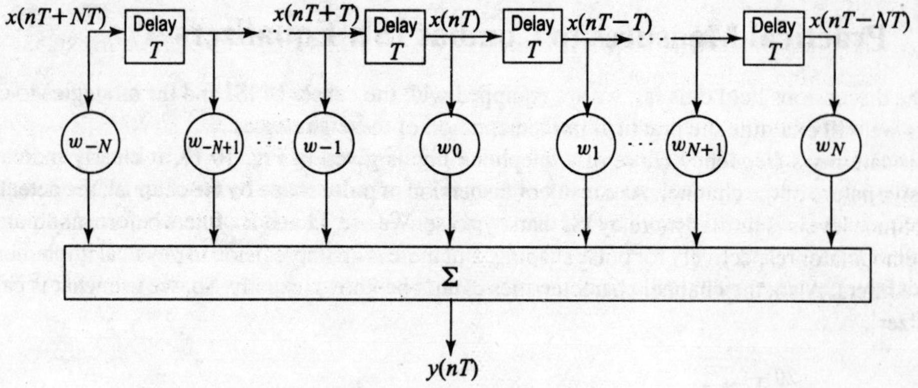

Fig. 10.19 *Tapped delay line equalizer.*

$$\sum_{k=-N}^{N} w_k \, C_{n-k} = \begin{cases} 1 & n = 0 \\ 0 & n = \pm 1, \pm 2, \ldots\ldots \pm N. \end{cases}$$

In matrix form, it appears as

$$\begin{bmatrix} c_0 & \cdots & c_{-N+1} & c_{-N} & c_{-N-1} & \cdots & c_{-2N} \\ \vdots & & \vdots & & & & \\ c_{N-1} & \cdots & c_0 & c_{-1} & c_{-2} & \cdots & c_{-N-1} \\ c_N & \cdots & c_1 & c_0 & c_{-1} & \cdots & c_{-N} \\ c_{N+1} & \cdots & c_2 & c_1 & c_0 & \cdots & c_{-N+1} \\ \vdots & & \vdots & \vdots & \vdots & & \vdots \\ c_{2N} & \cdots & c_{N+1} & c_N & c_{N-1} & \cdots & c_0 \end{bmatrix} \begin{bmatrix} w_{-N} \\ \vdots \\ w_{-1} \\ w_0 \\ w_1 \\ \vdots \\ w_N \end{bmatrix} = \begin{bmatrix} 0 \\ \vdots \\ 0 \\ 1 \\ 0 \\ \vdots \\ 0 \end{bmatrix}$$

10.10 Three Situations in Baseband Pulse Transmission System

Depending on the received signal-to-noise ratio, three different situations arise in baseband pulse transmission systems for channels with fixed characteristics:

1. Effect of ISI is neligible compared to that of noise: The procedure in this case is to use a matched filter (see Section 4.3) which is the optimum linear signal-to-noise ratio.
2. The received signal-to-noise ratio is high enough to ignore the effect of noise: In this case, ISI is to be taken care of. Control must be exercised over the shape of the received pulse. There are two approaches:
   - Use raised cosine spectrum for the overall frequency response of the baseband pulse transmission system
   - Use correlative level coding or partial response signalling that adds ISI to the transmitted signal in a controlled manner
3. ISI and noise are both significant: This situation calls for a joint optimization of the transmit and receive filters. In brief, a suitable pulse shape is first used to reduce ISI to zero at the

sampling instants and Schwartz's inequality is invoked to maximize the output signal-to-noise ratio at the sampling instant.

If the channel is random, the use of a fixed design based on average channel characteristics would not be adequate. An adaptive equalizer positioned after the receive filter in the receiver is used to compensate for variations in the frequency response of the channel automatically during data transmission. Adaptive equalizer is a tapped delay line equalizer following the LMS (linear mean squared) algorithm for control of tap weight.

Adaptive equalizers tackle the combined effect of ISI and receiver. It is a part of almost every modem in use today.

10.11 Adaptive Equalizers

The modern method for acheiving time domian equalization is to use a transveral filter (tapped delay line structure) as shown in Fig. 10.20. The unequalized baseband signal applied on the left, propagates through the delay line towards the right and the symbol-delayed versions are available at the taps for multiplication by a set of N weight coefficients $C_0,\ C_1, \ldots\ldots C_{n-1}$ before being summed. It can be shown that in order to control N points on the channel response, it takes about N taps. The most commonly used algorithm for controlling the taps, which converges to an equalized response (or an open eye), is the so-called linear mean squared (LMS) steepest descent algorithm, which for the nth tap gain is an iteration from time k to time $k+1$:

$$C_n\ (k+1) = C_n(k) - s(e_k)r(t-nt+kt),$$

where e_k is the error at time kt between the (partially) equalized output $z(t)$, and either the training sequence $T(t)$, or the output data $d(t)$, i.e.,

$$\begin{aligned} e_k &= z(kt) - t(kt) \qquad \text{(during training)},\\ e_k &= z(t) - d(kT) \qquad \text{(during tracking)}. \end{aligned}$$

The equalizer operates first in a training mode, whereby the transmitter sends a predetermined "pseudo-random noise" (PN) sequence which is also known to and generated at the receiver. The resulting error e_k is multipled by the delayed received signal $r(t-nt-kT)$ to determine the magnitude and polarity of the adjustment to the corresponding tap. A fixed constant 's' scales the 'step-size' of each adjustment in order to ensure smooth convergence. After the training mode is completed (about 20ms for 4800 bps modems to several hundred milliseconds for 14.4 kbps modems using a very large number of taps), the equalizer automatically switches to the tracking mode in postion (2) of Fig.10.20 and derives its reference from the reconstructed data $d(t)$ made available by making decisions on the equalized signal $z(t)$. Since the reconstructed data is now derived from an equalized signal with a high noise margin, it is a good margin, good enough to be used in place of the training PN sequences to derive the error, and the equalizer will now adapatively adjust to the slow changes in the channel. This is an example of decision feedback which is used in many other forms in digital communications. An LMS equalizer is stable, fast, and digitally implementable with either LSI chips or a special-purpose digital signal processing (DSP) microcomputer, or even on a general-purpose microcomputer if it is fast enough.

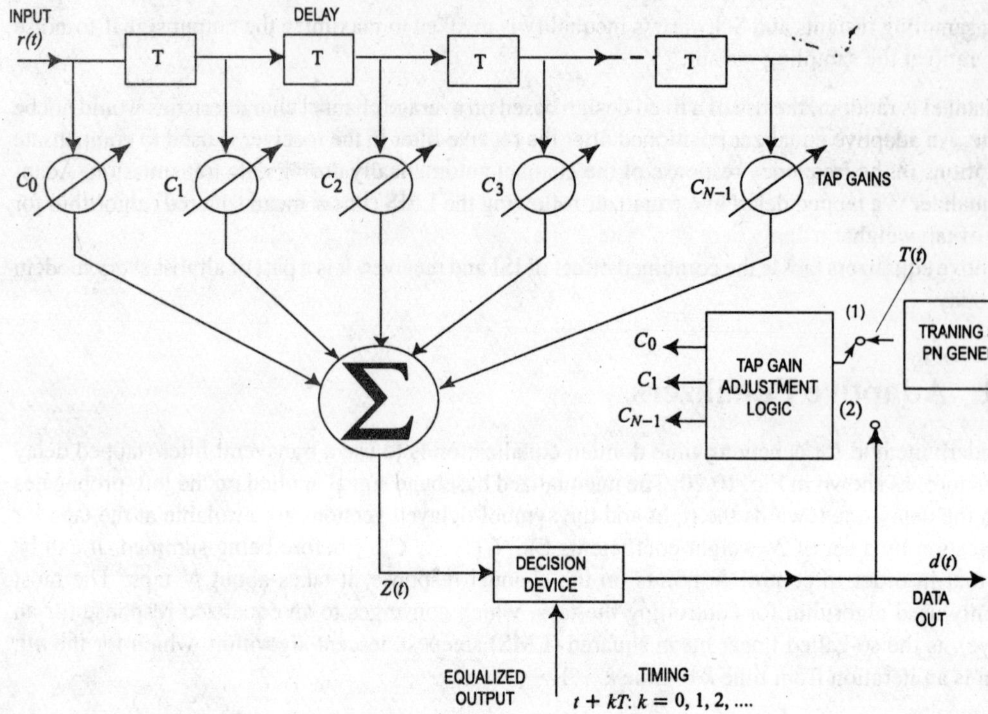

Fig. 10.20 *Adaptive equalizer.*

In modern high speed modems which are implemented almost entirely through digital processing, four adaptive equalizers are usually required to handle the modulated (non-baseband) signal, and the tap gains may require complex multiplication (product of two complex numbers involves four multiplication operations). The adaptive equalizer thus consumes the largest amount of processing resources in a modem. Without it however, the operation or the voice-grade lines above 2400 bps would not be possible.

10.12 Adaptive Echo Canceller

An impairment with fundamentally the same effect as inter-symbol interference due to pulse dispersion is the echo. Echoes are caused by reflections at mismatches in the transmission medium. In voice-grade channels it is at 4W/2W Hybrids. Echoes are particularly troublesome in full-duplex data communications since echo-supressors (voice- or signal-actuated switches) cannot be used. When satellite lines are part of the channel, echoes can cause significant deterioration of the error performance of a modem because of the very large delays (greater than 250 ms), and special techniques are required for mitigating this effect. One of the most promising of these techniques is that of an echo canceller. An echo canceller is an adpative filter which selectively eliminates echoes. It operates on the same principles as that of an adaptive equalizer using a tapped delay line transveral filter whose coefficients are adjusted by an LMS steepest descent algorithm (described earlier). The error for driving the algorithm is obtained from the difference of the echo simulated by the transversal filter and the actual echo.

10.13 Control of ISI in Passband Transmission

A digital MW radio network which is a passband transmission, uses band-limiting filters due to constraints of bandwidth. The non-ideal nature of these filters (as discussed in case of low-pass filters) causes channel dispersion and hence ISI.

Due to selective fading in the MW propagation pattern, the IF signal is distorted. Hence the techniques of adaptive equalization apply here also. The digital MW radio equipment uses 'transversal equalizer' to monitor and supress ISI at the input to PH-DEM module.

(*Note :* There are 2 other equalization operations not concerned with ISI:

- Delay equalization: Equalizes the group delay developed in RF branching (before Rx-IF module).
- Line equalization: Removes the distortion due to interconnection cable (part of Tx-DPU).

Chapter 11 is devoted to the study of these equalizations.)

10.14 Eye Pattern

Eye pattern is an experimental tool to evaluate the effect of impairments due to ISI and noise in an operational environment. It is the synchronized superposition of all possible realizations of the signal of interest viewed within a particular signalling interval. The signal of interest, in general is the received signal or receiver output.

The pattern resembles a human eye for binary waves, hence the name. The interior region of the pattern is called the eye-opening. The eye pattern interpretation is shown in Fig.10.21.
Eye patterns provide the following information:

- The width of the eye-opening defines the time interval over which the received signal can be sampled without errors due to ISI
- The preferred sampling instant is the maximum eye opening point
- The sensitivity of the system to timing errors is determined by the rate of closure of the eye as the sampling time is varied
- The height of the eye-opening at a specified sampling time defines the noise margin of the system
- When the effect of ISI is severe, traces from the upper portion of the eye pattern cross traces from the lower portion, with the result that the eye is almost closed
- In case of M-ary system, the eye pattern contains $(M - 1)$ eye openings, stacked vertically one up on the other
- Non-linearities in the communications channel cause asymmetries in the eye pattern; otherwise they are symmetrical

Eye pattern for received signal with no bandwidth limitation (for $M = 2$ and 4), and with bandwidth limited channels are shown Fig.10.22. Eye patterns before and after adapative equalization are shown in Fig.10.23.

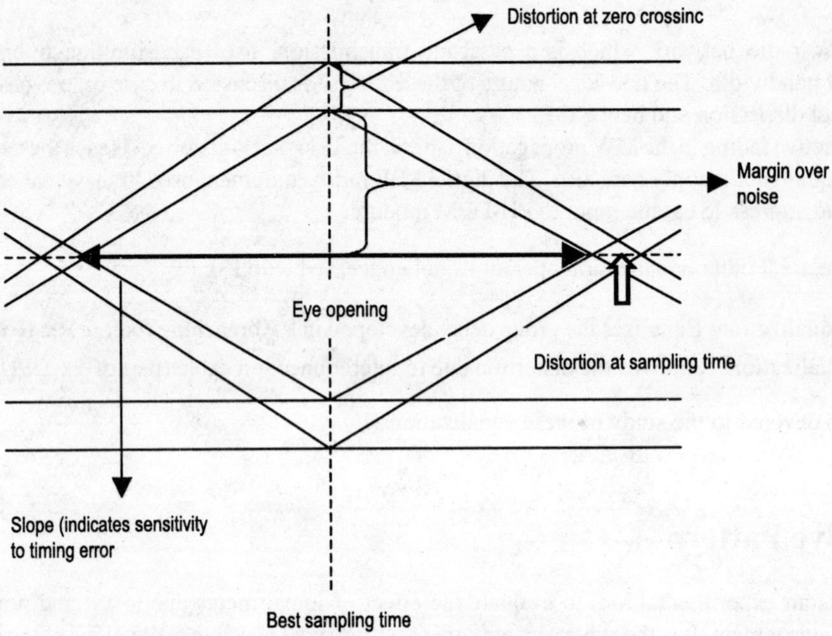

Fig. 10.21 *Interpretation of the eye pattern.*

10.15 Practical Applications of Duo-binary Techniques to Digital MW Radio

The discussion on duo-binary technique initiated in Section 10.7 is not complete without a peep into its practical applications in digital MW radio. Among duo-binary applications for modulations, the practically applicable ones are the co-relative APK, co-relative PSK and quadrature partial response signalling (QPRS). The scope of this book does not permit a full-fledged discussion on all these techniques. However, we shall study the co-relative QPRS which is extensively used in digital radios along with QPSK and 8 PSK modulation schemes.

Co-relative techniques find application in digital radio for four reasons (especially correlative QPRS):

(*i*) The implementation of co-relative QPRS is as simple as QPSK or 8 PSK radio

(*ii*) The problem of ISI is tackled well

(*iii*) QPRS can accomodate more bits per hertz in carrier modulation with only 3 dB more signal power than PSK

(*iv*) Digital service channels can be made completely independent of the main traffic bit stream. This is of great advantage because modulation and re-orientation process can be eliminated at non-dropping repeaters, thus greatly simplifying repeater arrangements

Fig. 10.22 (a,b). *Eye pattern of received signal for $M = 2$ (a) with no bandwidth limitation, and (b) with bandwidth limitation. The four different lines correspond to 4 different realizations of received signal in a baseband M-ary PAM transmission. In case (a), ISI is negligible; values are distinct at the sampling instant; symbol duration and bit duration are equal. In case (b) ISI is more, causing blurring of the pattern and reductio.. in eye opening.*

Fig. 10.22 (c,d). *Eye pattern of received signal for $M = 4$ (c) with no bandwidth limitation, and (d) with bandwidth limitation. The four different lines correspond to 4 different realizations of received signal in a baseband M-ary PAM transmission. Since $M = 4$, there are three eye-openings, symbol duration is twice the bit duration. In case (c), ISI is negaligible and the values are distinct at the sampling instant. In case (d) ISI is more, causing blurring of the pattern and reduction in eye opening.*

10.16 Correlative QPRS Modulation

The QPRS signal can be regarded as two AM-PSK 3-level duo-binary signals (discussed previously) in quadrature, as shown in Fig. 10.24. Each such signal has three states—the extremes at the tip of the vectors representing binary 0 and the centre state (absence of carrier) representing binary 1. When the in-phase and quadrature waveform are combined as in QPRS, the resultant is a set of nine possible states, as indicated in the signal–space diagram. The efficiency achieved is 2 bps/Hz.

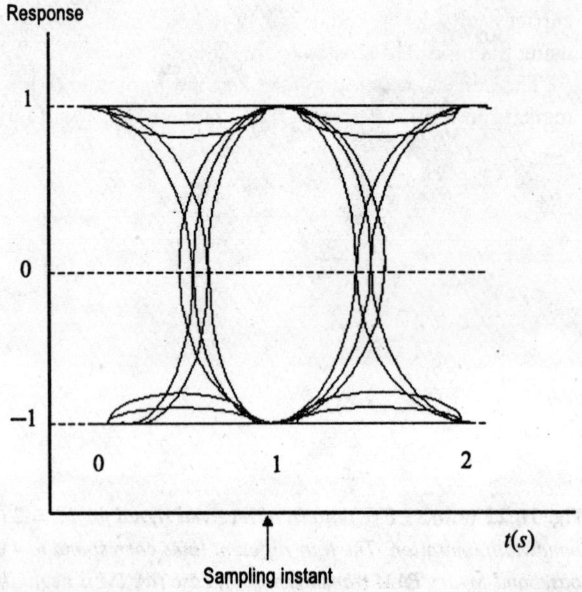

Fig. 10.23(a) *Eye pattern of received signal with adaptive equalization.*

There are inherent symmetries in the QPRS signal that are sufficiently significant in carrier acquisition and coherent domodulation. Such symmetries are apparent in Fig. 10.24. Remembering that the tips of the vectors represent "0", and the center "1", states 2, 4, 6, and 8 always represent the same pattern, namely (0, 0) on both I and Q channels. I and Q stand, respectively, for in-phase and quadrature channels. Further, states 1 and 5 represent (0, 1) and states 3 and 7 represent (1, 0). Thus, for the

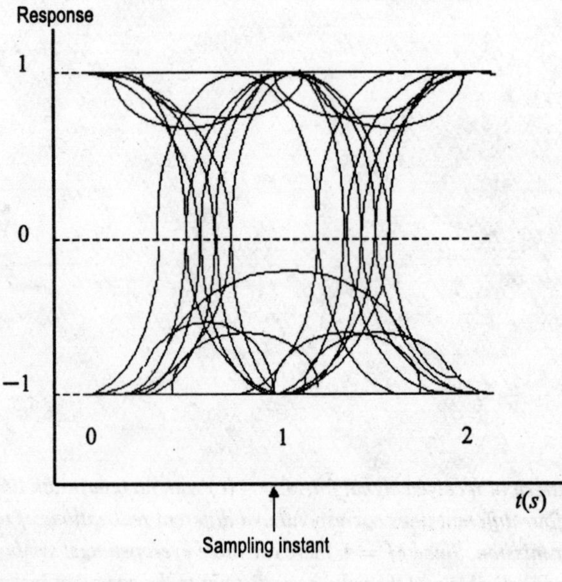

Fig. 10.23(b) *Eye pattern of received signal without adaptiveequalization.*

odd-numbered states (except state 9), a carrier ambiguity of 180° is immaterial. All even-numbered states are invariably (0, 0). Finally, the absence of carrier is represented by state 9 (1, 1). The coherent carrier is always locked to one of the odd-numbered states (except for state 9), and may be acquired by using the modified Costas loop.

The general block diagram and the implementation of QPRS system is very similar to the implementation of a four-phase QPSK modem described in the previous chapter (see also Fig. 10.25). The

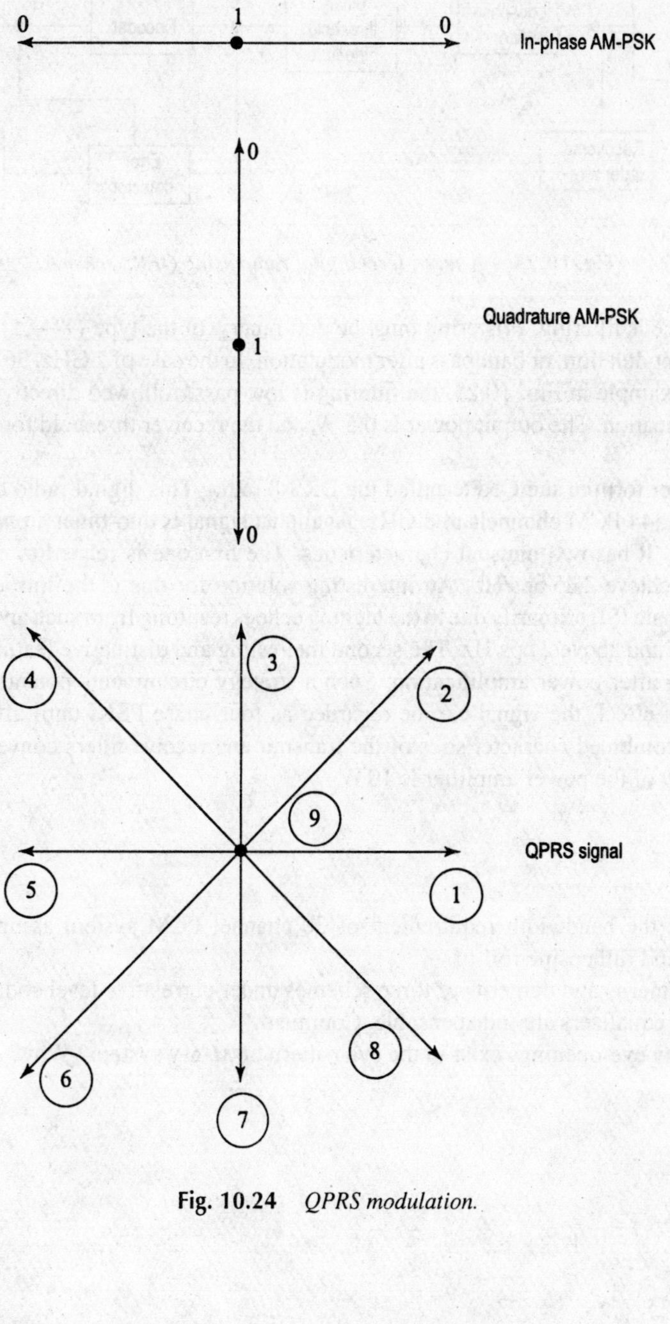

Fig. 10.24 *QPRS modulation.*

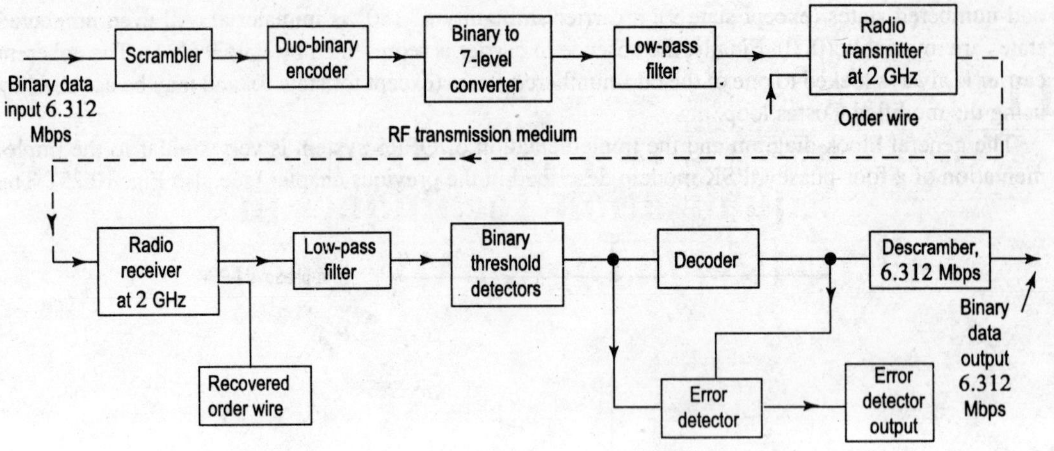

Fig. 10.25 *A seven-level digital radio using QPRS scheme.*

essential difference is filtering. Filstering must be duo-binary, of the type $(1 + \Delta)$, either low-pass in baseband before modulation, or bandpass after modulation. In the case of 2 GHz, 96 PCM channel radio presented as an example in Fig. 10.25, the filtering is low-pass, followed directly by RF modulation and power amplification. The output power is 0.5 W, and the receiver threshold for a 10^{-6} error rate is -82.7 dBm.

There is another form of the QRPS called the DRS-8 radio. This digital radio operates at 91 Mbs, accommodating 1344 PCM channels at 8 GHz. Again, its signal is duo-binary, phase modulated with I and Q channels. It has two unusual characteristics. The first one is related to effective bandwidth compression to achieve 2.25 bps/Hz. An interesting solution for this is the introduction of decision feedback to eliminate ISI, primarily due to the lagging echoes resulting from such an efficient bandwidth compression over and above 2 bps/Hz. The second interesting and distinctive feature is the duo-binary bandpass filtering after power amplification. Such a strategy circumvents potential problems due to non-linearities. In effect, the signal can be regarded as four-phase PSK, until after saturated power amplification of combined characteristics of the transmit and receive filters convert the waveform to QPRS. The output of the power amplifier is 10 W.

Exercises

1. Calculate the bandwidth requirement of 30 channel PCM system assuming ideal Nyquist channel and full cosine roll-of.
2. State the merits and demerits of three schemes under correlative level coding.
3. Adaptive equalizers are indispensable. Comment.
4. How many eye-openings exist in the eye pattern of M-ary system? Why?

11

Equalization Techniques in DMR-770 Digital MW Radio

11.1 Introdution

There are three equalization operations in all as far as the data stream in a DMR 770 digital MW radio is concerned. These are:

Delay equalization: Equalizes the group delay developed in the branching filter (phase velocity pertains to carrier, group velocity pertains to modulating envelopes (please also refer to Chapter 20))
Transversal equalization (at the input to PH-DEM): Suppresses ISI due to distortion in IF signal caused by selective fading
Line equalization: Removes the distortion due to interconnection cable

The equalizers performing these tasks are discussed in this chapter.

11.2 Delay Equalizer (DEL-EQL)

The DEL-EQL equalizes the IF (70 MHz) delay developed in an RF branching filter.

Performance characteristics

Input/output impedance :	75-ohm, unbalanced
Input/output return loss :	15 dB at 70 ± 10 MHz
Delay characteristics	As shown in Fig. 11.1

Principles of operation

The DEL-EQL is a latticed delay equalizer used in the IF (70 MHz) band. It equalizes the group delay developed in an RF branching filter. It is pre-mounted at the factory according to the delay characteristics developed.

Control

No control is provided. (The DEL-EQL is completely factory-adjusted and thus needs no adjustment at site during installation.)

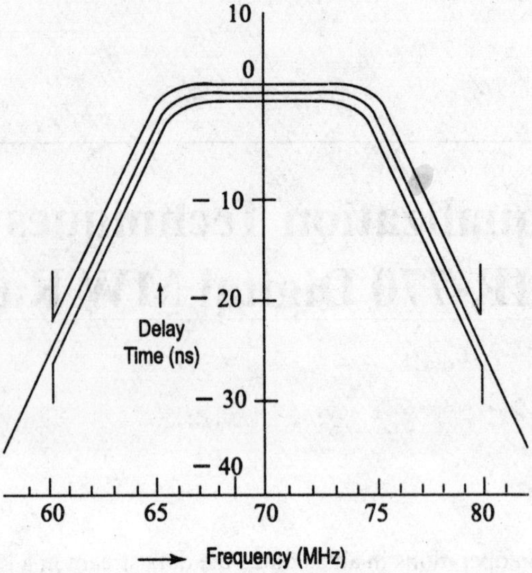

Fig. 11.1 *Delay characteristics of DEL-EQL.*

11.3 Transversal Equalizer (TRANSV-EQL)

The TRANSV-EQL module automatically compensates for both the amplitude and the delay distortion caused by selective fading in the microwave propagation path by monitoring and suppressing intersymbol interference at the PH-DEM module.

Performance characteristics

Transmission bandwidth	:	70 ± 15 MHz
Input/output frequency	:	70 MHz, nominal
Input/output level	:	−3 dBm, nominal
Input/output impedance	:	75 ohm, unbalanced

Circuit operation principles

The working of a transversal equalizer is based on adaptive amplitude and delay equalization by the transversal filter. See Fig. 11.2 for a functional block diagram.

The TRANSV-EQL module consists of a transversal filter circuit, a control signal generator (CONT-SIG-GEN), and an IF amplifier (IF-AMP). The input of the module is 70 MHz with a nominal level of −3 dBm and an impedance of 75 ohms, unbalanced.

The IF signal from the Rx-IF module is fed to the IF-IN jack.

Transversal filter circuit

The transversal filter circuit (variable bandpass filter) has four delay line units with three taps (signal outlet), eight multipliers which control tap weights, two signal adders and a directional coupler

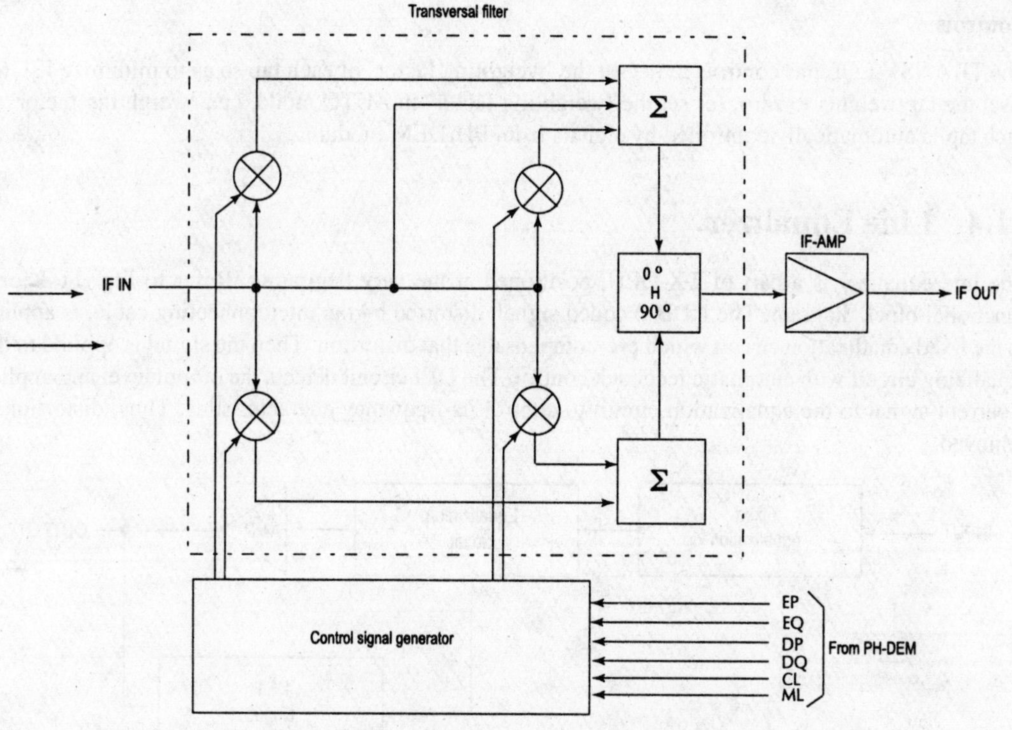

Fig.11.2 *Functional block diagram of TRANSV-EQL.*

(quadratic hybrid). The delay time of each of the delay line units is $1/f_C$ seconds, where f_C is the clock frequency.

In normal (undistorted) conditions, the IF signal traversing the central line path in Fig.11.2 shares the majority of the signal component. When the IF signal is distorted by selective fading in the transmission path, the upper part of the circuit eliminates the in-phase distortion, and the lower part compensates for distortion caused by the quadrature components. The tap weights are controlled with the help of signals from the control signal generator.

Control signal generator

The control signal generator supplies four control signals to the multipliers of the transversal filter. These control signals are generated by the data signals and the error signals of each *P* and *Q* channel from the PH-DEM module. The control signals are reset with a reset signal detected during carrier synchronization loss over maximum level error at the PH-DEM module.

IF amplifier

The equalized IF signal from the transversal filter circuit goes to the IF amplifier where the attenuated IF signal is amplified to −3 dBm at an impedance of 75 ohms, unbalanced.

Controls

The TRANSV-EQL has controls to (*a*) set the 'weighting factor' of each tap so as to minimize ISI, (*b*) reset the tap weights to zero, (*c*) set the 'weighting factor' in AUTO mode, i.e., 'weighting factor' of each tap is automatically controlled by signals from PH-DEM module.

11.4 Line Equalizer

The line equalizer is a part of TX-DPU, positioned at the very beginning. Refer to Fig. 11.3 for a functional block diagram. The HDB-3 coded signal, distorted by the interconnecting cable, is applied to the fixed equalization circuit which pre-compensates that distortion. Then the signal is applied to the equalizing circuit with automatic feedback control. The DEI circuit detects the output level and applies a current signal to the equalization circuit to control its frequency characteristics. Thus, distortion is removed.

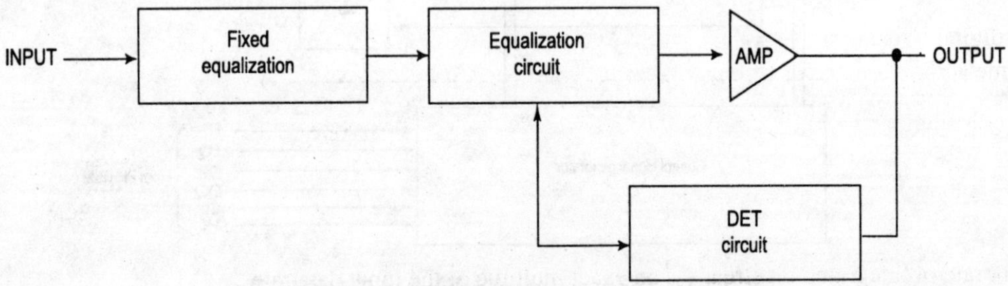

Fig. 11.3 *Functional block diagram of line equalizer.*

Reference

[1] DMR 770 34 Mb Digital Microwave Radio Manual, Chapter 6, Volume 1, M/s IT1 Ltd, 1996.

Exercises

1. Why does group delay develop in RF branching filter?
2. What is the reason for ISI (intersymbol interference) in IF signal?

12

Bit Stream Integration in Digital Transmission Systems

12.1 Introduction

In digital transmission systems, particularly the multiplexing equipment and the radio equipment, we come across three types of bit stream integration operations:

- Multplexing of synchronous digital signals
- Multplexing of asynchronous digital signals
- Integration of digital streams of different data rates

In the first type, the digital signals to be multiplexed are synchronised to one clock (master clock). The data rate of integrated bit stream is an exact multiple of the input data rate.

In the second type, tributaries, nominally at the same data rate but varying within specified limits are integrated. The data rate of the integrated bit stream is not an exact multiple of the input data rate.

In the third type, data streams at different data rates are integrated to form the master stream.

With the exception of the first type, bit stream integration is not a simple operation consisting of the straightforward multiplication or addition. The processes followed are complicated (essential to serve the purpose of data retrieval), and in this chapter we shall examine the various bit stream integration processes.

12.2 Multiplexing of Synchronous Data Signals

If the various bit streams to be integrated are synchronized to one clock (called the master clock) with a bit rate of T bps, then multiplexing n such bit streams gives an output bit stream with a rate of nT bps. The methods adopted for such a multiplexing are:

(*a*) *Block interleaving:* Information is taken as a block at a time from each tributary. The memory required is large.

(*b*) *Bit interleaving:* One bit is taken at a time from each tributary and fed to the output stream in a cyclic order. The memory required here is very small.

At the demultiplexer end it is necessary to recognize which bit of information belongs to which tributary. This is achieved by transmitting a fixed code called a 'frame' after a fixed number of information bits. This code is called the 'frame alignment signal'.

This method of multiplexing is easy but not reliable. If any deviation in the normal bit rate of the tributaries occurs, or there is any problem in data extraction from one or more of the formative tributaries, there would be a loss of information.

12.3 Multiplexing Asynchronous Signals

The bit streams being integrated in this process are at the same bit rate but varying within limits of tolerance, i.e., their corresponding significant instances occur nominally at the same rate, and any variation in rate is within the specified limits of tolerance.

Consider for example the formation of an 8 Mbps stream (as per CCIT Rec. G-742) from 2Mbps streams. The input streams are derived from different MUXes (i.e., different clock sources) at data rates of 2.048 Mbps $\pm$ 50 ppm. This means that although the input bit streams are at the same data rate with variation within tolerance limits, their corresponding significant events do not occur at the same instances. The bits stream are, in short, asynchronous. In such cases, it is not sufficient to just perform bit interleaving alone while forming the output stream—for ensuring that there is no loss of information the plural asynchronous streams have all to be synchronized to a common clock. The method adopted for achieving this is bit interleaving, together with 'justification'. Justification implies changing the bit rate of digital signals in a controlled manner. This process of integration is called integration of plesiochronous bit streams.

Justification can be of three types:

Positive justification: The common synchronization bit rate offered at each tributary is higher than the bit rate of individual tributaries.

Positive–negative justification: The common synchronization bit rate offered is equal to the normal value.

Negative justification: The common synchronization bit rate offered is less than the normal value.

In the integration processes associated with CCITT heirarchy based on 30 channel PCM (2.048 Mbps), positive justification is adopted.

12.4 Retiming by Justification

12.4.1 Justification process

In Fig. 12.1, T_1, T_2, T_3, and T_4 are the four plesiochronous bit streams to be integrated. The circuitry associated with T_1 alone is shown. The input pulse train T_1 is written in the memory MEM-1. The writing pulse train C is derived from T_1 using a clock extraction circuit. It has bit rate of f.

The written information is read-out with sufficient phase lag with respect to the time of writing in. Also, the reading clock has rate $f + \Delta f$ such that Δf is more than the allowable tolerance in f. In fact, we do not directly generate $f + \Delta f$. The output rate for 'n tributaries' equated is to $n(f + \Delta f)$, implying that the read-out rate per tributary from the memory is $n(f + \Delta f)/n$. We have a clock generator at frequency $n(f + \Delta f)$, which is divided by n to get the individual read-out clock for each tributary information.

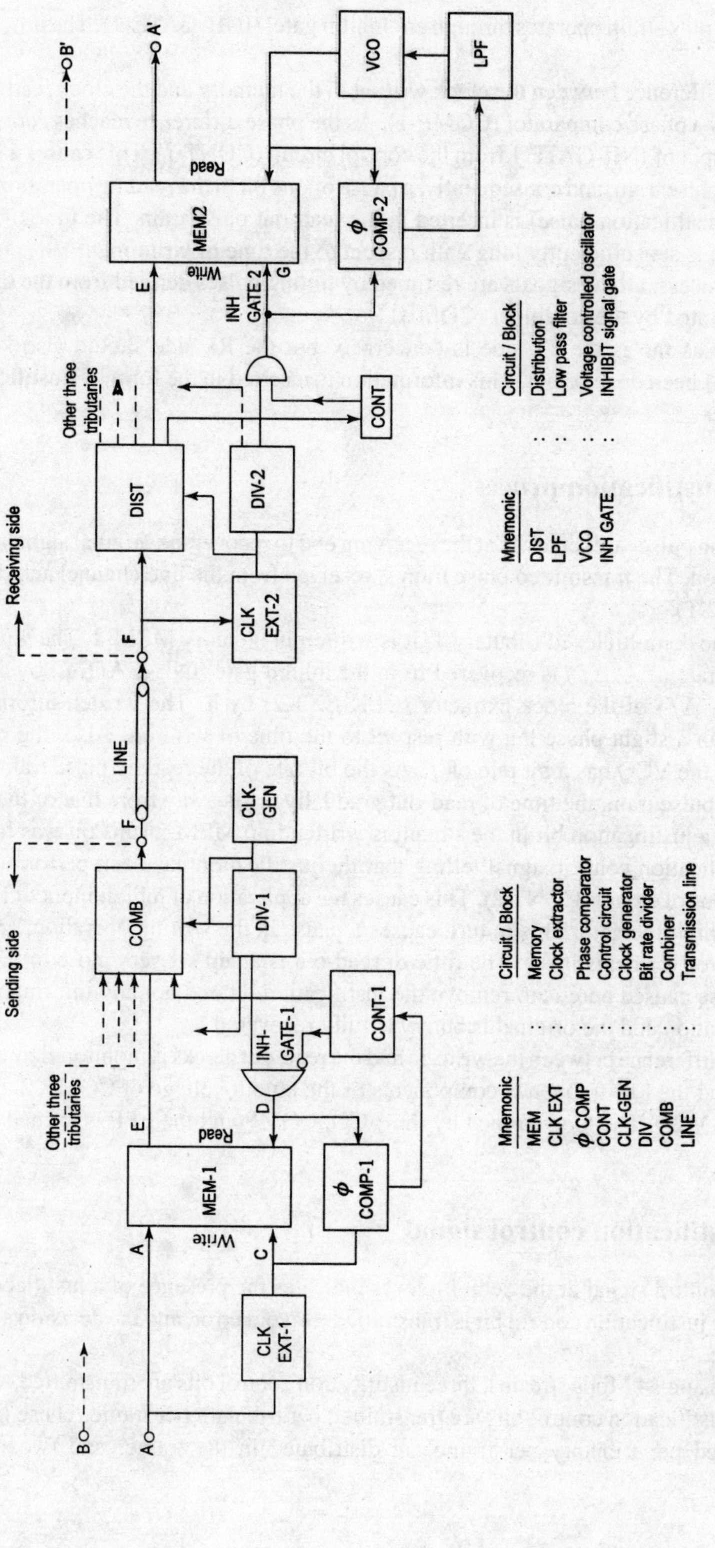

Fig. 12-1 *Justification process.*

The reading pulse train operates through an 'Inhibit gate' (INH-GATE-1). The process of justification is as follows.

The phase difference between the clock writing in the memory and the clock reading out of memory is monitored by a phase comparator (COMP-1). As the phase difference reaches zero, a pulse is applied to the inhibit input of INH-GATE 1 from the control circuit (CONT-1). This causes a pause of one cycle in the reading-pulse train, and consequently, a pause of one bit in the reading operation. An information-less pulse (or justification pulse) is inserted in the read-out pulse train. The time of read-out (i.e., the read-out clock) is set sufficiently long with respect to the time of write-in.

With this process, all the signals are re-timed by timing pulses derived from the CLK-GEN and can be easily integrated by the combiner (COMB).

This is fine as far as the Tx side is concerned, but the Rx side should also know whether any justification has been done or not. This information is inserted in the form of 'justification control bits' in the combiner.

12.4.2 Dejustification process

The justification pulses are removed at the receiving end to recover the original signals through a process of dejustification. The transmitted pulse train is received from the line/channel and demodulated at the distributor (DIST).

Consider the demultiplexed tributary 1. It is written in memory MEM-2. The writing clock for this operation at a rate $(f + \Delta f)$ is recovered from the inhibit gate (INH-GATE-2) by dividing the output bit rate $n(f + \Delta f)$ of the clock extractor (CLK-EXT-2) by n. The written information is read out of MEM-2 with a slight phase lag with respect to the time of write-in. Also, the reading pulse train obtained from the VCO has a bit rate of f. As the bit rate of the reading pulse train is lower than that of the writing pulse train, the time of read-out gradually drifts away from that of the write-in.

Just before a justification bit in the stream is written into MEM-2 (this bit was latched in at the Tx end), the justification control signal telling that the justification has been performed is applied from DIST to the control circuit (CONT-2). This causes the application of inhibit input of INH-GATE-2 from CONT-2 to inhibit the gate. This in turn causes a pause in the writing operation, and the justification pulse is removed (or dejustified). The time of read-out is again set very close to the time of write-in. Since the pause caused here is to remove the justification bit and not any information bit, there is no loss of information and the original tributary is fully recovered.

The phase difference between the write-in and the read-out clocks is monitored by a phase comparator (COMP-2), and the low frequency components of the output voltage of COMP-2 are applied through an LPF to the VCO. The loop formed by the VCO, COMP and the LPF is termed the 'phase control loop'.

12.4.3 Justification control signal

Justification control signal at the demultiplexer indicates the presence of a justifiable bit in the frame. More than one justification control bit is transmitted to avoid error, and the decision at the demultiplexer is by majority.

For 8 Mbps and 34 Mbps streams, three justification control bits are transmitted, and for a 140 Mbps system, five justification control bits are transmitted per tributary per frame. These justification control bits transmitted per tributary per frame are distributed in the frame, and two to three errors are required to cause false information of justification which will lead to a loss of frame alignment in lower hierarchies.

12.4.4 Summary of justification process

The information from a tributary is written in an elastic memory with the tributary clock derived from the incoming signal. The elastic memory is read-out by a clock faster than the write-in clock. The reading clock is derived from the common transmit clock by dividing it with the number of tributaries. Gaps are laced at locations where non-information bits occur in the frame structure, viz., the frame alignment bits, the justification control bits and the service digits.

Since the read clock is operating faster than the write clock, it is required to stop the read clock for a bit and insert the non-information bit (in the form of justification bit). The information when read out from the memory contains information bits, justification control bit and justification bit.

> *The justification bits are inserted at the instant when the linearly increasing phase difference crosses a threshold level. This threshold is selected in such a way that the average rate of read clock is equal to that of the write clock.*

In the demultiplexer, the clock timing of the input multiplex signal enables a control on the timing of operations. The detection frame alignment signals enable the receive frame to be aligned with the transmit frame, which, in turn enables the receiver to demultiplex the tributary information. The tributary information is written in an elastic memory as in the transmit tributary by the clock derived from the received clock. A phase-locked oscillator is used to read the elastic memory with a timing rate equal to that of the corresponding tributary signal at the input of the multiplexer.

12.5 The Perforated Clock

The concept of a perforated clock is of vital importance to the design of bit stream integration scheme. Let us examine it with the example of an 8 Mb MUX.

The frame structure for II-order digital MUX, i.e., 8 Mb MUX (CCITT Rec G.742) is shown in Fig. 12.2.

The 8 Mb MUX (integrating four 2 Mbps streams) has to first send twelve bits on the line as FAS and service bits. During the time these bits are sent by the MUX, no data should be read from the tributaries, i.e., no data should be read from the elastic memories. This is accomplished by suppressing the read clock at appropriate time intervals.

From the frame structure we know that the first twelve bits correspond to three bits of each tributary. Thus the clock is suppressed for a duration equal to that of the first three bits, and then, after 50 pulses, a 'justification control bit' (JC1) is due. The read clock is suppressed for 1-bit duration to facilitate the MUX to insert the JC bit. Again, after 52 bits, JC2 is due; after another 52 bits JC3 is due; then time-slot 155 (TS-155) appears. TS-155 carries a clock pulse, and data will be read from the elastic store if the JCs $(1, 2, 3) = 000$. Otherwise, the clock will be suppressed for TS-155 also (TS-155 corresponds to the justification-bit slot). Thus, the clock is perforated in slots corresponding to FAS service digits, JCs and justification bits.

The perforated clock is shown in Fig. 12.3.

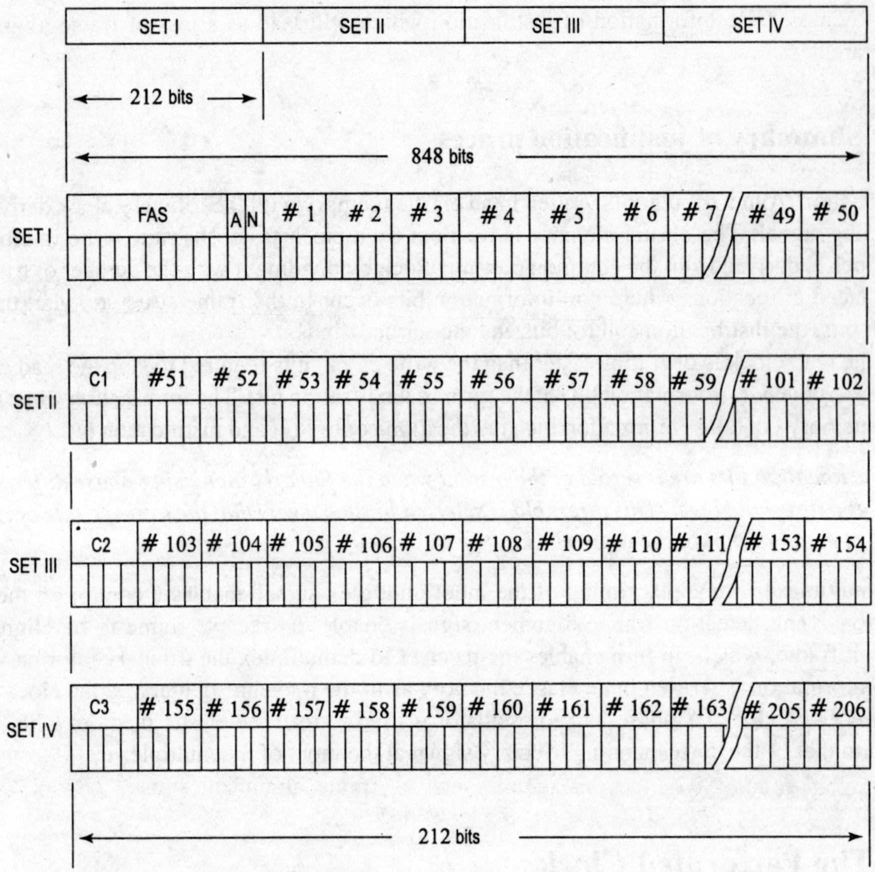

FAS:	Frame alignment signal (1111010000)
A:	Remote alarm bit (0 = Normal, 1 = Alarm)
N:	This bit is reserved for future use, always set to 1.
# 1 to # 206:	First bit of each group is assigned to Tributary 1, second bit of each group is assigned to Tributary 2, third bit of each group is assigned to Tributary 3, and the fourth bit of each group is assigned to Tributary 4.
C1 to C3:	Groups for carrying the justification control bits of each tributary.
# 155:	This group carries the justification bit of each tributary. If bits C1 to C3 of a tributary are 000, its bit in this group carries data. If they are 111, its bit in this group carries remote loopback information; a 0 indicates a request for loopback and a 1 indicates no request.

Fig. 12.2 *Frame structure for II-order digital MUX.*

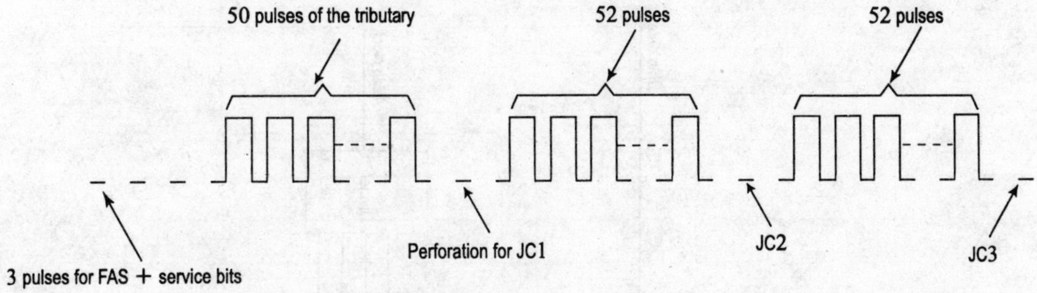

Fig. 12.3 *Perforated clock.*

12.6 Intergration of 2Mb Streams in II-Order Digital MUX

12.6.1 Integration of tributaries

Integration of four 2 Mbps (2.048 Mbps) tributaries to an 8.448 Mbps stream in II-order digital MUX and the retrieval of tributaries are discussed as examples of multiplexing asynchronous (plesiochronous) data streams. This would make the entire process clearer.

Refer to Fig. 12.4. An 8.448 Mbps clock is divided by four to supply a 2.112 Mbps clock for all the four tributaries. Each 2.112 Mbps clock is made out-of-phase by a quarter with respect to the previous one. These clock signals are used for reading data out of the 8-bit buffer memory. Only one channel is shown in Fig 12.4.

A gate through which the 2112 kHz clock signal passes is controlled by a multiplexer and a stuffing circuit. When overhead bits such as frame alignment signal, remote alarm bit, future use bit, justification control bit, and justification bit are inserted into the 8 Mb data stream, the pulses corresponding to these bits in the 2112 kHz clock signal are stopped. This results in an interrupted 2112 kHz clock, and is designated as the 2 Mb transmit read clock. Because the outputs of both transmit read counter and transmit write counter are compared, the 2 Mb transmit read clock has the same nominal information rate as the incoming data.

The 11-bit buffer memory is assigned to each 2 Mb tributary. The tributary data is called "2Mb data" and is written into the 11-bit buffer memory with a 2 Mb transmit write clock (2048 kHz $\pm$ 50 ppm) that is extracted from the same tributary data. The information is loaded and unloaded at different rates. The loading (writing) rate is constant while the unloading (reading) rate, even though it has a higher bit rate than the loading rate, is interrupted.

The 2 Mb transmit write clock and 2 Mb interrupted read clock have a relatively low accuracy (30 ppm for the read clock). In order to compensate for these variations, a phase comparator with a set threshold is used to detect the phase difference between the two clocks. Once the phase difference reaches the set threshold, the read clock is close to re-reading a previous bit. To avoid this, the phase comparator sends a stuff demand signal to the stuffing circuit when the phase difference reaches a set threshold. Upon its reception, the stuffing circuits sends "111" as the justification control bits (JCB) of the following frame. After the JCBs are sent, the stuffing circuit controls the rate so as to stop the 2112 kHz clock pulse for the time-slot # 155 (TS-155) of that tributary. It may be noted from Fig. 12.4 that each time-slot carries four bits—one bit for each tributary. TS-155 has data if the JCBs are 000, and justification bit

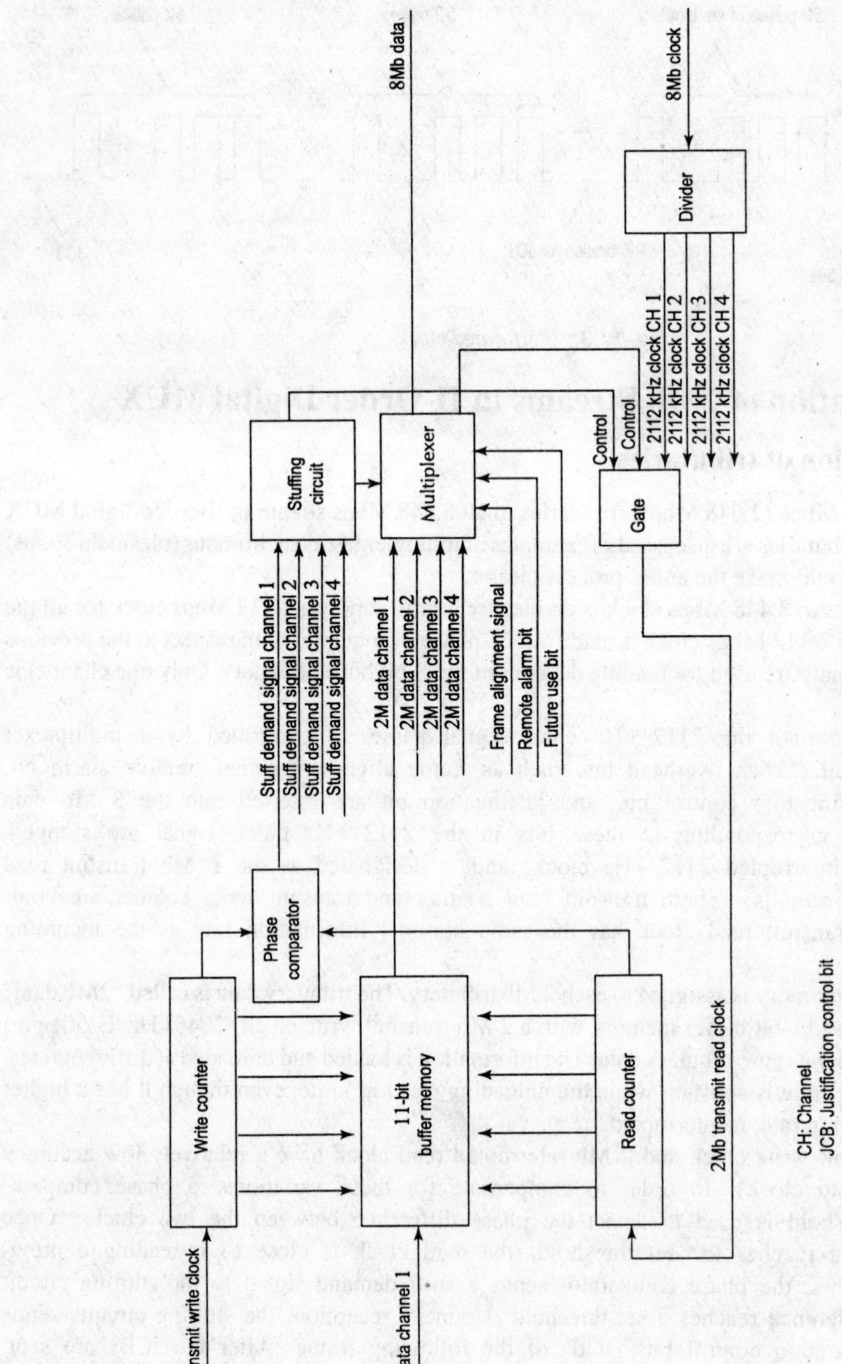

Fig. 12.4 *Intergration of 2Mb streams in II-order digital MUX.*

if JCBs are 111. This rule applies to each bit of TS-155 as per JCBs of the corresponding tributary.

12.6.2 Retrieval of tributaries

Refer to Fig. 12.5. The demultiplexer locates the synchronization word in the incoming 8Mbps signal, recovers the 8.448 Mpbs clock, and converts the input signal into four paralled streams of 2.112 Mbps each. The 2.112 Mbps data stream is written into an elastic store with the help of a write clock at 2.112 MHz (Rx clock/4).

The write clock is a 2.112 MHz perforated clock. The perforations are made at specific durations corresponding to FAS bit position, JC and justification bit slots. Since the clock is interrupted, the data is effectively written into the elastic store at a nominal rate of 2.048 Mbps even though the input to the elastic store is at a higher rate.

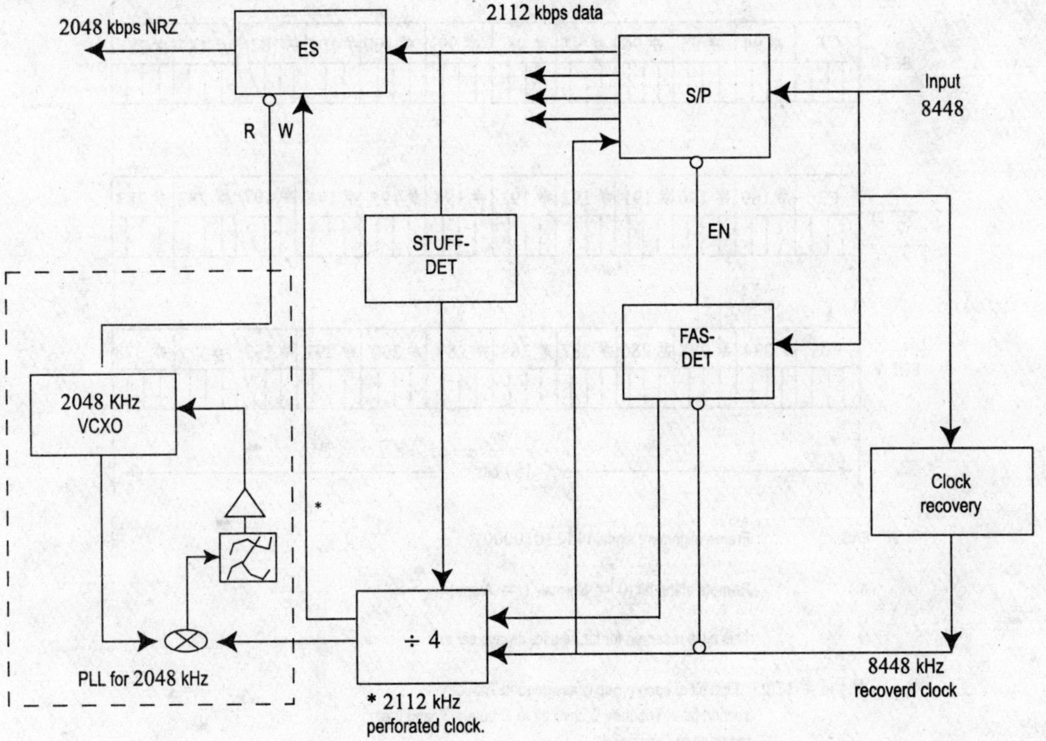

Fig. 12.5 *Retrieval of tributaries in II-order MUX.*

It may please be noted that the operation done here is the reverse of what was done at the multiplexer. In the multiplexer, data was read from the elastic store by providing perforations in the read clock.

The read clock for the elastic store is obtained from a 2.048 MHz VCXO which is phase-locked to the 2.112 MHz clock. The data can then be taken to the 2 Mbps interface for HDB-3 coding and onward transmission.

12.7 Integrating 8 Mb Streams in III-Order Multiplexer

The 34 Mbps (34.368 Mbps) frame structure of a III-order multiplexer is shown in Fig 12.6. Each frame has 1536 bits. Incoming tributaries are at 8 Mbps rate.

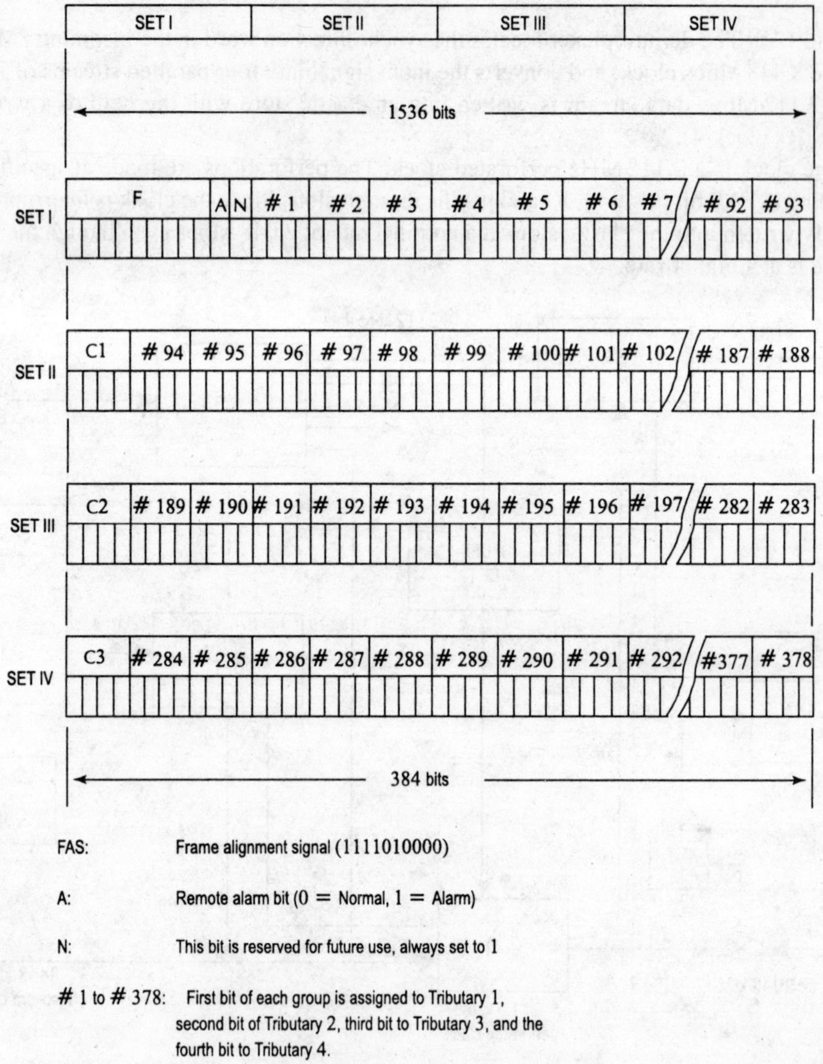

FAS: Frame alignment signal (1111010000)

A: Remote alarm bit (0 = Normal, 1 = Alarm)

N: This bit is reserved for future use, always set to 1

1 to # 378: First bit of each group is assigned to Tributary 1, second bit of Tributary 2, third bit to Tributary 3, and the fourth bit to Tributary 4.

C1 to C3: Groups for carrying the justification control bits of each tributary.

284: This group carries the justification bit of each tributary. If bits C1 to C3 of a tributary are 000, its bit in this group carries data. If they are 111, its bit in this group carries remote loopback Information; a 0 indicates a request for loopback and a 1 indicates no request.

Fig. 12.6 *Frame structure of III-order MUX.*

Bit stream integration process is similar to that in II-order MUX. Only the frame size and clock rates are different. The data read in clock is 8.592 MHz. Incoming data is at 8.448 Mbps ± 30 ppm. Each frame has 378 time-slots with 4 bits per slot; FAS and service digits are the first twelve bits of a frame; justification bits are in TS-284.

A block diagram of bit stream integration is shown in Fig 12.7.

It may be noted that TS-284 is also used for remote loopback at 8 Mb level (other than data and justification bits). If JC_1, JC_2 and JC_3, are 000; the bit in TS-284 carries data; if they are 111, the bit in TS-284 carries remote loopback information. A '0' indicates request for loopback and a '1' indicates no request.

12.8 Integration of Digital Streams of Different Data Rates

12.8.1 Where is it used in digital transmission?

This process is resorted to in digital radio equipment where we integrate 34 Mbps main data, 2 Mbps WS data and 72 kbps DSC data.

Please refer back to the complete block diagram of 34 + 2 Mbps DMR 770 digital MW radio equipment as shown as Fig 6.3(c) in Chapter 6. The Tx-DPU module integrates main 34 Mbps data, WS data and DSC data. Here also, elastic memory is used. The read clock is at a higher rate than the write clock. The read clock is perforated with a time gap ratio of 1 bit in every nine bits.

We know that in DMR 770 radio equipment, due to QPSK modulation the incoming 34.368 Mbps data is split as two 17.184 Mbps streams. So the write-in rate is 17.184 Mbps. Read-out rate is 17.184 × (1 + 1/8), which equals 19.332 Mbps.

The Tx-DPU has line equalizer, HDB-3-to-unipolar converter, timing extractor, multiplexer, transmitting switch, and alarm and control circuits. We will discuss only the multiplexer (and demultiplexer of Rx-DPU) in this section.

12.8.2 Multiplexer in Tx-DPU

The block diagram of multiplexer in Tx-DPU module shown in Fig. 12.8 (refer to Fig 6.3(c)also) has the following sub-units:

- Timing pulse generator
- Write/Read clock counter
- Stuff control circuit
- Buffer memory
- Scrambler
- Parity counter
- ID code selection

These units are described now.

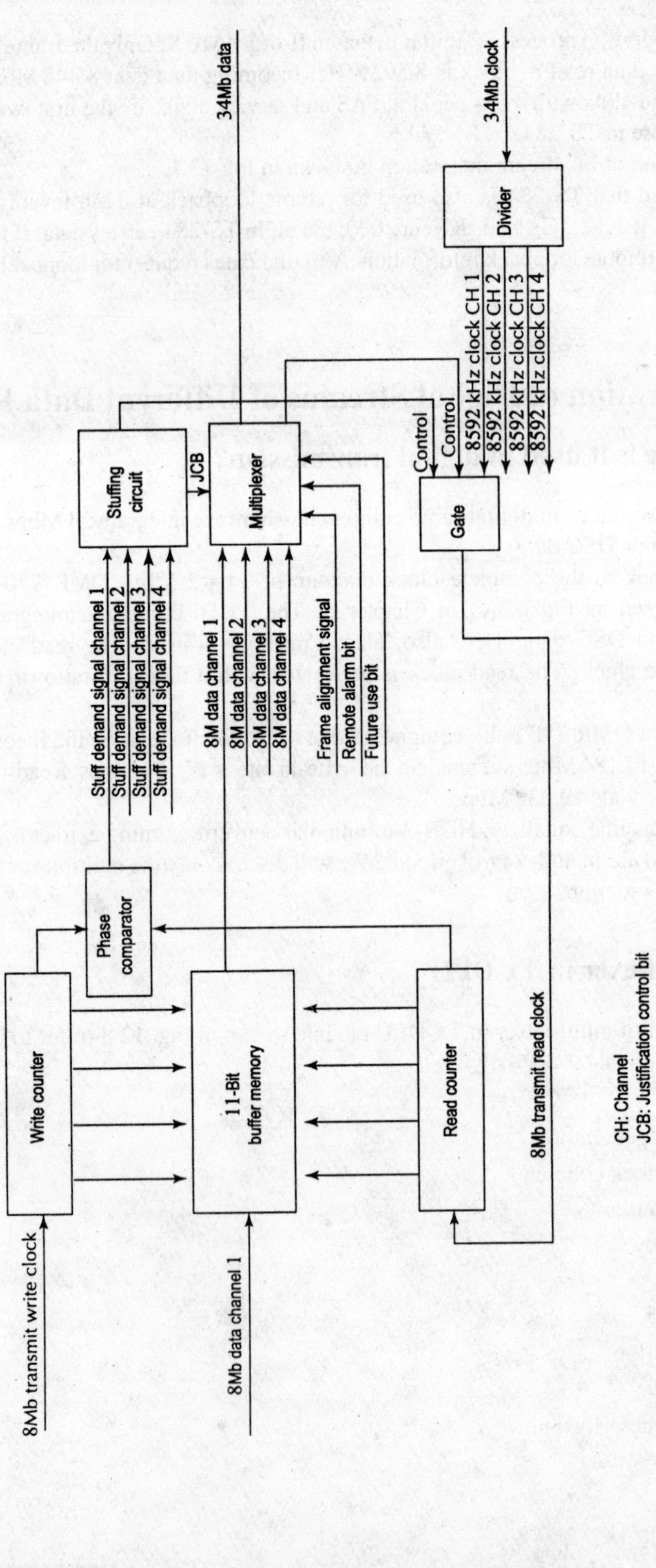

Fig. 12-7 *Integration of 8Mb streams in III-order MUX.*

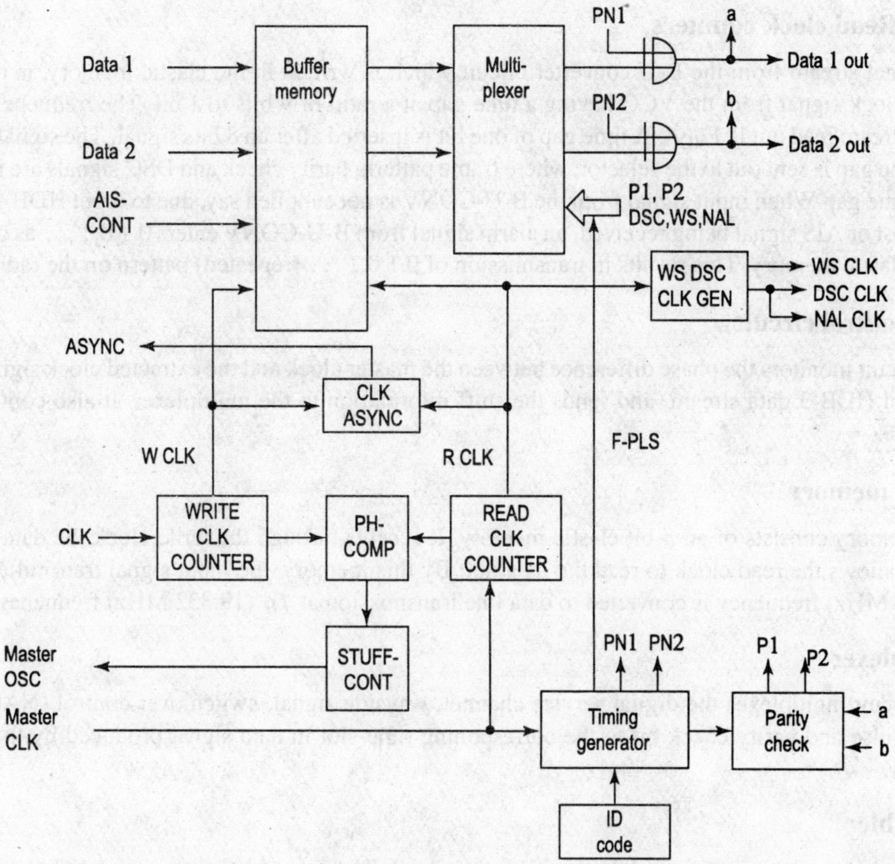

Fig. 12.8 *Block diagram of multiplexer in Tx-DPU.*

Timing pulse generator

The timing pulse generator section consists of a 432 bits random pattern generator, 1/2 frequency dividers and 1/6 frequency dividers. It produces the frame pattern signals inserted between time gaps of transmitted signals and the timing pulse needed in the time gap generator. The 432 bits random pattern generator produces a scramble pattern and subframe pulses (1/72). The 1/2 frequency divider produces control signals for parity check, and drives the 1/6 frequency divider. The 1/6 frequency divider outputs frame pattern signals FSI and FS2 (Fig. 12.9).

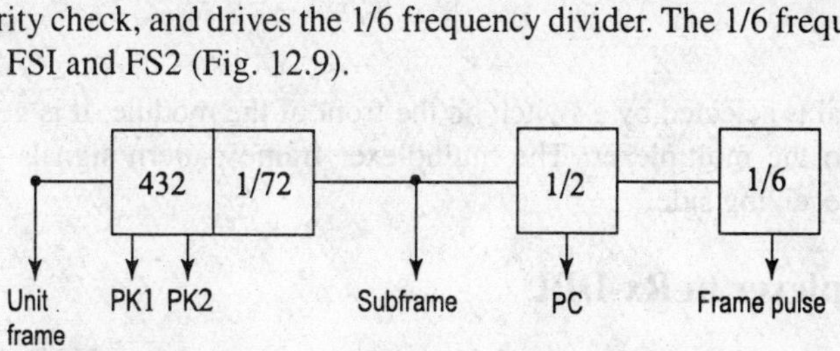

Fig. 12.9 *Timing pulse generator.*

Write/Read clock counters

The signal stream from the B-U converter circuit which is written in the elastic memory, is read out with a clock signal from the VCO having a time gap at a ratio of 9 bits to 1 bit. The frequency of the signal stream read out is F_{VCO}. A time gap of one bit is inserted after an 8 bits signal. The signal stream with time gap is sent out to the selector, where frame pattern, parity check and DSC signals are inserted in the time gap.When input signal from the B-U-CONV is not supplied say, due to input HDB-3 signal being lost or AIS signal being received, an alarm signal from B-U-CONV enters 0.1.0.1 . . . as contents of the elastic memory. This results in transmission of 0.1.0.1 . . . (repeated) pattern on the radio path.

Stuff control circuit

This circuit monitors the phase difference between the master clock and the extracted clock signal from the input HDB-3 data stream, and sends the stuff information to the multiplexer. It also controls the stuff rate.

Buffer memory

This memory consists of an 8-bit elastic memory. It accepts through the write clock the data written in, and allows the read clock to read the data out. By this memory, the input signal transmitted at f_L (17.184 MHz) frequency is converted to data line transmission at f_H (19.332 MHz) frequency.

Multiplexer

This circuit multiplexes the digital service channel, wayside signal, switch over control (NAL data), frame pulse and parity check bit to the corresponing time-slot in data signal produced by the buffer memory.

Scrambler

The scrambler consists of a modulo-2 arithmetic circuit. The scrambled pattern is a 432 bits random pattern produced by the timing pulse generator. The scrambled signal stream is sent out to the 4 PH-MOD.

Parity counter

By using the pulses supplied from the timing pulse generator, this counter calculates data in the modulo-2 for every bit of the original data signals simultaneously for two data streams, and multiplexes the result to the parity bit.

ID code selection

A frame pattern signal is selected by a switch on the front of the module. It is generated in the timing generator and sent to the multiplexer. The multiplexer frame pattern signals are used for channel identification at the receiving side.

12.8.3 Demultiplexer in Rx-DPU

The Rx-DPU comprises a demultiplexer, a delay adjustor, a unipolar-to-HDB-3 converter and alarm circuits. We will discuss the demultiplexer alone. The block diagram of demultiplexer of Rx-DPU is shown in Fig. 12.10. It has the following sub-units:

- Timing pulse generator
- Descrambler
- Write/Read clock counters
- Comparator
- Buffer memory
- AIS detector

These sub-units are described now.

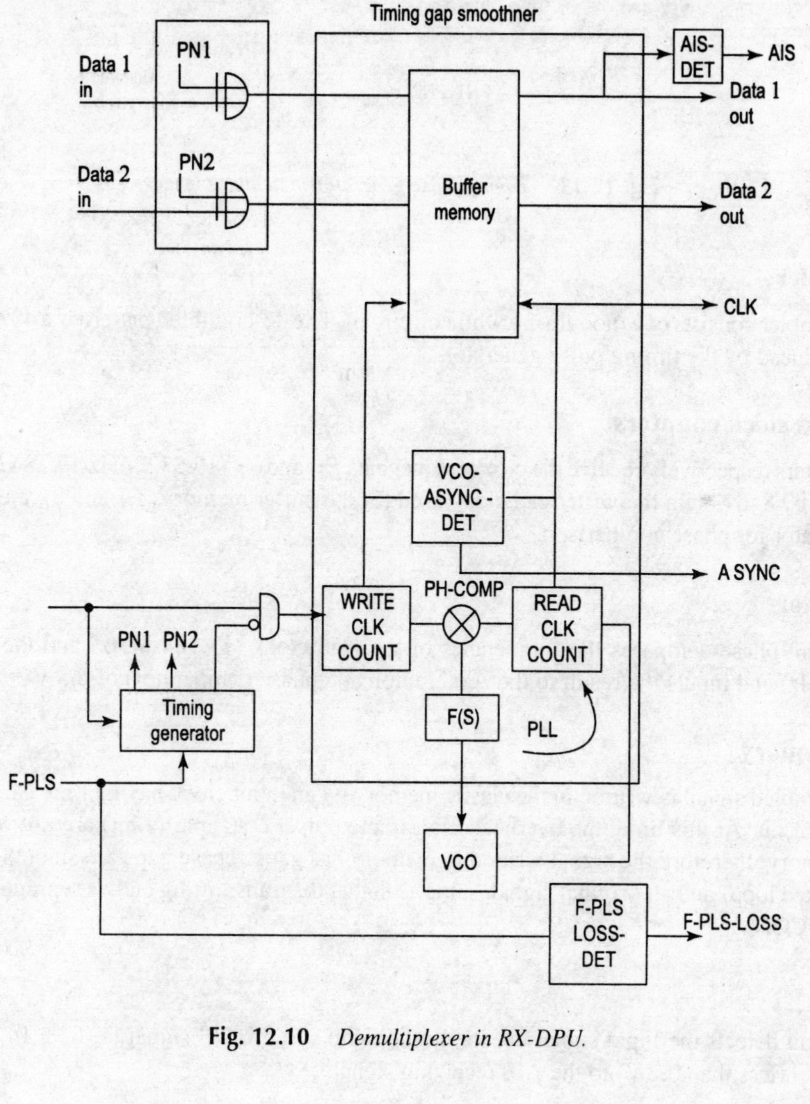

Fig. 12.10 *Demultiplexer in RX-DPU.*

Timing pulse generator

The timing generator has a 432-bit random pattern generator, controlled by an F-PLS signal from the PH-DEM module (or BIT-COMB module for hitless switching). It produces descrambled patterns (PN1 and PN2) and unit frame pulses (Fig. 12.11).

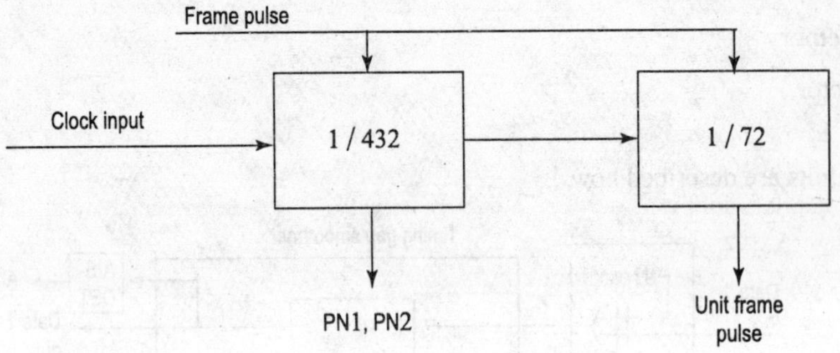

Fig. 12.11 *Timing pulse generator in demultiplexer.*

Descrambler

The descrambler consists of a modulo-2 arithmetic circuit. The descrambled pattern is a 432-bit random pattern produced by the timing pulse generator.

Write/read clock counters

These counters respectively receive the clock frequencies F_H and f_L (19.332 MHz/17.184 MHz) which are divided by 8 to obtain the write/read clock used for the buffer memory. f_H and f_L are also sent to the comparator for phase comparison.

Comparator

This sub-unit phase-compares the frequencies of the read clock (17.184 MHz) and the write clock (19.332 MHz) and inputs the result to the APC (automatic phase comparator) of the VCO.

Buffer memory

The descrambled signal is written in the elastic memory by an input clock having time gaps at the ratio of 9 bits to 1 bit. At this time, the overhead bits (frame pulse, DSC, parity bit) are not written in the elastic memory; therefore the signal written in memory has gaps. These gaps are smoothed by a PLL (phase-locked loop) and an original signal, same as that at the transmitting end, is reproduced and sent to the U-B-CONV.

AIS-DET

This sub-unit detects the input signal becoming the radio system AIS signal (1, 0, 1, 0), and outputs the low level (less than 0.2 V) to the AIS receiving signal.

References

[1] CCITT Rec. G. 742 — Frame structure for II-order digital multiplexing.

[2] CCITT Rec. G. 751 — Frame structure for III-order digital multiplexing.

[3] *DMR 770. 34 Mb Digital MW Radio Manual*, Vol.1, Chapter 6, M/S ITI Ltd., Bangalore, 1996.

Exercises

1. What is the maximum number of perforations in the clock used to read data from elastic memory in II-order MUX? What is the minimum number?
2. What is the maximum number of perforations in the clock used to read data from elastic memory in III-order MUX? What is the minimum number?
3. What is the purpose of scrambler in Tx-DPU?
4. What are the input and output data rates associated with buffer memory in the multiplexer of Tx-DPU?

13

Line Codes

13.1 Introduction

In a digital transmission system, the data streams originating from primary MUX equipment are first connected to higher order MUX equipment and then to the radio. Thus the encoded output of the PCM equipment is transmitted over VF cable pair or unbalanced wire before entering the transmission medium. On these interconnecting cables the signal is likely to undergo high frequency attenuation, distortion and cross-talk. Moreover, the signal has a strong dc content and thus prevents the use of ac-coupled circuits.

For distortion-free transmission, the encoder output should be converted into a suitable code which will match the characteristics of the medium. This code is called the "line code" and the signal converted to the line code is called the line signal.

13.2 Characteristics of Line Codes

A practically useful line code shall have the following characteristics:

- Restricted bandwidth
- Low energy in the upper part of the signal spectrum to reduce attenuation distortion
- Low energy in the lower part of the spectrum to reduce cross-talk
- No dc component so that transformers can be used for coupling purposes
- Contain adequate timing information
- Have an in-built error monitoring capability

13.3 Types of Line Codes

The following codes are used in various working systems

- NRZ binary
- RZ binary
- Bipolar (AM)
- HDB-3
- CMI
- 5B6B

13.4 NRZ Binary Code

NRZ stands for 'non-return to zèro' code. Suppose we have a code 100110111001. In the pulse form this would appear as shown in Fig. 13.1. Here it may be seen that whenever a '1' is continuously transmitted, the output continues at '1' level for a duration equal to the number of bits transmitted. For example, in a 30 channel PCM system, the bit duration is 0.488 μs. If three 1s are transmitted, the output signal is a pulse which is $3 \times 0.488\ \mu$s wide, i.e., 1.464 μs wide.

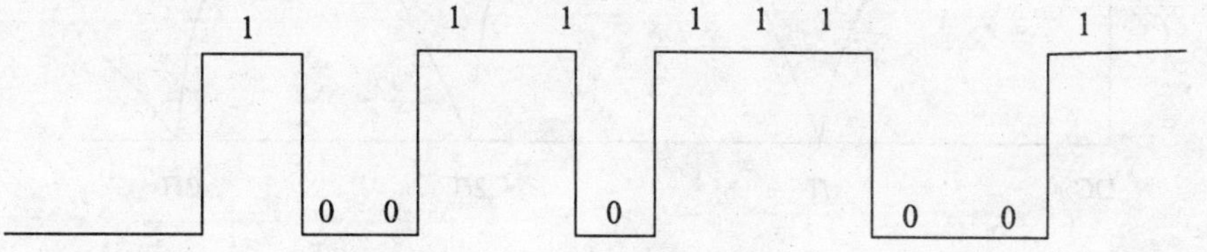

Fig. 13.1 *NRZ binary waveform.*

In this example the signal has only one polarity. A '0' is 0 V and a '1' is positive (say +5V). Sometimes, a '1' can be denoted by a positive voltage and '0' by a negative voltage. With this code also, when there is a string of 1s to be transmitted the output continues at '1' for as many bits as are continuously transmitted.

In both cases the output does not return to zero after every '1' bit when a number of 1s are transmitted. For this reason, this type of code is called a non-return to zero or NRZ binary. When the signal has only one polarity, as in the first example, the code is called unipolar or unbalanced NRZ binary, and when the signal has dual polarity as in the second example, it is called a balanced NRZ binary or bipolar NRZ.

13.4.1 Limitations of NRZ binary code

Figure 13.2 shows the spectrum of an NRZ binary signal. From the spectrum it can be seen that

(*i*) there is a strong dc component;

(*ii*) there is a large low frequency content. This may result in cross-talk;

(*iii*) there is no frequency component at $1/T, 2/T$, etc. It means that there is no component corresponding to the clock frequency. This makes efficient recovery of timing pulses very difficult;

(*iv*) the high dc component does not permit the use of transformers for coupling.

Thus the simple NRZ binary code does not satisfy the requirements of a line code.

13.5 RZ Binary Code

This is a modification of the NRZ code and stands for 'return to zero' binary. In this, a '1' bit is represented by a pulse of half-bit duration as shown in Fig. 13.3 (b). The spectrum for this code is shown in Fig. 13.3 (c). Figure 13.3 (a) shows the digital signal in NRZ binary form.

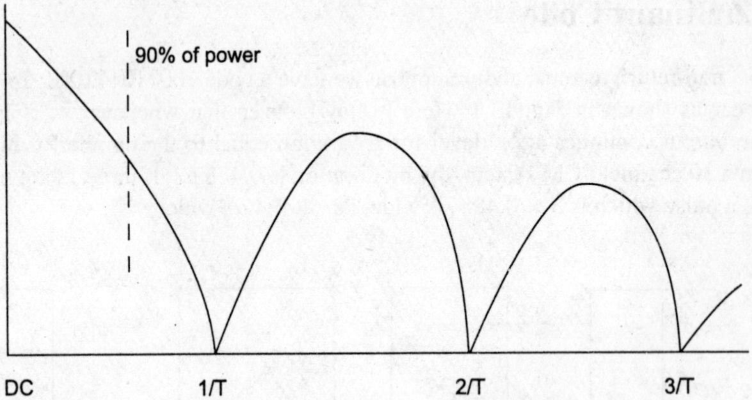

Fig. 13.2 *Spectrum of an NRZ signal.*

Here the '1' bit pulses have only 50% duration. From the spectrum we can see that a recovery of timing information at the receiver is possible. But still, because of the strong dc component and low frequency content, this code is also not suitable for transmission.

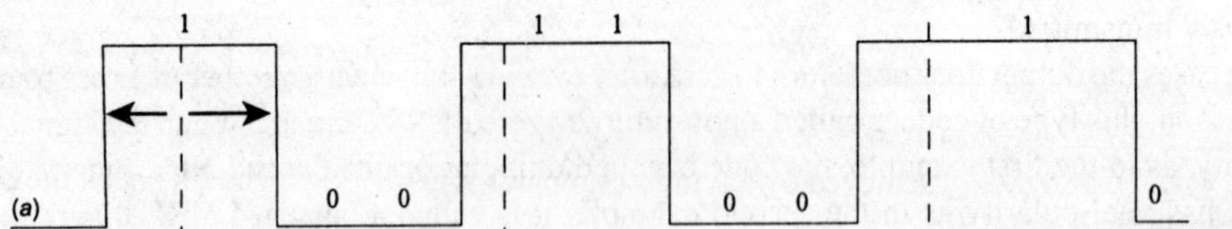

Fig. 13.3 *RZ binary waveform and spectrum.*

13.6 Bipolar Coding (AMI Code)

AMI stands for 'alternate mark inversion'. This code solves the dc content problem. Here, a logic '0' is represented by 0 volt and logic '1' is alternately encoded with positive and negative voltages. Therefore, the average voltage is maintained very close to zero and hence there is no dc component. Under steady-state conditions, only a low dc of the order of 0.4 to 0.8 volts remain.

The waveform of an AMI code is shown in Fig. 13.4.

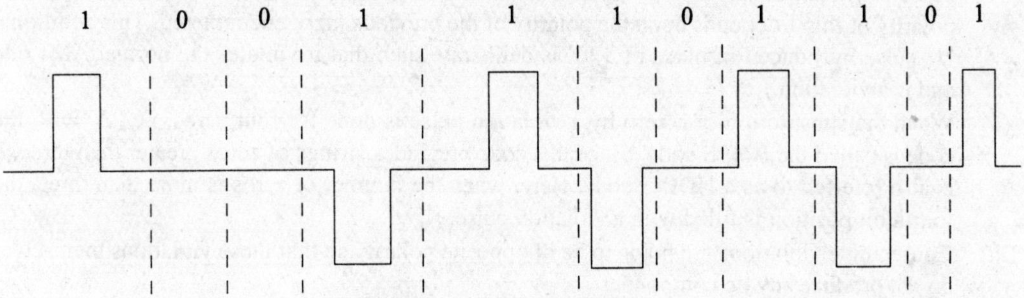

Fig. 13.4 *AMI code signal waveform.*

(*i*) From the AMI waveform it can be seen that this code has a built-in error monitoring facility. Since alternate marks (or 1s) are to be inverted, any deviation from this would mean an error. This can be practically achieved by having a comparator network which will check the polarity of the 1s received.

(*ii*) Though AMI code subdues the dc component and has large low frequency content, the timing information may be lost if a series of '0's are to be transmitted. The HDB-3 code which will be discussed in the following section, overcomes this problem. The spectrum of an AMI code signal has the shape shown in Fig. 13.5.

(*iii*) From the spectrum for the AMI code it can be seen that the maximum power is centred around the half-bit rate, i.e., $1/2T$, and that there is no dc component. The AMI code is used in 24 channel PCM systems in USA, Japan and Canada.

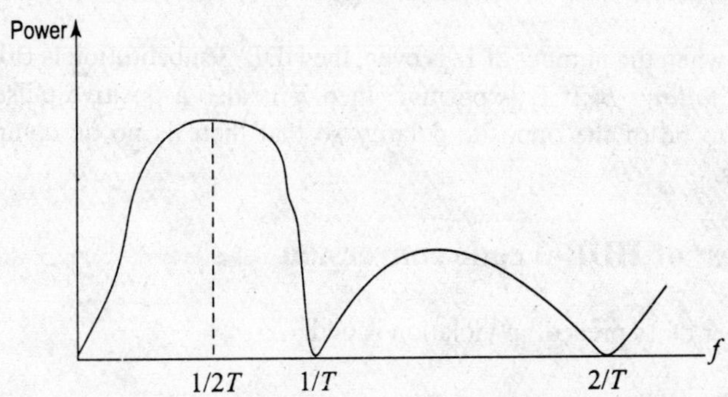

Fig. 13.5 *Spectrum of AMI signal.*

13.7 HDB-3 Code

HDB stands for 'high density bipolar' code. It belongs to the BNZS (binary N-zero substitution) family of codes. The coding rules are summarized below:

(*i*) The PCM signal is usually transmitted according to the AMI code, but when a string of N zeros is encountered, the N zeros are replaced by a special code which will deliberately introduce a bipolar deviation or violation. (Normally in the AMI code, if there are N zeros, they will be transmitted as such. But in the BNZS method, a '1' pulse is introduced deliberately. The polarity of this 1 depends upon the polarity of the previous mark encountered. This additional '1' pulse introduced in place of a '0' is deliberate such that it violates the normal AMI rule and is a violation.)

(*ii*) When the substitution of a zero by a violation pulse is done for four zeros, i.e., $N = 4$, the code is called the BNZS code. Since this code precludes strings of zeros greater than three, it is also referred to as a HDB-3 code. Here, when the number of zeros is more than three, the fourth bit position is filled with a violation pulse.

(*iii*) Consecutive violations are made to be of opposite polarity so that these violations themselves do not produce any dc component.

(*iv*) The violation pulse is always placed in the last bit position. Suppose there are 4 zeros coming in a row, then the HDB-3 code for this would of the form BCONV in general when V is the violation pulse. The polarity of this depends on the polarity of the last '1' and the number of 1s encountered prior to the four zeros.

(*v*) The first of the code was shown as B in *(iv)* above. B is set to '0' if the number of 1s encountered prior to the violation is odd. If it is even or zero, then the 'B' bit is filled with a '1' whose polarity is in accordance with the AMI code, i.e., if the previous '1' was positive, then B is '1' with negative polarity and vice versa.

(*vi*) The substitution rules stated above are tabulated below:

Polarity of previous '1'	Number of 1s since last violation: Odd	Even
Negative	000 V^-	$B^+00\,V^+$
Positive	000 V^+	$B^-00\,V^-$

It can be seen that when the number of 1s is even, the HDB-3 substitution is B00V: in this, *B follows the AMI code* and *V follows B*. If B is positive, then V is also a positive pulse. Thus, consecutive violations are made to be of the opposite polarity so that there is no dc component added by the violations themselves.

13.7.1 Examples* of HDB-3 code conversion

Condition 1: Number of 1s preceding violation is odd.

---

* Examples presented here are taken from the hand-out on PCM prepared for the S & T Department of S.C Railways.

Example 1

Consider the NRZ binary waveform given in Fig. 13.6 (*a*). Assume that there is no previous violation. Figure. 13.6 (*b*) is the RZ binary form for Fig. 13.6 (*a*) and Fig. 13.6 (*c*) is the corresponding HDB-3 code.

(*i*) Notice that up to pulse Z, HDB3 code follows the AMI code.
(*ii*) After pulse Z, we have four consecutive zeros. This calls for a violation.
(*iii*) Prior to the arrival of these zeros, three 1s were encountered, i.e., the number of 1s preceding the violation is odd.
(*iv*) This means that the HDB-3 substitution for the zeros will be of the form 000V.
(*v*) Also, the polarity of the last '1' before the arrival of zeros is positive. Therefore the violation pulse will also be a positive pulse, as shown shaded in Fig. 13.6. (*c*).
(*vi*) Then the fourth pulse P arrives, and it is converted according to AMI code as shown in the figure.

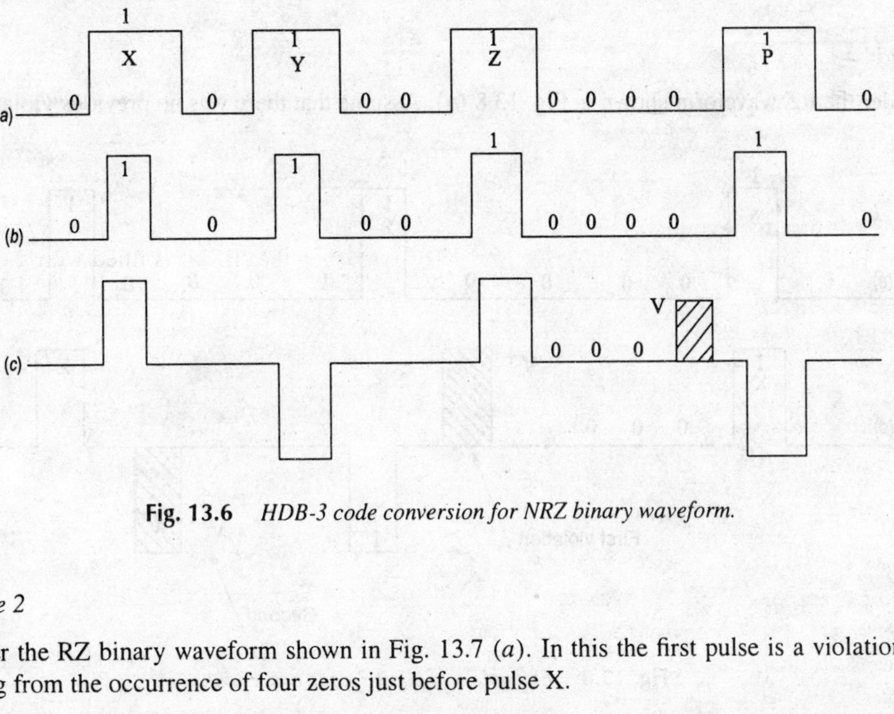

Fig. 13.6 *HDB-3 code conversion for NRZ binary waveform.*

Example 2

Consider the RZ binary waveform shown in Fig. 13.7 (*a*). In this the first pulse is a violation pulse resulting from the occurrence of four zeros just before pulse X.

(*i*) Here, the first pulse is a positive violation pulse. The next pulse (i.e., pulse X) is converted in accordance with AMI code and is therefore a negative pulse (Fig. 13.7 (*b*)).
(*ii*) After pulse X, we get four zeros.
(*iii*) Now, the total number of 1s since the violation is one, i.e., odd. Therefore the substitution is of the form 000V.
(*iv*) As the polarity of the last '1' before the arrival of zeros is negative, the violation pulse is also negative which is shown as a shaded pulse (V^-).

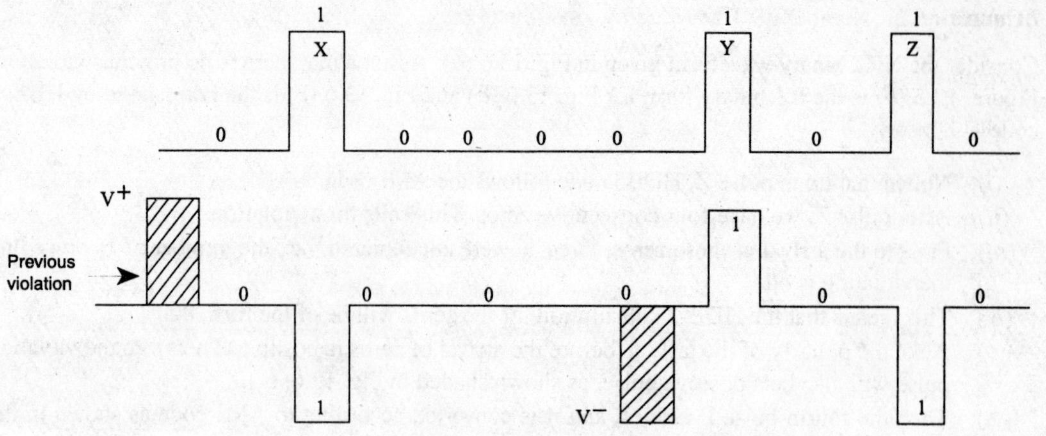

Fig. 13.7 *Example 2 for HDB3 code conversion.*

Example 3

Consider the RZ waveform shown in Fig. 13.8 (*a*). Assume that there was no previous violation.

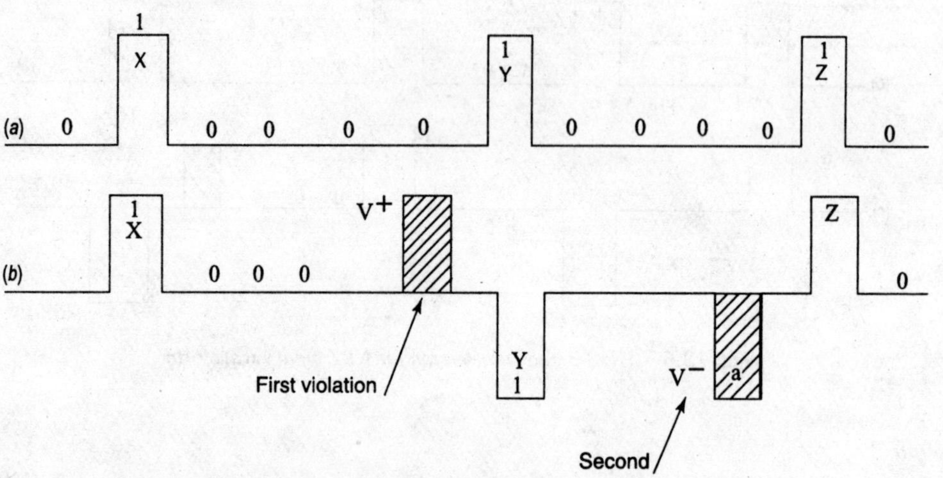

Fig. 13.8 *Example 3 of HDB-3 code conversion.*

(*i*) Here, after pulse X we have four zeros and again after pulse Y, four more zeros.

(*ii*) The first pulse goes as a positive pulse. The next four zeros are substituted by 000V because the polarity of the last '1' is positive and the total number is odd.

(*iii*) Then pulse Y is converted into a negative pulse according to AMI code.

(*iv*) The next 4 zeros are substituted by 000V since in this case the total number of 1s is again odd and the polarity of the last '1' is negative.

Condition 2: Number of '1's preceding violation is even.

Example 4

Consider the RZ binary shown in Fig. 13.9 (*a*) and assume that there was no previous violation.

(*i*) In this case, the pulses X and Y are converted according to AMI code as shown in Fig. 13.9 (*b*).

(*ii*) Four zeros are encountered after pulse Y. Here, the number of 1s prior to these zeros is *even* and therefore the substitution is of the form B000V.

(*iii*) Since the last '1' is a negative pulse, the substitution is B00V.

(*iv*) The HDB-3 substitution for the four zeros is shown in Fig. 13.9(*b*) as shaded positive pulses.

(*v*) The next pulse Z is converted as a negative pulse in accordance with AMI code.

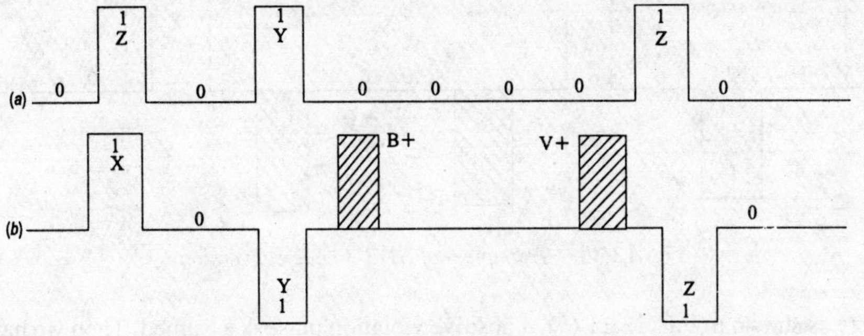

Fig. 13.9 *Example 4 of HDB-3 code conversion.*

Example 5

Consider the RZ waveform shown in Fig. 13.10 (*a*). Assume there was no previous violation.

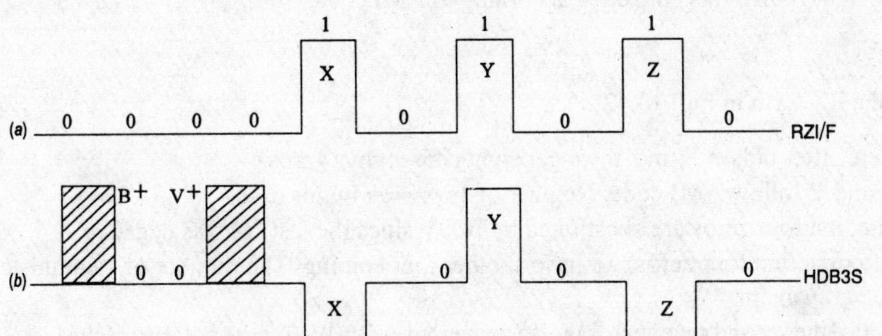

Fig. 13.10 *Example 5 of HDB-3 code conversion.*

(*i*) In this case, the waveform begins with four zeros. There are no previous violations. The number of 1s preceding the string of 0s is zero, i.e., *even*.

(*ii*) Therefore, the substitution is of the form B00V.

(*iii*) The (AMI) coding network is so designed that the very first bit is always a positive pulse. Hence the HDB code for Fig. 13.10 (*a*) would be B00V which is as shown in Fig. 13.10 (*b*).
(*iv*) The following 1s (X, Y and Z) are converted according to AMI code.

Example 6

Consider the RZ waveform in Fig. 13.11 (*a*). Assume a positive violation pulse to start with.

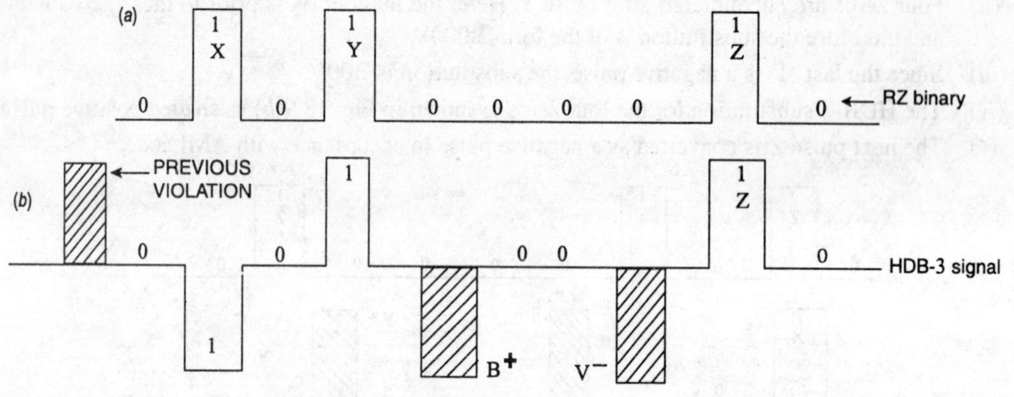

Fig. 13.11 *Example 6 of HDB-3 code conversion.*

(*i*) Here as shown in Fig. 13.11 (*b*), a positive violation pulse is assumed. Then we have pulses X and Y which are converted according to AMI code.
(*ii*) After pulse Y, we get four zeros. These should be substituted by B00V since the number of 1s is even.
(*iii*) Further, since the polarity of the last '1' is positive, the code would be B00V. This is shown as shaded pulses in Fig. 13.11(*b*).
(*iv*) Then the pulse Z is converted according to AMI code.

Example 7

Consider the RZ form in Fig. 13.12(*a*).

(*i*) Here, after pulses X and Y we get eight consecutive zeros.
(*ii*) X and Y follow AMI code. Number of 1s is *even* in this case.
(*iii*) The first four zeros are substituted by B00V since the last '1' was negative.
(*iv*) After the first four zeros, we have another four coming. The number of 1s in this case is zero, i.e., *even* again.
(*v*) Hence the second set of zeros is also converted as B00V. But the polarity of the last '1' (although it was a violation pulse), was positive. Hence, the second set of zeros is converted to B00V.
(*vi*) Pulse Z follow the AMI code.
(*vii*) The substitution pulses are shown shaded in Fig. 13.12(*b*).

But for the insertion of violation pulses, the HDB-3 coding is similar to the AMI code.

The spectrum for the HDB-3 code is shown in Fig. 13.13. From the spectrum it can be seen that there is no dc component and that maxium power is around $0.16/T$. It means that the power in the lower and

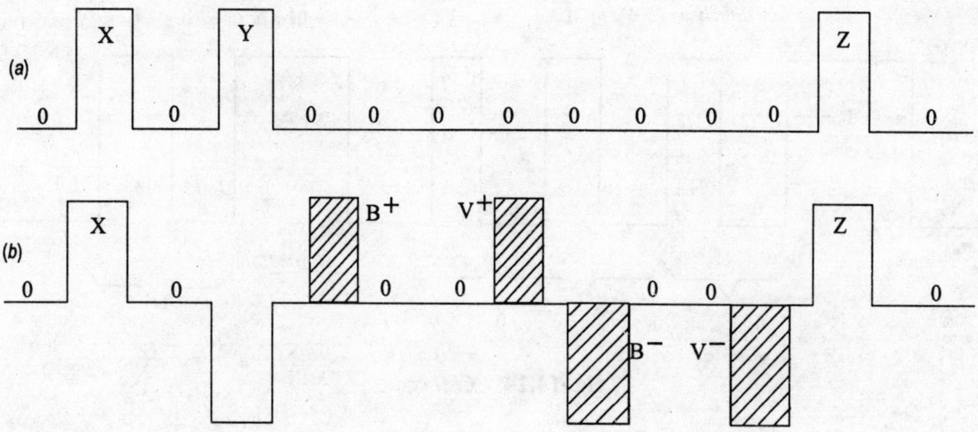

Fig. 13.12 *HDB code conversion—example 7.*

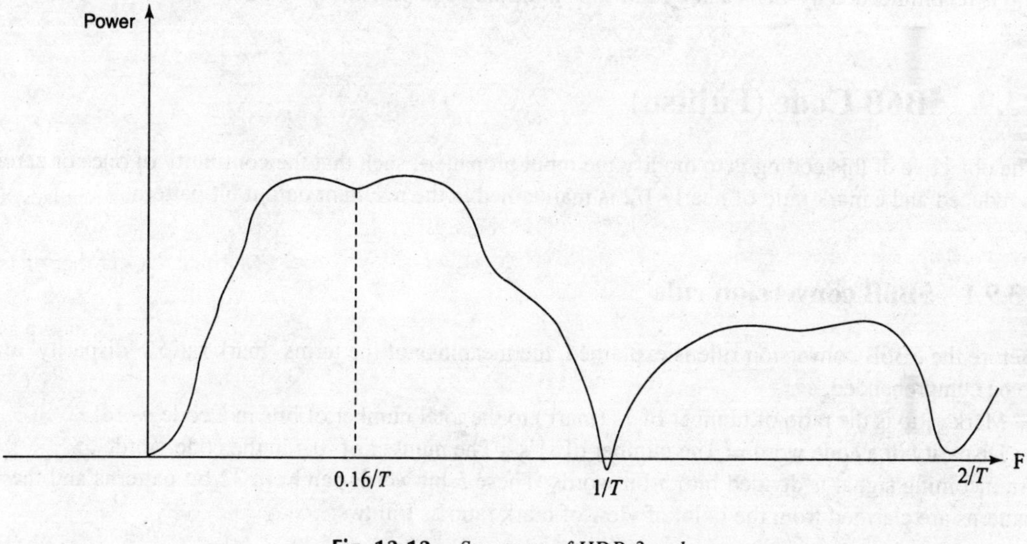

Fig. 13.13 *Spectrum of HDB-3 code.*

upper limits of the spectrum is low. This would minimize high frequency attenuation and cross-talk. Although this spectrum has nulls at $1/T, 2/T$, etc., because of the violation pules introduced, timing is not lost when a long string of zeros is encountered. The HDB-3 code satisfies all the requirements of a line code and is therefore specified by CCITT for 30 channel PCM systems having 2048 kbps clock.

13.8 CMI Code (Coded Mark Inversion Code)

This is a two level NRZ code in which a binary '0' is coded as 01 and the binary 1s are coded alternatively as logic '0' or '1'. In the case of a binary '0', the two CMI bits '0' and '1' are for half clock duration only, whereas for binary 1s the '0' and '1's are for full clock duration. This is illustrated in Fig. 13.14.

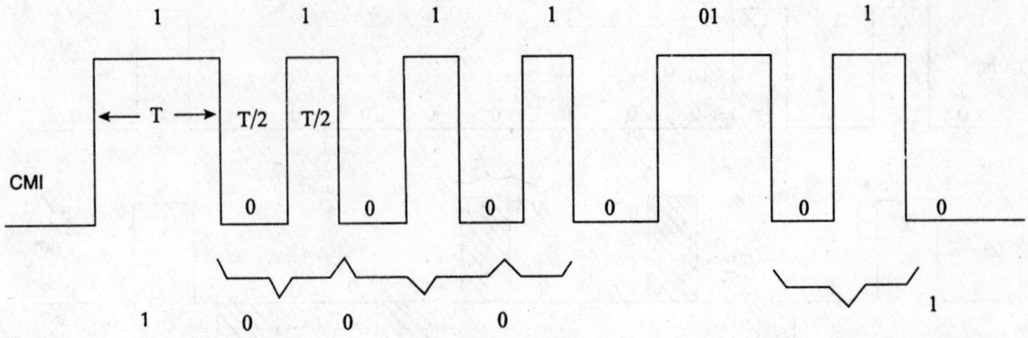

Fig. 13.14 *CMI code.*

This is basically a binary code and the bit rate of the code is twice the bipolar AMI code. For this reason CMI code is grouped with 1B2B family of line codes. The CMI code has a high clock content and is recommended by CCITT for 140 Mbps multiplex equipment.

13.9 5B6B Code (Fujitsu)

The objective of this coding is to modify the input bit pattern such that the continuity of ones or zeros is reduced and a mark ratio of nearly 1/2 is maintained in the resultant output bit pattern.

13.9.1 5B6B conversion rule

Before the 5B6B conversion rule is explained, the meanings of the terms 'mark ratio', 'disparity' are to be comprehended.

Mark ratio is the ratio of number of 1s (mark) to the total number of bits in a code word.

Disparity of a code word = The number of '1's – The number of '0's in the code word.

An incoming signal is divided into 5-bit words. These 5-bit words can have 32 bit patterns and these patterns are clarified from the point of view of mark ratio as follows:

2/5 mark ratio	:	10 patterns
3/5 mark ratio	:	10 patterns
5/5, 4/5, 1/5, 0/5 mark ratio	:	12 patterns

Each 5-bit input word is converted to a 6-bit word using 5B6B code to obtain 3/6 mark ratio.

For a 5-bit input word of mark ratio 2/5, a '1' is added to it to get mark ratio of 3/6 (disparity '1'). Similarly for a 5-bit input word having mark ratio of 3/5, a '0' is added to make the mark ratio of the output word 3/6 (disparity '0').

Out of the 64 bits pattern available for a 6-bit word, only 20 have mark ratio of 3/6 (disparity '1'). Hence output words with mark ratio of 5/5, 4/5, 1/5 and 0/5 cannot be converted to words with 3/6 mark ratio straightaway. Therefore, from the remaining patterns, the ones having 3/6 and 4/6 mark ratio patterns are alternatively used to obtain 3/6 mark ratio as the mean value. 6-bit

5B6B code conversion rule

	Input signal (5 bit)	5B6B Code S1 (−1)	5B6B Code S2 (+)
Mark ratio : 2/5	1 1 0 0 0	x x x x x 1	Disparity 0 Mark ratio: 3/6
	1 0 1 0 0		
	1 0 0 1 0		
	1 0 0 0 1		
	0 1 1 0 0		
	0 1 0 1 0		
	0 1 0 0 1		
	0 0 1 1 0		
	0 0 1 0 1		
	0 0 0 1 1		
Mark ratio : 3/5	0 0 1 1 1	x x x x x 0	Disparity 0 Mark ratio: 3/6
	0 1 0 1 1		
	0 1 1 0 1		
	0 1 1 1 0		
	1 0 0 1 1		
	1 0 1 0 1		
	1 0 1 1 0		
	1 1 0 0 1		
	1 1 0 1 0		
	1 1 1 0 0		
		Disparity + 2 Mark ratio : 4/6	Disparity − 2 Mark ratio : 2/6
Mark ratio : 5/5	1 1 1 1 1	1 1 1 0 1 0	0 0 0 1 0 1
Mark ratio : 2/5	1 1 1 1 0	1 1 0 1 1 0	0 0 1 0 0 1
	1 1 1 0 1	1 0 1 1 1 0	0 1 0 0 0 1
	1 1 0 1 1	1 1 1 0 0 1	0 0 0 1 1 0
	1 0 1 1 1	1 1 0 1 0 1	0 0 1 0 1 0
	0 1 1 1 1	1 0 1 1 0 1	0 1 0 0 1 0
Mark ratio : 1/5	1 0 0 0 0	0 1 1 1 0 1	1 0 0 0 1 0
	0 1 0 0 0	1 1 0 0 1 1	0 0 1 1 0 0
	0 0 1 0 0	1 0 1 0 1 1	0 1 0 1 0 0
	0 0 0 1 0	0 1 1 0 1 1	1 0 0 1 0 0
	0 0 0 0 1	1 0 0 1 1 1	0 1 1 0 0 0
Mark ratio : 2/5	0 0 0 0 0	0 1 0 1 1 1	1 0 1 0 0 0

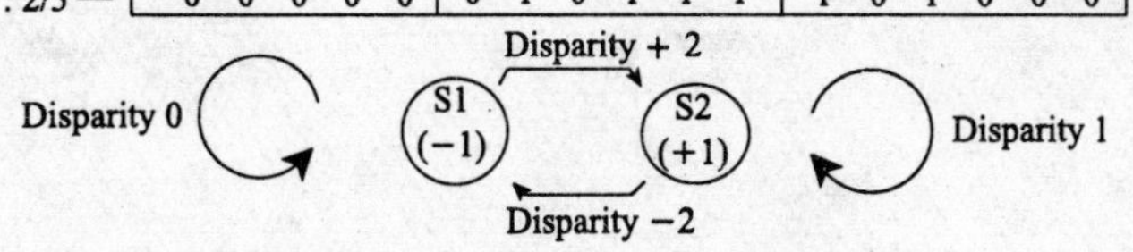

Fig. 13.15 *Transition diagram for a 6-bit pattern.*

words with mark ratio 2/6 are said to be in State 1(S1) and those with 4/6 are said to be in State 2(S2).

State 1: State 1 is that state where a 6-bit coded word has more ones than zeros. This is given a value of −1.

State 2: State 2 is that state where a 6-bit coded word has more zeros than one. This is given a value of +1.

When the mark ratio of input word is 5/5 or 4/5 or 1/5 or 0/5, the corresponding 6-bit output word is selected from either State 1 or State 2. The next state to be used is selected based on the information regarding the present state and the disparity of the 6-bit word that is being transmitted currently.

For example, to start with, if State 1 is selected, the input 5-bit data is converted into corresponding 6-bit word in State 1. Now if the disparity of this output word is +2, the control is transferred to State 2 for the next word. If the disparity of the next coded word is zero, the control remains with this state itself, viz., S2. Instead, if the disparity is −2, the control transfers to State 1 (refer to transition diagram shown in Fig. 13.15). The same can be explained with the help of following illustration. The transmission of next state is decided by adding disparity of the present coded word to the value of the present state.

Suppose the input 5-bit pattern has a mark ratio of 0/5 when the control is in S1 (−1). An output of 5/6 mark ratio pattern of S1 (disparity +2) is obtained and the state transfers to S2 (+1).

Next, if a 2/5 mark ratio pattern is input, an output of 3/6 mark ratio pattern (disparity 0) is obtained, and the state is not transferred and remains in S2.

$$+1 + 0 = +1.$$

Exercises

1. Why is CMI code recommended for 140 Mbps interface by CCITT?
2. Whys is 5B6B recommended for high data rate interfaces in OFC systems
3. What is the biggest limitation of AMI coding?
4. How are high frequency alternation and cross-talk minimized in HDB-3 coding?

14

Sources of Jitter in Digital Transmission Systems

14.1 Introduction

The most unwanted and yet widely present component in digital transmission system are noise, intersysmbol interference (ISI) and jitter. They contribute to bit error. Noise is of primary concern in passband (bandpass) systems and ISI is of primary concern in baseband systems. Jitter is of concern to both systems for various reasons.

Noise and ISI have been dealt with in earlier chapters. We now turn our attention to jitter.

Jitter is the time shift in the symbol retiming clock in any regenerator with respect to the actual symbol clock. By now we are all familiar with the function of regeneration—it has to do with making binary decision as to whether the incoming bit stream is above or below a threshold called the decision threshold, and re-timing the 'sliced' bit stream so that the signal stream leaving the circuit is identical to the originating bit stream.

In passband systems, the re-timing is by symbol synchronization. In Chapter 9 we saw that carrier synchronization and symbol synchronization are performed simultaneously in a digital radio equipment. The discussion on early–late gate symbol synchronizer where clock recovery follows carrier recovery was reserved for this chapter. The block diagram of the same is shown in Fig. 14.1. As can be seen, it has a phase-locked loop and correlators.

The correlators integrate over the entire symbol interval in two different paths with time lag (hence the name 'early–late'). The difference between the correlators' output is given as input to the PLL to achieve symbol synchronization.

In the MUX equipment also, PLL is used to extract the clock (Fig. 14.2). A closer look at these PLLs reveals the following problem:

Jitter manifests itself as phase modulation because jitter is due to difference in recovered timing and symbol timing; this can vary from 0 to a maximum of one clock period, i.e., a phase difference of 0 to 2π radians. Use of high loop-gain in clock recovery ensures almost perfect timing due to the tracking properties of PLL. If the PLL's loop parameters are not chosen properly, the loop response can gain at some frequencies which will amplify any phase modulation such as jitter. This will appear as variation in the triggering time of the recovered clock.

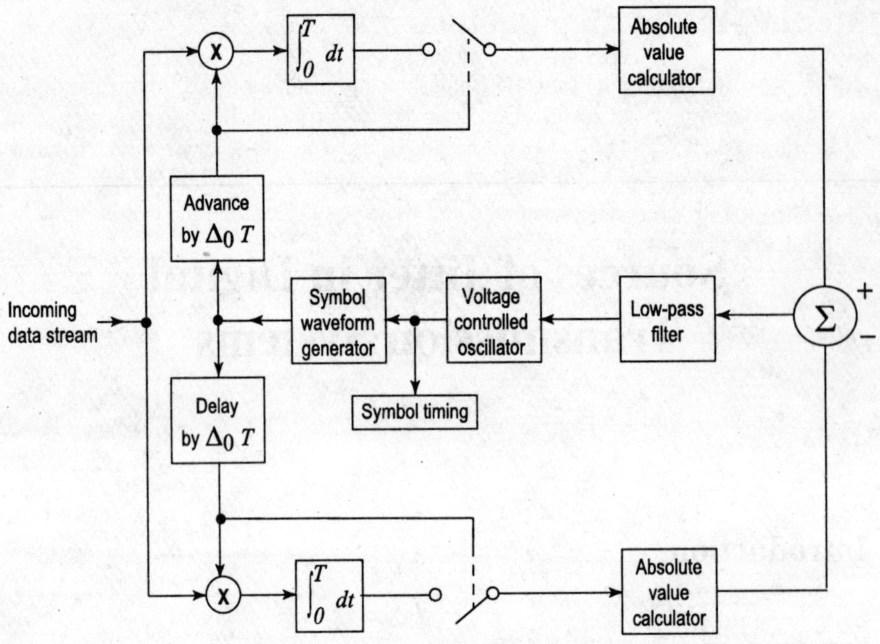

Fig. 14.1 *Early–late gate symbol synchronizer.*

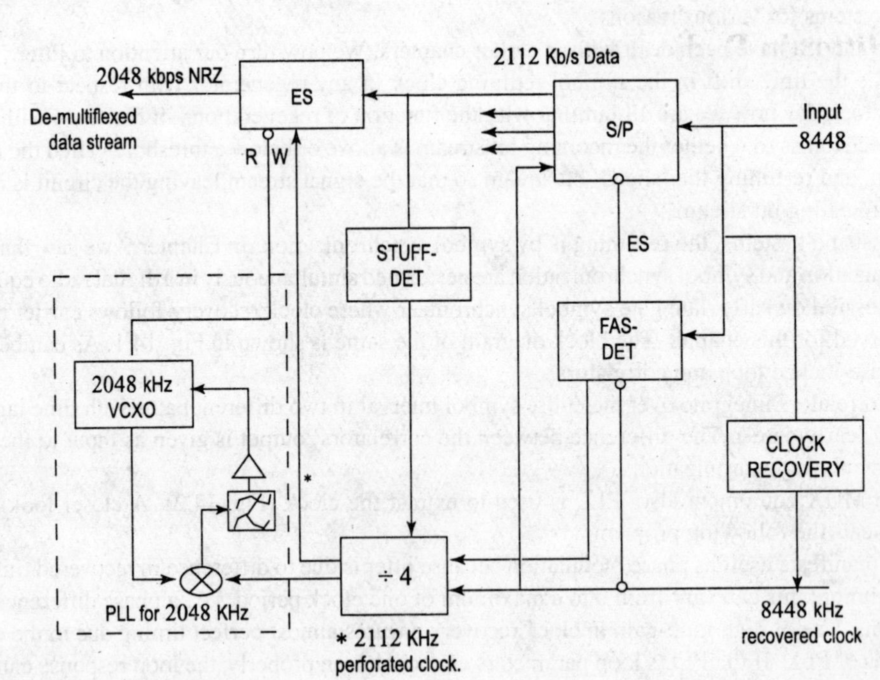

Fig. 14.2 *PLL in MUX equipment clock recovery. (ES : Elastic storage.S/P : Serial to parallel)*

14.2 Jitter in Multiplexing Equipment

In Chapter 12, we saw that during the process of intergrating plesiochronous bit streams in the MUX equipment, FAS, service digits and justification control bits were inserted. Also, justification bits in TS-155 of 8 Mb frame and TS-284 of 34 Mb frame (maximum four bits—one for each tributary) were inserted. Further, there exists a phase difference between write and read clocks which varies from frame to frame. The time difference between read and write clocks is called the *waiting time*. All these contributes to jitter, and thus jitter in a MUX can be:

- jitter introduced due to insertion of FAS, service digits and justification control bits,
- jitter introduced by justification,
- waiting time jitter due to variation in waiting time from frame to frame.

Since FAS and service digits are 12 per frame, their clocking rate is a fraction equal to 12/848 of an 8 Mb frame and 12/1536 of a 34 Mb clock (the frame size is 848 bits for an 8 Mb MUX and 1536 bits for a 34 Mb MUX). Similarly, justification bits are clocked at fractions equal to 4/848 and 4/1536 respectively of 8 Mb and 34 Mb frames.

Hence, the first two jitter components are at high frequencies compared to the passband of the PLL and are thus filtered out. Waiting time has a low frequency component and cannot be filtered out by PLL at the demultiplexer output. Waiting time difference builds up due to build-up of phase difference throughout the read/write operations associated with elastic memory. So, we have to live with the waiting time jitter of multiplexing equipment.

14.3 Jitter in Radio Equipment

The bit stream integration process in the digital radio equipment has been discussed in Chapter 12. It may be recalled that in Tx-DPU's multiplexer, the read-out data stream has the time of one bit for every eight bits of read-out data to facilitate insertion of framing, WS and DSC data (Fig. 14.3). Also, since the buffer memory has eight bits, it fits in with the plan of insertion of a 1-bit gap for every eight of information. The jitter due to this insertion is at high frequencies compared to the timing clock, and are hence filtered out by the PLL at the demultiplexer output.

The essential differences between the operations from the elastic memory here and that in the case of II/III-order multiplexing equipment are:

Multiplexing in TX-DPU	*Multiplexing in II/III order*
It is the integration of data streams of different rates.	It is the integration of data streams of same nominal rate but with differences within specified tolerance.
Only one read clock is derived.	Read clocks are derived from master clock (4 each at 2112 kHz for II-order MUX and 8592 kHz for III-order MUX). Each clock has a phase difference of a quarter and is used for different tributary data read-out.
The phase difference between read and write clocks varies systematically from 0 to 2π but does not vary from frame to frame.	The phase difference between read and write clocks varies from frame to frame, i.e., variable waiting time from frame to frame.

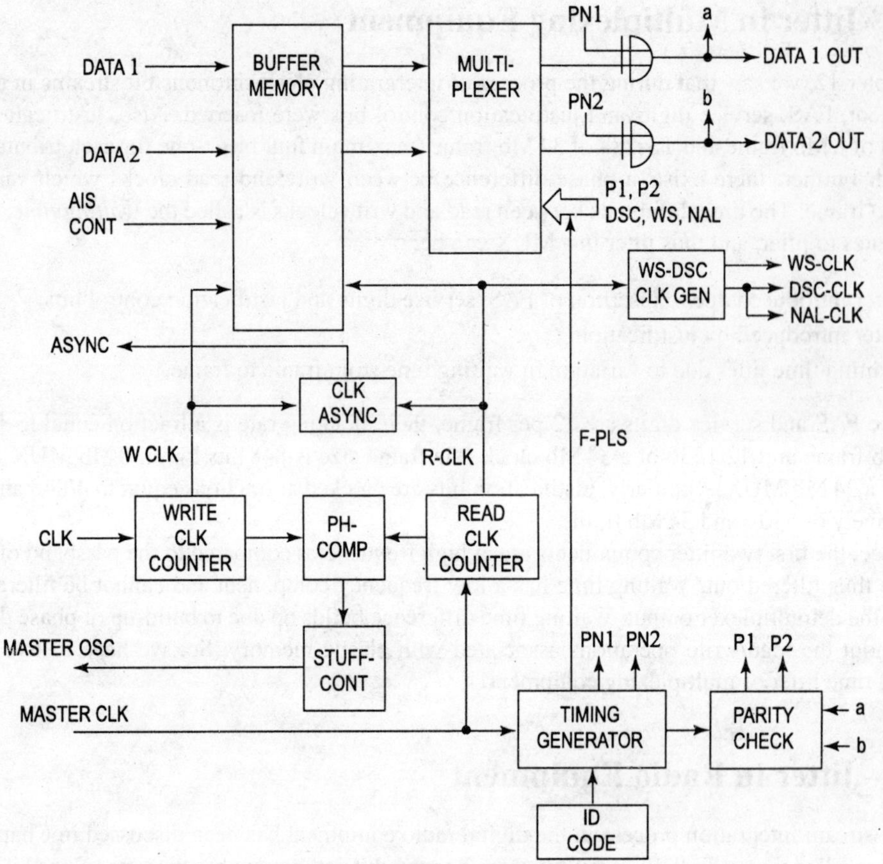

Fig. 14.3 *TX-DPU's multiplexer.*

Hence, there is waiting time jitter in the case of MUX equipment but not in the case of radio equipment.

14.4 Jitter Reduction using Dejitteriser

The schematic of a dejitteriser is shown in Fig. 14.4. Clock recovered from the incoming digital sequence (say frequency f) is used to clock the incoming sequence into a shift register A. Another clock of rate f/n is used to transfer the contents of shift register A to an intermediate buffer B (since register length is n cells).

The n bits in B are transferred to the output register C in the middle of the frame sequence under the control of f/n clock. The recovered clock of rate f is used to clock the bits out of register C. The clock recovery circuit operating on the incoming sequence has a high Q.

The length n of the registers is determined by the peak jitter to be handled by the dejitteriser circuit. For low jitter frequencies, n is large. So for very low frequency jitter components, n has to be excessively large. One has to strike a balance, and the very low frequency components of jitter may be ignored

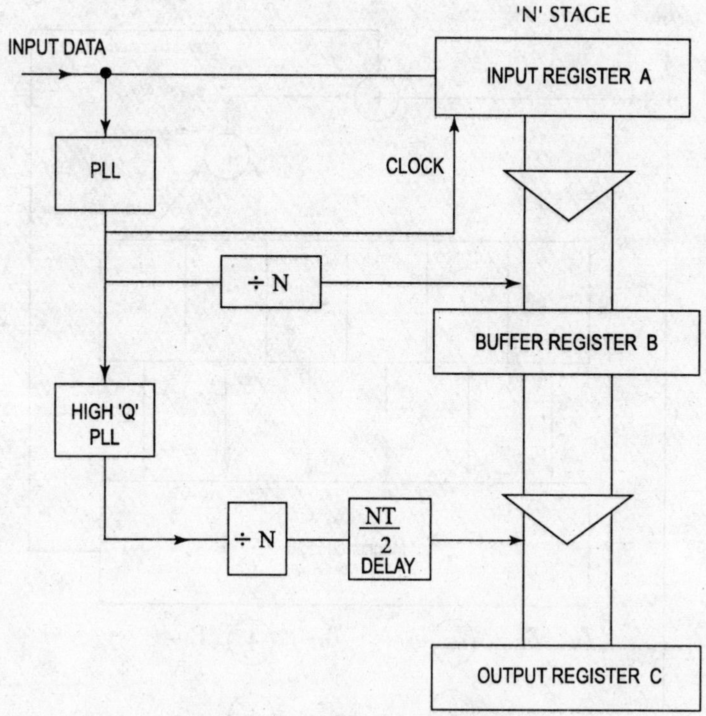

Fig. 14.4 *Dejitteriser schematic.*

as long as the accumulation of jitter over the entire route (i.e., over all the regenerators in-between) is within tolerable limits.

14.5 Jitter Reduction using Scrambling

If the incoming data sequence has repetitive 0s or 1s, the PLL of the clock recovery circuit may not maintain its narrow-band features. One way out of this difficulty is to monitor the incoming sequence, and if the repetition is more than a specified limit, force a sign change and do the reverse operation at the receiver. This requires the insertion of sign change and information regarding the sign change. These operations, if it happens frequently, are themselves sources of jitter. So, data scrambling at the sending end is resorted to randomise the incoming sequence. At the receiving end a reverse operation is performed.

Refer to Fig. 14.5 illustrating the scrambling operation. The output of the shift register at certain stages (say j, k) are taken, and an EX-OR operation performed on them. The result of the EX-OR operation is given as an input to the register so that the ouput is a pseudo random sequence of length $2n - 1$ for an n-bit sequence. The descrambler has to undertake converse operations to get back the actual sequence.

If I_n is the incoming sequence, T_n the transferred sequence, and R_n the recovered sequence, we have,

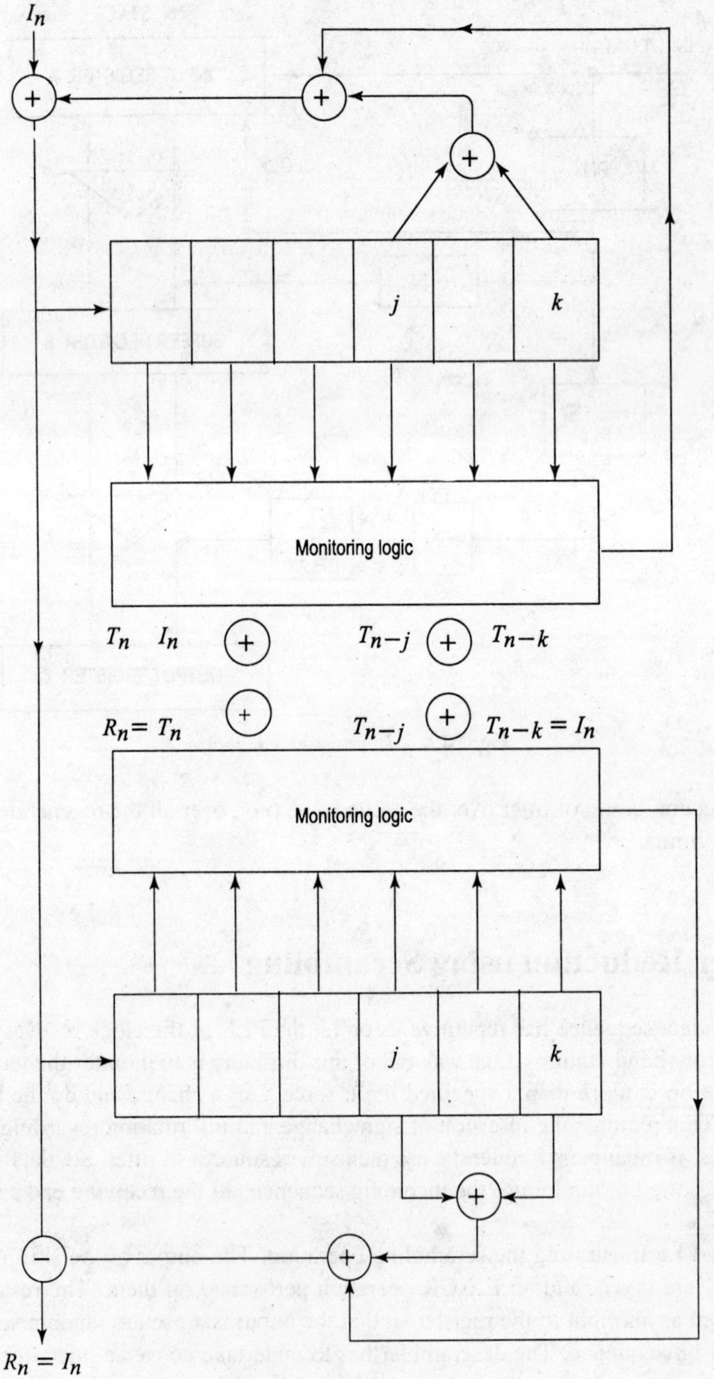

Fig. 14.5 *Scrambling operation.*

$$T_n = I_n \oplus T_{n-j} \oplus T_{n-k},$$
$$R_n = T_n \oplus T_{n-j} \oplus T_{n-k} = I_n.$$

Waiting time jitter may not be a problem for radio equipment as seen in Section 14.3. But a pattern-sensitive jitter (due to repetitive '0' and '1's) is definitely a problem. Hence the radio equipment uses a scrambler in the Tx-DPU module and a descrambler in Rx-DPU module.

The scrambler scrambles each data stream with a 432-bit pattern generated by the time signal generator of the Tx-DPU's multiplexer. With such scrambling, the transmission data mark ratio between the radio stations is averaged, and the transmit output power spectrum is kept constant irrespective of the mark ratio of the output signal from the higher order MUX equipment at the originating station.

Reference

[1] DMR 770 34 Mb Digital MW Radio Manual, Vol I, Chapter 6, M/S ITI Ltd, Bangalore, 1996.

Exercises

1. How can you say that jitter introduced due to justification process is insignificant?
2. How can you say that insertion of FAS and service digits is insignificant?
3. Can you conclude that waiting-time jitter is non-existent in radio equipment in spite of main and WS data?

15

Reliabilty Considerations for Long Haul Digital MW Link

15.1 Introduction

In the systems design of a digital MW link, the following must be ensured:

- The link satisfies regulatory requirements
- The link is compatible with the digital hierarchy
- The link performs well

The regulatory requirements are concerned with the SACFA (Standing Advisory Committee for Frequency Allocation) clearances and the WPC (wireless planning and coordination committee) license for operation of the link. CCIR recommendations are followed for frequency allocations. The digital hierachy followed in a country should be adopted for standard interfaces with exchanges and other links.

The third requirement of the link is subjective. In this chapter, we will be focusing on the performance issue.

15.2 Overall P_e of Multi-hop Digital MW Link

While discussing the advantages of a digital communication system over its analog counterpart in Chapter 2, we have seen that because of the regeneration of data stream at every repeater, there is no cumulative effect of noise. So, ideally speaking the performance of a digital link is independant of the number of repeaters and the system length and topology. But in reality, transmission quality is not completely independent of the length of the system or the number of repeaters.

The following example makes the issue clear.

Consider a five-hop digital microwave system. Assume the hops have equal length, say 40 km. The total one-way system length is therefore 200 km (= 5 × 40). Refer to Fig. 15.1. Evaluate the error probabilites of the individual hops with a pseudo random binary sequence (PRBS) generator and error detectors on a hop-by-hop basis. If the error probabilites of individual hops $P_{h(e)}$ are equal and very small, then the overall error probability is

$$P_{0(e)} = nP_{h(e)},$$

where n, the number of hops is not too large.

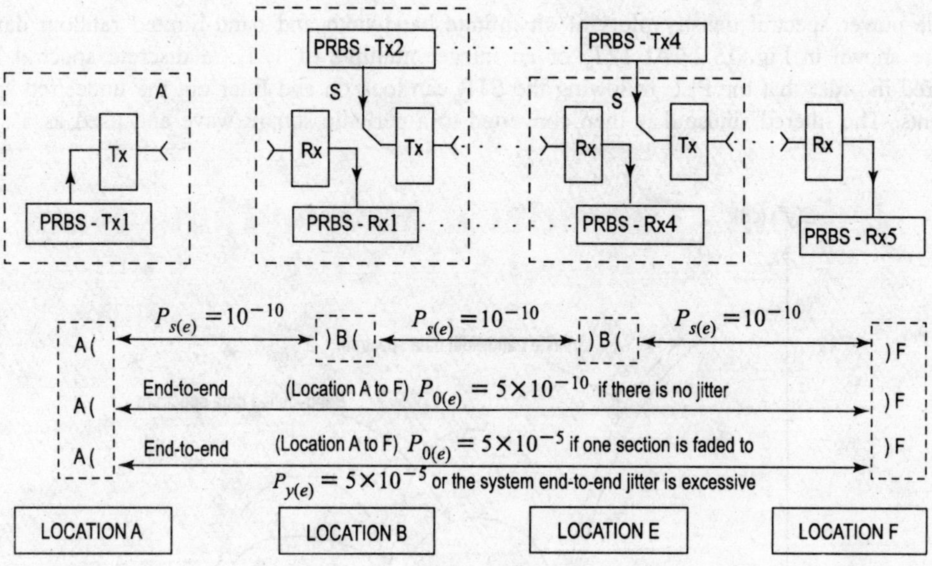

Fig. 15.1 *Five-hop digital MW link.*

In the present example,

$$P_{h(e)} = 10^{-11}, \quad n = 5$$

$$P_{0(e)} = 5 \times 10^{-11}.$$

Most of the time there is no fading in a well-designed system, and hence $P_{h(e)} = 10^{-11}$. Deep fading, which causes performance degradation, is unlikely to occur as equal depths fade simultaneously on two or more hops. So the overall P_e is controlled by the hop having the deepest fade.

Let the deepest fade hop have $P_e = 10^{-5}$ and the moderately fade hops have $P_e = 10^{-7}$.

$$P_{0(e)} = (3 \times 10^{-11}) + 10^{-7} + 10^{-5} = 10^{-5}.$$

Hence, it is clear that the equation $n P_{h(e)}$ is not applicable during deep fades. In such a situation, $P_{0(e)} = P_{w(e)}$, where $P_{w(e)}$ is the worst faded section's P_e.

15.3 Effect of Jitter on P_e

The example presented in Section 15.2 on P_e calculation does not take into account the error due to jitter accumulation. For low values of n, say $n = 5$, the jitter accumulation is negligible, or not so much as to cause 'bit skipping'.

The data signal on the digital link is random—it is a multiplexed stream of various voice data channels and is band-limited around the carrier. On the receive side (of every repeater), the STR (symbol timing clock recovery) circuit extracts the clock information from the data signal, i.e., the band-limited random data signal. The random data pattern does not contain discrete spectral lines at the symbol rate. Hence, non-linear processing is required in the STR circuit.

The power spectral density plots of an infinite bandwidth and band-limited random data signal are shown in Fig. 15.2. At $1/T_s$ or an integer multiple of $1/T_s$, a discrete spectral line is required in order that the PLL following the STR can lock on and filter out the undesired spectral contents. The filtered sinusoid is then converted to a periodic square wave and used as a symbol clock.

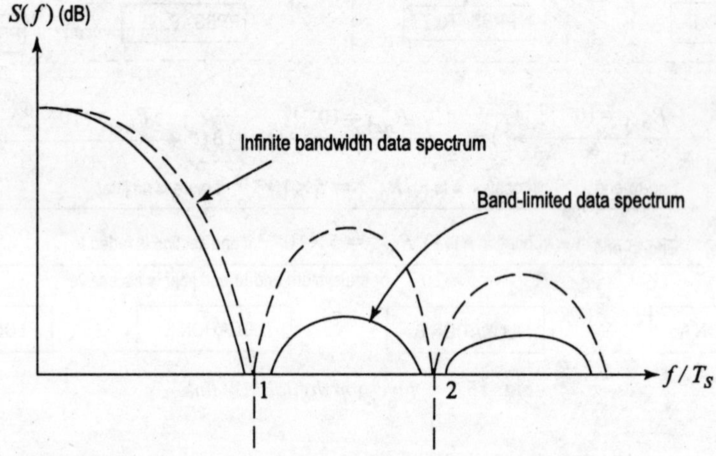

Fig. 15.2 *Power spectral density of infinite bandwidth and band-limited random data signals.*

But, a. already mentioned, the random data pattern does not have discrete spectral lines at $1/T_s$, and hence we resort to non-linear processing in STR circuits. To understand this non-linear processing, let us go back to time domain from frequency domain.

The time domain representation of a data signal is shown in Fig. 15.3. Jitter-free symbol clock is set to sample the eye diagram at its maximum opening. The traces of the eye diagram cross the threshold level at various time instants. The data transition peak jitter J_{Dp} is the largest possible shift of the data threshold crossing point from the nominal crossing point $T_s/2$.

The time domain operation in Fig. 15.3, followed by a PLL would give us a spectrum as shown in Fig. 15.4. We no doubt got the spectral lines, but with it also got the 'haze' pertaining to phase noise caused by jitter as well as channel noise. This will lead to a degradation in performance. The non-linearly processed signal exhibits a timing peak clock jitter J_{CP}.

For the 5-hop digital link example of Section 15.2, J_{CP} introduced in the first hop does not enter into the second hop as the second hop is fed by a jitter-free reference clock provided by the PRBS data transmitter. The situation is the same for subsequent hops also.

$$P_{0e} = 5P_{he} = 5 \times 10^{-11}.$$

If the 5-hop link is through end-to-end, without any interposing PRBS transmitters, the accumulated peak clock jitter $J_{5CP} > T_s/2$ (see Fig. 15.3). This excessive clock jitter degrades P_e performance of the sampled bits and is the cause of occasional bit-skipping.

In any digital link, it must be ensured that the overall peak-to-peak clock jitter in the worst case is less than 10% the symbol duration.

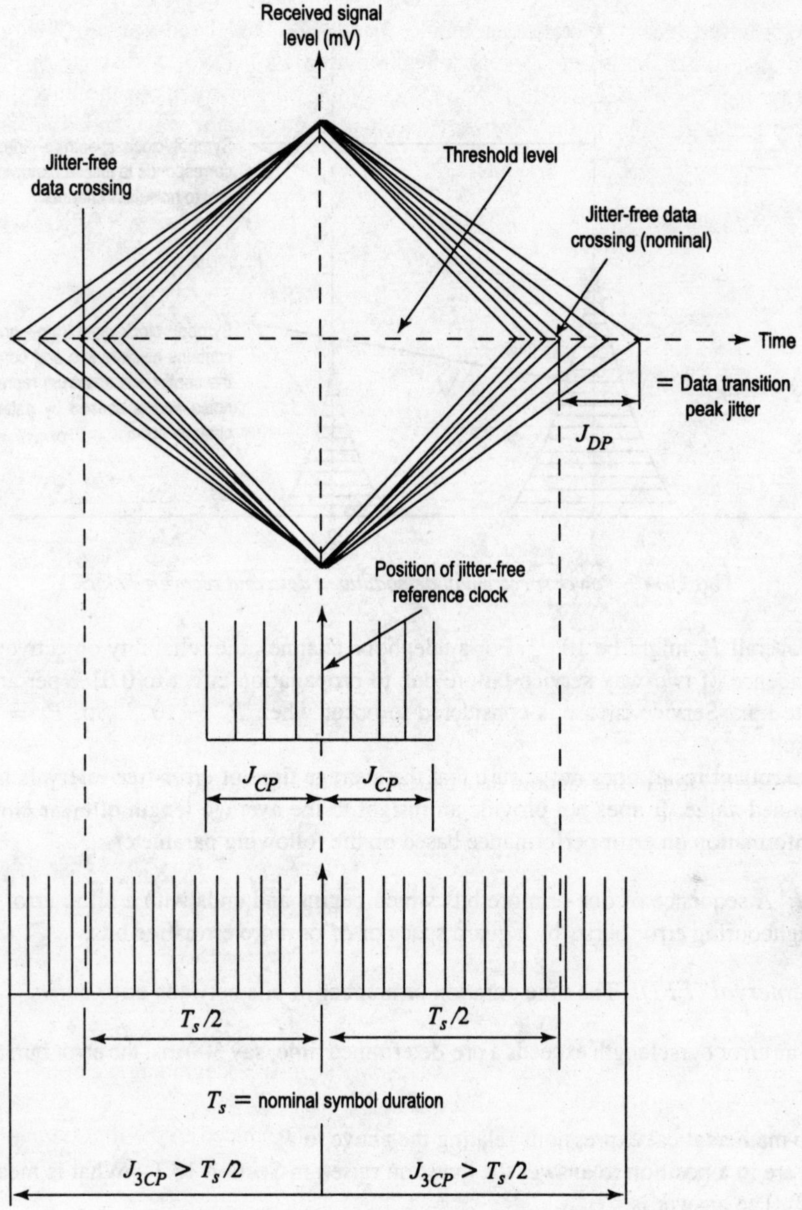

Fig. 15.3 *Sampling with a jittery reference clock.*

15.4 Reliability Objectives

The probality of error P_e represents a statistical average of the ratio of average number of bits in error to the average number of transmitted bits in a sufficiently long measurement interval. In an unfaded

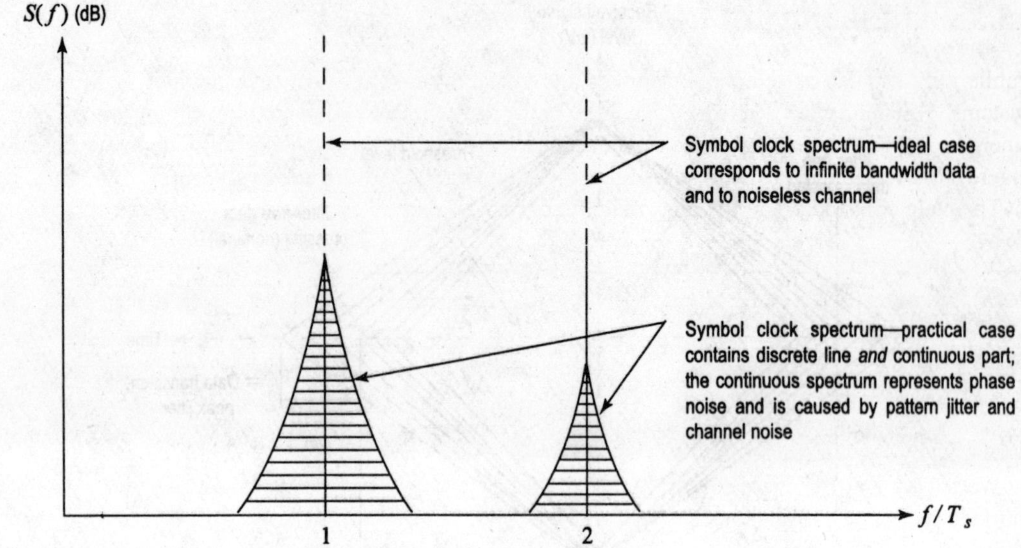

Fig. 15.4 *Power spectrum of demodulated data and recovered clock.*

system, the overall P_e might be 10^{-11}. For a telephone channel, the reliability objective would be to limit the incidence of two-way service failure due to propagation effect to 0.01% per annum over a 400 km route link. Service failure is considered to occur when $P_e > 10^{-6}$. So, $P_e = 10^{-11}$ is an excellent result.

But this excellent result does not ensure that the average time of error-free intervals is larger than a pre-determined value. It does not provide an insight to the average length of error burst. We need additional information on error performance based on the following parameters:

Error burst: A sequence of one or more bits which begins and ends with a bit in error is separated from the neighbouring error bursts by a guard space of M or more error-free bits.

Error-free-interval (EFI): The time duration or number of bits between error bursts.

Outage: If an error burst length exceeds a pre-determined time, say 300 ms, the error burst is redefined as an outage.

There are no mathematical expressions relating the above to P_e.

Now, we are in a position to answer the question raised in Section 15.1—What is meant by 'well-performing?'. The answer is

- The digital link should have a P_e better than 1×10^{-6}, i.e., $p_e < 1 \times 10^{-6}$
- Worst-case overall peak-to-peak clock jitter < 10% of symbol duration
- EFI shall be 99.9999%
- The availability shall be 99.98% for unprotected, non-diversity system and 99.9998% for protected, diversity system.

15.5 Availability of Link

Public utilities such as telephone services and information system networks require highly reliable systems. For the Indian Railways, the MW network is the backbone of its communications during emergencies. The digital microwave link should, therefore, achieve the reliability objective specified in terms of availability of link or maximum allowable time of failure due to all causes.

The main contributors to the unavailability of network are:

(*a*) those controllable by equipment and route design (fading),
(*b*) those caused by operational activities (maintenance, system change-over, expansion, partial cut-over, etc.),
(*c*) catastrophies such as fires, tilting of antennas by severe cyclonic storms, problems with feeders and also theft of feeder; this is indeed possible in India as experienced on a 100 m tower Eluru MW station of the South Central Railway in 1995).

We have a control on factors under (*a*), and hence can quantize the availability and relate it to the fade margin requirement. But on factors under (*b*) we have only partial control and on factors under (*c*), we do not have any control. Hence, we cannot quantize or provide any mathematical relations for these. However, these factors can be kept under check as will be seen later in this chapter.

As regards (*a*), the allowed unavailability is 0.02% for a two-way long haul digital MW link. Of this 0.01% is apportioned to equipment failure and 0.01% to fading. These are the specifications for a non-diversity, unprotected system. The specifications for a diversity system are 0.0001% unavailability, i.e., a hundred-time improvement is expected by the provision of diversity.

15.6 Fade Margin

15.6.1 Requirements for specified system availability

The starting point for all microwave communications systems design is the aspect of 'system gain'. It incorporates many parameters of interest to the designer, and hence determines the allowable trade-off between power, bandwidth, complexity of equipment and cost of towers. The simplest definition of system gain is

It is the difference between transmitter output power and receiver threshold sensitivity for a given bit error rate.

Mathematically,

$$G_s = P_{Tx} - C_{\min}(> F_m) + L_p + L_f + L_b - G_{TX} - G_{RX} \ldots, \tag{15.1}$$

where,

G_s = system gain in dB,

P_{Tx} = Tx power output (dBm) excluding the antenna branching network,

$C_{\min}$ = the received carried level (dBm) for a minimum quality objective (usually BER of 1×10^{-6}). This is also termed the receiver threshold,

$L_p =$ path loss $= 92.4 + 20 \log d + 20 \log f$ (15.2)

where d is the path length and f is the frequency,

$L_f =$ feeder loss, about 9.5 dB, considering a 100 m waveguide, elliptical at both ends in the 7 GHz band,

$L_B =$ branching circuit loss = 3 dB,

$G_{Tx}, G_{Rx} =$ Gains of Tx and Rx antennas = 44.8 dB for a 3 m metallic antenna in the 7 GHz band,

$F_m =$ fade margin in dB for non-diversity system.

Applying Barnett–Vigant reliability formula[1], F_m for unprotected, non-diversity system is given by

$$F_m = 30 \log d + 10 \log(ABf) - 10 \log(1 - R) - 70, \quad (15.3)$$

where,

$1 - R =$ reliability objective (one-way),

$A =$ roughness factor (4 for smooth terrain and water; 1 for average terrain; 1/4 for mountains and rough terrain),

$B =$ factor to convert the worst month probability to annual probability (1/2 for lakes and humid areas; 1/4 for inland areas; 1/8 for mountains and dry areas).

The reliability objective is given below.

Operator	*Case*	*Reliability objective* (1 − R)
Common carrier	Two-way	0.0001d
	One-way	400
Utilities	Worst month (without diversity)	0.0001
	Worst month (with diversity)	0.000001

It may please be noted that if R is reliability expressed as fraction but not percentage (99.99% reliability means R is 0.9999), $1 - R$ is taken as reliability objective. This notation helps in easier formulation of outage of the link and also easier correlation with fade margin.

Observe from the equation for F_m that if $(1 - R)$ is increased by 10 times, F_m increases by 10 dB. This means that an order of magnitude of improvement in reliability requires 10 dB improvement in F_m. Hence, the fade margin is of prime concern in the system design.

From (15.1), (15.2) and (15.3),

$$G_s = 50 \log d + 30 \log f + 10 \log(6A) + 62.4 + L_F + L_B - G_{TX} - G_{RX}.$$

A plot of G_s versus path length d is plotted in Fig. 15.5 for an F_m of 37.5 dB and a reliability objective $(1 - R)$ of 0.01 % for a 7 GHz system.

In this context, we present the results of an important experiment carried out by the South Central Railways at Vijayawada on the BZA–GDR digital MW link:

> *If we increase the system gain by about 5 dB, for the same number of hops, smaller antenna dishes can be used, thus reducing the wind loads. In yet another method, the mounting height can be reduced for the same 3 m metal antenna. Both these methods can help save costs on tower. The results of the experiment are discussed in detail in Chapter 19.*

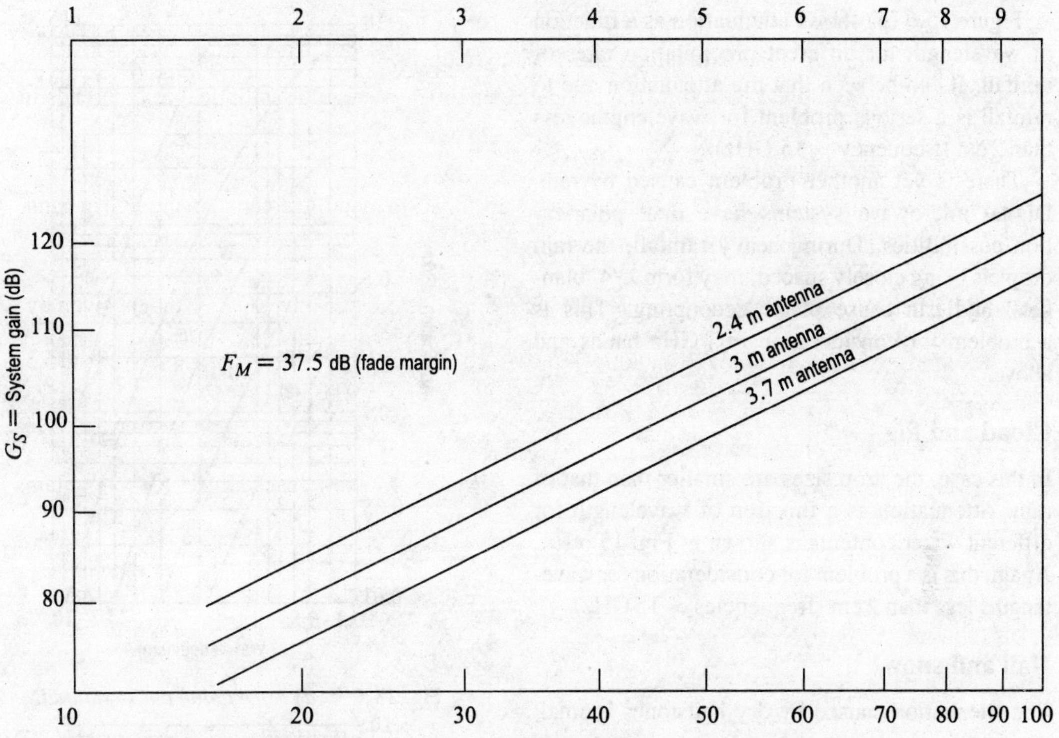

Fig. 15.5 *G_s versus path length for F_m of 37.5 dB.*

15.6.2 System engineering considerations

Fade depth is the total manifestation of fading and signal loss due to propagation problems. It is a time-varying term. The usual causes of fading and signal loss are:

- multipath propagation
- rainfall
- cloud and fog
- hail and snow
- anomalous propagation disturbances
- reflections
- obstructions and insufficient path clearances

The problems of obstructions and insufficient path clearances can be avoided by careful system design and observance of sighting clearances. Multipath fading is dealt with seperately later.

Rainfall

Attenuation due to rainfall is dominant in the 18 GHz band where the rain drop size is of the order of wavelength of radio carrier. In the 7 GHz band, the attenuation on this account is negligible, and moreover, raindrop attenuation and multipath fading will not occur simultaneously.

Figure 15.6 (*a*) shows attenuation as a function of wavelength for different precipitation rates of rainfall. It can be seen that the attenuation due to rainfall is a serious problem for wavelengths less than 2 cm (frequency > 15 GHz).

There is yet another problem caused by rain. Digital microwave systems have dual polarization possibilities. During heavy rainfall, the rain droplets being closely spaced, may form $\lambda/4$ 'blankets' and can cause phase decoupling. This is a problem for consideration in 7 GHz bands and above.

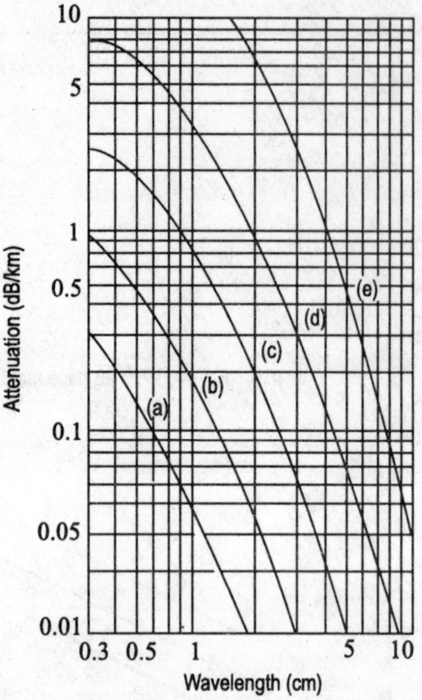

Fig. 15.6 (*a*) *Attenuation due to rainfall.*

Cloud and fog

In this case, the drop sizes are smaller than that of rain. Attenuation as a function of wavelength for different water contents is shown in Fig. 15.6 (*b*). Again, this is a problem for consideration for wavelengths less than 2 cm (frequencies > 15 GHz).

Hail and snow

The attenuation caused by dry hailstones is small compared to that caused by rain of equivalent water content. The same observation holds good for snow also.

Anomalous propagation disturbance

The anomalous propagation disturbance relevant for MW frequency range is called 'atmospheric ducting'. The tropospheric refractive index depends on the air temperature and pressure and on the amount of water vapour present. This dependence is given by the following equation:

$$(N-1) = (80/T)\times(10^{-6})[P+(4800\times w/T)],$$

where N is the refractive index, P the atmospheric pressure in millibars, w the partial vapour pressure in millibars, and T the absolute temperature.

If we differentiate the expression w.r.t. height, we get the variation of N w.r.t. height as

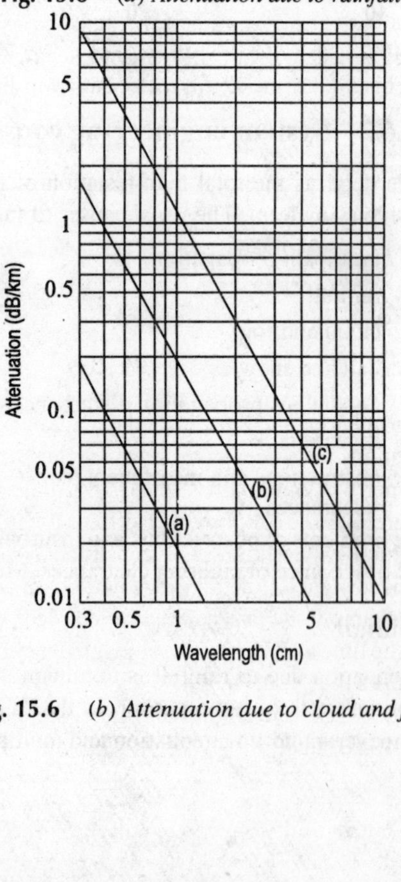

Fig. 15.6 (*b*) *Attenuation due to cloud and fog.*

$$\left(\frac{dN}{dh}\right)10^6 = \frac{80}{T}\frac{dP}{dh} - \frac{80}{T^2}\left(P + \frac{9600w}{T}\right)\frac{dT}{dh} + \frac{80 \times 4800}{T^2}\frac{dw}{dh}.$$

The first term on the RHS is always negative since P decreases with h. The sign of the second and third terms depend on atmospheric conditions.

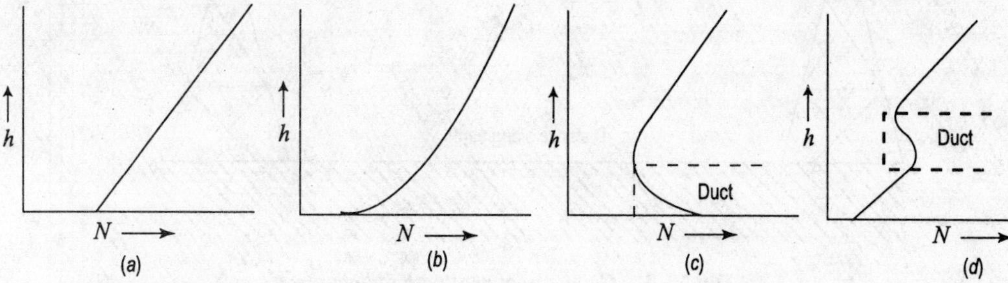

Fig. 15.7 *Variation of refractive index with height.*

Refer to Fig. 15.7 for a plot of h v/s N. Under certain atmospheric conditions, dT/dh and dw/dh differ greatly from standard values. (When warm dry air passes over cool sea surface, the air close to the water will be cooled and thus an increase in temperature with height will occur—the opposite of normal case. In addition, water vapour content will decrease with height more rapidly than usual.)

This would make dn/dh negative in the region close to surface (see Fig. 15.7 (c)) resulting in a 'surface duct'. Sometimes, the duct can be elevated as in Fig. 15.7 (d). These conditions which cause dN/dh to drop below normal value are called 'super-standard' conditions. The result is super refraction or trapping (d) and abnormally high field strength regions. The electromagnetic wave gets 'trapped' in this region.

Note : 'Sub-standard' conditions causing increase of dN/dh above normal value as in Fig. 7.8 (b) may also occur. This causes reduction in field strength.

15.7 Multipath Fading

Atmosphere is the transmission medium for electromagnetic radiation. It is non-homogeneous as spatial variations in temperature, pressure, humidity, and turbulance cause variation in refractive index. The geometry of multipath fading is illustrated in Fig. 15.8. The two rays (Ray 1 and Ray 2), because of differential refraction, cover different path lengths and have a phase difference. It can lead to either destructive interference or constructive interference, and hence, there would be fluctuations in the received signal. If the aggregate of phase angles of multipath arrivals at the receiver is such that the net signal amplitude is zero, it is called 'deep multipath fading'.

The time during which a signal fades below a level L is called the duration of fade. Refer to Fig. 15.9. It shows the duration of a 40 dB fade. The fade for level L relative to normal level is given by $20 \log L$. For $L = 10^{-2}$, i.e. 100 times poorer signal than the normal signal, the fade is $20 \log(10^{-2}) = -40$ dB, or a 40 dB fade.

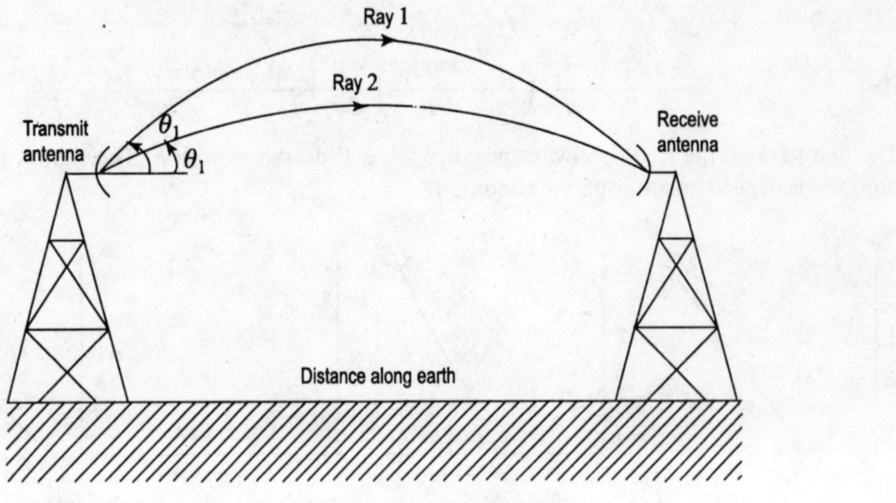

Fig. 15.8 *Geometry of multipath fading.*

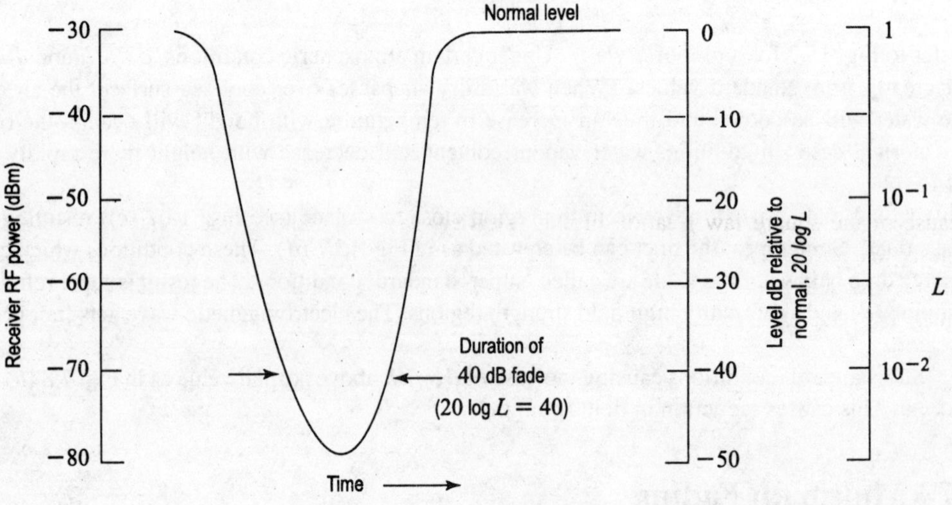

Fig. 15.9 *Duration of fade.*

The average duration of fading is given by

$$t_f = 410\,L \text{ seconds } (L \leq 0.1).$$

The plot of probability of fade, $P(V \leq L)$ against fade depth (20 log L) is shown in Fig. 15.10. Over water, $P(V \leq L) \leq L$.

In non-diversity systems on ordinary routes,

$$P(V \leq L) \propto L^2.$$

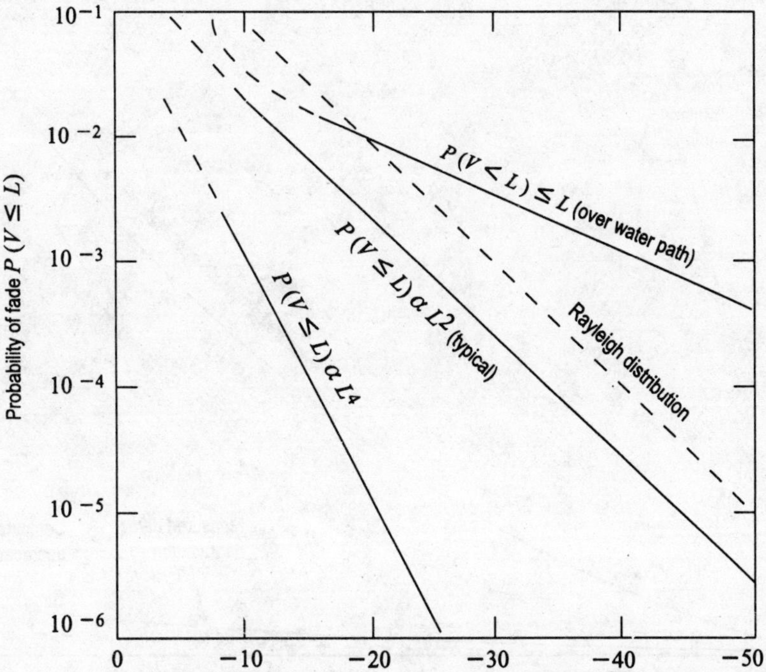

Fig. 15.10 *P*(*V* ≤ *L*) *against fade depth.*

Because of the square law relationship, it is called Rayleigh fading. The square law relationship suggests that the fade is deeper, which is obvious, since it is a non-diversity system. Multipath fading causes amplitude distortion and group delay distortion due to its frequency selective nature (Fig. 15.11). This causes ISI, which increases the threshold BER limit of the receiver by 10 to 15 dB, thus reducing the overall system gain. As a counter-measure, adaptive equalization techniques are used (refer to discussion on transversal equalizer in Chapter 10). In addition, we have diversity techniques in reception.

15.8 Diversity Techniques

In radio reception, two diversity techniques are employed—frequency diversity and space diversity.

Refer to Fig. 15.12. In frequency diversity (FD), the same digital information is fed on two carriers generated in Tx-1 and Tx-2. The carrier spacing should be at least 4% to have the benefits of frequency diversity ($\Delta f/f = 0.04$).

In space diversity, there is vertical separation of the two receiving antennas (minimum 6 m for discernible results).

Because of the overwhelming demand for carriers, the number of carriers allotted by the regulatory authority in the frequency plan of the link is limited. Morever, in digital MW, the number of hops is more due to reduced repeater spacing (about 40 km compared to the analog system's 60 km), with or without diversity, and a hot standby equipment is available. If the standby equipment also operates

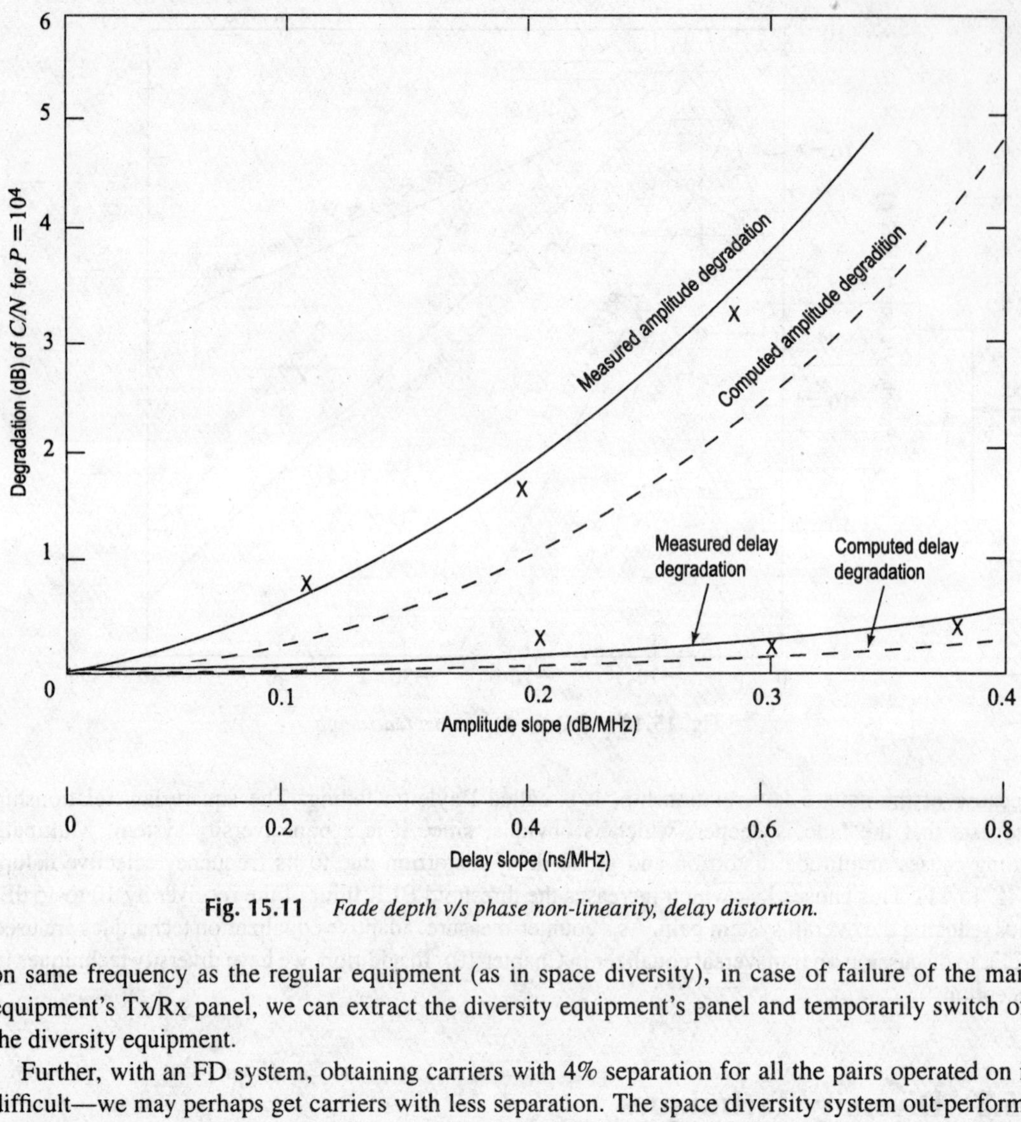

Fig. 15.11 *Fade depth v/s phase non-linearity, delay distortion.*

on same frequency as the regular equipment (as in space diversity), in case of failure of the main equipment's Tx/Rx panel, we can extract the diversity equipment's panel and temporarily switch off the diversity equipment.

Further, with an FD system, obtaining carriers with 4% separation for all the pairs operated on is difficult—we may perhaps get carriers with less separation. The space diversity system out-performs the frequency diversity system in terms of availability (Fig. 15.13). Hence, space diversity is preferred for digital MW networks.

Link availability measurements on BZA–GDR–RU digital MW link of the South Central Railway are presented in Chapter 19.

15.9 Hitless Switching in Hot Standby Systems

Switch-over from regular to protection equipment and vice-versa, with or without diversity, should not introduce any error burst or discontinuity in the data path. Hence, such switching is performed at baseband on the regenerated data streams. This is known as 'hitless' switching.

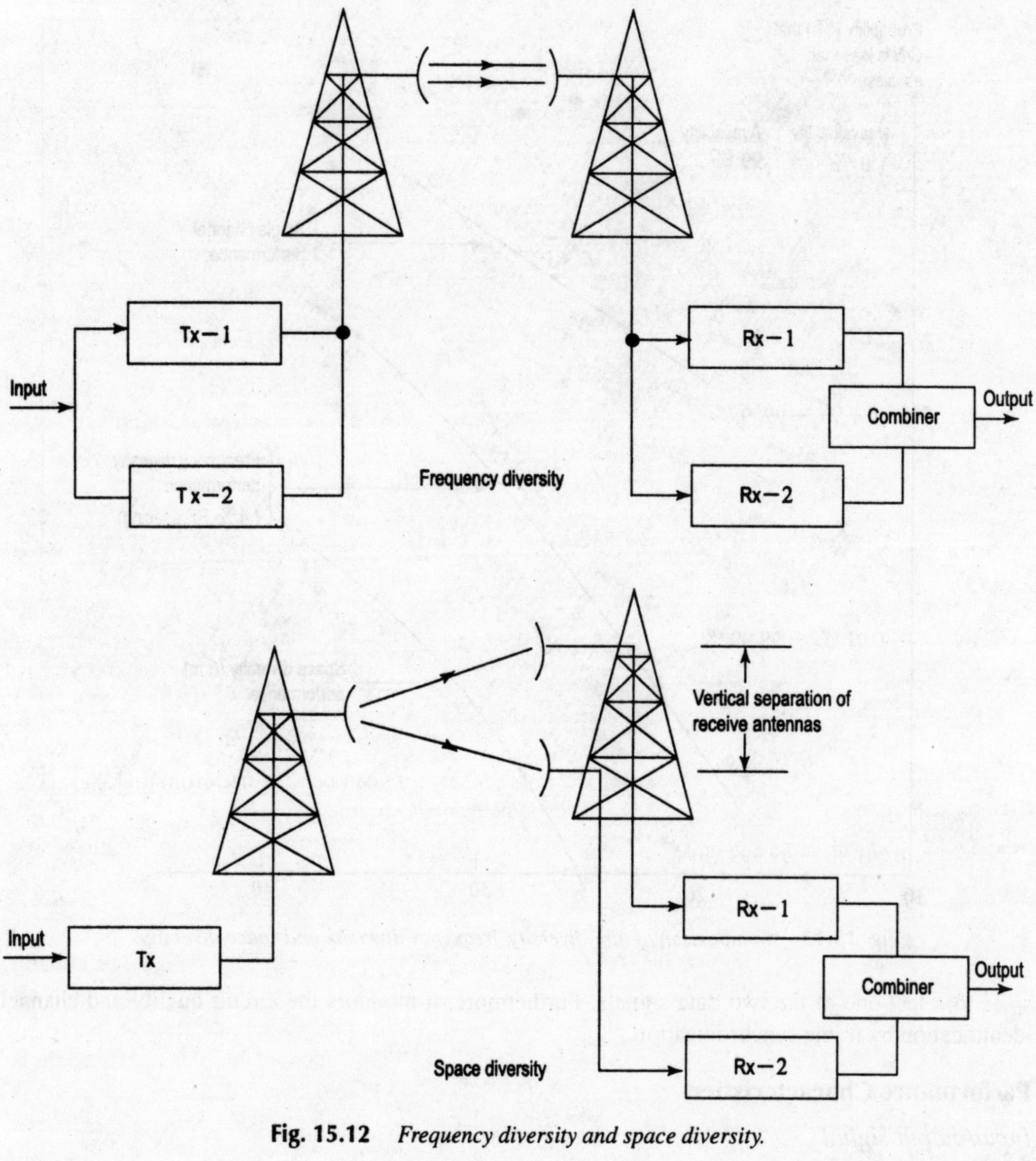

Fig. 15.12 *Frequency diversity and space diversity.*

If we refer to the block diagram of digital radio DMR 770 (Fig 6.3 (*c*)), we will note that the PH-DEM outputs of regular and protection paths are given to BIT-COMB of both paths which is instrumental in hitless switching. It is described in detail in the following section.

15.9.1 Hitless switching in DMR 770 using BIT-COMB

The BIT-COMB (bit combiner) corrects the delay timing of each input data stream in the regular and protection channels, and accomplishes hitless switching by triggering the four switching mode signals

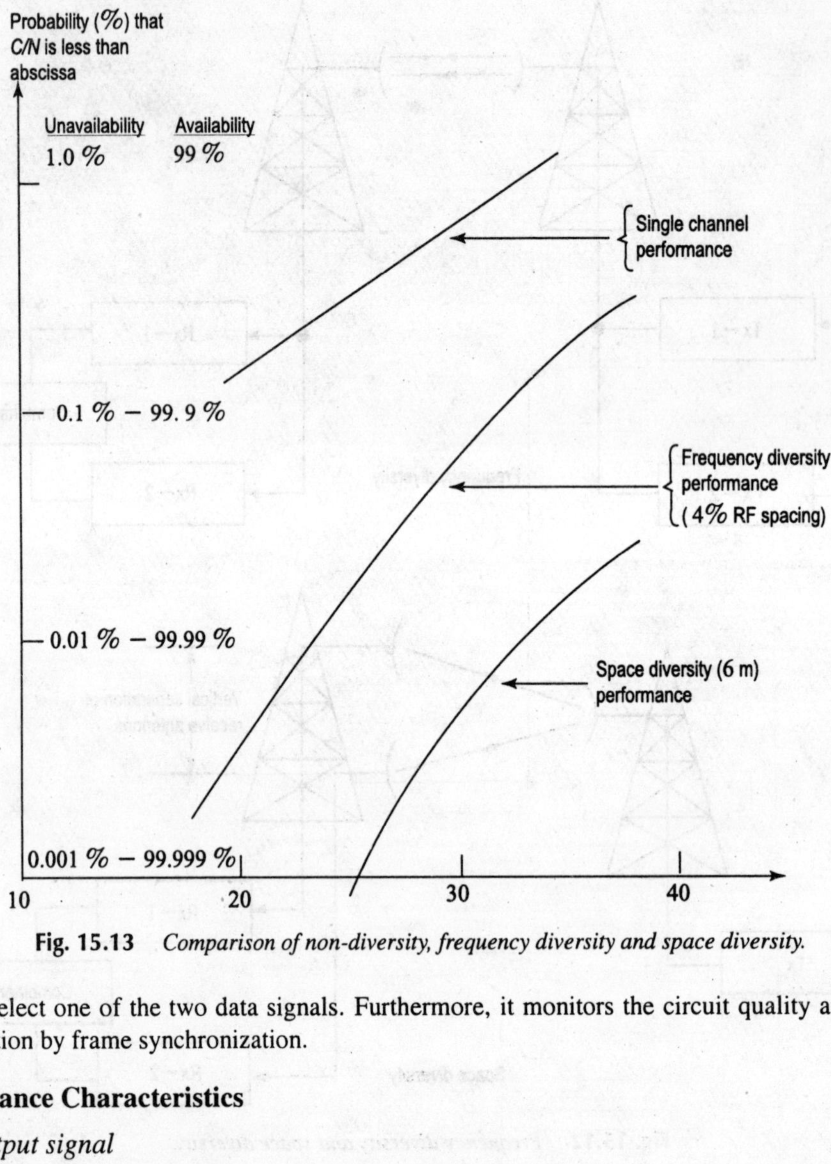

Fig. 15.13 *Comparison of non-diversity, frequency diversity and space diversity.*

so as to select one of the two data signals. Furthermore, it monitors the circuit quality and channel identification by frame synchronization.

Performance Characteristics

Input/output signal

Code format	:	NRZ
Bit rate	:	19.331 Mbps
Level	:	0 to 2 V (75 ohm, unbalanced)

Circuit operation principles

The BIT-COMB consists of a PROT/REG data selector, delay adjustor-1, delay adjustor-2, data line selector, switching logic circuit, frame synchronizer and alarm control circuit. Block diagram of BIT-COMB is shown in Fig. 15.14.

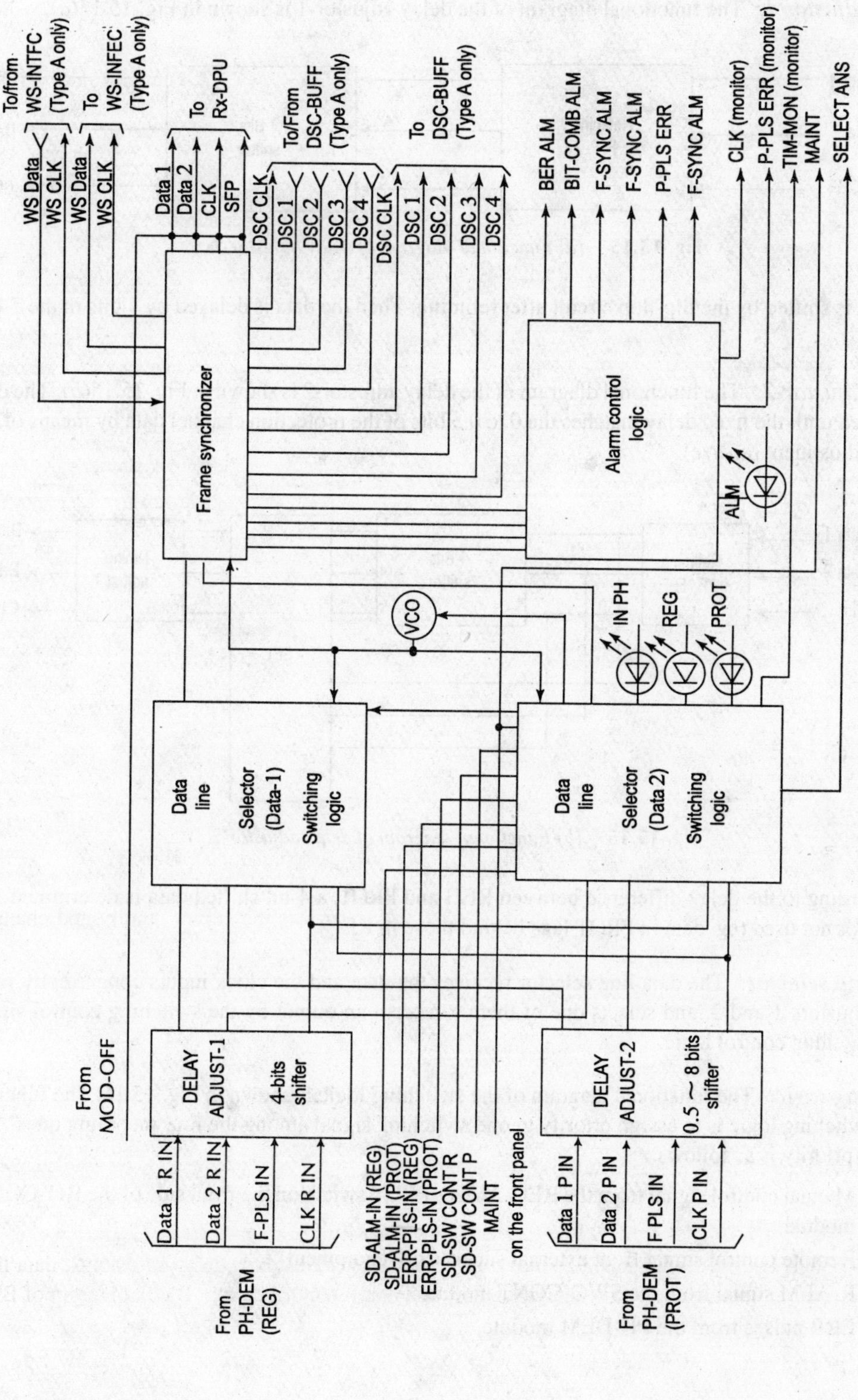

Fig. 15.14 *Block diagram of BIT-COMB.*

Delay adjustor 1: The functional diagram of the delay adjustor-1 is shown in Fig. 15.15(*a*).

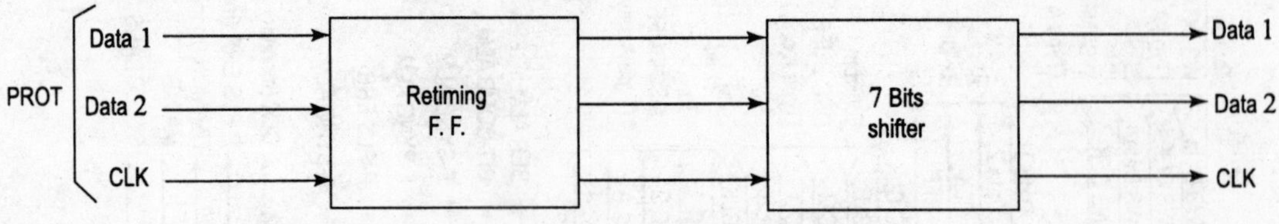

Fig. 15.15 *(a) Functional diagram of delay adjustor-1.*

The data is shifted by the flip-flop circuit after retiming. Then the data is delayed by 7 bits in the 7-bits shifter.

Delay adjustor-2: The functional diagram of the delay adjustor-2 is shown in Fig. 15.15(*b*). The data is adjusted until the fixed delay matches the 0 to 0.5 bits of the protection channel data by means of the timing adjustment (coarse).

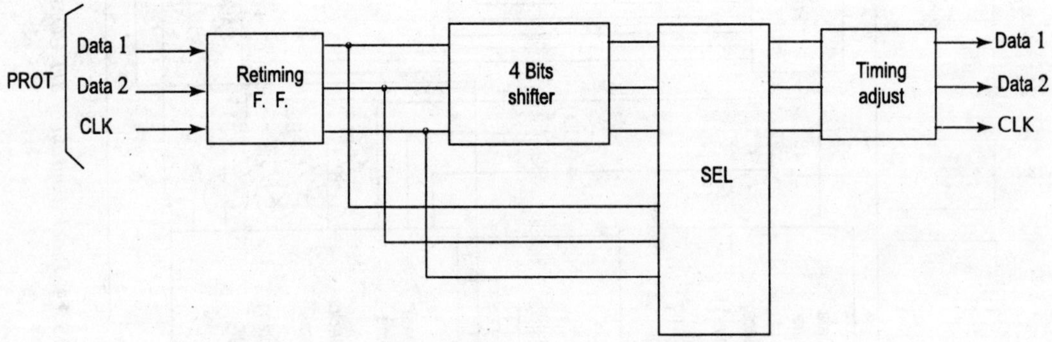

Fig. 15.15 *(b) Functional diagram of delay adjustor-2.*

According to the delay difference between REG and PROT, a 4-bit shifted data is determined—to be used or not used (eg. data in PROT lags behind those in REC).

Data line selector: The data line selector receives the data and the clock inputs concurrently from delay adjustors 1 and 2, and selects one of them to accept an output by the switching control signal from switching control logic.

Switching logic: The functional diagram of the switching logic is shown in Fig. 15.16. The function of the switching logic is to assign priority to one switching signal among the four incoming ones. The order of priority is as follows:

1. Manual control signal from the REG-AUTO-PROT switch on the front side of the BIT-COMB module
2. Remote control signal from external supervisory equipment
3. R-ALM signal from the SWO-CONT module
4. ERR pulses from the PH-DEM module.

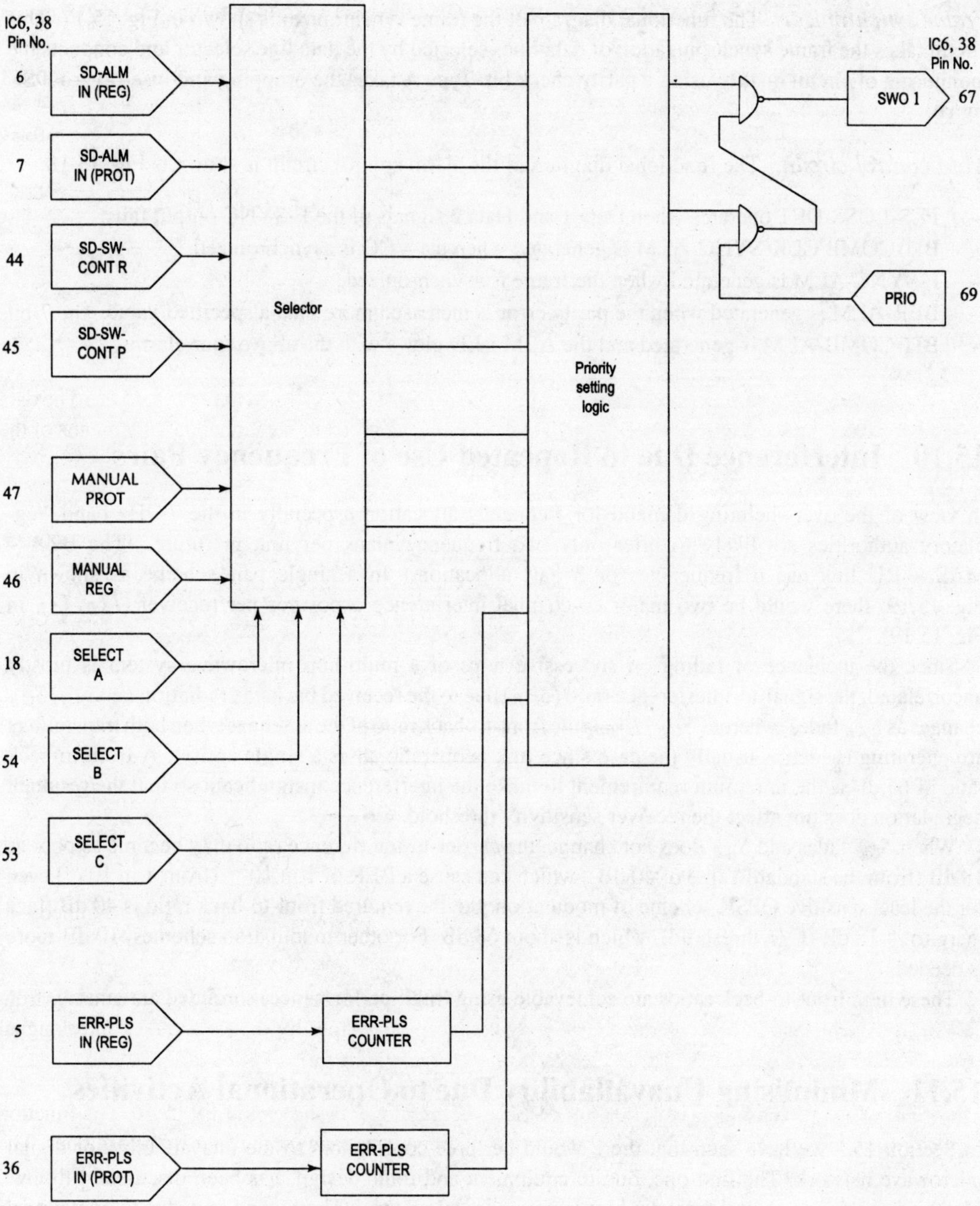

Fig. 15.16 *Functional diagram of switching logic.*

Frame synchronizer: The functional diagram of the frame synchronizer is shown in Fig. 15.17. This circuit takes the frame synchronization of data lines selected by the data line selector and conducts the monitoring of circuit quality using a parity check bit. Type A takes the dropping and insertion of DSC signal.

ALM control circuit: The functional diagram of the alarm control circuit is shown is Fig. 15.18.

- PLS-LOSS-DET operates when Data 1 and Data 2 signals of the F-SYNC output fails.
- BIT-COMB CLK-SYNC-ALM is generated when the VCO is asynchronized.
- F-SYNC-ALM is generated when the frame is asynchronized.
- BER-ALM is generated when the parity error is increased more than a specified value.
- BIT-COMB-ALM is generated and the ALM LED glows with the above four alarms.

15.10 Interference Due to Repeated Use of Frequency Pairs

In view of the overwhelming demand for frequency allocation especially in the 7 GHz band, regulatory authorities are likely to offer only two frequency plans per link in future. (The BZA – MAS – RU link has 6 frequencies or 3 pair allocation.) In a single pair scheme as shown in Fig. 15.19, there would be two major co-channel interference exposures per receiver (I_{AB}, I_{CB} in Fig. 15.19).

Since the incidence of fading on successive hops of a multi-hop microwave system is mostly uncorrelated, the signal-to-interference ratio (S_{CB}) due to the received backside radiation i.e. S_{AB}/S_{CB} changes as S_{AB} fades, whereas S_{CB}/I_{CB} is the front-to-back ratio of the antennas when both transmitters are operating (which is usually the case since link is operational as a single entity). A front-to-back ratio of 65 dB is the minimum requirement to make the interference insignificant so that the resultant degradation does not affect the receiver sensitivity threshold.

When S_{CB} fades and S_{AB} does not change, the carrier-to-interference ratio may become as poor as 18 dB (from the standard value of 40 dB), which can cause a BER of 1 in 10^{-5} (from 1 in 10^{-6}) even for the least sensitive QPSK scheme of modulation. So, the required front-to-back ratio is 40 dB (face margin) + 18 dB (C/I threshold), which is about 68 dB. For other modulation schemes, 10 dB more is needed.

These high front-to-back ratios are achievable using (high-performance) shrouded antennas.

15.11 Minimising Unavailability Due to Operational Activities

In Section 15.5 we have seen that there would be three contributors to the unavailability of digital microwave network. The first one, due to equipment and route design, has been discussed till now. The third one due to natural catastrophies is not under our control. The second one, due to operational activities, is partially under our control. The following tips would help in this regard:

- Proper maintenance of power plant
- Regular audit of power plant and protection system
- Ring earth for lightning protection (refer Appendix)

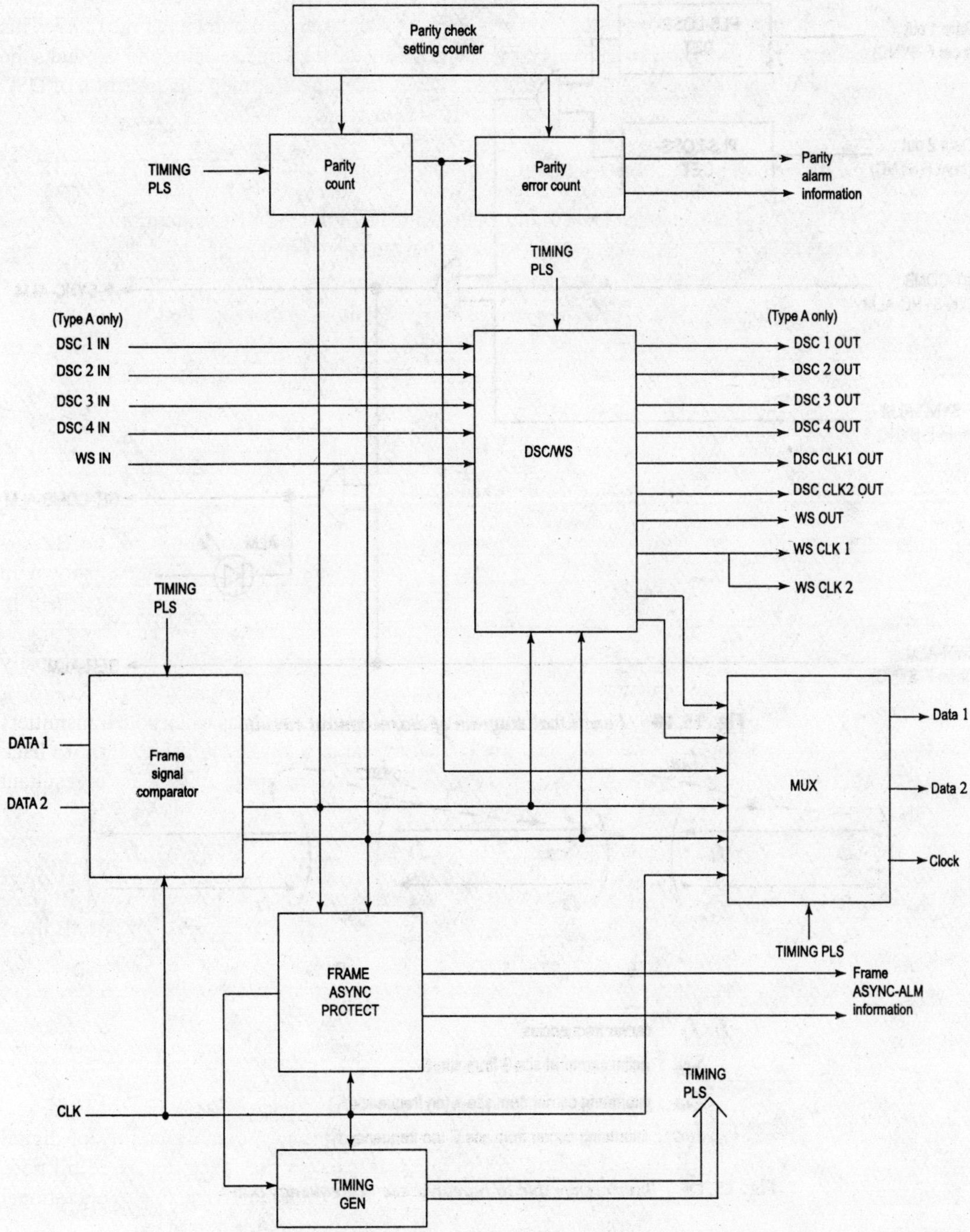

Fig. 15.17 *Functional diagram of frame synchronizer.*

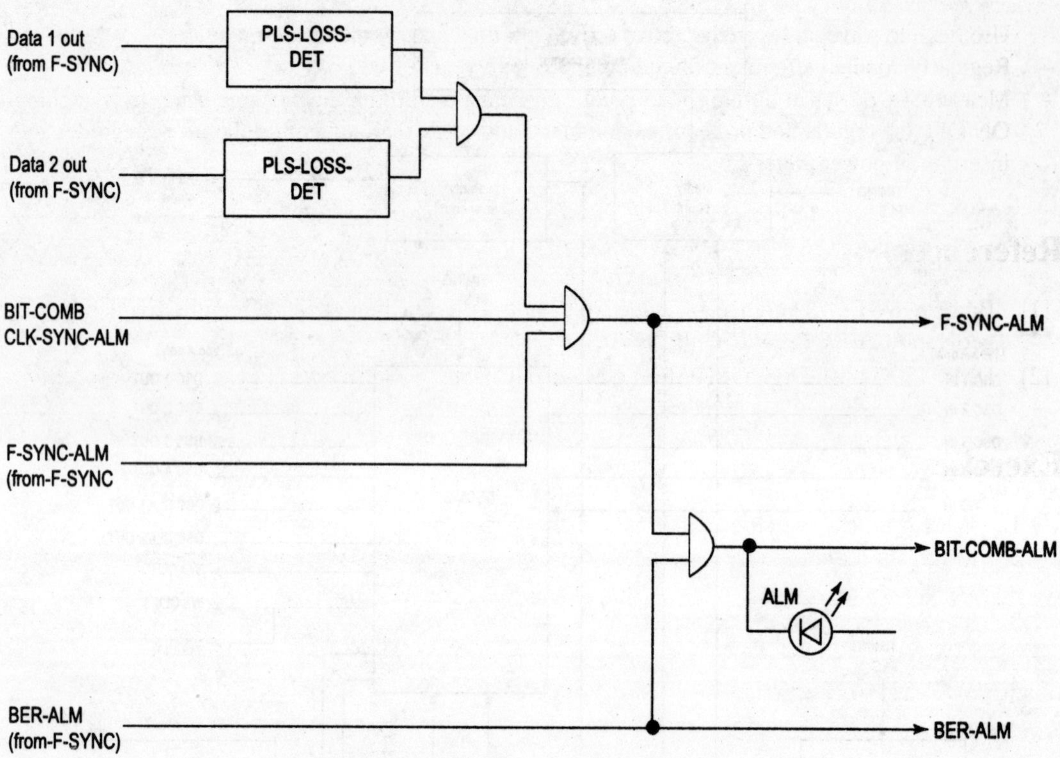

Fig. 15.18 *Functional diagram of alarm control circuit.*

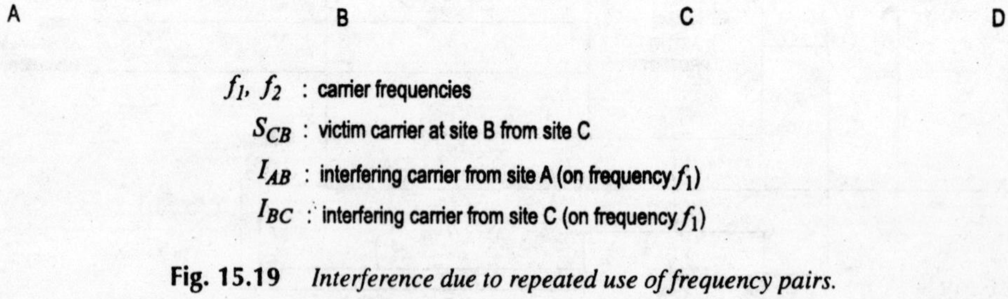

f_1, f_2 : carrier frequencies

S_{CB} : victim carrier at site B from site C

I_{AB} : interfering carrier from site A (on frequency f_1)

I_{BC} : interfering carrier from site C (on frequency f_1)

Fig. 15.19 *Interference due to repeated use of frequency pairs.*

While changing over from analog to digital by partial coverage method, first install (for digital link) antennas at lower heights than the existing antennas, then operate digital link, transfer the circuits, and thereafter remove the analog equipment. After that, connect the top antennas (old ones) to main ports and the newly installed bottom antennas to SD ports.

- Use flexible waveguides in the above activity till final arrangements are made.
- Regularly conduct alignment of equipment (once a year).
- Measure Tx-power at dummy load point rather than disturbing the working setup by switching ON/OFF the regular and protection equipment, and while removing of miniature waveguides and insertion of power meters.

References

[1] Barnett, W.T., Occurence of selective fading as a function of path length, frequency and geography, *IEEE Proc. ICC-70*, 1970.
[2] DMR 770 34 Mb Digital MW Radio Manual, Chapter 6, Vol 1, M/s ITI Ltd., Bangalore, 1996.

Exercises

1. What is hitless switching? How is it achieved in space diversity equipment?
2. Multipath fading reduces overall system gains in a digital MW link. Comment in analytical sequence.
3. How do you reduce interference due to repeated use of frequency-pairs on long haul digital MW link?
4. Why is space diversity preferred over frequency diversity?

16

Control of Jitter in Long Haul Digital Networks

16.1 Introduction

In chapter 6, while exploring jitter in multiplex systems, we concluded that the following are the different types of jitter introduced in the system:

- Jitter due to the routine insertion of frame alignment words, service digits and justification instructions
- Justification jitter
- Waiting time jitter

Further, we have seen that data transition jitter gets introduced in the symbol timing recovery circuits that extract clock from the demodulated baseband signal in the radio equipment. The Main/WS/ DSC data streams get combined and segregated respectively in the multiplexer and demultiplexer of the radio equipment, and if proper control is not exercised, under certain circumstances, jitter can accumulate to such an extent that the following impairments arise:

- Increase in probablity of introducing errors in digital signals at points of signal regeneration as a result of timing signals being displaced from their optimum postion in time
- Introduction of uncontrolled slips in digital signals through store spillage and depletion in certain types of terminal equipment incorporating buffer stores and phase comparators
- Degradation of digitally encoded analog information as a result of phase modulation of reconstructed samples in the digital-to-analog conversion device at the end of connection

Hence, it is imperative that proper control should be exercised on jitter. In Chapter 15, we looked at an example of long haul digital MW network and emphasised that a system's worst case overall peak-to-peak clock jitter should be less than 10% of the symbol duration. Keeping this objective of system design in mind, in this chapter we proceed to examine CCITT recommendation regarding control of jitter and jitter accumulation for cascaded equipment on long haul network. Finally, we will examine some of the design techniques to keep jitter under check.

16.2 Jitter Control Philosophy

The jitter control philosophy should provide answers to the following three questions:

1. What is the maximum network limit that should not be exceeded by jitter at any hierarchical interface?

2. What are the consistent specifications of individual digital equipment?
3. What are the guidelines to measure and control jitter accumulation in any network configuration?

These issues are now examined one by one.

16.3 Network Limits for Maximum Output Jitter and Wander

16.3.1 Jitter v/s wander

Jitter is the time-shift of the recovered clock from the original symbol clock. Its sources are the equipment's timing circuits, or rather the role played by them in the extraction of timing information and insertion of bits. Wander is the long-term variation of significant instants of digital signal from its ideal postion in time. It arises due to changes in propagation delay in transmission media and equipment. Wander causes slips in transmission if not accomodated properly at the input ports of the equipment.

16.3.2 Network limits for jitter at hierarchical interface

CCITT Rec.G.823 gives the maximum permissible levels of jitter at hierarchical interfaces within a digital network. These limits should be met for all operating conditions, regardless of the number of equipment preceding the interface. The limits specified here are compatible with the minimum tolerance to jitter required to be provided by ports of all equipment.

The jitter limits and values of jitter cut-off frequencies as per CCITT Rec.G.823 for different hierarchial levels are given in Table 16.1 as per CCITT Rec.G.823. The frequency response of the filters associated with the measurement apparatus should have a roll-off of 20 dB/octave.

For systems in which the output signal is controlled by an autonomous clock (e.g. the crystal oscillator), more stringent output jitter values than those of Fig. 16.1 may be defined in the equipment specifications. For e.g., CCITT Rec.G.735 specifies maximum peak-to-peak output jitter as 0.05UI (UI stands for 'unit interval') for the MULDEX (multiplexer–demultiplexer equipment).

16.3.3 Network limits for wander at hierarchical interface

CCITT Rec G.823 does not define the maximum network limit for wander at all hierarchical levels. But the actual magnitude of wander, which is dependent on fundamental propagation characteristics of the transmission media and the ageing of the clock circuitry, can be predicited. Studies on telcom networks worldwide have shown that input ports can tolerate wander in accordance with the input tolerance limits of a particular hierarchical level (± 50 ppm for 2 Mb, ± 20 ppm for 34 Mb), and slips introduced as a result of exceeding input tolerance will be rare.

CCITT Rec.G.823 does however define the limits for wander for interfaces to network nodes (Fig. 16.1). These limits state that the MTIE (maximum time interval error) over a period of S seconds shall not exceed $(10^{-2}S + 10000)$ ns for $S > 10^4$. The study is still under way for $S < 10^4$.

16.3.4 Jitter and wander consideration for synchronised networks

In a synchronised network, it is assumed that the digital equipment at different nodes accomodate permitted phase deviations on the incoming signal (with jitter and wander). Thus, under normal conditions no 'slip' occurs.

Table 16. 1 *Maximum permissble jitter at a hierarchical interface*

Parameter value / Digit rate (kbps)	Network limits		Measurement filter bandwidth		
	B_1 unit interval peak-to-peak	B_2 unit interval peak-to-peak	Bandpass filter having a lower cut-off frequency f_1 and f_3, and an upper cut-off frequency f_4		
			f_1	f_3	f_4
64*	0.25	0.05	20 Hz	3 kHz	20 kHz
2.048	1.5	0.2	20 Hz	18 kHz	100 kHz
8.448	1.5	0.2	20 Hz	3 kHz (80 kHz)	400 kHz
34.368	1.5	0.15	100 Hz	10 kHz	800 kHz
139.264	1.5	0.075	200 Hz	10 kHz	3500 kHz

* For the codirectional interface only.

Note – The frequency values shown in parenthesis apply only to certain national interfaces.

The unit interval (UI) for various data rates are:

For 64 kbps	1 UI = 15.6 μ s
For 2048 kbps	1 UI = 488 ns
For 8448 kbps	1 UI = 118 ns
For 34 368 kbps	1 UI = 29.1 ns
For 139 264 kbps	1 UI = 7.18 ns

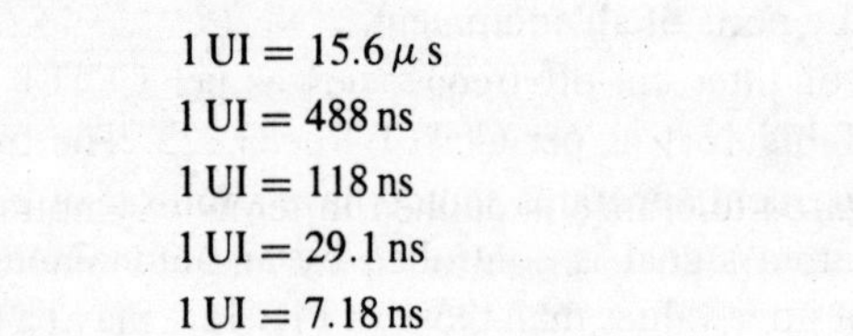

Fig. 16.1 *MTIE v/s observation period S for the output of a network node.*

However, due to performance degradation failure conditions, maintenance actions and other events, the TIE (time interval error) between incoming signal and the interval timing signal of the terminating equipment may exceed the wander and jitter tolerance of the equipment. Such a situation leads to controlled slip.

At nodes interconnecting independently synchronised networks (or plesiochronous operation as in national networks), the TIE between incoming signal and the internal timing signal of the terminating equipment may exceed the wander and jitter tolerance limits leading to 'slip'.

The maximum permissible long-term mean controlled slip rate resulting from the above situation is 1 slip in 70 days (CCITT Rec.G.711).

16.4 Jitter Limit Specifications for Individual Digital Equipment

For individual digital equipment, jitter performance is specfied in three respects:

- Jitter and wander tolerance of digital input ports
- Maximum output jitter in the absence of input jitter
- Jitter and wander transfer characteristics

16.4.1 Jitter and wander tolerance of digital input ports

For converting any equipment to recommended hierarchical interface within a network, it is necessary that the input ports of all equipment be capable of accommodating levels of jitter up to the maximum limits specified in Section 16.3.

To be more precise, the required tolerance is defined in terms of amplitude and frequency of the sinusoidal jitter, which when modulating a test pattern, should not cause any significant degradation in the operation of the equipment.

Thus, all the digital input ports of the equipment should be able to tolerate a digital signal having electrical characteristics in accordance with the requirements of Rec.G.703, but modulated by a frequency relationship as defined by Fig. 16.2. Table 16.2 shows the appropriate limits for different hierarchical

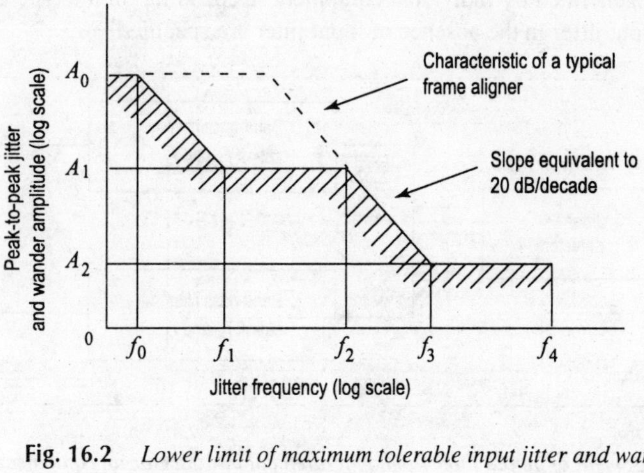

Fig. 16.2 *Lower limit of maximum tolerable input jitter and wander.*

levels. These requirments should be met, regardless of the information content of the digital signal. For test purposes, the equipment binary content of the signal with jitter modulation should be a pseudo-random bit sequence as tabulated in the last column of Table 16.2.

It may please be noted that:

(*a*) The test condition in itself is not intended to be representative of the type of jitter that may be found in practice in a network. The purpose of the test specified is to ensure the following:

- The Q-factor associated with the timing signal recovery of the equipment input circuitry is not excessive
- Adequate amount of buffer storage (elastic buffer for read/write operations bit streams) has been provided

(*b*) The wander effects are predominant below f_1. Many transmission equipment such as digital line systems and synchronous MULDEXes using justification techniques are transparent to these very low frequency changes in phase which are not amenable to simple practical evaluation. Hence these limits should be taken care of at the design stage of the equipment.

(*c*) The part of mask between f_1 and f_4 reflects the maximum permissible jitter magnitude in a digital network. But the part of the mask below f_1 does not represent the maximum permissible wander that might occur in practice. Below f_1 the mask is derived such that wherever necessary, provision of this level of buffer storage at the input of an equipment facilitates the accomodation of wander generated in a majority of real situations.

(*d*) A short term reversal of the relative TIE between the incoming signal and the internal timing signal of the terminating equipment shortly after the occurence of a controlled slip, should not cause another slip. In order to prevent such a slip, the equipment should be designed with suitable hysteresis for this phenomenon. This hysteresis should be at least 18 μs (refer to Table. 16.2 for values of f_1, f_2, f_3 and f_4 for various data rates).

16.4.2 Maximum output jitter in the absence of input jitter

The arrangement for measuring output jitter is shown in Fig. 16.3. In the absence of input jitter, we measure only the jitter generated by individual equipment. Depending on the type of equipment, the maximum levels of output jitter in the absence of input jitter are specified.

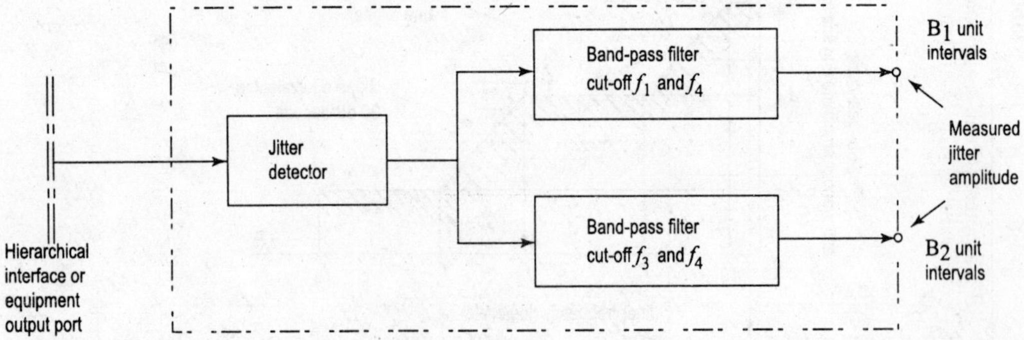

Fig. 16.3 *Measurement of output jitter from a hierarchical interface or an equipment output port.*

Table 16.2 *Parameter values for input jitter and wander tolerance*

Parameter value / Digit rate (kbps)	Peak-to-peak amplitude unit interval			Frequency					Pseudo-random test signal
	A_0	A_1	A_2	f_0	f_1	f_2	f_3	f_4	
64 (Note 1)	1.15 (18 μs)	0.25	0.05	1.2×10^{-5} Hz	20 Hz	600 Hz	3 kHz	20 kHz	$2^{11} - 1$ (Rec. O. 152)
2.048	36.9 (18 μs)	1.5	0.2		20Hz	2.4 kHz (93 Hz)	18 kHz (700 Hz)	100 kHz	$2^{15} - 1$ (Rec. O. 151)
8.448	152 (18 ms)	1.5	0.2	1.2×10^{-5} Hz	20 Hz	400 Hz (10.7 kHz)	3 kHz (80 kHz)	400 kHz	$2_{15} - 1$ (Rec. O. 151)
34.368	*	1.5	0.15	*	100 Hz	1 kHz	10 kHz	800 kHz	$2^{23} - 1$ (Rec. O. 151)
139.264	*	1.5	0.075	*	200 Hz	500 kHz	10 kHz	3500 kHz	$2^{23} - 1$ (Rec. O. 151)

* Values under study.

Note 1 – For the codirectional interface only.

Note 2 – For interfaces within national network the frequency values (f_2 and f_3) shown in parenthesis may be used.

Note 3 – One Unit Interval (UI) for various data rates are:

For 64 kbps 1 U1 = 15.6 μs
For 2048 kbps 1 U1 = 488 ns
For 8448 kbps 1 U1 = 118 ns
For 34368 kbps 1 U1 = 29.1 ns
For 139 264 kbps 1U1 = 7.18 ns

The value of $A_o(18\mu s)$ represents a relative phase deviation between the incoming signal and the interval timing local signal derived from the reference clock. This value of A_0 corresponds to an absolute value of 21 μs at the input to a node (i.e. equipment input port) and assumes a maximum wander of the transmission link of 11 μs between two nodes. The difference of 3 μs corresponds to the 3 μs allowed for long-term phase deviation in the national reference clock (Rec. G.811, §3c).

For example, for 34 Mb MUX, it is specified that the peak-to-peak output jitter at the tributary output in the absence of input jitter should not exceed 0.25 UI (UI stands for unit interval) when measured in the frequency range up to 100 kHz. When measured with an instrument incorporating a bandpass filter having a lower cut-off frequency of 18 kHz, a roll-off of 20 dB/octave and an upper limit of 100 kHz, the peak-to-peak output jitter should not exceed 0.05 UI, with a probability of 99.99% during a measurement period of 10.

Coming to the multiplexer signal output jitter, in the case of transmitting signal being derived from an internal oscillator, the peak-to peak jitter at the 34368 kbps output should not exceed 0.05 UI when it is measured within the frequency range from $f_1 = 100$ Hz to $f_4 = 800$ kHz. When measured with an instrument incorproating a bandpass filter having a lower cut-off frequency of 3 kHz, a roll-off of 20 dB/ decade, and an upper limit of 400 kHz, the peak-to-peak output jitter should not exceed 0.05 UI, with a probability of 99.99% during a measurement period of 10 s.

16.4.3 Jitter and wander transfer characteristics

Jitter transfer characteristics define the ratio of output jitter amplitude to input jitter amplitude for various jitter frequencies for a given bit rate. When jitter is present at the digital input port of the equipment, some portion is transmitted to the output port, and only the higher frequency jitter components get filtered. To control jitter in cascaded homogeneous digital equipment, it is important to restrict the value of the jitter gain.

Figure 16.4 indicates the general shape of jitter transfer characteristic. The appropriate values for x, y, f_1, f_5, f_6 and f_7 can be obtained from the relevant recommendation.

In a synchronous digital equipment, the bandwidth of the phase smoothening circuits is above 10 Hz. Hence, wander on input signal may appear virtually unattenuated on the output. However, in certain particular digital equipment (for example nodal clock), it is necessary to attenuate wander. For such synchronous equipment, wander transfer characterstics are also specified in CCITT recommendations.

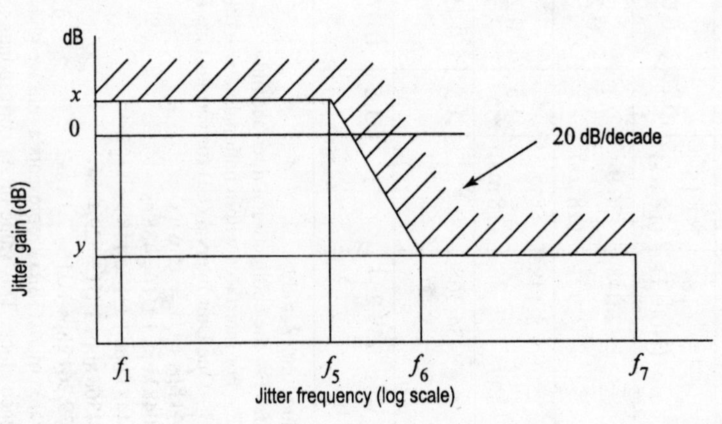

Fig. 16.4 *Jitter transfer characteristics.*

16.5 Measurement of Jitter

Jitter measurement can be classified into two categories:

- Measurements using an undefined traffic signal which may be considered as quasi random (this is applicable under operational condition of the digital link)
- Measurements using specific test sequences (these are applicable during laboratory, factory and commissioning conditions)

16.5.1 Measurements using an undefined traffic signal

Because of the quasi random nature of jitter and its possible dependence on traffic loading, accurate peak-to-peak measurements need to be made in operational networks over long periods of time. In practice, it is expected that with the experience gained from working with particular systems it would be possible to identify abnormalities measured over shorter measurement periods, which would go on to indicate that the maximum permissible limit might be exceeded over longer intervals of measurement.

The network limits specified in Section 16.3 are so defined that the probability of exceeding such levels is very small. Any practical observation of such a measurement interval is not possible. So, a smaller interval with relatively greater probability limit is chosen. The test conducted on BZA–MAS–RU digital MW link is presented in Chapter 19.

16.5.2 Measurement using a specific test sequence

It is convenient to test jitter performance of digital equipment with PRBS (pseudo random binary sequence) in factory tests and commissioning tests. But in operational situations the information content of the signal is more random (e.g. telephone channel). Hence, a correction factor is used to relate traffic-based measurement to PRBS-based measurement. The correction factor applied should also take care of shortcomings in PRBS technique due to inadequate spectral content of PRBS within jitter bandwidth of the system on which measurement is being taken. These are elaborated next.

16.5.3 Relationship between random traffic-based measurement and PRBS- based measurement

In PRBS technique, we are basically emulating the random type traffic signal by pseudo random sequence. The results of PRBS technique, in optimistic sense, should be identical with those of random traffic signal technique. But, in reality, there is a disparity. The reason is that in random traffic signal, an almost unrestricted range of jitter amplitudes is possible, whereas in PRBS technique, only a finite range is possible.

The results of the two techniques can be related by a correction factor based on operational experience. It is very difficult to prescribe a formula for this factor as we have to establish a relation for every conceivable practical situation with several parameters—the measurement period, presence/absence of scramblers, system length, value of Q of timing recovery circuit, sequence length, etc.

A good practical choice of correction factor for most of the situations are given below:

- 1.5 at 2048 kbps and 8448 kbps (based on the use of $(2^{15} - 1)$ PRBS generated in accordance with CCITT Rec.0.151)
- 1.3 at 34368 kbps and 139264 kbps (based on the use of $(2^{23} - 1)$PRBS generated in accordance with CCITT Rec.0.151)

The estimated jitter amplitude when transmitting random traffic signal } = (correction factor) × (measured jitter amplitude with PRBS technique).

16.6 Spectral Content of PRBS

PRBS is cyclical and hence has power spectrum with spectral lines occurring at regular intervals. For achieving results with acceptable measurement error, it is essential that the PRBS used when measuring output jitter has adequate spectral content within the jitter bandwidth of the system being measured. The bandwidth of the jitter spectrum at the output of a chain of digital regenerators is a function of the Q factor of the timing recovery circuit and the number of regenerators in tandem.

$$\text{Jitter bandwidth} = \frac{f_1}{Q \times n} \text{ Hz for large } n \text{ where,}$$

f_1 = frequency of the timing signal extracted from the incoming signal by the timing recovery circuit,

Q = Q factor of one repeater,

and n = number of cascaded repeaters.

$$\text{PRBS repetition rate} = f/L \text{ Hz,}$$

where f is the bit rate and L is the sequence length.

For adequate spectral content, the pattern should be less than $1/y$ of the jitter bandwidth of the system under test.

$$\text{i.e,} \quad \frac{f}{L} \leq \frac{f_1}{y \times Q \times n}.$$

Value of y is under study by various research groups.

16.7 Jitter Accumulation Relationships for Cascaded Equipment

In this section the expressions for accumulated jitter when homogenous (same type) regenerators are in cascade are presented. Jitter accumulatation in MULDEX (multiplexer–demultiplexer) is also discussed.

16.7.1 Cascaded homogeneous regenerators

Most of the repeaters are fully regenerative and self-timed, i.e., the output signal is retimed under the control of a timing signal derived from the incoming signal. We study here

- pattern-dependent jitter,
- random pattern jitter,
- two repetitive pattern jitter.

Pattern-dependent jitter

The most significant component of jitter in a regenerator is the waiting-time jitter which is pattern dependant. If this jitter is truly random, the total rms jitter J_N on a link with N regenerative repeaters is

$$J_N \cong J \times \sqrt[4]{N} \text{ for large value of } N,$$

where J = rms jitter of single regenerator due to controlled source.

Since the pattern-dependent jitter is not random, the above expression is modified as

$$J_N \cong J_1 \times \sqrt{2N} \text{ for large values of } N.$$

Based on operational experience,

$$0.4 \text{ UI} \leq J_1 \leq 1.5 \text{ UI}.$$

Use of PLL in the timing recovery circuit modifies the expression further as

$$J_N \cong J_1 \times \sqrt{2NA},$$

where A is a factor dependent on the number of repeaters and the damping factor of PLL.

From the above expressions, the following inferences can be drawn:

– Pattern-dependent jitter accumulates more rapidly than non-pattern-dependent jitter as the number of regenerators increased
– The amplitude of jitter produced by a chain of regenerators increases without limit as the number of regenerators is increased

Jitter due to random pattern

The jitter produced by random pattern is itself random. Its amplitude has a Gaussian probability distribution function. Hence, for a given standard deviation (rms amplitude), the probability of exceeding any chosen peak-to-peak amplitude can be calculated. Normally, the probability of exceeding peak-to-peak rms ratio of 15 is very low.

Jitter due to two repetitive patterns

When the signal being transmitted is composed of two repetitive patterns alternating at low frequency, the jitter appears as a low frequency repetitive wave with amplitude proportional to the number of regenerators.

The maximum peak-to-peak jitter is given by

$$J_{NP} = N \times PSJ,$$

where PSJ stands for pattern sensitive jitter, produced by a single regenerator when it is subjected to alternating repetitive patterns.

This situation is unlikely in normal operations, but to rule out the slighest possibility, digital scramblers are used.

16.7.2 Jitter accumulation in MULDEX equipment

Waiting-time jitter is a significant contribution from a MULDEX equipment. Accumulation is in between $\sqrt[4]{N}$ and $\sqrt[2]{N}$, where N is the number of MULDEX pairs.

16.8 Jitter Control Techniques for Digital Radio and Optical Fibre System

The following techniques are in vogue to keep jitter under check in cascaded digital radio and optical fibre systems:

– Use of scrambler/descrambler
– Use of jitter reducers

These have been discussed in detail in Chapter 14.

16.8.1 Jitter reduction with scrambler/descrambler

Inclusion of a scrambler/descrambler in the digital network at each regenerative repeater ensures that there is no correlation in the jitter contributed, i.e., pattern-dependent or repetitive pattern jitters are reduced.

The jitter accumulated is

$$J_m = J_0 \times \sqrt[4]{KM},$$

where J_m = jitter on the M digital hops,

J_0 = rms jitter on single equipment,

and $1 < K < 2$. As M increases $K \to 2$.

16.8.2 Jitter reducer

With jitter reducer the accumulation relationship given above applies, but in place of J_0 we have J_s.

$$J_s = 2MJ_0\sqrt{f_c B},$$

where M = no. of cascaded repeaters,
f_c = cut-off frequency of jitter reducer,
B = half-bandwidth of the regenerative repeater.

References

[1] CCITT Rec. G.823–Network limits for jitter at hierarchical interfaces in digital networks.

Exercises

1. Distinguish between jitter and wander.
2. Which method of jitter measurement is preferred in (a) equipment test situation, (b) operational conditions. What are the reasons?
3. Why do we apply correction factors when we measure jitter using PRBS test sequence?
4. Where is jitter accumulation more?

   – Using scrambler / descrambler
   – Using jitter reducer

   Justify your answer quantitatively.

17

Transmultiplexer

17.1 Introduction

A transmultiplexer provides direct, through-connection of analog channels on FDM to digital channels on TDM and vice versa. Frequency division multiplexing (FDM) and time division multiplexing (TDM) were discussed in detail in Chapter 2. In FDM, 60 channels constitute one supergroup (SG); SG-1 spans 60–300 kHz, SG2 spans 312–552 kHz, etc. Each SG has a 240 kHz band (60 channels × 4 kHz per channel). In TDM, 30 channels constitute one E-1 frame or a tributary, with data rate of 2048 kbps.

A 60 channel transmultiplexer carries out conversion between two 30 channel PCM signals at 2048 kbps and a 60 channel analog SG in the 312–552 kHz band. A 24 channel transmultiplexer carries out conversion between a 24 channel PCM signal at 1544 kbps and two 12 channel analog primary group in the 60–108 kHz band.

In this chapter, we deal with a 60 channel transmultiplexer.

17.2 Transmultiplexing Schemes

A transmultiplexer's function can be realised in three ways:

- Pseudo transmultiplexing
- Analog filtering transmultiplexing
- Digital filtering transmultiplexing

17.2.1 Pseudo Transmultiplexing

In this method, the analog and digital MUX equipment are placed back to back, and the channels are patched at VF level in the distribution frame (Fig. 17.1). There is no transmultiplexer (transmux) involved, really speaking. The equipment to be installed are more, and there is lot of wiring (jumpering) work involved.

If spare MUXes and channel cards are available, this method can be followed to save on the cost of transmux. Also, problems can be easily isolated to either analog or digital portion in this method.

17.2.2 Analog filtering transmultiplexing

The functional diagram of an analog filtering transmultiplexer is given in Fig. 17.2.

In the digital-to-analog direction, the 64 kbps digital signal corresponding to a given channel is first extracted from the 2048 kbps PCM signal by digital demultiplexing, and then converted to analog

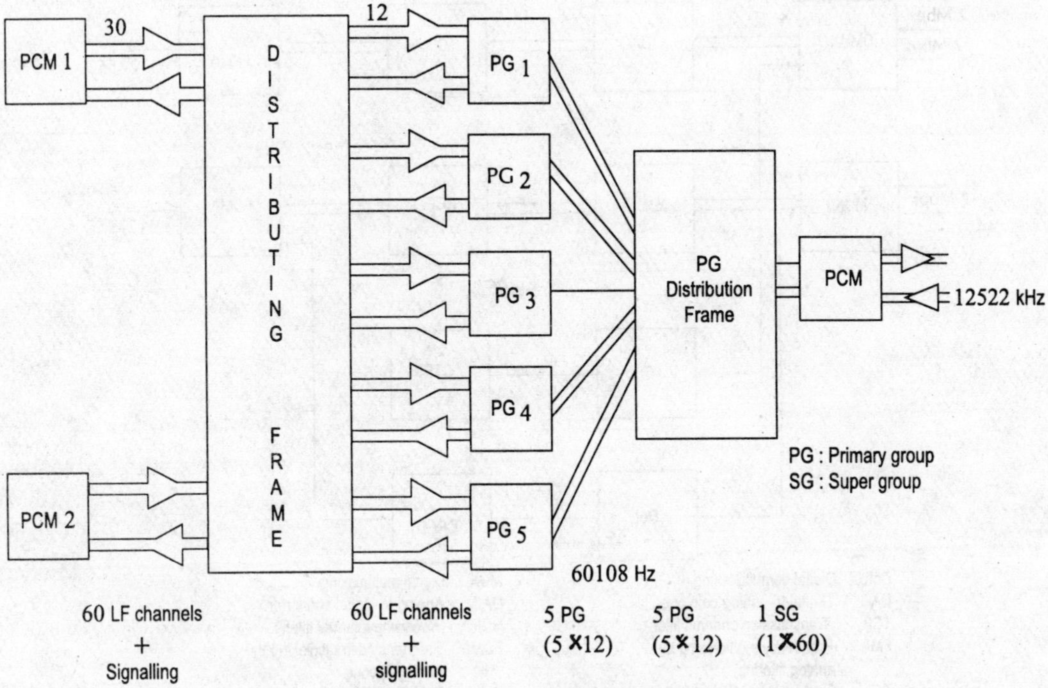

Fig. 17.1 *Pseudo transmultiplexing.*

signal by digital-to-analog conversion (sampled at 8 kHz). This analog signal is then 'frequency division multiplexed' with the signals corresponding to the other channels to form the standard basic supergroup. In the analog-to-digital direction, the operations are the reverse of the above operation.

In case analog multiplexing follows out-of-band signalling (R2 code with signalling frequency 3825 Hz), it is necessary to limit the 0–4 kHz band to 300–3400 Hz, and extract the 'signalling' signals from the PCM frames.

Four types of filters are used in the MUX–DEMUX operations of this scheme:

TCF - Transmission channel filter (VF signal is cleaned around signalling frequency)

RFC - Receive channel filter (eliminates residues of signalling signals from speech signals)

CTF - Call transmission filter (limits the spectrum of signalling signal)

CRF - Call receive filter (allows extraction of signalling tone from received signal)

Analog filtering transmultiplexing is easy from the design point of view, but not elegant; sophisticated digital processing techniques available are not exploited. 'Digital filtering transmultiplexer' utilizes digital signal processing (DSP) techniques which are described next.

17.2.4 Digital filtering transmultiplexing

In this method, A/D and D/A conversions are done at FDM–TDM level using digital signal processing techniques. Digital filtering transmultiplexer is compatible with analog and digital MUX equipment, in accordance with CCITT recommendations. It has three sub-units, functionally:

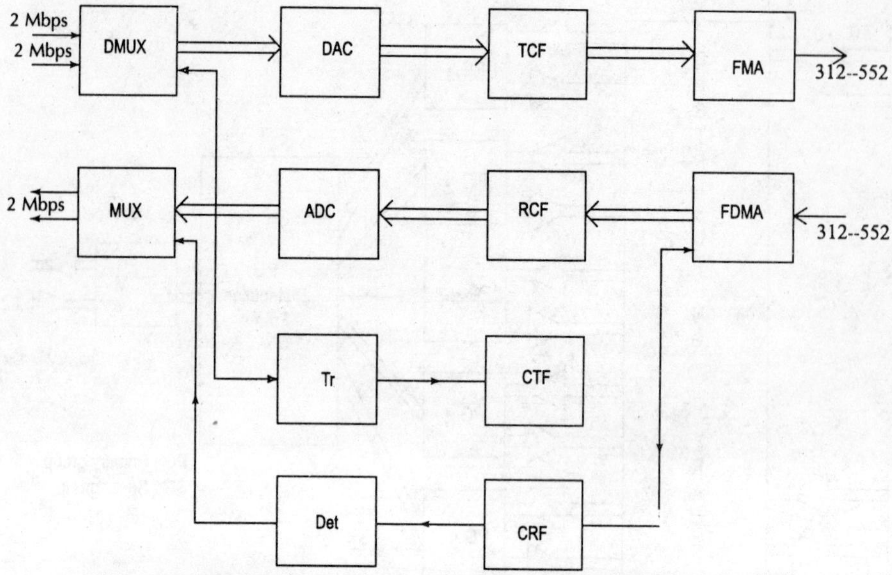

DMUX	Digital demultiplexing	MUX	Digital multiplexing
DAC	Digital-to-analog converter	DAC	Analog-to-digital converter
TCF	Transmission channel filter	RCF	Receiving channel filter
FMA	Frequency multiplexing by analog filtering	FDMA	Frequency demultiplexing by analog filtering
Tr	Transmission of signalling	Det	Detection of the signalling
CTF	Call transmission filter	CRF	Call receiving filter

Fig. 17.2 *Analog filtering transmultiplexer.*

– PCM interface unit
– Digital processing computer
– Analog interface unit

A block diagram of a digital filtering transmux is shown in Fig. 17.3, and its functional diagram in Fig 17.4. Henceforth, in this chapter, the term T-MUX will be used to refer to digital filtering transmultiplexer.

17.3 Functional Description of T-MUX

A T-MUX of M/s PCL make (TM-7800 MI) is taken as an example and discussed in detail to illustrate the operation of a T-MUX equipment. The complete block diagram of TM-7800 M1 T-MUX is shown in Fig 17.5. In the PCM-to-FDM direction, the signal first flows through the PCM interface units, then the tone generator, signalling mapping, discrete cosine transform (DCT), modulator weighting network (WN), and D/A converter units. In the FDM-to-PCM direction, the signal flows through the A/D converter, demodulator WN, DCT, tone detector, signalling mapping, and the two PCM interface units. The plug-in unit descriptions follow the signal flow in the same order.

Group and supergroup pilot frequencies are shown in Table 17.1.The relationship of PCM channels to FDM frequencies is shown in Table 17.2.

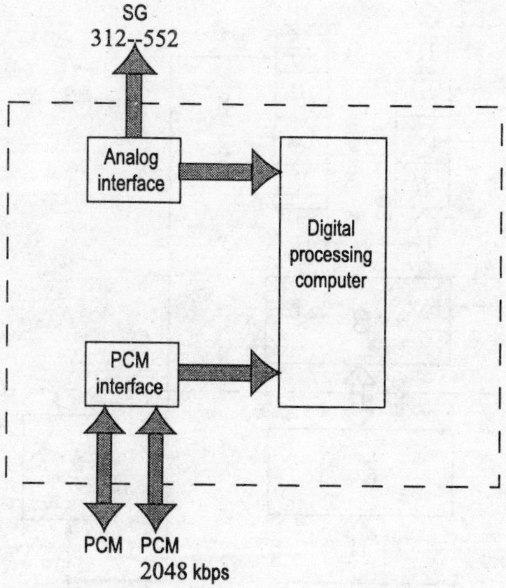

Fig. 17.3 *Block diagram of a T-MUX.*

Table 17. 1 *Supergroup and group pilots*

PCMI	Channel 6	Group 1 pilot at 335.92 kHz
PCMI	Channel 18	Group 2 pilot at 383.92 kHz
PCMI	Channel 25	Supergroup pilot at 411.92 kHz
PCMI	Channel 30	Group 3 pilot at 431.92 kHz
PCM2	Channel 12	Group 4 pilot at 479.92 kHz
PCM2	Channel 24	Group 5 pilot at 527.92 kHz

Although the signal processing is discussed in analog terms, digital techniques are used for all functions except those on the FDM side of the actual D/A and A/D conversions.

The following parts describe the operations of T-MUX at the functional level. The signal processing is followed through each plug-in unit, first in the PCM-to-FDM direction, then in the FDM-to-PCM direction. Support functions are discussed later.

17.3.1 PCM-to-FDM direction

PCM interface

The incoming PCM signal enters via connectors and backplane and goes to the PCM interface unit (Fig. 17.6). Test points on the front panel allow the incoming signal to be monitored. The signal then goes through test jack J1 at the front of the unit. Signals normally pass through J1, but when the PCM

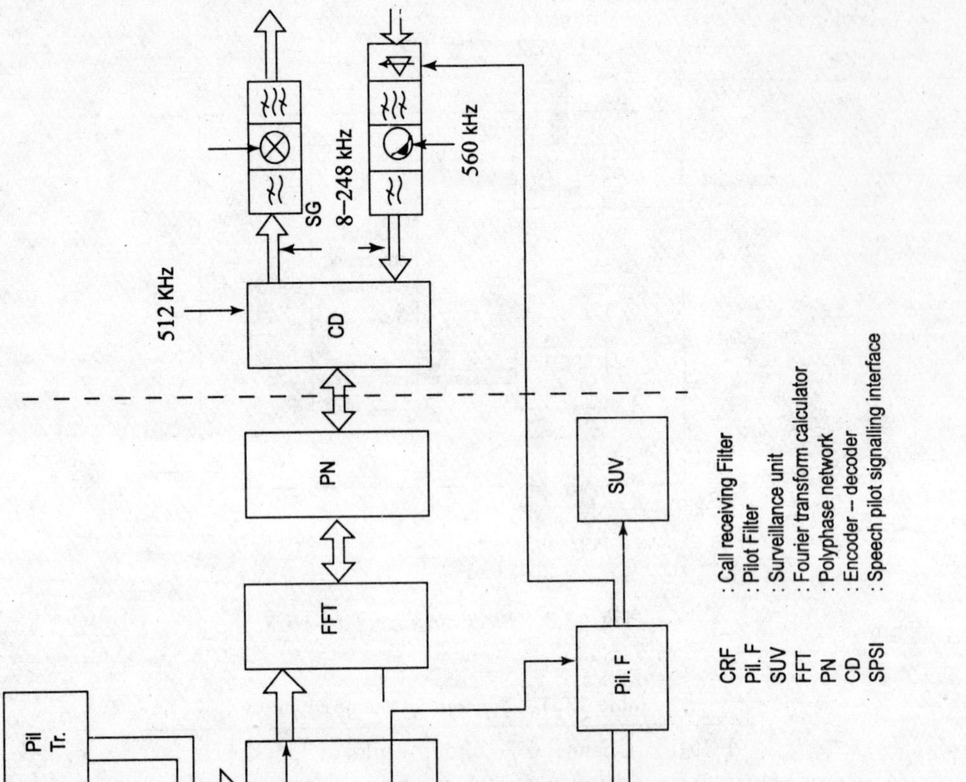

Fig. 17.4 *Functional diagram of a T-MUX.*

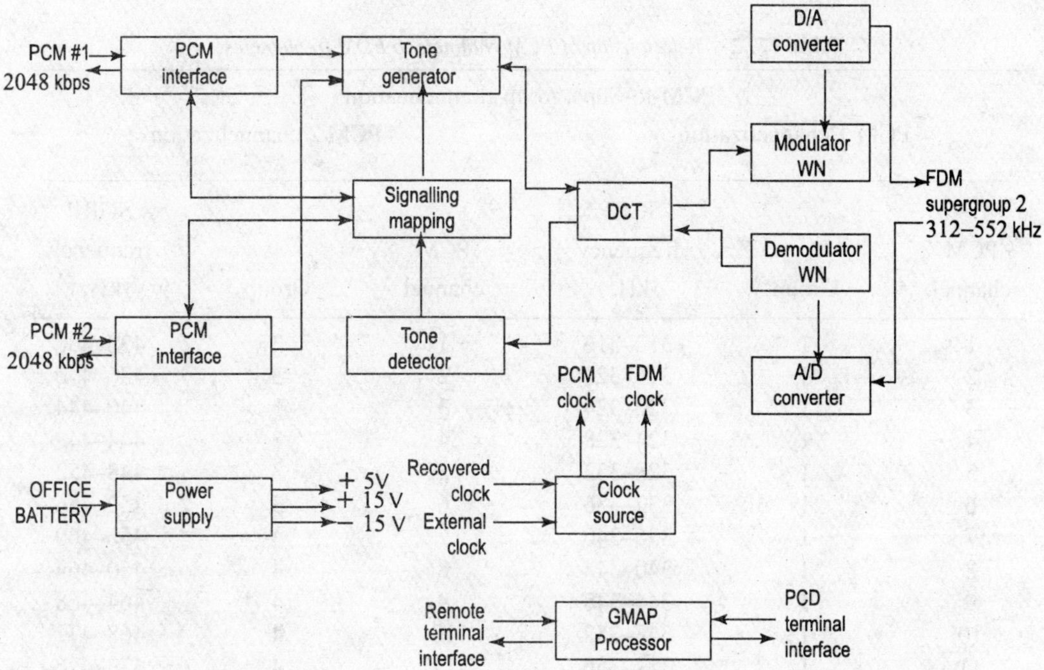

Fig. 17.5 *Block diagram of TM-7800 M1 T-MUX.*

test cable assembly is plugged into the jack, the signal path is broken. This provides access to the line and equipment side of the signal path.

For a 75-ohm operation, an unbalanced matching pad made of two resistors is strapped into the signal path. For a 120-ohm operation, a 6-dB balanced pad made of four resistors is strapped into the signal path.

The automatic line buildout (ALBO) automatically adjusts for variations in the incoming signal level. The HDB-3 decode circuit converts the HDB-3 coded bipolar signal into a decoded unipolar signal. When a loop signal is received from the GMAP processor, the HDB-3 decoder uses the transmit signal as an input instead of the incoming PCM signal.

The clock recovery circuit recovers a 2048 kHz clock from the incoming bit stream. This becomes the clock for the receive (incoming) side of the PCM interface unit.

The frame and signalling detector finds both the frame and the multiframe words in the incoming signal so that individual channels and signalling information can be identified. If a frame or a multiframe is lost, signals are sent to the GMAP processor for alarm processing. Signalling information is also sent to the signalling mapping unit.

The serial-to-parallel converter converts the serial 2048 kbps data into 8-bit parallel words representing each channel.

Before leaving the unit, the data goes through a test tone combiner circuit. This circuit, when selected, injects a test tone into one of the channels in place of the incoming data in that channel. The test tone control signal is originated by the GMAP processor. The voice data then exits the PCM interface unit en route the tone generator unit.

Table 17.2 *Relationship of PCM channels to FDM frequencies*

PCM-to-Supergroup channelization					
PCM 1 channelization			PCM 2 channelization		
PCM channel	Group	SGRP. frequency (kHz)	PCM channel	Group	SGRP frequency (kHz)
1	1	312–316	1	3	432–436
2	1	316–320	2	3	436–440
3	1	320–324	3	3	440–444
4	1	324–328	4	3	444–448
5	1	328–332	5	3	448–452
6	1	332–336	6	3	452–456
7	1	336–340	7	4	456–460
8	1	340–344	8	4	460–464
9	1	344–348	9	4	464–468
10	1	348–352	10	4	468–472
11	1	352–356	11	4	472–476
12	1	356–360	12	4	476–480
13	2	360–364	13	4	480–484
14	2	364–368	14	4	484–488
15	2	368–372	15	4	488–492
16	2	372–376	16	4	492–496
17	2	376–380	17	4	496–500
18	2	380–384	18	4	500–504
19	2	384–388	19	5	504–508
20	2	388–392	20	5	508–512
21	2	392–396	21	5	512–516
22	2	396–400	22	5	516–520
23	2	400–404	23	5	520–524
24	2	404–408	24	5	524–528
25	3	408–412	25	5	528–532
26	3	412–416	26	5	532–536
27	3	416–420	27	5	536–540
28	3	420–424	28	5	540–544
29	3	424–428	29	5	544–548
30	3	428–432	30	5	548–552

Tone generator

The VF data from the two PCM interface units enters the tone generator (Fig. 17.7) and goes to an elastic store (buffer). The purpose of the elastic store is to allow for possible difference between the

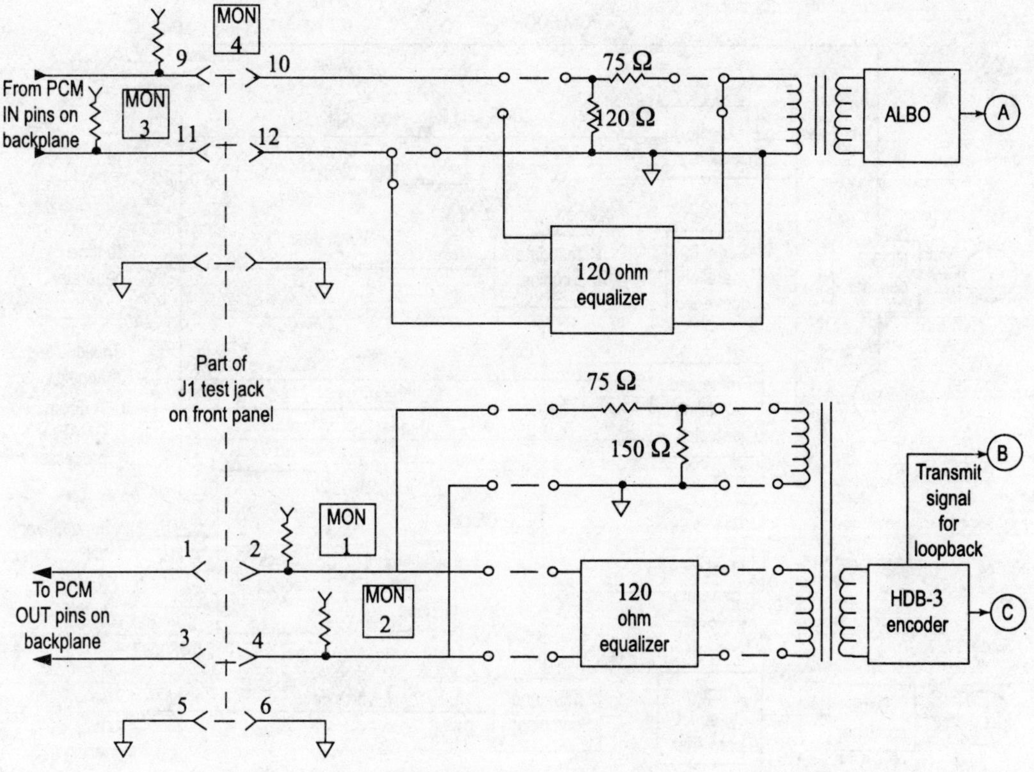

Fig. 17.6(*a*) *PCM interface unit (Part I).*

PCM and FDM clock rates. The data is entered into the elastic store at PCM clock rate and taken out of the elastic store at FDM clock rate.

If data is entered into the elastic store at a rate different from the rate at which data is taken out, a controlled slip occurs. A slip is one frame of data being lost or repeated. The slip detector notifies the GMAP processor of each slip.

The signal then goes through an A-law linear conversion. A-law is a companded PCM encoding method not usable by an FDM equipment, so the signal must be linearized. The signal is filtered by a low-pass filter to make sure that the 3825 Hz signalling slot is clean.

The final function performed by the tone generator unit is to add the signalling information, which is the presence or absence of 3825 Hz. The tone insertion circuit adds 3825 Hz from the 3825 Hz tone generator to the VF data, as directed by the signalling mapping unit.

Signalling mapping unit

The signalling mapping unit (Fig. 17.8) accepts the signalling information from both PCM interface units. The PCM-to-FDM mapping module examines the incoming PCM signalling and maps the signalling on a per-channel basis into a tone ON/OFF data as required by the tone generator unit. The signalling format for each channel is stored in the GMAP processor, as programmed by the program, control and display (PCD) terminal.

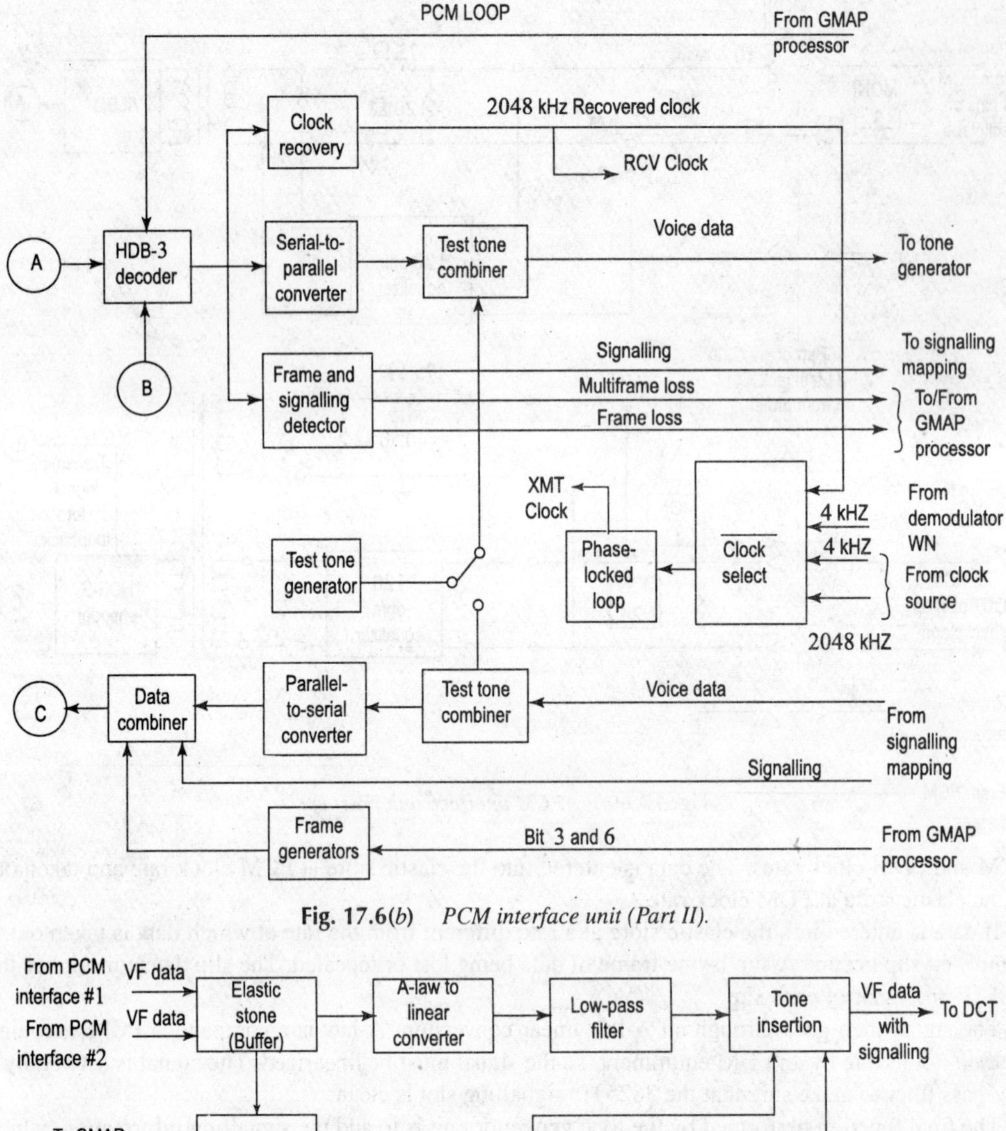

Fig. 17.6(*b*) *PCM interface unit (Part II).*

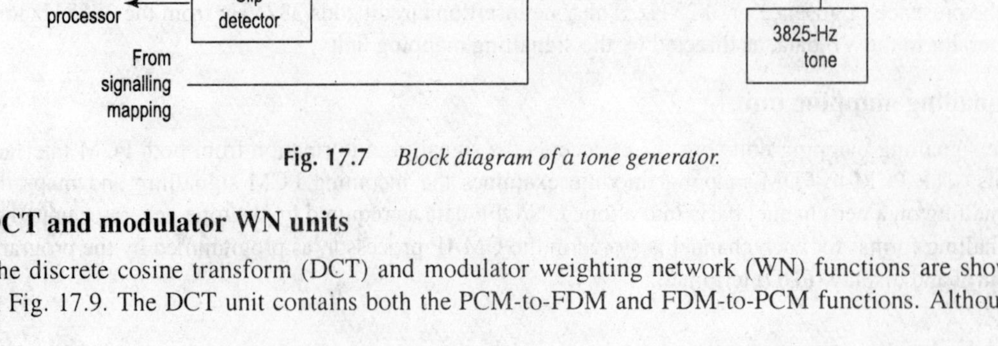

Fig. 17.7 *Block diagram of a tone generator.*

DCT and modulator WN units

The discrete cosine transform (DCT) and modulator weighting network (WN) functions are shown in Fig. 17.9. The DCT unit contains both the PCM-to-FDM and FDM-to-PCM functions. Although

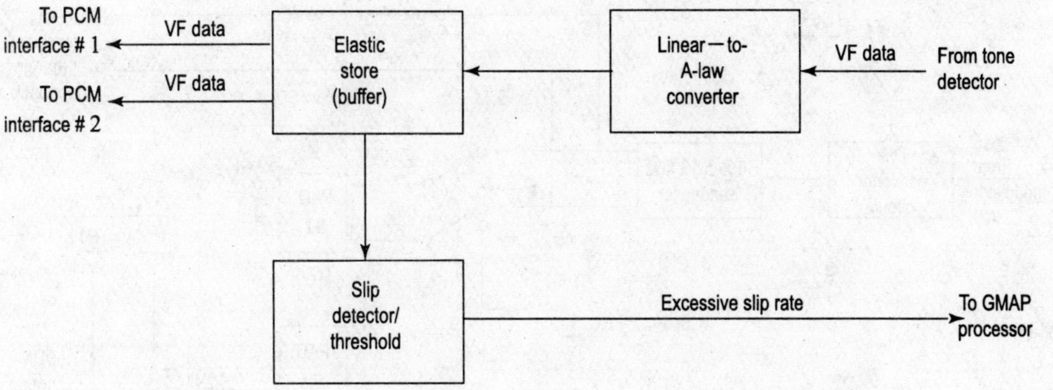

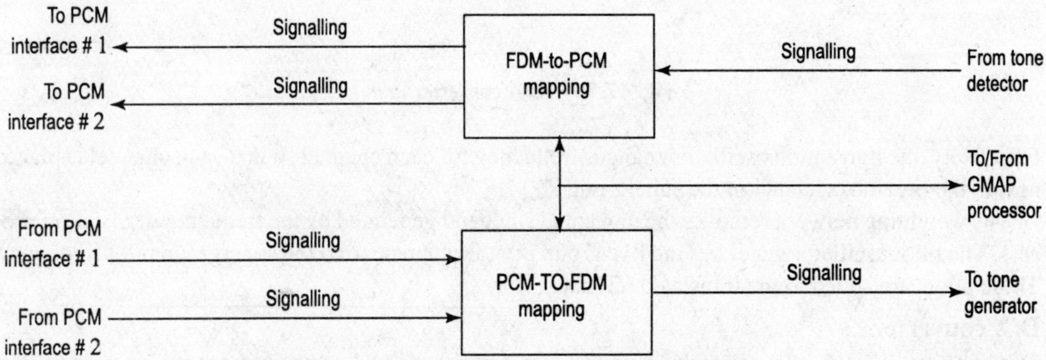

Fig. 17.8 *Block diagram of signalling mapping unit.*

the signal goes through the DCT first, and then the modulator WN, the plug-in units work together to accomplish the functions shown.

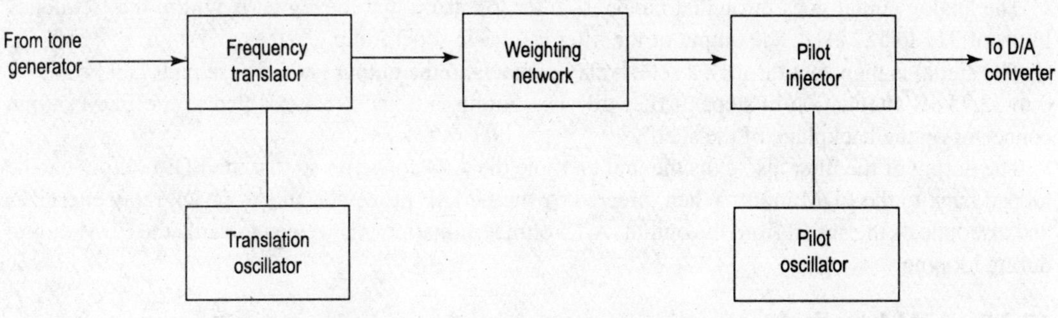

Fig. 17.9 *DCT and modulator WN units.*

The VF data for each channel occupies the spectrum from 0 to 4 kHz. The frequency translator shifts the frequency of each channel to the proper position in the FDM supergroup (see Table 17.2).

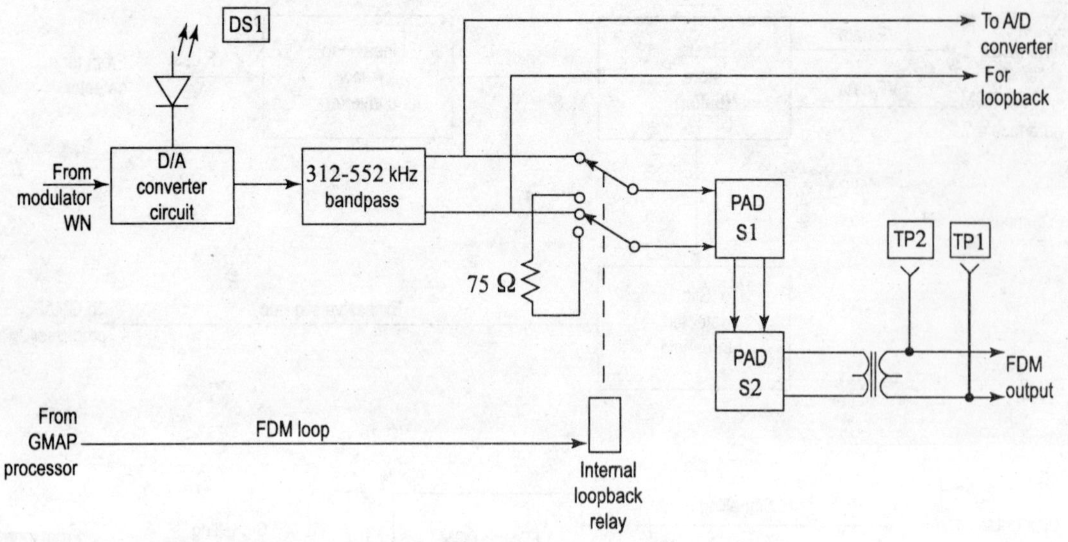

Fig. 17.10 *D/A converter unit.*

Effectively, the translation oscillator changes frequency for each channel so that each channel is placed next to the previous channel in the supergroup.

The weighting network removes the unwanted sideband generated by the frequency-translation process. The pilot oscillator generates the five group pilot frequencies and the supergroup pilot frequency. These pilots are added to the translated VF data.

D/A converter

The circuitry in D/A converter unit (Fig. 17.10) converts digital input into an analog signal. The D/A converter circuit also contains a power output monitor. If the output of the D/A converter circuit goes above a particular threshold, the output turns off and a lamp (DS1) on the front panel lights. This protects the analog equipment (e.g. the FM radio) from overload in the case of malfunction.

The analog signal goes through a bandpass filter to ensure that the signal is within the frequency limits of 312 to 552 kHz. The output of the filter is a basic supergroup.

The signal is then sent through a set of relay contacts to the output pads. These pads can be set for 0 to 33.75 dB attenuation in steps of 0.25 dB. The output is transformer-coupled to the FDM output connector on the backplane of the shelf.

The output of the filter also exits the unit en route the A/D converter so that the FDM output can be looped back to the FDM input. When directed by the GMAP processor, the loopback relay energizes and disconnects the signal from the output. A 75-ohm termination is provided towards the FDM output during looping.

17.3.2 FDM-to-PCM direction

A/D converter

The FDM signal enters the A/D converter unit (Fig. 17.11) and is transformer-coupled to the attenuator pads. These pads provide 0 to 33.75 dB of attenuation in steps of 0.25 dB. The output of the pads goes

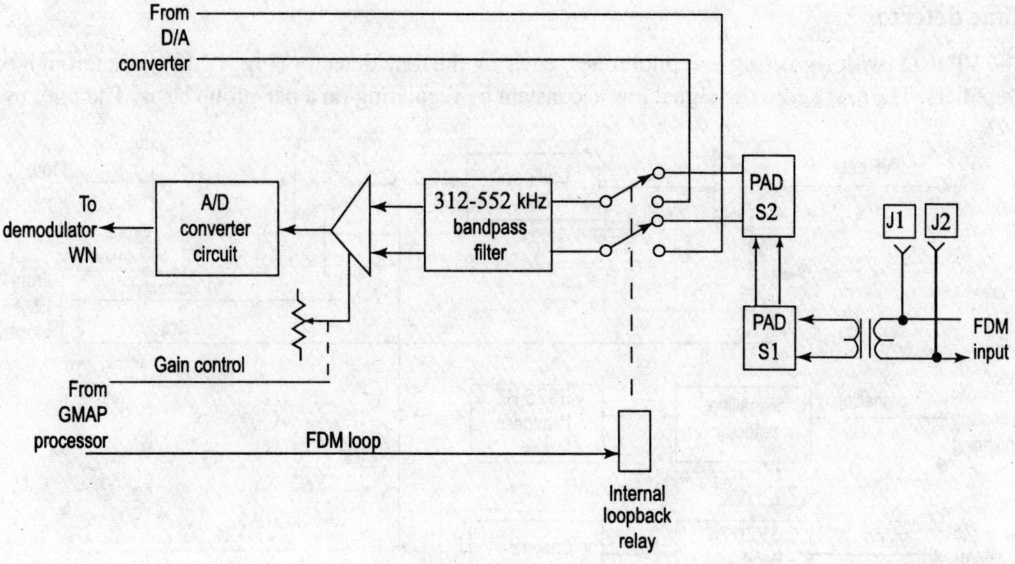

Fig. 17.11 *A/D converter.*

through a set of relay contacts to an amplifier. The gain of the amplifier is controlled by the GMAP processor unit for pilot regulation. The FDM signal is then filtered to reduce out-of-band noise and applied to the A/D converter circuit. The output of the A/D converter is a digital signal that is ready for demodulation.

When directed by the GMAP processor unit, the loopback relay is energized and the input to the amplifier becomes the filtered output of the D/A converter.

Demodulator WN and DCT

The demodulator WN and DCT units work together to accomplish the functions shown in Figure 17.12. The signal goes first through the demodulator WN and then the DCT unit.

The digital signal first goes through the weighting network which allows only one channel at a time to pass. As each channel enters the frequency translator, the translation oscillator changes to the correct frequency to shift the supergroup channel down to a 0–4 kHz baseband signal.

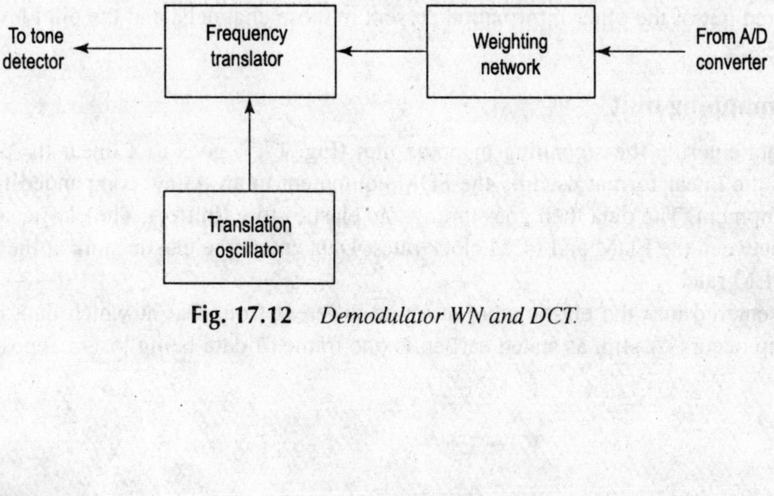

Fig. 17.12 *Demodulator WN and DCT.*

Tone detector

The VF data (with signalling and pilot tones) entering the tone detector (Fig. 17.13) goes through two amplifiers. The first keeps the signal levels constant by regulating on a per-group basis. The pilot level

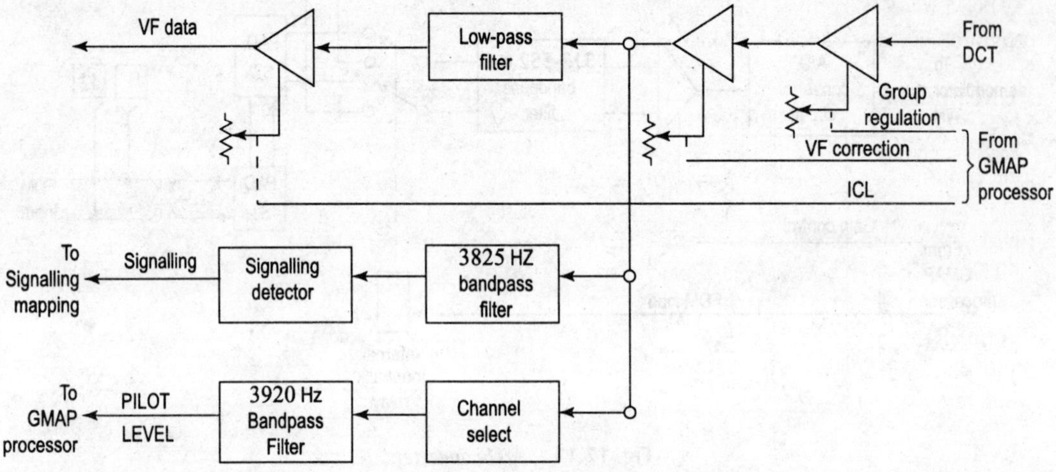

Fig. 17.13 *Tone detector.*

information is sent to the GMAP processor unit, which returns a control signal to vary the amplifier gain as required. The gain of the second amplifier is called VF correction level, and is programmed on a per-channel basis using the PCD connected to the GMAP processor unit. The amplifier gain can be adjusted from − 3.0 dB to + 3.0 dB in steps of 0.1 dB. The VF data then goes through a low-pass filter that eliminates the signalling and pilot information from the voice path. The level of the VF data is then adjusted on a per-channel basis, as directed by the GMAP processor. This level is programmed into the GMAP processor unit by the PCD as inserted connection loss (ICL). This loss can be varied from 0 to −6 dB in steps of 0.1 dB. The VF data is then sent to the signalling mapping unit.

After going through the VF correction amplifier, the signals also go to a 3825 Hz bandpass filter. This filter passes only the signalling information. After filtering, the signalling detector determines whether the signalling tone is ON or OFF and passes this information to the signalling mapping unit. The output of the VF correction amplifier also goes to the channel selector that allows information only in those channels carrying the selected supergroup pilot and the five group pilots to pass. The pilots are then filtered out of the other information present in those channels, and the pilot levels sent to the GMAP processor.

Signalling mapping unit

The voice data entering the signalling mapping unit (Fig. 17.7) goes to a linear to A-law converter that converts the linear format used by the FDM equipment to an A-law companded format used by the PCM equipment. The data then goes through an elastic store (buffer). The elastic store allows for differences between the FDM and PCM clock rates. Data enters the elastic store at the FDM rate and exits at the PCM rate.

If data is entered into the elastic store at a rate different from that at which data is taken out, a controlled slip occurs. A slip, as stated earlier, is one frame of data being lost or repeated. If the slip rate exceeds one slip in two and a half seconds, the slip detector/threshold notifies the GMAP processor unit.

The signalling information entering from the tone detector goes through the FDM-to-PCM mapping module which maps the signalling on a per-channel basis into an '*a*' or '*b*' bit signalling required by the PCM interface units. The signalling format for each channel is stored in the GMAP processor as programmed by the PCD.

PCM interface unit

Voice data enters the transmit side of the PCM interface unit (refer to Fig. 17.6) and then goes to a test tone combiner which injects a test tone from the test tone generator into the selected channel when a test tone has been selected. The test tone generator can inject a tone into either the receive side (toward FDM) or the transmit side (toward PCM), but not both at the same time.

The data is then converted from 8-bit parallel words into a 2048 kbps serial bit stream by a parallel-to-serial converter. In the data combiner, the voice data is combined with signalling and framing information. The framing information comes from the frame generators which generate frame and multiframe words, with alarm information inserted during bit 3 of time-slot 0 and bit 6 of time slot-16, when required.

The signal then goes through an encoding circuit which formats the signal to an HDB-3 line code, and is then transformer-coupled to the output circuits. To compensate for various cabling lengths, an unbalanced line matching pad or an equalizer is strapped onto the signal path respectively, for a 75 ohm or a 120 ohm operation.

The front panel test points allow the outgoing signal to be monitored. The signal proceeds through test jack J1 on the front panel, then goes to the PCM output connector via the backplane.

The clock selects one of the following circuit outputs: (*i*) a 2048 kHz clock recovered from PCM signal, (*ii*) a 4 kHz clock from the demodulator WN unit (synchronous to the FDM clock), or (*iii*) a 2048 kHz clock generated external to the transmux. The clock selection is controlled by the GMAP processor unit as programmed by the PCD. A phase-locked loop locks onto the selected clock, and the output of the PLL is the transmit clock.

17.3.3 Support functions

There are some plug-in units which do not handle signal flow directly, but supply functions required for the system operation. These are described next.

Clock source unit

In the clock source unit (Fig. 17.14), the synchronization source for the 9792 kHz FDM clock is selected. The clocks entering the clock select circuit are the 2048 kHz clock recovered from the PCM #1 signal, the 2048 kHz clock recovered from the PCM # 2 signal, an external 2048 kHz (TTL), or a 4 kHz TTL.

The GMAP processor, as programmed, directs the PLL to either lock onto one of the three clock inputs or run free. When free-running, the PLL is a very stable internal clock source. The output of the PLL is a 9792 kHz clock used for all FDM functions.

The PLL output goes through a divider to obtain either 24 kHz or 96 kHz, which is then sent to the power supply. This signal synchronizes the switching-type power supply to reduce noise in the system. The frequency is factory-selected to match the type of power supply used.

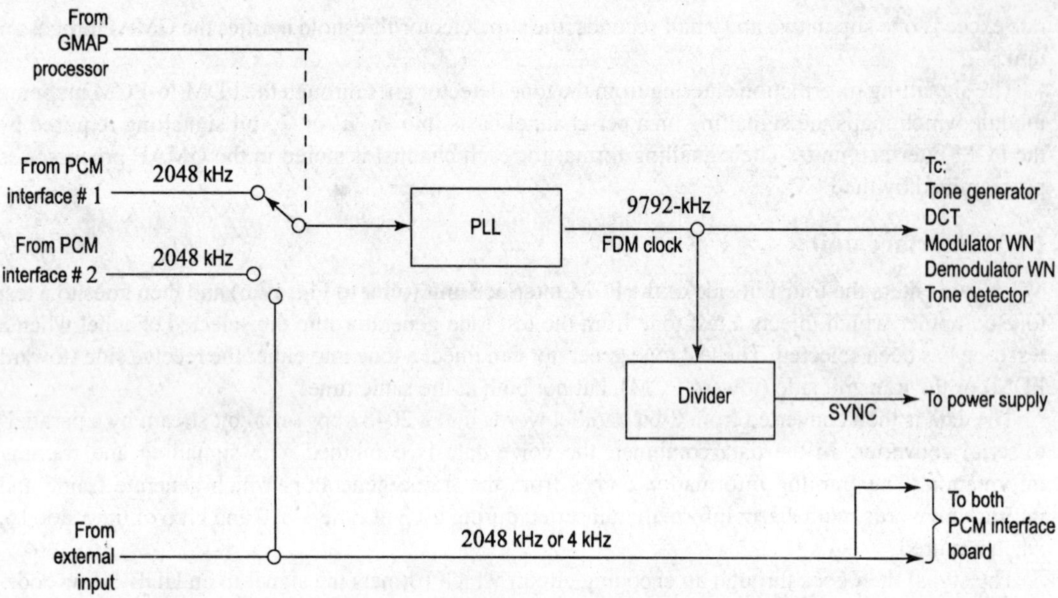

Fig. 17.14 *Clock source unit.*

The external clock, when used, passes through the clock source unit to the PCM interface units. The output is on two lines so that the PCM interface units can choose one line for a 2048 kHz external clock and the other line for a 4 kHz external clock. The choice of PCM interface synchronization must be consistent with the external clock supplied.

GMAP processor

The microprocessor on the GMAP unit (Fig. 17.15) communicates with all the other units in the system. Incoming and outgoing alarm and signalling information is exchanged with other units. Alarm and signal processing are as per CCITT G.793, under Type 1 setting. Type 2 offers a different set of alarm handling features. Option settings are sent to those units as required.

If the microprocessor fails, the MPFAIL lamp lights. The microprocessor is reset on power-up or when the RESET is activated (activating the RESET features does not erase option settings, but will interrupt traffic).

Options are entered into the system from the PCD through the front panel connector. The options entered are stored in an electrically alterable, read-only memory (EAROM) module. It the GMAP processor must be replaced, the EAROM module can be unplugged and transferred to the replacement unit, eliminating the need for reprogramming.

The power monitor circuit monitors the voltages +5, +15, and −15 volts. If any voltage fails, the PWR-ALM lamps lights and the microprocessor is notified. If the fuse that supplies +5 volts to the PCD blows, the PCD FUSE lamp lights.

The microprocessor receives status signals from both PCM interface units and sends signals to the lamp drivers to light the required lamps. Any trouble on an incoming PCM signal causes the local alarm lamp to light. An incoming alarm (indicating trouble at the far-end PCM equipment) causes the remote alarm lamp to light.

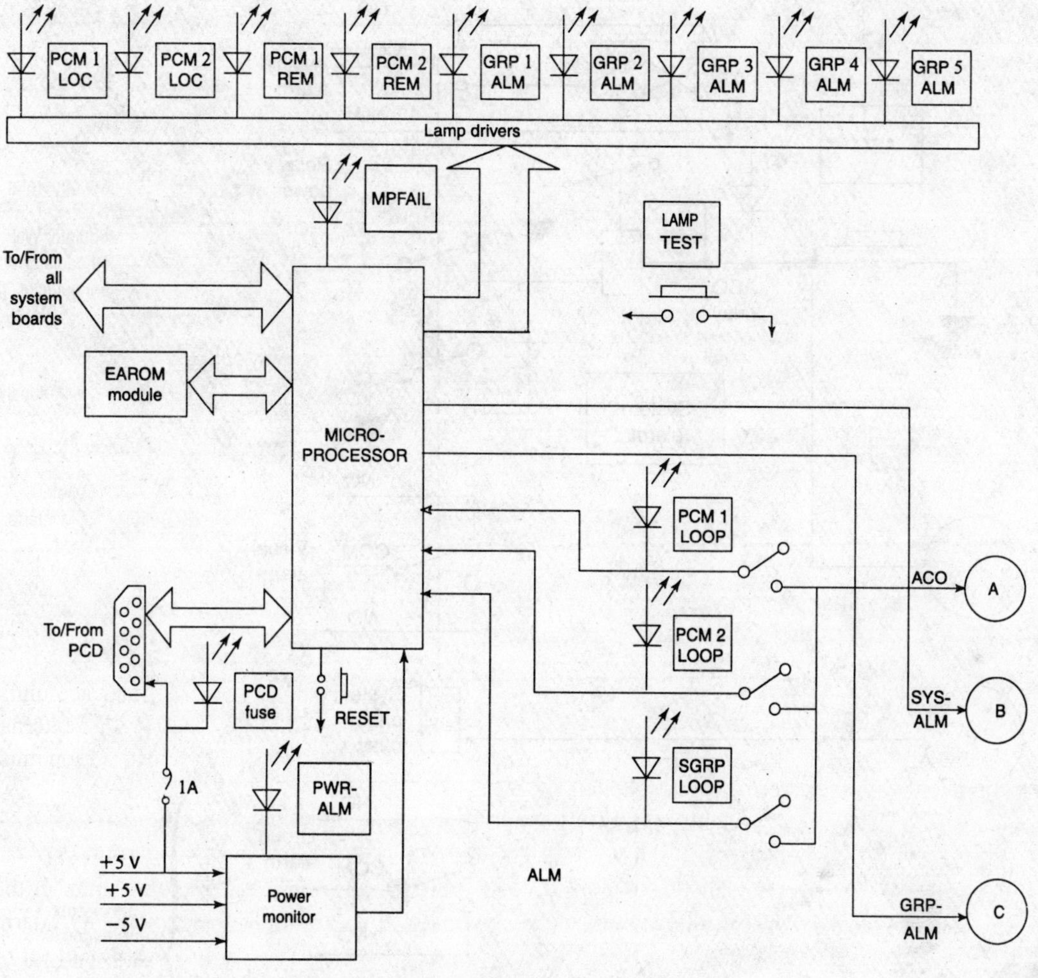

Fig. 17.15(*a*) *GMAP processor (Part I).*

Pilot level information is supplied by the tone detector unit. If a group pilot level goes above or below the nominal by more than 3 dB, a group alarm lamp lights and the corresponding group alarm relay provides contacts for office alarms. If the supergroup pilot is off by more than 3 dB, all the group alarm lamps flash and all the group alarm relays energize.

The GMAP processor uses the supergroup pilot level from the tone detector to adjust the gain of the receive signal in the A/D converter. The group pilot levels, also supplied by the tone detector, are used to adjust the gain in the tone detector.

The slip indications from the tone generator unit are monitored, and if the slip rate exceeds one in two and a-half seconds, a status message is generated and indicated on the PCD. A signal from the signalling mapping unit indicating an excessive slip rate also causes a status message to be displayed.

A system alarm is activated when any of the following occur:

- PCM local alarm
- Group alarm

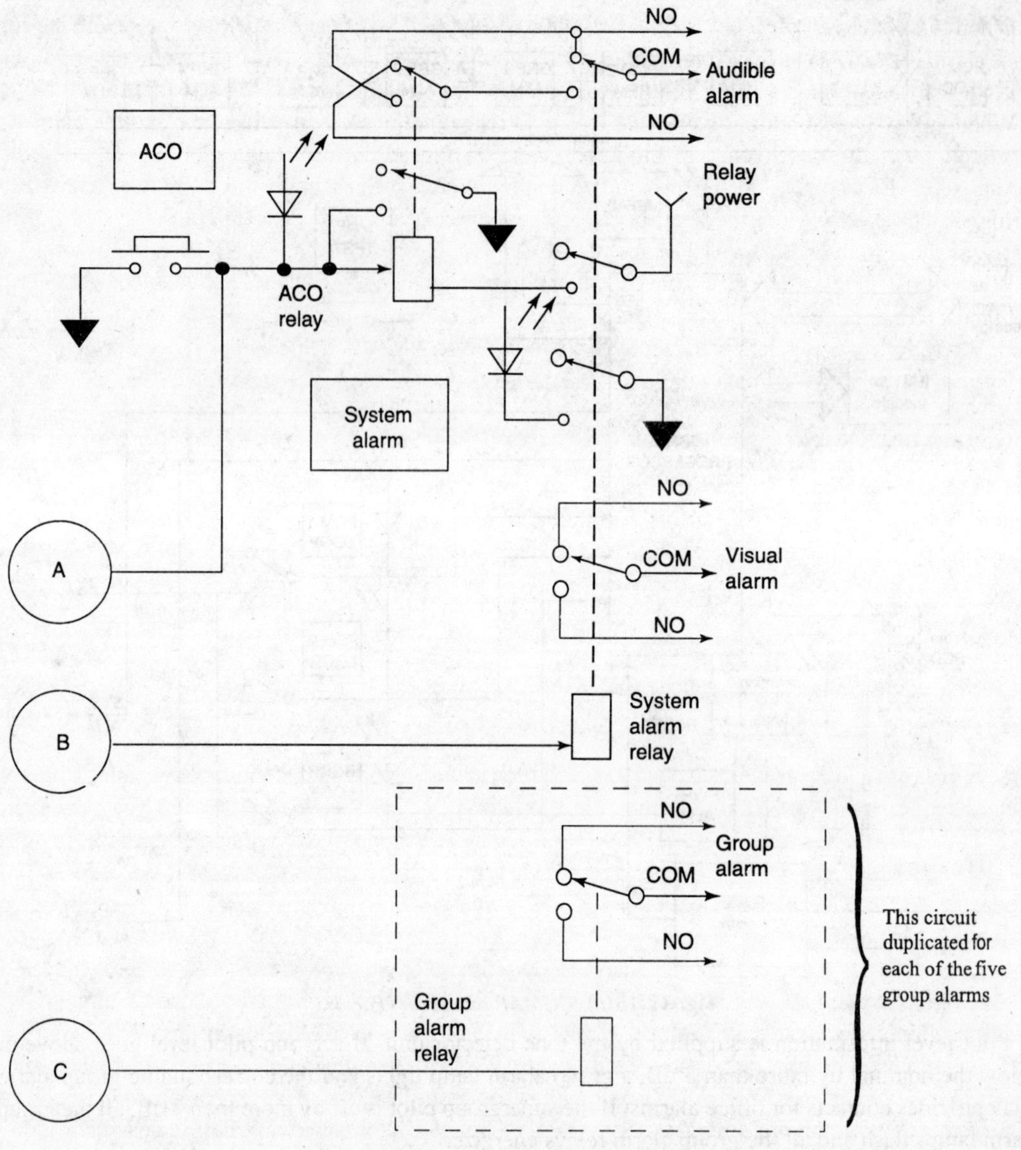

Fig. 17.15(*b*) *GMAP processor (Part II).*

- Power alarm
- FDM output failure
- Synchronization loss (PCM or FDM)
- Unit missing

The system alarm relay is normally held energized. During a system alarm, the system alarm relay releases. This allows a system alarm to be indicated during complete power failure.

When the system alarm relay is released, both audible and visual alarm contacts are available for office alarm. Both these have normally-open and normally-closed contacts. Also, a system alarm causes the SYSTEM-ALM lamp to light, and provides relay power to the alarm cut-off (ACO) relay.

When the ACO pushbutton is pressed, the ACO relay energizes. This causes the audible alarm to effectively change states so that an office bell can be silenced during troubleshooting. In addition, another set of ACO relay contacts provide a latching function which holds the ACO relay energized during system alarm. When the condition that causes the system alarm is cleared, the system alarm energizes (goes to the no-alarm state) and the ACO relay releases.

When the ACO relay energizes, the ACO lamp lights and the loop switches are enabled. Therefore, there must be a system alarm and the ACO must be activated before the front panel loop switches can be used. When the loop switch is activated, the microprocessor is notified. The microprocessor maintains the system alarm so that signals can be looped for testing when no alarm condition exists.

Power supply

The 48 volts dc (standard) enters the power supply unit (Fig. 17.16) through a fuse and an ON/OFF switch. Voltage is applied to a startup circuit that slowly raises the +15 volt line. A time delay keeps the oscillator OFF until the +15 volt line is sufficiently high.

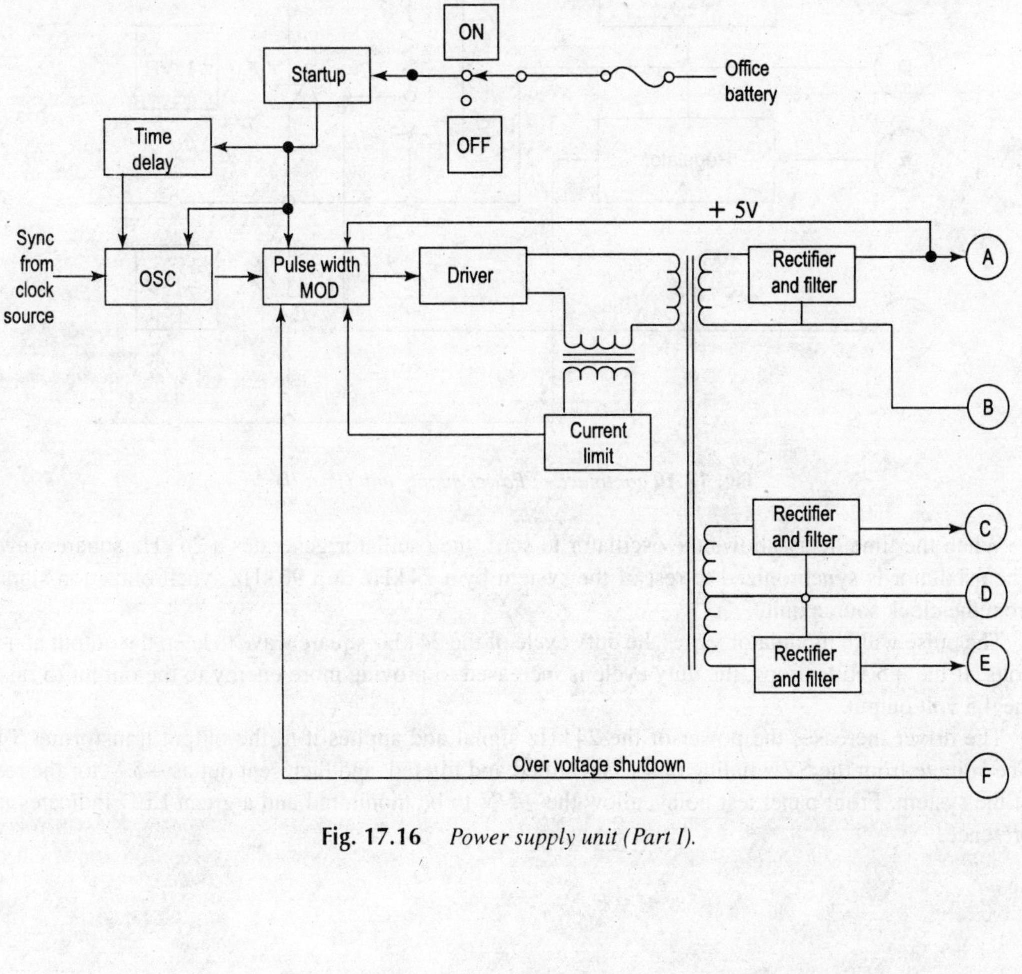

Fig. 17.16 *Power supply unit (Part I).*

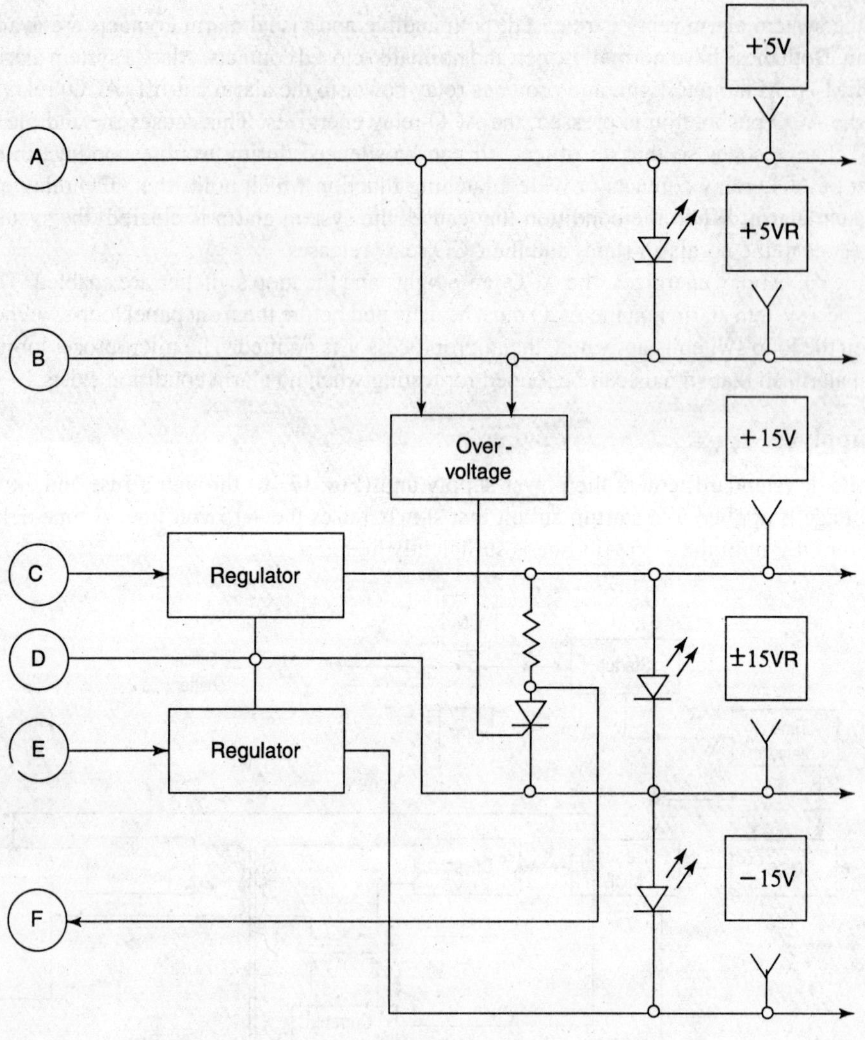

Fig. 17.16 *continued* *Power supply unit (Part II).*

When the time delay allows the oscillator to start, the oscillator generates a 24 kHz square wave. The oscillator is synchronized to rest of the system by a 24 kHz or a 96 kHz synchronization signal from the clock source unit.

The pulse width modulator varies the duty cycle of the 24 kHz square wave to keep the output at +5 volts. If the +5 volts drops, the duty cycle is increased to provide more energy to the output to raise the +5 volt output.

The driver increases the power of the 24 kHz signal and applies it to the output transformer T1. The voltage from the 5V winding of T1 is rectified and filtered, and then sent out as +5 V for the rest of the system. Front panel test points allow the +5 V to be monitored and a green LED indicates its presence.

Another winding on T1 provides voltage for the +15 and −15 volt outputs. The voltage from each half of the winding is rectified and filtered. Both +15 and −15 volt lines go through a series regulator before being sent out. Test points allow these voltages to be monitored and LEDs indicate their presence.

Two safety circuits are included in the power supply: (*i*) The current limit circuit monitors the current through T1. If the current exceeds a present limit, a control signal causes the pulse width modulator to reduce the output. (*ii*) The over-voltage circuit monitors the +5 V output. If the output voltage exceeds a present threshold, a silicon-controlled rectifier is tuned to short the +15 V as an over-voltage shutdown signal to the pulse width modulator.

17.4 T-MUX Specifications

The PCM interface, FDM interface and channel characteristics specifications for T-MUX are given below.

General

Channel capacity	:	60 channels
Voltage	:	–40 to 64 V_{dc} ($-48V_{dc}$ nominal)
Current	:	2.5 A maximum at $-48V_{dc}$
Power	:	80 W typical, not including hand-held PCD terminal

17.4.1 PCM interface

Compatibility	:	Compatible with CCITT G.703
No. of interfaces	:	2
Line rate	:	2048 kbps ±50 ppm
Line code	:	HDB-3
Impedance	:	Standard : 75 ohms ±5% unbalanced
		Optional : 120 ohms ±5% balanced
Transmit pulse amplitude	:	75 ohms : 2.37 V ±0.237 V
		120 ohms: 3V ±0.3 V
Receive pulse amplitude	:	75 ohms : 1.06 V to 2.61 V
		120 ohms: 1.35 V to 3.30 V
Time-slot sequence	:	0, 1, 2, ... 31
Frame organization	:	32 eight-bit time-slots (TS).
Coding law	:	CCITT G. 711

Note: TS-0 is allocated to framing information; TS-16 is allocated to signalling information: TS1-15 and TS17-13 are available for VF channels.

17.4.2 FDM interface

Compatibility	:	Compatible with CCITT G.233
Frequency	:	312 to 552 kHz (basic supergroup 2)
Impedance	:	75 ohms, unbalanced
Transmit level	:	−42.5 to −22.5 dBm (dBr) in 0.25 dB steps
Receive level	:	36.4 to 5 dBm (dBr) in 0.25 dB steps.
Channelization	:	60 channels, 4 kHz spacing, upright sideband, suppressed carrier
Group pilot	:	84.08 kHz
Supergroup pilot	:	411.92 kHz

17.4.3 Channel characterists (Loopback mode)

Frequency gain	:	300 to 400 Hz	:	0.6 to −1.7 dB
Relative to 0.81 kHz	:	400 to 600 Hz	:	0.6 to −1.0 dB
		600 to 2400 Hz	:	0.6 to −0.6 dB
		2400 to 3000 Hz	:	0.6 to −1.0 dB
		3000 to 3400 Hz	:	0.6 to −2.4 dB
Carrier leak	:	−50 dBmo maximum		
Intelligible cross-talk	:	−65 dBmo maximum		
Unintelligible cross-talk	:	−60 dBmo maximum		
Idle noise	:	−65 dBmo maximum		
Loaded noise	:	−62.5 dBmop maximum, with +6.1 dBmo noise loading of supergroup as per CCITT G 792.		
Signal-to-quantizing	:	33dB minimum for an input signal to 810 Hz, 0 dBmo.		
Distortion ratio	:			
Linearity	:	±0.5 dB with input signal at 810 Hz (−40 at +3 dBm)		
Standard filters	:			
Delay	:	3.0 ms absolute delay,		
Delay distortion	:	500 to 600 Hz	:	2.0ms
		600 to 1000 Hz	:	1.5 ms
		1000 to 2600 Hz	:	0.5 ms
		2600 to 2800 Hz	:	2.0 ms
Inserted connection				
Loss (ICL)	:	0 to −6 dB in 0.1 dB steps, programmable		
Adjustment:				
VF level adjustment	:	−3 to +3 dB in 0.1 dB steps, programmable		

17.4.4 Synchronization

PCM source	:	Programmable
External clock	:	Recovered PCM frame clock, independent go and return clocks
External clock	:	2048 kHz (TTL) or 4kHz (TTL)
Internal clock	:	Internally generated precision clock

17.4.5 Signalling conversion

Selection	:	Programmable
PCM signalling	:	a, b bits in time-slot 16 (TS–16)
FDM signalling	:	3825 ± 4 Hz, -20 dBmo, idle = tone ON or tone OFF
Optionals	:	Digital R_s (1 or 2 bit), analog R_2.
		Digital E & M, extraction /insertion of TS-1

References

[1] Transmultiplexing Equipment TM-2800 Manual, M/s Punjab communications Ltd., Chandigarh, 1996.

Exercises

1. Why is digital filtering transmultiplexing superior to analog filtering transmultiplexing?
2. In FDM to PCM conversion in transmux, how is out-of-band noise reduced?
3. How is the gain of receive signal adjusted in A/D converter?

18

Common Problems in the Working of Transmux

18.1 Introduction

The following problems were reported in the working of a T-MUX from various installations (Railways and DOT networks):

– Call-dropping (disconnection of calls during conversation) when the decibel level of the speaker is high
– Excessive heat in T-MUX bay
– Intermittent noise in T-MUX
– Loss of channel while accommodating group pilot
– Loss of sync
– Signalling distortion
– Call disconnection due to improper signalling bias
– Excessive metering
– Signalling option selection

The above problems and their recommended solutions are discussed in this chapter. The most important performance characteristic of a T-MUX as per CCITT Rec.G.791, G.792, G.793 are discussed at the end of this chapter.

18.1.1 Call dropping when the level of speaker is high

Many field units have noted the problem of disconnection of calls during conversation (with T-MUX in the communication link) when the voice level of speaker is high. This problem can be easily investigated by feeding the tone at different levels and varying the frequency from 300 Hz–3400 Hz at the input of PCM-MUX (connected to T-MUX) and monitoring the level of the 3825 Hz signalling frequency in 'tone-off' mode.

The results of such an experiment are presented below:

(*i*) Not all input frequencies produce 3825 Hz at the output in overload condition
(*ii*) Only the following input frequencies produce 3825 Hz tone at output–
1180, 1200, 1210, 1270, 1390, 2200, 2830, 3090 (Hz)
(*iii*) The level of signalling tone (3825 Hz) obtained at the output port for most of the input frequencies given in (*ii*) is about −56 dBm in overload condition

(*iv*) When the input signal is below the overload level, the level of 3825 Hz received is much below−90 dBm in all frequencies in VF band (300 Hz–3400 Hz) including the frequencies mentioned in (*ii*)

It may be noted that subscriber loops are designed such that the average talker 'level' is around−12 dBm. If such subscriber loop standards are met, the peak signal of a telephone conversation will not exceed +3.14 dBmo. Considering these aspects, the send signal level at 0 dBmo is adopted as 0 dBmo for 2 W channel and −3.5 dBmo for 4 W channel. Transmission systems are tested with respect to −3.5 dBmo for line-up.

From these observations, it follows that the T-MUX generates 3825 ± 25 Hz when the input signal is a subharmonic of 3825 Hz, and the level exceeds 3.14 dBmo (−0.36 dBm in 4-wire channel) in the overload limit. Now, let us analyse why at all this happens, i.e., why T-MUX should generate 3825 ± 25 harmonic?

The overload limit is 3.14 dBm for 2W and −0.36 dBm for 4W channels as already seen. When the VF level exceeds this overload limit, continous 11111111 and 01111111 codes are received from PCM MUX at the digital input port of the T-MUX. The T-MUX input digital filter is designed with the presumption that + 3.14 dBmo is not to be exceeded. When this is exceeded, the continuous 11111111 and 01111111 patterns cause square-wave distortion of input signal (as taken in by the T-MUX), which generates even and odd harmonics of the input signal. This means that any frequency which is a sub-multiple of 3825 Hz, produces a signalling frequency of 3825 Hz when the level is sufficiently high. This pseudo tone creates a 'clear forward' condition and disconnects the call.

T-MUX is designed as per CCITT recommendations which assumes that the input level at the digital port will not exceed the overload limit of VF signal. But in our networks (Railways or the DOT), the VF level often exceeds the overload limit. In fact, the 1275 ± 100 Hz frequency component in a speaker's voice is the prime contributor to this problem.

This problem can be present in the case of PCM MUX–T-MUX chain or DTAX–T-MUX chain. The solutions recommended for each case are given below.

PCM MUX–T-MUX interface

In this case, the T-MUX is connected to the PCM-MUX of the local exchange. The solution to call-dropping problem is to introduce a 4 dB pad in the send side of the VF portion of the PCM-MUX and remove the 4 dB pad in the D/A unit of the T-MUX (this ensures maintenance of the same level at the SG output port).

Implementation

The T-MUX is tested by feeding 0 dBmo (−3.5 dBm in the case of 4 W channel) from the PCM simulator or the PCM MUX in one of the channels, and measuring the signal levels at the output port of the T-MUX. The measured level should be −34.8 dBm. With the removal of the 4 dB pad, it should be −30.8 dBm. Since this pad is common to all channels, the level of all 60 channels is up by 4 dB.

Coming to PCM-MUX, all the MUXes have a strappable pad of about 15 dB in both transmit and receive side. These pads are located in the circuit just before the A/D convertor codec and after the input transformer. Firstly, a self-loop is created by looping back the send and receive digital port. Then a −3.5 dBmo signal is fed at the VF port of the PCM MUX. The receive level at the output VF port of the same channel is then measured. The level received should be −3.5 ± 0.5 dBm. Introduce a 4 dB pad from the strappable pad at the send side and conduct level measurement again. The level should be −7.5 ± 0.5 dBm. Pads can be similarly adjusted for all channels.

Digital TAX–T-MUX interface

Digital TAX or DTAX (digital trunk-to-automatic exchange) may not have the provision to accept a local subscriber, and the input ports are generally 2Mbps digital junctions. PCM-MUXes are situated at the remote local exchanges which carry subscribers of both local and trunk calls. The solution in the form of introduction of a 4 dB pad in the send of VF portion of PCM-MUX and the removal of the 4 dB pad in the D/A unit of T-MUX can be implemented in this case also. However, if the 2 Mbps is coming from E-10B local exchange, use of the 4 dB pad is not possible. In such cases, the T-MUX has to be programmed to work in 'tone-off' idle condition. Necessary changes are to be made in the far-end FDM equipment also.

18.1.2 Excessive heat in T-MUX bay

The T-MUX bay when equipped to full capacity, will house six T-MUXes. A T-MUX is rated for +5°C to +45°C operation. For better heat dissipation, the rack cover and the side plates are to be removed. But this would invite dust and rodents in spite of well-maintained equipment rooms.

The power dissipation is about 80 W per T-MUX, i.e., 480 W for the entire bay. With proper ventilation on the bay top and also in equipment room, this problem can be solved (the problem is severe for the top-most T-MUX). If the ac arrangements are not satisfactory, equip only three to four T-MUXes in each bay, leaving two slots free in-between.

18.1.3 Intermittent noise in T-MUX

Intermittent noise (ISG noise level of −35 dBmo against tolerable level of −40 dBmo) may be generated in a T-MUX. ISG noise means noise at inter supergroup frequencies, i.e., at 308 kHz (between SG1 & SG2) and 556 kHz between (SG2 & SG3).This is wrongly attributed to BPF at the output of the modulator. But the actual reason is that properly screened cables for 2 Mbps digital ports connections with PCM MUX or DTAX are not used. Earthing of screened cable should also be checked.

It can easily be established that ISG noise is not due to the T-MUX. Since T-MUX works in supergroup 2(312 kHz−552 kHz), feed −20 dBmo at the line point and measure the line output at ISG frequencies, viz. 308 kHz and 556 kHz. The levels received would be less than −150 dBmo. This implies that there is no contribution to ISG noise by a T-MUX.

18.1.4 Loss of channel while accomodating group pilot

Some T-MUXes have only provision to transmit both group pilot and SG pilots together, there is no provision to transmit SG pilot alone. If group pilot is transmitted, channel # 6 is not used for VF channel in every group (in practically all channelling equipment) and is hence lost.

The solution to this is to re-program the EPROMs to enable transmission of SG pilot only, i.e., the following options should be made available:

- SG pilot only
- Group pilot only
- SG and group pilot
- No pilot

This requires a change in the mod weighting network and the GMAP.

18.1.5 Loss of sync

The synchronization options of T-MUX are:

(*i*) Clock recovered from incoming PCM-1 signal
(*ii*) Clock recovered from incoming PCM-2 signal
(*iii*) External 2048 kHz / 4 MHz clock
(*iv*) Internal (precision) clock 19.584 MHz ± ppm

If network synchronization scheme is implemented, either option (*i*) or (*ii*) is to be selected. Else, option (*iv*) is recommended.

18.1.6 Signalling distortion

For T-MUX, nine options are available for AGC at the analog input port. If AGC is not used, signalling distortion would result in much as a 3ms/dB level variation of the received pilot signal due to a change in signalling bias.

When SG pilot alone is received, code 8 of AGC is to be used, and when group pilot is also received, code 7 of AGC is to be used.

18.1.7 Call disconnection due to improper signalling bias

The signalling tone transmitted by T-MUX in PCM-FDM direction is −20 dBmo. Since channelling equipment used to work earlier on analog transmission network, with the introduction of T-MUX signalling bias required adjustment. This has to be done at one time for all channels.

It is recommended that the bias of signalling in the FDM equipment be adjusted to zero.

18.1.8 Excessive metering

This problem exists only for DOT networks. In Railways, telecom is free of cost. If metering is introduced on all inter-Railway and Railway STD/Trunks, the revenue would be to the tune of Rs. 3 billion. The S & T department is saving Rs. 3 billion worth of DOT bill per anum for the Railways with its telecom network!

Coming to the point, the causes of excessive metering are similar to the causes of call-dropping. If the T-MUX is connected in multi-exchange area and allowed to handle the multimetering pulse, any unwanted frequency tone appearing on and off might be treated as a metering pulse, resulting in excessive metering. The solution for this is the same as suggested in Section 18.1.1.

18.1.9 Signalling option selection

A T-MUX in the network interfaces either two PCM-MUXes at the digital port, or two E-1 frames from the digital TAX.

PCM-MUXes in general use R2 signalling (2-bit) as line signalling in TS-16. DTAX supports R2 signalling and R1 signalling (1-bit). Hence, there can be a dilemma in choosing the signalling option. This dilemma can be resolved using the following guidelines:

PCM-MUX connected to digital ports of T-MUX

- Option 15 (1-bit transparent tone-on) if tone-on idle condition is desired
- Option 16 (1-bit transparent tone-off) in case tone-off idle condition is desired.

D-TAX connected to digital ports of T-MUX

- Option 16 (1-bit transparent tone-on) if tone-on idle condition is desired
- Option 15 (1-bit transparent tone-off) if tone-off idle condition is desired
- URM of DTAX to be programmed to select 1-bit R1 signalling.

URM is the PCM interface between EIOB exchange and remote connection units or circuits from satellite exchanges. URM is an abbreviation for a French term meaning 'multiplex connection unit'. Each URM can have up to 8 modules and each module can support 4 PCM links, i.e., one URM can handle 32 PCM links.

18.2 CCITT Recommendations for T-MUX Performance

The most important performance characteristics of a 60 channel T-MUX as per CCITT recommendations G.791, G,792, G.793 are given below.

Idle channel noise

When a PCM signal corresponding to amplitude number 1 for the A-law in all channels of the T-MUX is applied to the digital input of the T-MUX, the psophometric noise measured over any channel at the digital output should not exceed −65 dBmo.

Total distortion

When a sinusoidal signal of frequency f (700 Hz < f <1100 Hz) is applied to the channel at the digital input of the T-MUX, the ratio of signal-to-total distortion power measured with appropriate noise weighting should be within the limits prescribed in Fig. 18.1.

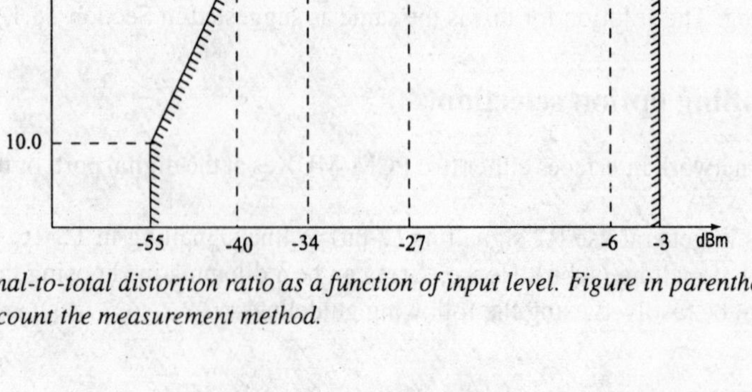

Fig. 18.1 *Signal-to-total distortion ratio as a function of input level. Figure in parentheses takes into account the measurement method.*

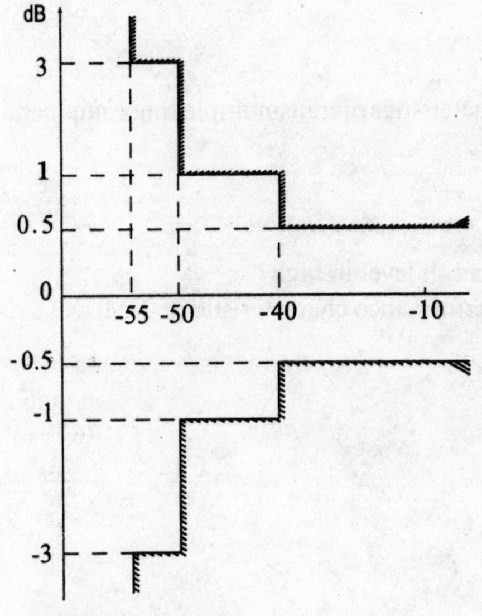

Fig. 18.2 *Variations in gain with input level.*

Variation of gain with input level

When a sinusoidal signal in the frequency range 700–1100 Hz is applied to the digital input port of any channel at a level between −55 dBmo and +3 dBmo, the gain variation of that channel relative to the gain at an input level of −10 dBm should be within the limits prescribed in Fig. 18.2.

Attenuation/Frequency distortion

The variation of attenuation with frequency in any channel should be within the mask shown in Fig. 18.3.

Interchannel cross-talk

The cross-talk ratio measured between any pair of channels for any frequency f (700 Hz < f < 1110 Hz) at 0 dBmo level should be greater than 62 dB.

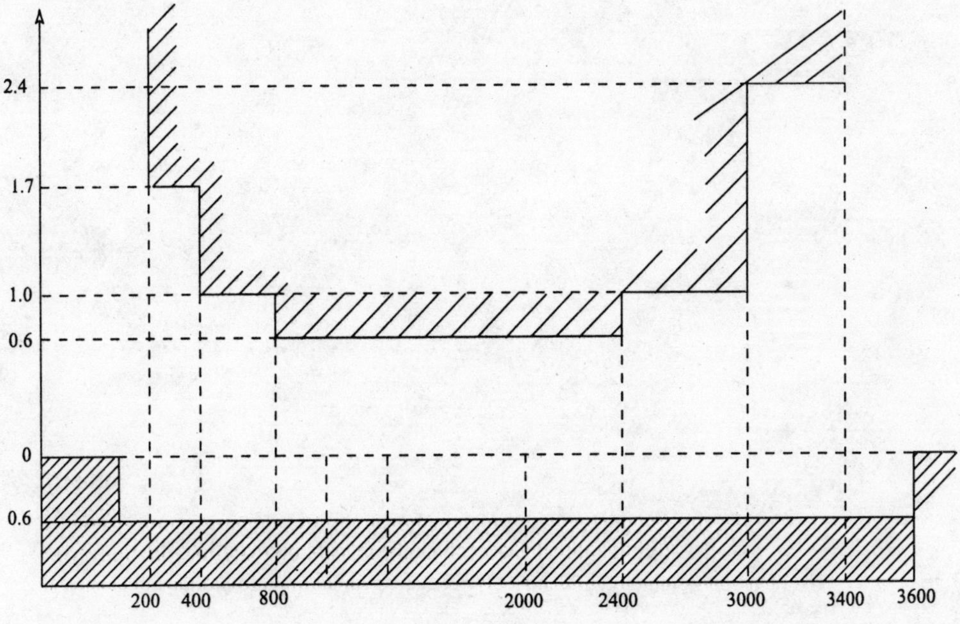

Fig. 18.3 *Attenuation distortion as a function of frequency (for all channels).*

References

[1] CCITT Rec. G.791, G.792, G.793 — Performance characteristics of transmultiplexing equipment.

Exercises

1. Why does 'call-dropping' occur in T-MUX when speech level in high?
2. What are the CCITT recommendations regarding performance characteristic of T-MUX?

19

Some Experimental Results

19.1 Introduction

The BZA–MAS–RU (Vijayawada-Madras-Renigunta) digital MW link commissioned by the South Central Railway has seventeen MW stations over a route length of 450 km approximately (refer Fig 5.1). Among these stations, three are terminals (BZA, MAS, RU), fourteen are repeaters (the MW station at Naidupet is branching station).

Results of the following experiments[1] conducted on this link are presented in this chapter:

- Simulation of tandem working of number of III-order MUXes
- Measurements confirming improvement in the availability of link with space diversity
- Hoisting of antennas at reduced heights—fade margin and availability trade-off
- Jitter accumulation measurements

19.2 Simulation of Tandem Working of III-Order MUXes

The BZA–MAS–RU digital MW link has III-order MUXes of the terminal type at BZA, MAS and RU, and repeater type MUXes at GNT (Guntur), BTTR (Bitragunta) and NYP (Nayudupeta).

A tributary developed from BZA to RU via GNT, BTTR and NYP goes through eight III-order MUXes and thirteen radio repeaters. If the tributary is looped back at the far end (RU), we have a setup of 16 III-order MUXes and 26 radio-repeaters. This is a good setup to investigate the P_e performance (see Table 19.1). The prime contributors to errors would be noise and jitter (mostly waiting-time jitter).

19.3 Measurement of Link Availability with Space Diversity

The BZA–MAS–RU digital MW link has space diversity on eight hops. The link was first commissioned without space diversity. Later, space diversity was introduced on eight hops as per the initial design of the link.

[1]Details of experiments can be obtained from the author.

Availability of link (calculated over a three-month period, with non-availability due to maintenance operations ignored) without space diversity	:	99.99%
Availability of link (calculated over a three-month period, with non-availability due to maintenance operations ignored) with space diversity	:	99.9999%

19.4 Hoisting Antennas at Reduced Heights: Fade Margin and Availability Trade-off

At OGL, BTTR, and N.KONDA, antennas were hoisted at lower than specified (5 m to 10 m) height. The link was operating with the fade margin reduced by 15 dB. The availability of link was 99.99% (unavailability due to fading alone). These results point to the possibility of saving on tower costs (due to reduced height) while trading off on availability.

19.5 Jitter Accumulation Measurements

Jitter accumulation due to a number of MUXes in cascade was measured at various frequencies simulating the number of MUXes as was done in experiment. The results are in line with the mathematical predictions under CCITT Rec. G-283.

Table 19.1 *Performance of digital MW link*

		Normal condition	**Faded condition*
P_e without loopback (8 III-order MUXes, 13 repeaters)	:	1×10^{-11}	1×10^{-11}
P_e with first loopback (simulation of 16 III-order MUXes, 26 repeaters)	:	1×10^{-10}	1×10^{-9}
P_e with second loopback (simulation of 32 III-order MUXes, 52 repeaters)	:	1×10^{-9}	1×10^{-7}
P_e with third loopback (simulation of 64 III-order MUXes, 104 repeaters	:	1×10^{-7}	1×10^{-5}

* *Faded condition: Fading on one hop physically, but, the number of faded hops to be taken as per the number of loopbacks.*

Part - III

20

Guided Waves and Waveguides

20.1 Introduction

There are two types of transmission lines commonly used—balanced line (parallel wire) and unbalanced line (coaxial line). Where balanced properties are required, parallel wire line is employed and where unbalanced properties are required, coaxial line is used.

For example:

Parallel line usage	-	Connection of folded-dipole antenna to a TV receiver
	-	Connection of rhombic antenna to HF transmitter
Coaxial line usage	-	Connection of broadcast transmitter to its grounded antenna
	-	Interconnection of UHF and MW equipment panels, to avoid risk of radiation from the line itself

Any system of conductors is likely to radiate if the conductor separation approaches half-wavelength at the operating frequency. Thus, in parallel wire transmission the amount of energy lost in radiation is negligible for frequencies upto 100 MHz. Most of the electromagnetic energy travels in the space closely surrounding the parallel wires. For frequencies higher than 100 MHz, the radiation is considerable, hence parallel wires cannot be used at such frequencies for transmission

In a coaxial line, the outer conductor surrounds the inner conductor and is invariably grounded. Hence energy is guided along the space between the two coaxial conductors and there is no scope for loss due to radiation. However, there is loss due to resistance of the conductors which increases with frequency. There is an additional loss due to leakage in the insulating materials used for supporting the inner conductor. These losses assume a major proportion at frequencies of the order of 3 GHz.

The obvious question that comes up is, Why not guide the electromagnetic energy by confining it inside a single hollow tube so that losses due to conductor resistances and insulating material usage become negligible? This is indeed feasible, and precisely the method of transmission through waveguides. But this cannot be adopted for all frequencies. The cross-sectional dimensions of the waveguide must be of the order of the wavelength. This means that for frequencies below 3 GHz, the dimensions of waveguide (if used as transmission line) are inconveniently large. Thus, waveguide transmission is restricted to SHF (microwave) band. In this band also, it is used as a subsystem connecting the antenna to the equipment because although line-of-sight propagation is feasible in this band, waveguide loss is large for too long spans. Figure 20.1 indicates the typical loss in a waveguide for various wavelengths.

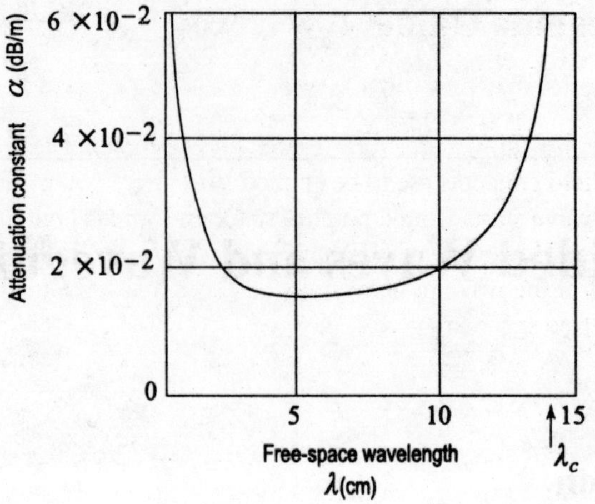

Fig. 20.1. *Attenuation versus wavelength for a rectangular waveguide (copper waveguide of inside dimensions* 7.1 cm × 3.35 cm).

20.2 Boundary Conditions

There are two approaches to understand how a waveguide guides the wave. One is direct derivation from Maxwell's equations for electromagnetic waves and the other is by applying boundary conditions (of conducting surfaces which are the waveguide walls) to plane waves. The former method is cumbersome but serves as a proof. The latter method is an analytical one and gives an insight to what is happening physically. Thus, it provides clues for many practical considerations in the design of a waveguide subsystem. Hence the latter method is discussed in the following sections. The mathematical treatment is given in the Appendices at the end of the book for interested readers.

Before we commence discussion on how a waveguide guides the wave, the laws governing the electric field $\boldsymbol{E}$ and the magnetic field $\boldsymbol{H}$ of an electromagnetic wave intercepted by conduction surfaces are discussed.

When dynamic electric and magnetic fields of an electromagnetic wave come across a perfectly conducting surface, they obey certain laws at such boundaries. These laws which are readily deducible from Maxwell's equations are termed 'boundary conditions'. The four boundary conditions are:

1. (a) Inside a perfect conductor, electric field is zero since there cannot be a potential gradient in a perfect conductor.
   (b) If there is charge Q just outside the conducting surface, an electric field normal to the surface will be set up outside but there will be no field inside. The charge moves to the surface to balance the field.
2. An electric field tangential to a perfectly conducting boundary vanishes at the boundary.
3. A magnetic field perpendicular to perfectly conducting boundary vanishes at the boundary.
4. A magnetic field parallel to a perfectly conducting boundary is associated with a surface current I in the conductor whose magnitude is given by

   $$I = |H_t| \quad \text{A/m},$$

   where H_t is the tangential magnetic field.

20.3 How Waveguide Guides the Wave

20.3.1 Plane waves

For a waveguide, there is no 'flow and return' circuit and one cannot talk of currents and voltages. Hence, electromagnetic field concepts are to be applied.

If an electromagnetic wave starts from a point in space and spreads out uniformly in all directions, it forms a spherical wave. However, an observer situated at a great distance from the source will observe only a small part of the wave in his vicinity and sees it as a plane wave. Properties of a plane electromagnetic wave in free space are given below.

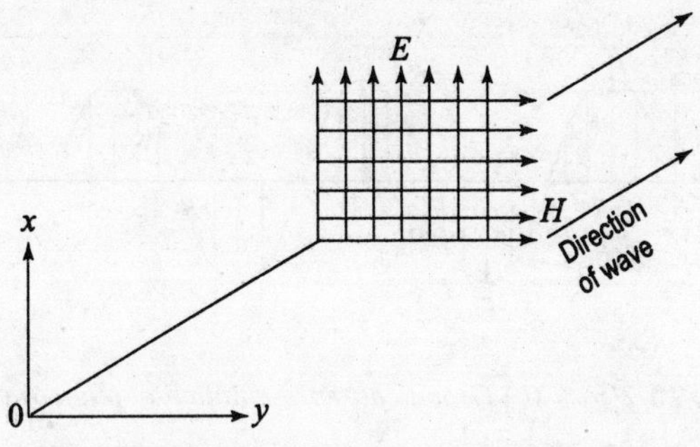

Fig. 20-2(a) *E and H in a plane wave.*

1. The elelctric field ***E*** and magnetic field ***H*** are at right angles to each other and to the direction of propagation (Fig. 20.2*a*).
2. The velocity of propagation of the wave in free space is given by

$$v = \frac{1}{\sqrt{k_0\,\mu_0}} \text{ m/s,}$$

where k_0 is the permitivity of free space and μ_0 the permeability of space.

$$k_0 = \frac{1}{36\pi} \times 10^{-9} \quad \text{Farad/m.}$$

$$\mu_0 = 4\Pi \times 10^{-7} \quad \text{Henry/m.}$$

Thus, it can be found that

$$v = \frac{1}{\sqrt{(1/36\pi \times 10^{-9} \times 4\pi \times 10^{-7}}} = 3 \times 10^8 \text{ m/s.}$$

This is the value used in all calculations but the exact value is 2.99790×10^8 m/s.

3. ***E*** and ***H*** always oscillate in phase, and the ratio of their amplitudes is always constant and equal to

$$\sqrt{\mu_0/k_0} = 120\pi(\approx 377).$$

4. All electrmagnetic waves in free space travel at the velocity mentioned in (2) above. Light is also composed of electromagnetic waves although the frequencies are much higher than those of radio waves. It also travels at the velocity of 3×10^8 m/s. The velocity of light is always represented by the symbol c.
5. The propagation of a plane wave in free space along x-direction can be grapically represented as in Fig. 20.2(b). $\boldsymbol{E}$ lies in plane xoz and $\boldsymbol{H}$ is in plane yoz. The length of each vector represents its value at that distance along oz. They vary with distance from maximum to minimum to maximum, completing a cycle over a distance 'λ' called the 'wavelength.' It is obvious that $f\lambda = c$. To an observer at a fixed point on oz, $\boldsymbol{E}$ and $\boldsymbol{H}$ appear to vary rapidly as the field rushes past him.

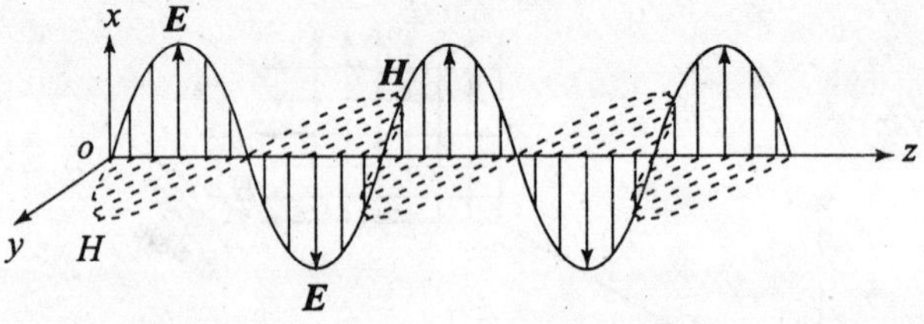

Fig. 20-2(b) (b) *Graphical representation of plane wave.*

Figure 20.2(b) shows the wave only at one instant of time. As the e.m. wave propagates, the planes containing $\boldsymbol{E}$ and $\boldsymbol{H}$ may rotate, but at any time the orthogonality between them is maintained. If these are made to reside in fixed planes, the wave is said to be polarized.

6. A plane wave is transverse if its $\boldsymbol{E}$ and $\boldsymbol{H}$ field vectors lie in a plane which is transverse to the direction of travel. At any instant $\boldsymbol{E}$ and $\boldsymbol{H}$ are constant in magnitude and direction over the whole surface of any plane set up perpendicular to the direction of propagation. such a plane is known as wavefront.

20.3.2 Boundary conditions imposed on plane waves

Suppose two plane conducting sheets are placed in the positions parallel to yoz plane as shown in Fig. 20.3. Here the extent of plane waves is limited to the x-direction.

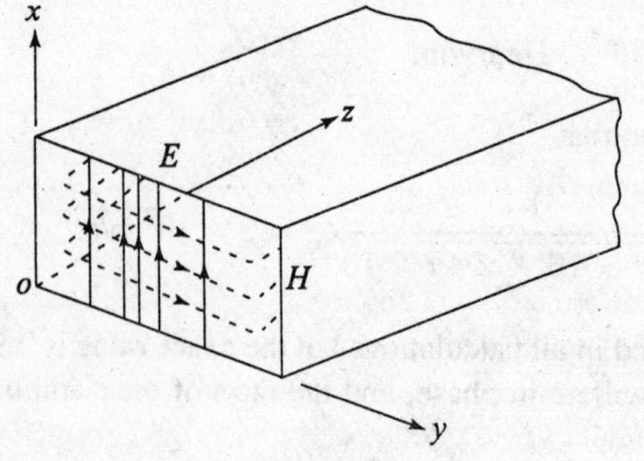

Fig. 20-3 *Lines of* E *and* H *in rectangular waveguide.*

In the space between the plates the electric field is normal to the plates and hence sastisfies the boundary condition for electric lines at the surfaces. The magnetic field is tangential and hence it also satisfies the boundary condition for magnetic lines. Thus the electric lines terminate on the plates and a plane wave can travel between them, remaining undistorted but limited to the space between the plates.

By inserting two more conducting plates parallel to xoz (this forms a rectangular hollow tube) and applying the boundary conditions, it can be seen that with the new boundaries the electric field would be tangential and the magnetic field would be normal. This is a contradiction to the result obtained by inserting conducting plates parallel to yoz plane. This means that a plane wave cannot be propagated in a rectangular tube.

However, if by superposing waves travelling in two different directions the wave pattern is modified, wave propagation is possible. In this situation, $\boldsymbol{E}$ has a maximum at the centre of the tube, falling to zero on the side walls. The $\boldsymbol{H}$ lines curve round so that they pass down the tube tangential to the side walls.

20.4 Rectangular Waveguides

In the previous section, it was shown how boundary conditions do not allow a rectangular waveguide round a plane wave but permit it round two plane waves superposed. More details of this superposition are given below.

- Figure 20.4(a) represents a section of a plane wave travelling in the diection of arrow. The wave is polarized such that the magnetic vector lies in the plane of the paper. The full lines correspond to the magnetic field lines at successive maxima in alternate directions (shown by arrows on them). The dashed lines are the zero magnetic field positions. The distance between two successive maxima is half wavelength. The electric field is normal to the plane of the paper and in phase with the magnetic field so that the positions of electric lines of maximum intensity entering and leaving the plane of the paper are as shown by the dots and the circles.
- Figure 20.4(b) shows a section of a wave similar to that in Fig. 20.4(a) but travelling in a different direction.
- The two waves of Fig. 20.4 (a) and Fig. 20.4 (b) are superposed in Fig. 20.4 (c) coinciding the two points A. The $\boldsymbol{H}$ lines of composite wave form loops and the $\boldsymbol{E}$ lines add/subtract to give the maxima and the minima. For example, along AA′ in Fig. 20.4 (c), $\boldsymbol{E}$ falls to a zero midway, and again rises to a maximum.
- By inserting two parallel conductiong plates, we get the situation in Fig. 20.4 (d). CC′ is tangential to $\boldsymbol{H}$, and along CC′ plane $\boldsymbol{E}$ is zero.

Thus boundary conditions are satisfied and wave propagation is possible.

To summarize, a wave pattern which is the resultlant of a combination of two plane waves of equal amplitude travelling at an angle to each other, can propragate freely down a rectangular tube. In practice, we do not 'bring' two such plane waves and arrange their superposition. Such a specific wave can be set up by particular arrangement of conductors and transmitted through the waveguide. This wave and all its properties can be derived directly from Maxwell's equations by applying appropriate boundary conditions. Since in this book the emphasis is more on physical visualization of the phenomenon, the

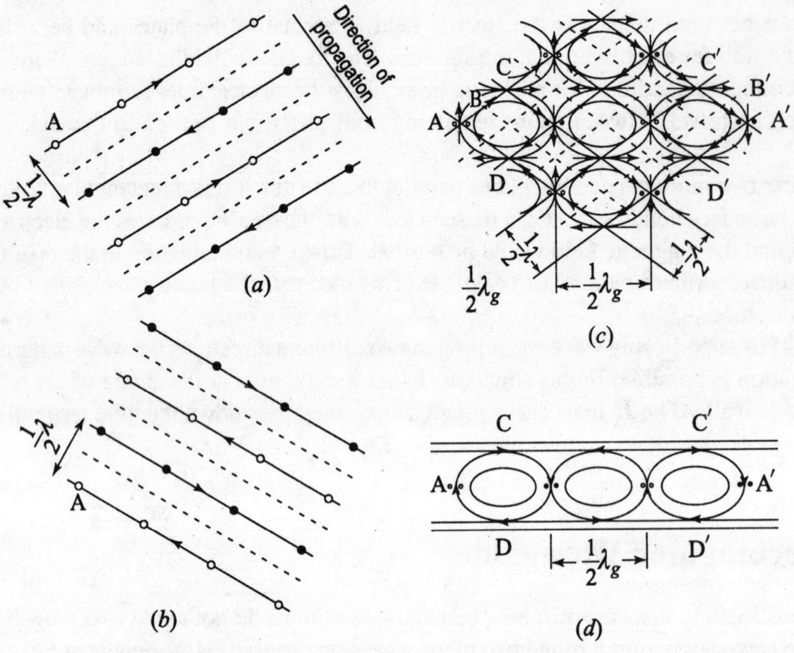

Fig. 20-4 *Superposition of plane waves forming waveguide modes.*

derivation is presented in Appendix. Interested reader may study this and tally the propagation modes supported in a waveguide.

20.4.1 Propagation modes in rectangular waveguide

Fundamental mode

Figure 20.4(*d*) is re-drawn as Fig. 20.5 with *x*-axis along the broadside of the waveguide. It can be observed that

- all ***E*** lines are transverse to the direction of the wave; they go directly across form side to side,
- ***H*** lines have longitudinal componetnts since they lie partly along the direction of motion,
- the closeness of spacing of the lines or dots indicate the the strength of the field,
- while passing along *x*-axis, lines have a maximum at the centre and fall to zero at the ends,
- while going along y-aaxis, no ***E*** lines come across.

Since ***E*** is transverse and only ***H*** has longitudinal component (i.e., component along direction of propagation), it is called TE or H wave. Since there is one field maximum along *x*-axis and none along *y*-axis, the wave is called TE_{10} or H_{10} wave.

20.4.2 Higher modes

H_{10} is not the only possible mode of propagation, higher modes are also possible. In H_{10} wave, one row of ***H*** loops are enclosed between side walls. Two or more rows could also be enclosed. In Fig. 20.6, two

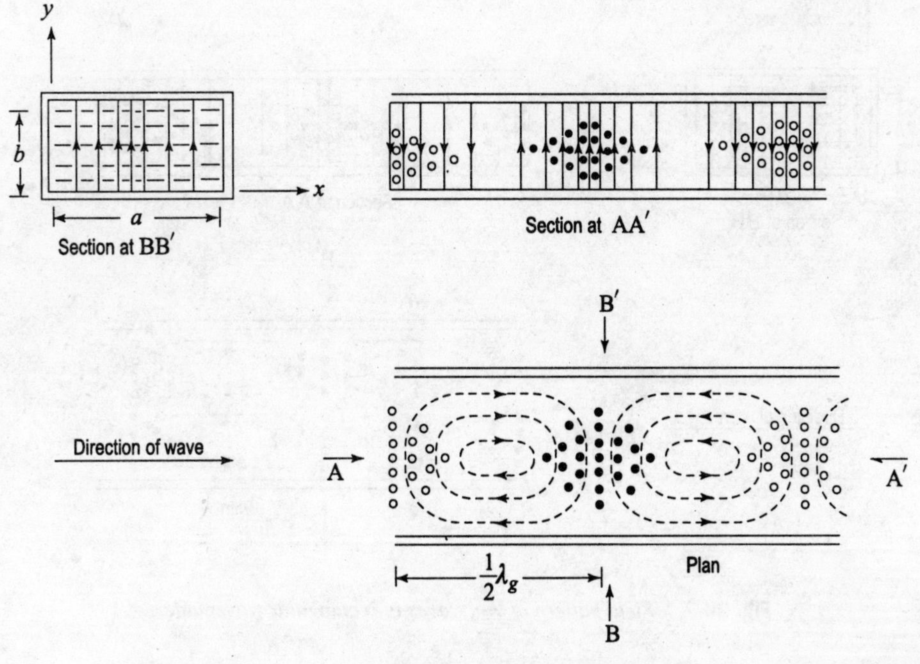

Fig. 20-5 *Field pattern of* H_{10} *wave in rectangular wave guide.*

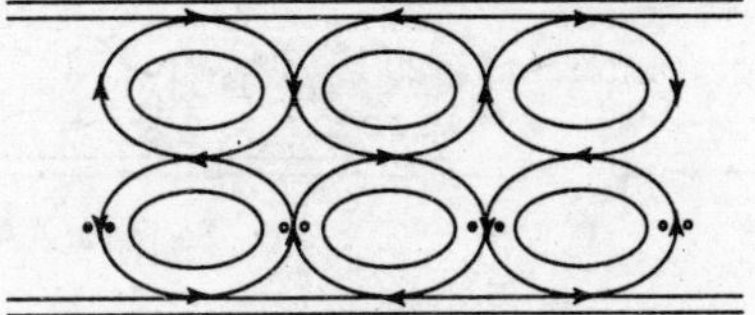

Fig. 20-6 *Two rows of H loops in waveguide.*

are shown. There are two maxima along x-axis and none along y-axis. Thus, it is called H_{20} or TE_{20} wave. Similarly, H_{11} mode can be visualised with one maxima along x- and y-axis (see Fig. 20.7).

E-modes

By superposition of plane waves, it can also be shown that E wave or TM wave is possible. In this wave, the magnetic field is completely transverse and the electric field has a component in the direction of propagation. E_{10} or TM_{10} mode is not possible. If we try to construct E_{10} wave, the pattern in a longitudinal section parallel to x-axis would be as in Fig. 20.8. Then in the xz plane, $\boldsymbol{E}$ lines would be tangential to the wall which violates the boundary conditions.

In E wave or TM wave, modes E_{11} or TM_{11} onwards are possible (see Fig. 20.9). A 3-dimensional view of H_{10}, E_{11}, H_{11} are shown in Plates 1(*a*–*c*). Generation of different modes and their applications are discussed in Section 20.7.

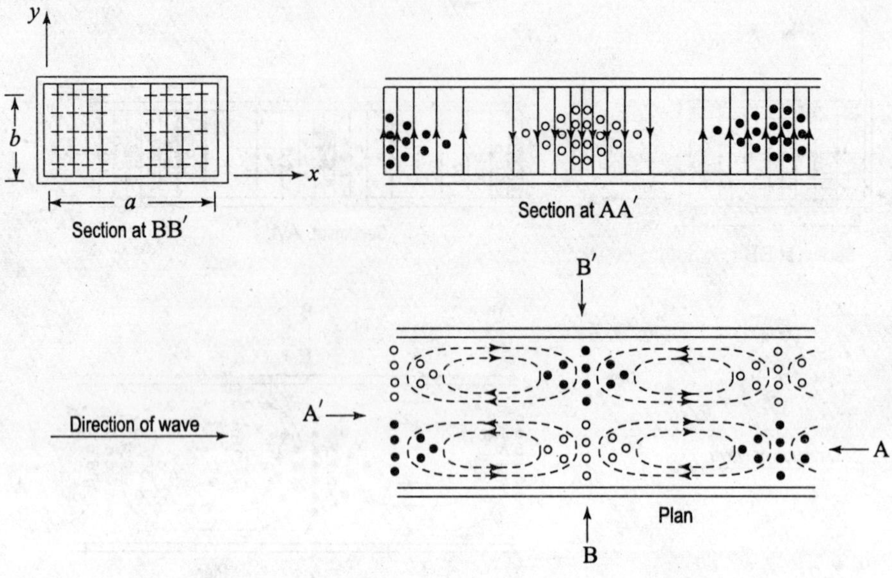

Fig. 20-7 *Field pattern of* H_{20} *waves in rectangular waveguide.*

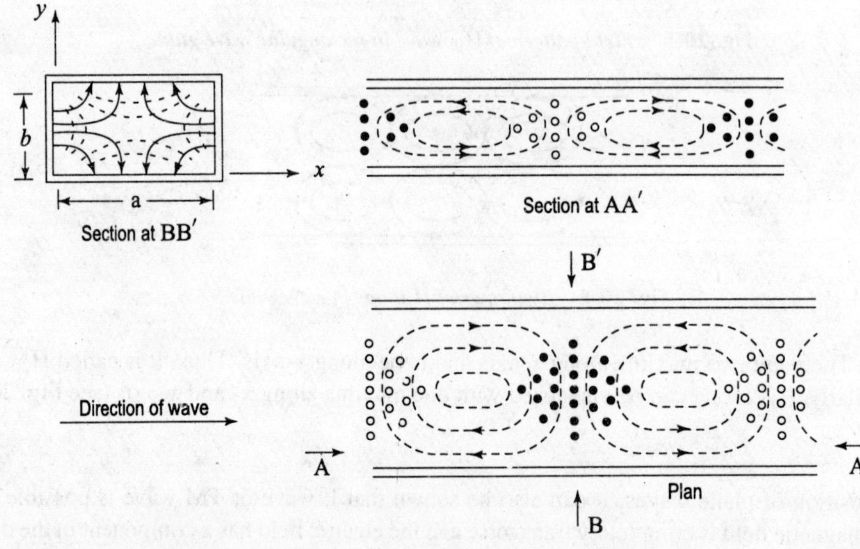

Fig. 20-8 *Field pattern of* H_{11} *waves in rectangular waveguide.*

Velocity and wavelength

As already seen, there are two constituent plane waves in an H_{10} wave. Let lines AC, AD represent the wavefronts of these two constituent waves. These wavefronts move at velocity c along CE, DE which are perpendicular to AC, AD respectively (Fig. 20.10).

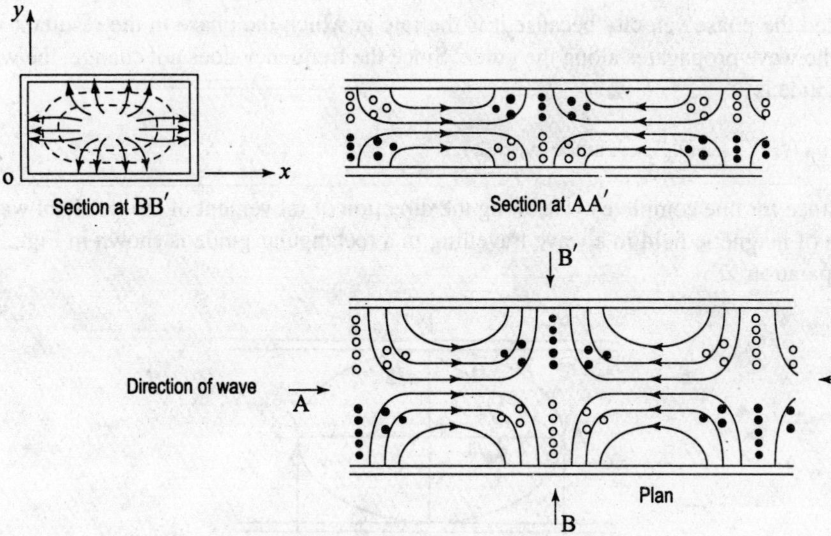

Fig. 20-9 *Field pattern of* E_{11} *waves in rectangular waveguide.*

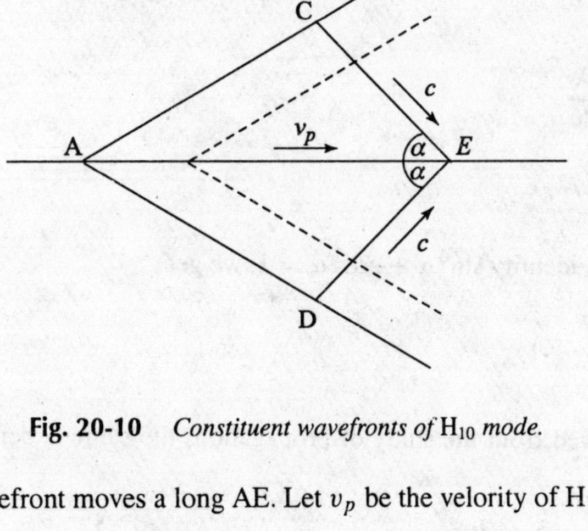

Fig. 20-10 *Constituent wavefronts of* H_{10} *mode.*

The resultant H_{10} wavefront moves a long AE. Let v_p be the velority of H_{10} wavefront. Then,

$$\frac{c}{v_p} = \frac{\text{CE}}{\text{AE}} = \cos\alpha \qquad \left(\alpha = \frac{1}{2}\right),$$

where $\alpha = \angle AED$, i.e.,

$$v_p = \frac{c}{\cos\alpha}.$$

Since $\cos\alpha < 1$, $v_p > c$.

v_p is called the phase velocity because it is the rate at which the phase in the resultant wavefront changes as the wave propagates along the guide. Since the frequency does not change, the wavelength in the waveguide is

$$\lambda_g = v_p/f \qquad (\lambda_g > \lambda \text{ since } v_p > c).$$

λ_g is the distnce for one complete cycle along the direction of movement of the resultant wavefront.

One loop of magnetic field in a wave travelling in a rectangular guide is shown in Fig. 20.11 with side wall separation 'a'.

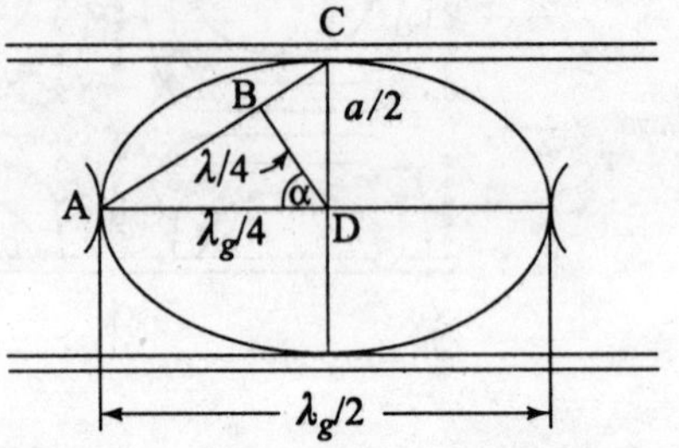

Fig. 20-11 *One loop of H field in* H_{10} *wave.*

From this figure,

$$\sin\alpha = \frac{BD}{CD} = \frac{\lambda}{2a},$$

$$\cos\alpha = \frac{BD}{AD} = \lambda/\lambda_g.$$

Using the trigonometric identity $\sin^2\alpha + \cos^2\alpha = 1$, we get

$$\left(\frac{\lambda}{2a}\right)^2 + \left(\frac{\lambda}{\lambda_g}\right)^2 = 1.$$

This expression is derived from the study of propagation; the same is derived in Appendix 8 using Maxwell's equation.

Critical or cut-off wavelength

If α is increased, $\sin\alpha$ increases. When $\lambda = 2a, \alpha = 90°$ and $\sin\alpha = 1$, the wave is reflected back. The wavelength at which this occurs is called the critical wavelength λ_c. The corresponding frequency is $f_c = v_p/\lambda_c$, f_c is called the criticl frequency; a wave of frequency less than f_c cannot be transmitted.

Phase velocity and group velocity

According to the theory of relativity, energy can never be propagated at speeds greater c, the speed of light. Then how could we get $v_p > c$?

The answer is that v_p is not the actual velocity of energy transmission. It is the rate of change of phase or velocity of phase travel. The actual progress of energy is at group velocity v_g (Fig. 20.12) which is given by

$$v_g = c \cos \alpha = c \frac{\lambda}{\lambda_g},$$

$$v_p = c / \cos \alpha = c\lambda_g / \lambda.$$

Hence $\quad v_p v_g = c^2.$

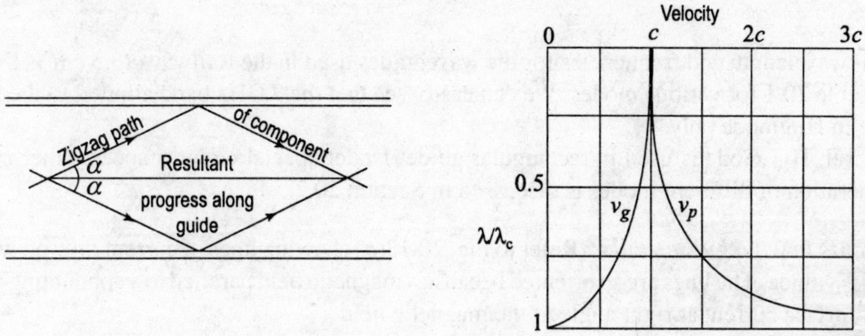

Fig. 20-12 *Relation between phase and group velocity.*

In all practical applications, waves carry intelligence (imposed on it by modulation). The modulation envelope travels at group velocity whereas the carrier travels at phase velocity in the waveguide.

Properties of higher modes

(*a*) *Critical wavelength:* Earlier, we discussed critical wavelength w.r.t. fundemental mode H_{10}. For E_{mn} or H_{mn} mode in general, and for a waveguide of dimensions $a \times b$,

$$\frac{1}{\lambda^2} = \left(\frac{m}{2a}\right)^2 + \left(\frac{n}{2b}\right)^2,$$

$$\frac{1}{\lambda_g^2} = \frac{1}{\lambda^2} - \frac{1}{\lambda_c^2},$$

where, λ_g = wavelength in the guide,

λ = wavelength in free space,

λ_c = cut off wavelength.

For H_{10}, $m = 1, n = 0$ and $\lambda_c = 2a$

$$\frac{1}{\lambda_g^2} = \frac{1}{\lambda^2} - \left(\frac{1}{2a}\right)^2$$

as already seen.

Table 20. 1 *Critical wavelengths and frequencies for the waveguides used in Indian Railways. Dimension of the waveguide is 3.5 cm × 1.6 cm*

Mode	λ_c Propagation possible for $\lambda < \lambda_c$	f_c Propogation possible $f > f_c$
H_{20}	7 cm	4.28 GHz
H_{10}	3.5 cm	8.57 GHz
H_{01}	3.2 cm	9.375 GHz
H_{11}, E_{11}	1.455 cm	20.6 GHz

Critical wavelength and frequencies for the waveguides used in the Railways (3.5 cm × 1.6 cm) are given in Table 20.1 for various modes. We can easily see that the 7 GHz band allotted to the Railways is feasible in H_{10} mode only.

In general, H_{10} mode is used in rectangular guide. Under special circumstances, other modes are used. Generation of different modes is discussed in Section 20.7.

(*b*) *Currents in waveguide walls:* Refer to Fig. 20.13(*a*) showing lines of current flow on waveguide walls for H_{10} mode. The lines are so oriented because a magnetic field paralled to a conducting boundary induces a surface current at right angle to the magnetic field.

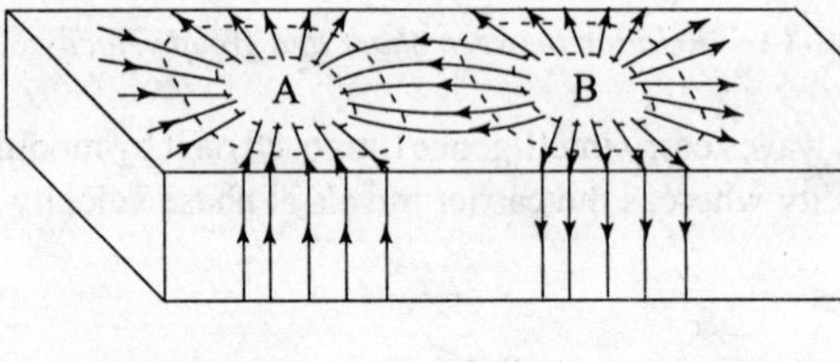

Fig. 20-13(*a*) *Current flow on waveguide walls.*

The pattern of current flow travels with the wave at phase velocity v_p. Any slot not interrupting the current lines has negligible effect on the propagation characteristics. Therefore, in Fig. 20.13(*b*) slots P and Q do not have any efect on the propagation (these slots are used for probes), while slots R, S and T impede the wave.

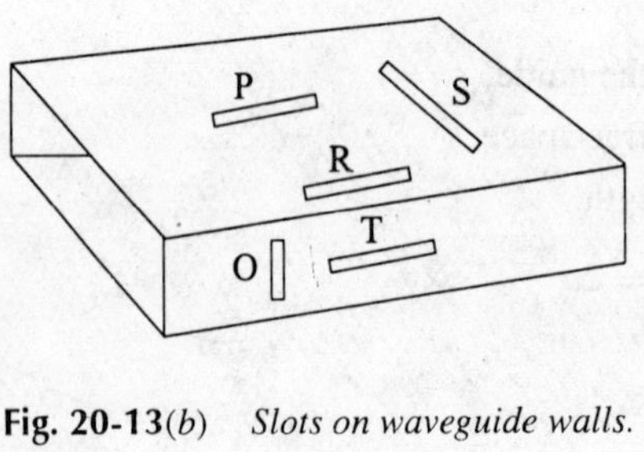

Fig. 20-13(*b*) *Slots on waveguide walls.*

(*c*) *Power carrying capacity:* In waveguides, there is no loss in the air between the wall of the guide. The power dissipated in the wall is not much. The limit is imposed by the breakdown voltage in the air. Dry air at atmospheric pressure breaks down at 30 kV/cm.

Then, one may wonder — in MW communications systems we talk powers of the order of 1 watt. Why discuss breakdown electric fields at all?

Digital MW systems are pulsed systems, and carriers are gated as per modulation scheme chosen. At GHz frequencies, the pulse-duration is in nano seconds and average pulse power applied over a pulse duration is sufficient to cause a breakdown. The maximum value of the mean power which can be handled without breakdown (for H_{10} wave) is

$$P = E_{max}^2 \frac{ab}{480\pi} \sqrt{\{1 - \left(\frac{\lambda}{2a}\right)^2} \text{ watt},$$

where E_{max} is the electric field at $x = a/2$, i.e. half way across the waveguide (refer to Appendix A11 for the derivation).

Because of the possibility of a breakdown voltage, the Tx powers of MW systems, though adjustable, are limited to a maximum value specified by the manufacturer.

A breakdown voltage of 30 kV/cm is at normal atmospheric pressure. At high altitudes the atmospheric pressure is less, so the power carrying capacity is reduced. This can be avoided by 'pressurising' the waveguide using dehydrators. Dehydrators not only keep the waveguide dry but also maintain pressure. In dry waveguides the breakdown voltages are less. So, dehydrators are to be taken seriously. For example, at Undavalli, Panchadarala and Narayanagiri MW stations of the South Central Railways located at high attitudes, dehydrators are employed even though the waveguide used is only of the order of few meters.

(*d*) *Attenuation:* Since currents are not along the waveguide walls when a wave is propagated, there would be some attenuation. The plus point is that the RF signals do not penetrate deep into the walls but go only skin depth (for copper it is 0.001 mm at $\lambda = 10$ cm, i.e., 3 GHz). For a conductor of permeability μ and resistivity ρ, the skin depth δ (at frequency f) is given by

$$\delta = \sqrt{\frac{\rho}{\pi \mu f}},$$

surface resistivity $R_s = \rho/\delta$ ohms,

$$\text{and attenuation } \alpha = \frac{R_s}{120\pi b}\left\{1 + \frac{2b}{a}\left(\frac{\lambda}{2a}\right)^2\right\}\frac{1}{\sqrt{1-(\lambda/2a)^2}} \text{ nepers/m.}$$

See Appendix A9 for a derivation.

Attenuation for various modes is shown in Fig. 20.14. It is lowest for H_{10}.

Why H_{10} mode in rectangular waveguide?

- Attenuation is lowest
- Cut-off wavelength is maximum
- It is easy to generate H_{10} mode

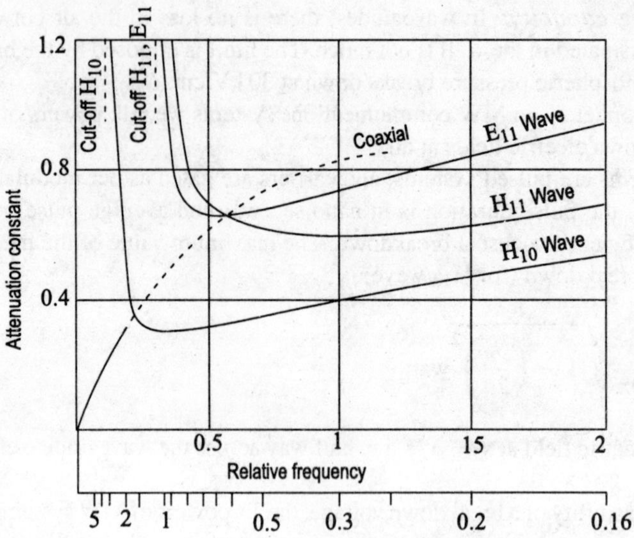

Fig. 20-14 *Attenuation for various modes.*

(*e*) *Waveguide dimensions:* Waveguide dimensions are so chosen that only H_{10} mode propagates. (For special applications, other modes are chosen). Dimensions are determined on the following considerations:

- Desired mode
- Critical wavelength of the chosen mode
- Attenuation
- Power-carrying capacity

Let $a \times b$ be the dimensions of the waveguide and λ the operating wavelength.

$$\lambda_c = 2a \quad (\text{for } H_{10});$$
$$\lambda_c = a \quad (\text{for } H_{20});$$
$$\lambda_c = 2b \quad (\text{for } H_{01}).$$

So if λ is given and only H_{10} is to be propagated, choose a, b such that

$$\lambda/2 < a < \lambda; \quad \text{and} \quad 2b < \lambda.$$

For a chosen dimension 'a', the mean operating wavelength for H_{10} is

$$\lambda_{\text{mean}} = \frac{2}{3}\lambda_c,$$

$$\text{i.e, } \lambda_{\text{mean}} = \frac{2}{3} \times 2a = \frac{4}{3}a.$$

The operation band is limited to 20% on either side of the mean (i.e., 4/3a), and b is usually chosen close to $a/2$ because this gives the lowest attenuation and highest breakdown level (while avoiding H_{01} mode).

20.5 Circular Waveguides

In a rectangular waveguide, by applying the boundary conditions on plane waves one can easily study the propagation modes which can be tallied with the derivations using Maxwell's equations and bounday conditions. Such a simple approach is not feasible in the case of circular waveguides because of the difficulty in fitting plane waves with boundary conditions on circular surface. But we can extrapolate from the propagation modes of rectangular guide.

Imagine a square waveguide (special case of rectangular waveguide with $a = b$). Distort it gradually to take the shape of a circle (see Fig. 20.15). We have the following results:

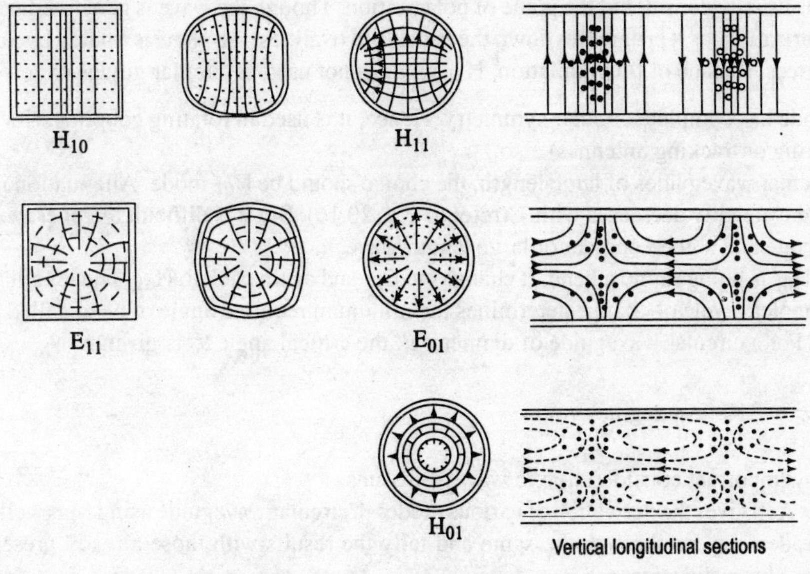

Fig. 20-15 *Development of circular waveguide modes from rectangular waveguide modes.*

- Circular guide can also carry E and H waves.
- The field in a longitudinal section is of the same form as the corresponding rectangular guide modes.
- H_{11} is the fundamental mode in circular waveguide having the longest λ_c.
- E_{01}, H_{01} modes are other fundamental modes.
- H_{01} has no counterpart in rectangular guide.
- Supscript '0' in E_{01} and H_{01} indicates that there is no variation of the field in moving round the circumference of the guide. (In the case of rectangular guides, '0' subcript means absence of a particular field in the direction of motion.)
- The value of second subscript determines the complexity of the field variations along the radius. Only cases with second subscript = 1 are of importance.
- $v_p = c\dfrac{\lambda_g}{\lambda}; \quad v_g = c\dfrac{\lambda}{\lambda_g}; \quad v_p v_g = c^2;$

  as in rectangular waveguide.

– Attenuation α for a circular waveguide of diameter d for H_{01} mode is given by

$$\alpha = \frac{R_s}{60\Pi d}\left(\frac{\lambda}{\lambda_c}\right)^2 \frac{1}{\sqrt{1-(\lambda/\lambda_c)^2}} \quad \text{neper/m},$$

where, λ_c is the critical wavelength, and R_s = surface resistance.

Plates 2(*a-c*) shows H_{11}, E_{01}, H_{01} modes in a circular waveguide.

The reason for not using H_{11} mode in a circular waveguide though it has the longest λ_c is that H_{11} has no plane of polarization in the strictest sense. A plane passing through the axis and coinciding with the central line is usually considered the plane of polarization. Though the wave is generated with the same plane of polarization, as it progresses down the bends and ovalities, the plane is rotated by unknown and variable degrees. Becaue of this limilation, H_{11} mode is not used in circular guides over long lengths.

– E_{01} mode has complete circular symmetry. Hence, it is used in rotating couplings(for connection to rotating or tracking antennas) also.
– For circular waveguides of large length, the choice should be H_{01} mode. Attenuation is lowest for H_{01} and it steadily decreases with λ (refer to Fig. 20.16). But it is difficult to generate. E_{01} is easy to generate and a good choice for large length waveguides.
– When H_{01} is going round a bend, it changes to E_{11} and again back to H_{01}. The length required for this complete cycle of change determines the minimum required angle of bend called the 'critical angle.' For a circular waveguide of diameter d, the critical angle θ_c is given by

$$\theta_c = 310 \times \frac{\lambda}{d} \text{ degrees.}$$

This is a very important consideration in waveguide runs.

Appendix A10 gives the derivation of various modes in circular waveguide using Maxwell's equation. Interested readers can go through the same and tally the results with those already presented in this section using a heuristic approach.

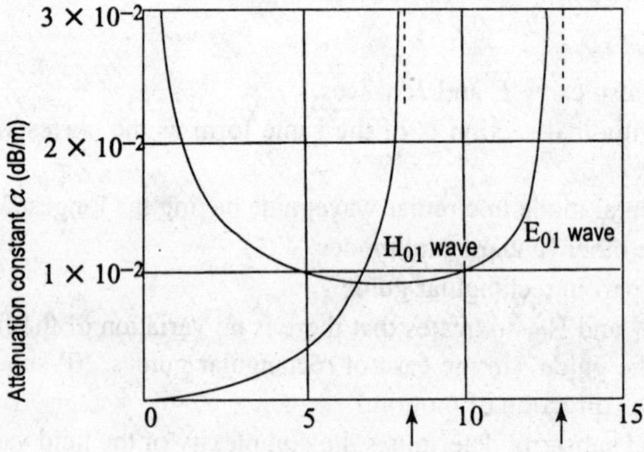

Fig. 20-16 *Attenuation in circular waveguide for various modes.*

20.6 Elliptical Waveguides

Distortion of square waveguide (special case of rectangular waveguide with $a = b$) as shown in Fig. 20.17 gives the modes in an elliptical waveguide. These modes are H_{11}, E_{01}, H_{01}.

The most commonly used mode is the H_{11} mode. The reasons are:

- H_{11} has longest λ_c.
- H_{11} comes from H_{10} during the transition from rectangular to elliptical waveguide (equipment side to antenna side).

Critical wavelength λ_c

For an elliptical waveguide, we can get an approaximate idea of the critical wavelength λ_c by considering the rectangular case. In a square tube of side a, the H_{10} wave has a critical wavelength of $2a$. Assuming the same ratio of critical wavelength to perimeter for H_{10} of square (rectangular) waveguide and corresponding H_{11} of elliptical waveguide,

$$\frac{\lambda_c \text{ (square)}}{\text{Perimeter (square)}} = \frac{\lambda_c \text{ (elliptical)}}{\text{Perimeter (ellipse)}}$$

$$\frac{2a}{4a} = \frac{\lambda_c \text{ (elliptical)}}{\text{Perimeter (ellipse)}}$$

$$\therefore \quad \lambda_c \quad \text{(elliptical)} = \frac{\text{Perimeter of ellipse}}{2} = \frac{\pi\sqrt{(a^2+b^2)/2}}{2} = \frac{\pi\sqrt{a^2+b^2}}{2\sqrt{2}}$$

But this is to be modified by Bessel constant which gives

$$\lambda_c = \frac{\pi\sqrt{a^2+b^2}}{2\sqrt{2}S},$$

where $S = 1.841$ for H_{11} mode.

Attenuation

This can be extrapolated from the attenuation of circular waveguide (Section 20.5) as

$$\text{Attenuation } \alpha = \frac{R_s}{60\pi\sqrt{\frac{a^2+b^2}{2}}}\left(\frac{\lambda}{\lambda_c}\right)^2 \frac{1}{\sqrt{1-(\lambda/\lambda_c)^2}},$$

where a and b are the major and minor axis of the ellipse respectively.

20.7 Generation of Different Modes in Waveguides

20.7.1 Excitation of EM field pattern

A wave is launched in a waveguide by setting up the desired mode of electromagnetic fields at one end. The excitation is done using probes in waveguides (and loops in cavity resonators). The probe lies in the direction of the desired electric field and the loops encircle the desired lines of magnetic field.

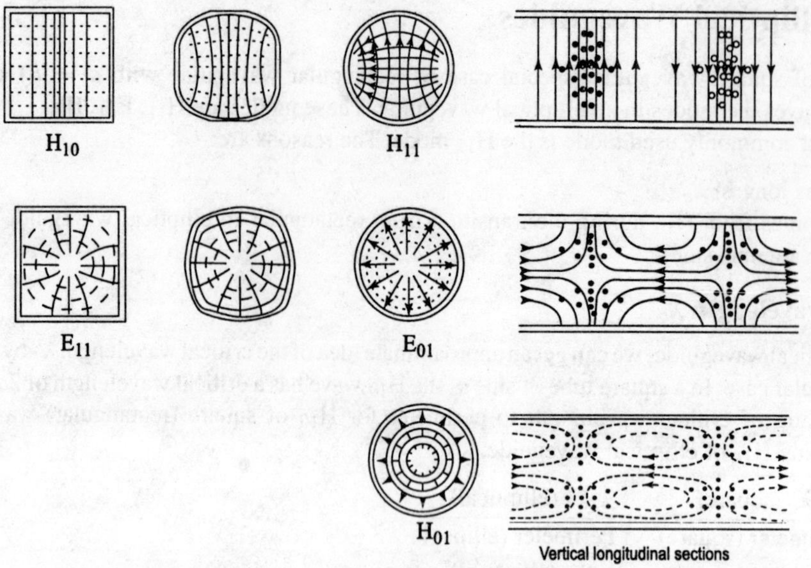

Fig. 20-17 *Development of various modes in elliptical waveguide.*

Rectangular waveguide

The mostly used mode in rectangular waveguide is H_{10} as already discussed in Section 20.4.

Generation of H_{10} by means of simple probe, crossbar transformer and doorknob transformer are illustrated in Fig. 20.18 (probe on the broad face for TE wave or H wave and narrow face for TM wave or E wave). Generation of H_{11}, E_{11}, H_{20} and E_{21} mode are illustrated in Fig. 20.19.

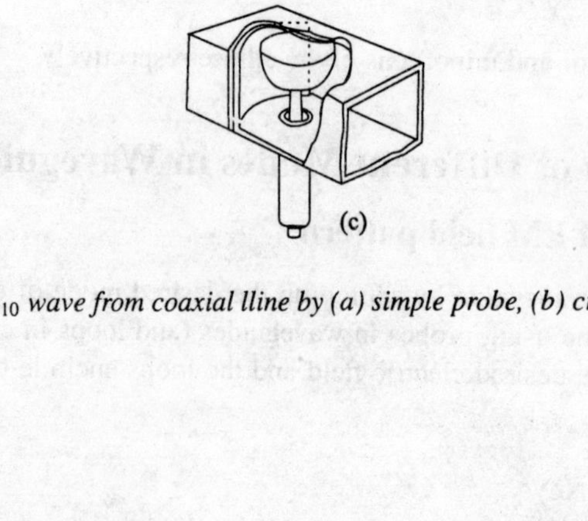

Fig. 20-18. *Generation of* H_{10} *wave from coaxial lline by (a) simple probe, (b) cross bar transforme, (c) doorknob transformer.*

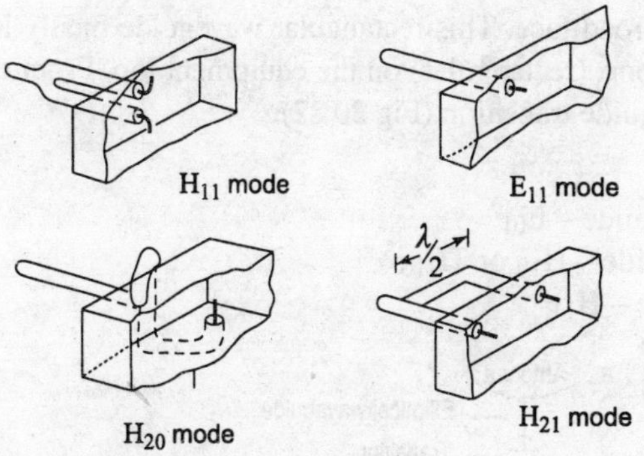

Fig. 20-19 *Generation of different modes in rectangular waveguides.*

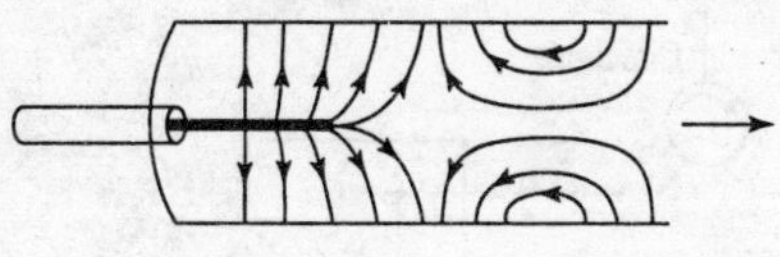

Fig. 20-20 E_{01} *mode generation.*

Circular waveguide

E_{01} is mostly used for circular waveguides. It is excited by a centrally located probe (Fig. 20.20). H_{01} is derived from TE_{20} in a rectangular guide using transitions (see Fig. 20.21).

Elliptical waveguide

H_{11} wave comes from H_{10} of a rectangular guide through a transition. A transition is a waveguide component interposed between the rectangular port of an equipment and the elliptical waveguide for a reflection-free transition in the guided path of an electromagnetic wave.

20.7.2 Case study

The digital MW radio equipment (DMR 770) has a miniature circular waveguide connecting the Tx panel to the circulator section. This circular waveguide is probe-excited centrally. It is terminated on

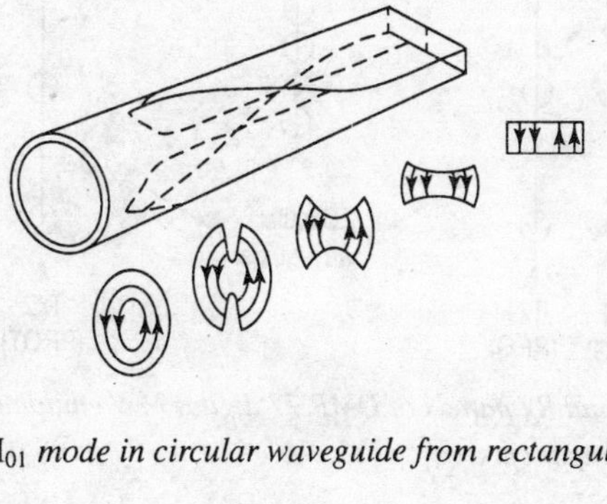

Fig. 20-21 H_{01} *mode in circular waveguide from rectangular waveguide.*

a rectangular guide on its broad face. This rectangular waveguide finally leads the Tx signal through a circulator to the antenna port (rectangular) on the equipment top. From there it goes to an elliptical waveguide through a waveguide transition (Fig 20.22).

The modes involved are:

- In miniature circular guide—E_{01}
- In rectangular waveguide—H_{10} or TE_{10}
- In elliptical waveguide—H_{11}

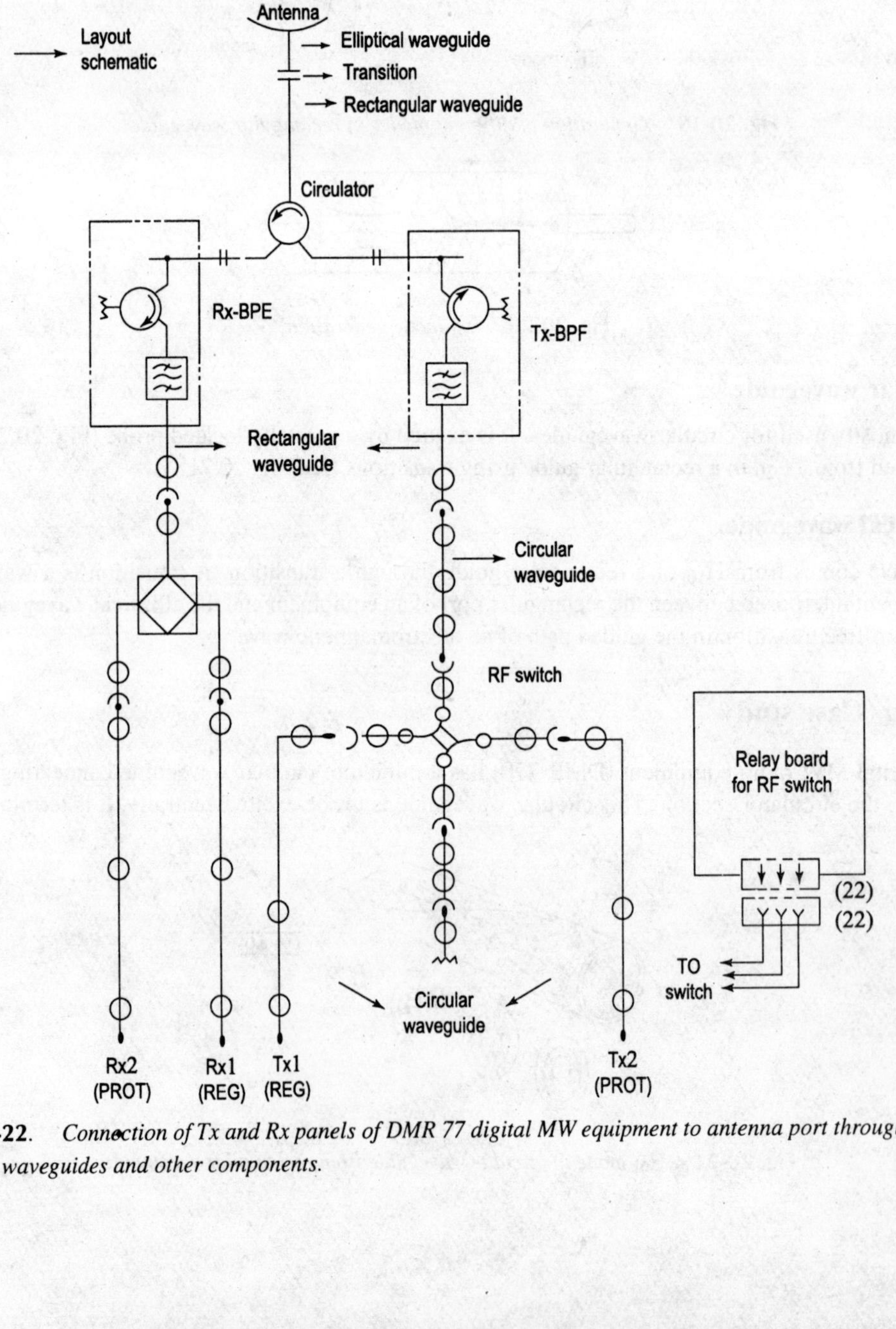

Fig. 20-22. *Connection of Tx and Rx panels of DMR 77 digital MW equipment to antenna port through various waveguides and other components.*

20.8 Comparision of Different Waveguides

Rectangular waveguides have the lowest attenuation, followed by elliptical waveguides (copper) and elliptical waveguides (aluminium) in the 7 GHz band.

Circular waveguide (H_{01})	:	1 dB/100 m
Rectangular waveguide	:	4 dB/100 m (E Cu 2y type)
Elliptical waveguide (aluminium)*	:	7 dB/100 m

*WEA-7-L/H-HQ inner dimensions: Major axis 42.4 mm, minor axis 27.4 mm).

Rectangular waveguides need sealing to prevent moisture entry at every joint. Also, it is costlier. Hence, elliptical waveguide is the most preferred waveguide despite its higher attenuation. Also, elliptical waveguide can be easily hoisted.

λ_c	f_c	Waveguide type	Size (major axis × minor axis)
5.605 cm	5.35 GHz	WEA-7-L-HQ Al(Aluminium) waveguide	42.4 × 27.4 mm
5.462 cm	5.49 GHz	Cu (copper) waveguide (E Cu 2y)	4.4 × 2.2 mm

Exercises

1. Though H_{11} has the longest λ_c, it is not used in a circular waveguide. Why?
2. $v_p > c$ and $v_g < c$. What is the physical significance?
3. In a rectangular waveguide of dimensions $a \times b$, b is chosen close to $a/2$. Why?
4. For E_{mm} or H_{mn} mode in a rectangular waveguide of dimension $a \times b$, show that $1/\lambda^2 = (m/2a)^2 + (n/2b)^2$.

21

Waveguide Components and Accessories

21.1 Introduction

The waveguide span from equipment to antenna, together with the components and accessories, is called the 'waveguide run'. There are several components and accessories along the waveguide run, each of which serves a definite pupose. These components and accessories are discussed in this chapter.

Waveguide bends, corners, tapers, twists, flexible guides and loading elements are waveguide components. Waveguide flanges, couplings, clamps, earthing kits, wall feed-throughs, bending tools come under accessories.

21.2 Waveguide Bends

Waveguide bends facilitate change of direction in a waveguide run. These bends are similar to elbows in a water pipeline.

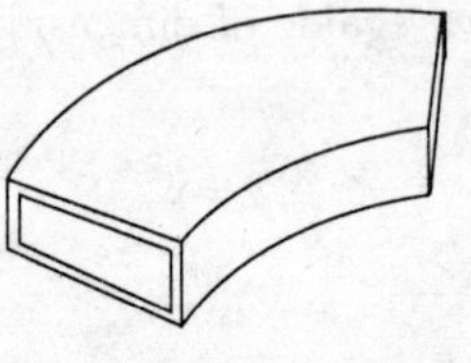

Fig. 21.1 *H-bend.*

H-bend : A waveguide bend whose plane is parallel to the plane of magnetic field vector is called an H-bend (Fig. 21.1). An H-bend can rest only on its broad face, the narrow face is curved. So its plane is the broad face which is parallel to the H vector. Hence it is an H-bend. Physically, the bend is on the narrow side of a rectangle, i.e., hard to bend. So, an easy way of remembering H-bend is to associate with the clue 'hard to bend'.

E-bend : A waveguide bend whose plane is parallel to the plane of electric field vector is called an E-bend (Fig. 21.2). An E-bend can rest only on its narrow face (the broad face is curved). So its plane is the narrow face which is parallel to the E vector. Hence, it is an E-bend. Physically, the bend is on the broad side of rectangle, i.e., easy to bend. So an easy way to remember an E-bend is to associate with the ciue 'easy to bend'.

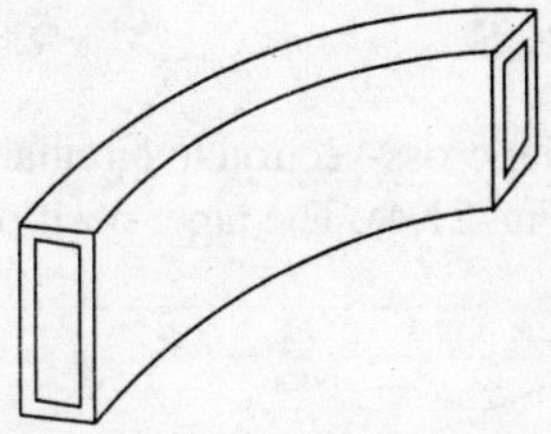

Fig. 21.2 *E-Bend.*

Problems with the bends

Since the velocity of propagation in a bent guide differs slightly from that in a straight guide, every bend introduces some discontinuity and hence causes reflection of some energy. Though it can be minimized by keeping the radius of curvature large compared to the guide dimensions, it cannot be totally avoided. Hence, use of too many bends in the waveguide run should be avoided. The placement of equipment with respect to waveguide entry in the equipment room, and also the order of different waveguides (for different direction hops) should be such that the bends used are minimum.

21.3 Waveguide Corners

A mitred corner (single mitred or double mitred) as shown Fig. 21.3 (*a*), (*b*), (*c*) can be used in place of bends. These corners can be made almost reflectionless by suitably choosing the corner width d. The value of d depends on the wavelength and the guide dimensions.

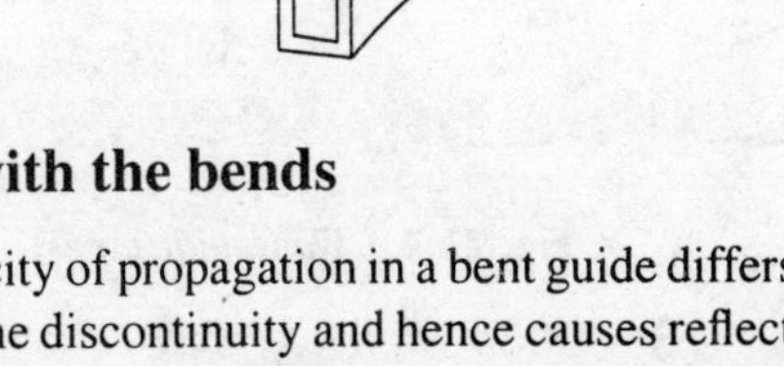

(*a*)

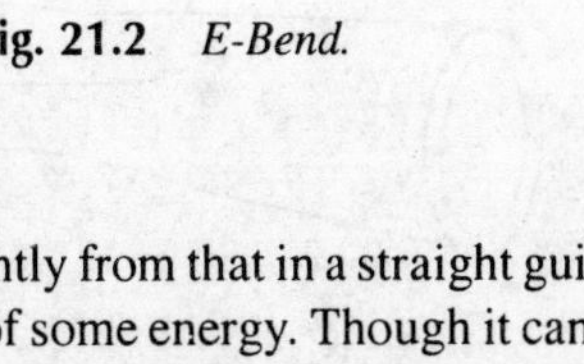

(*b*)

Fig. 21.3(*a*) *Single mitred H-corner.* Fig. 21.3(*b*) *Single mitred E-corner.*

In a single mitred corner, d is $0.9a$ for a 90° H-corner and $8.85b$ for a 90° E-corner, where a and b are the broad and narrow sides of a waveguide respectively.

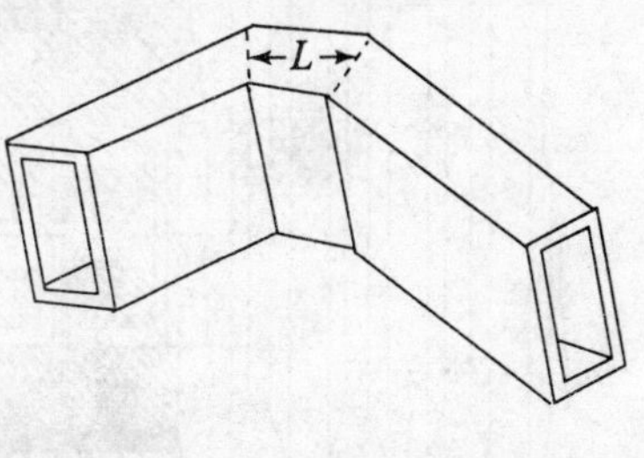

Fig. 21.3(*c*) *Double mitred E-corner.*

The double mitred corner in Fig. 21.3 (*c*) allows a wider frequency range of operations than a single mitred type. The length L is chosen such that the reflections from both the corners cancel each other (hence $L = \lambda_g/4$). Thus, a reflectionless corner is a double mitred corner.

21.4 Waveguide Taper

When it is needed to change over from rectangular cross-section to circular (or elliptical) cross-section and vice versa, a waveguide taper is used (refer Fig. 21.4). The taper position is usualy $2\lambda_g$ to minimize losses due to reflections.

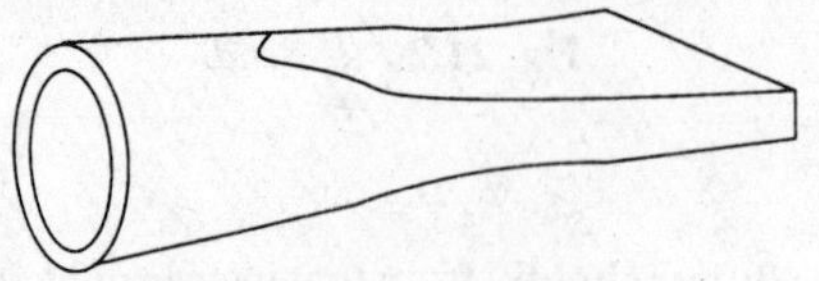

Fig. 21.4 *Waveguide taper.*

21.5 Waveguide Twist

When it is required to change the direction of polariztion of a guided wave, a waveguide twist is used. In almost all cases the change is by 90°. The twist is extended over a minimum length of $2\lambda_g$ to avoid losses due to reflections (Fig. 21.5).

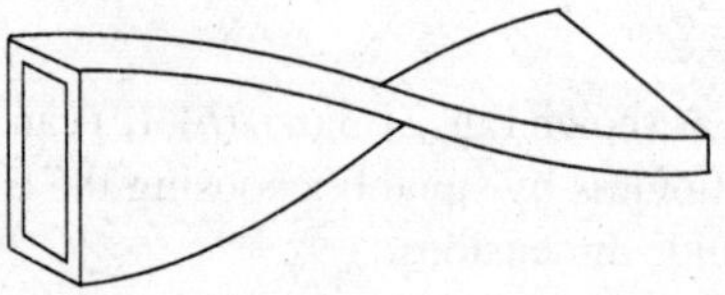

Fig. 21.5 *Waveguide twist.*

21.6 Flexible Waveguide

Metal hose-type flexible waveguides consist of a continuous spiral of interlocking metal strip (formed in the same way as flexible metal hose) of rectangular cross-section with a moulded rubber jacket (Fig 21.6).

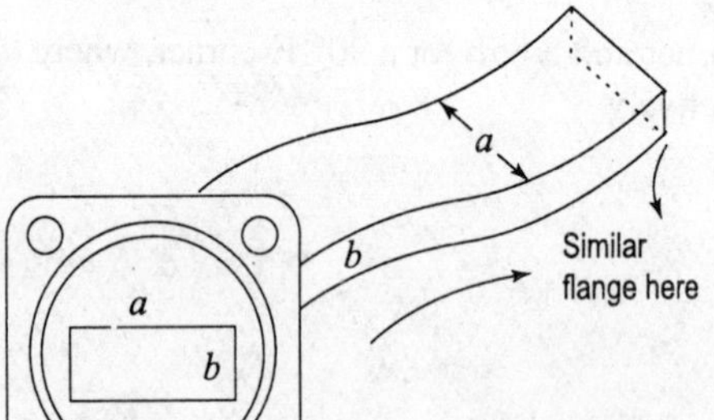

Fig. 21.6(*a*) *Flexible waveguide (metal hose).*

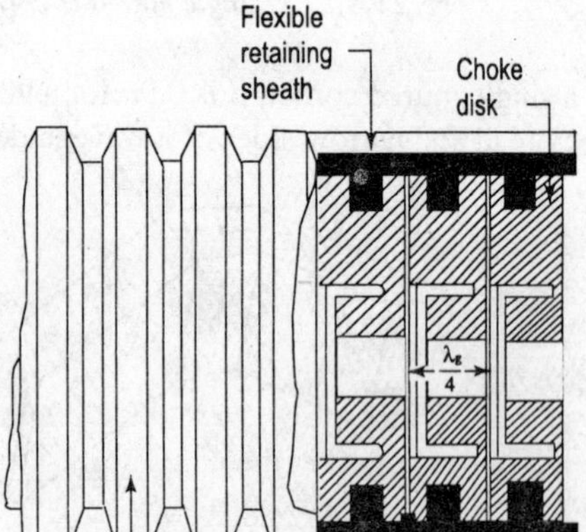

Fig. 21.6(*b*) *Flexible waveguide (vertebral).*

A vertebral type flexible waveguide consists of a series of discs, each pierced with a rectangular hole (of the same size as the waveguide) retained within a ribbed rubber or polythene jacket. The ribs maintain small spacing between the discs, and each disc has choke grooves in one face the dimensions of which are such that pairs of discs act as choke couplings. The combined length of rib spacing and segment is λ_g to minimise reflections.

Metal hose type waveguides are available upto a length of 0.5 m and vertebral types are available in 1 m or 1.5 m lengths.

21.7 Loading Elements

In this section, we will discuss matched terminations, attenuators, irises, apertures and filters. All these components have a common property—they 'load' the transmission through waveguide.

Matched termination

In a closed-ended waveguide, reflections from the closed end causes formation of standing wave pattern. E_y is zero at the closed end and maximum at lengths equal to odd multiples of $\lambda_g/4$ from the closed end. H_x is zero at points where E_y is maximum. Thus, at $\lambda_g/4$, E_y/H_x is infinite which implies that the input impedance of $\lambda_g/4$ section with closed end is infinite (in practice, a very large impedance value). Similarly, input impedance of a $\lambda_g/2$ section with closed end is zero.

If a resistive film is placed at $\lambda g/4$ from the closed end, all the power propagated is absorbed by it (since it is parallel with an infinite impedance) and hence it becomes a matched termination (Fig. 21.7(a)).

In practice, an abrupt film is not used. A length of lossy material that can fit inside by sliding and having a tapered end is used. This ensures a gradual change in the waveguide form and thereby reflections are minimized. The material used is wood for higher powers, and iron dust or lossy ceramic material for ordinary powers.

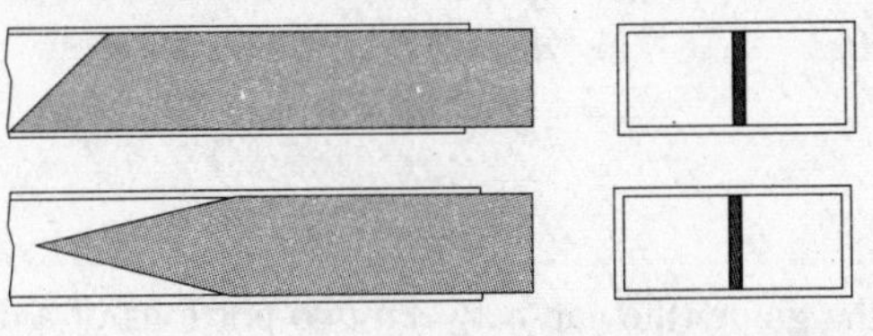

Fig. 21.7(a) *Matched termination.*

Resistive attenuator: A resistive film mounted in a waveguide as in Fig. 21.7(b) acts as an attenuactor. It can be made variable by changing its tranverse orientation or 'cut-in' area. Values of attenuation upto 40 dB are possible, but for higher values the design has to be precise to avoid coupling through the 'cut-in' slot.

Lumped reactances: Any waveguide component introduces a certain amount of mismatch. The answer for this problem is a matching device. A matching device can be (i) waveguide stub of adjustable length, (ii) lumped reactance. Lumped reactances are in the form of metallic obstacles of various shapes placed in the guide. The different types are (refer to Fig. 21.7(c&d)):

- Inductive diaphragm or inductive iris
- Capactive diaphragm or capacitive iris

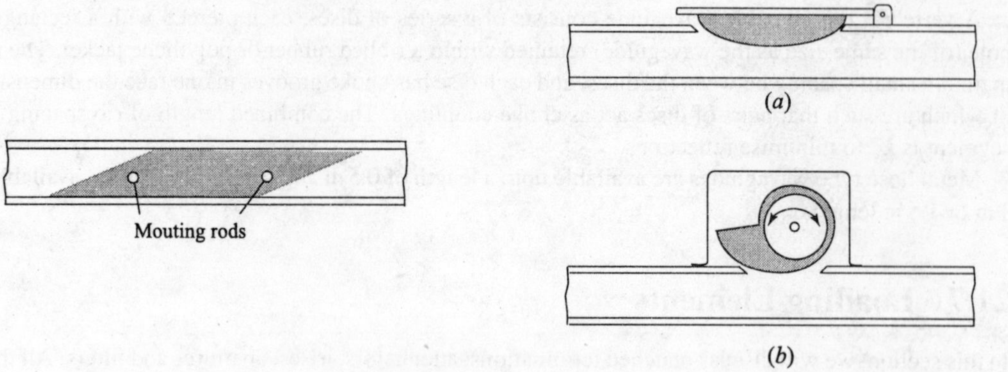

Fig. 21.7(*b*) *Resistive attenuators.*

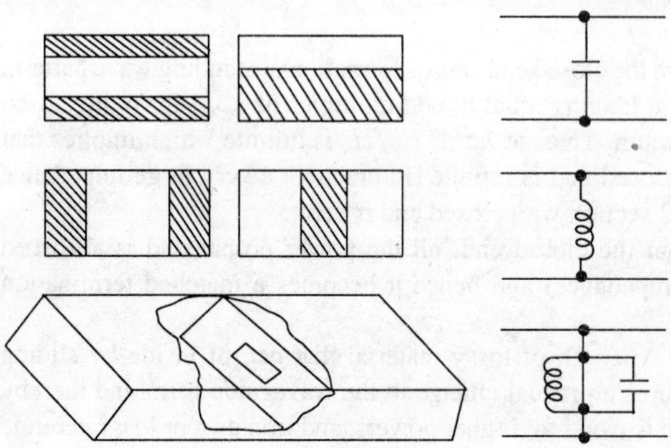

(*c*)

Fig. 21.7(*c*) *Lumped reactances-Irises.*

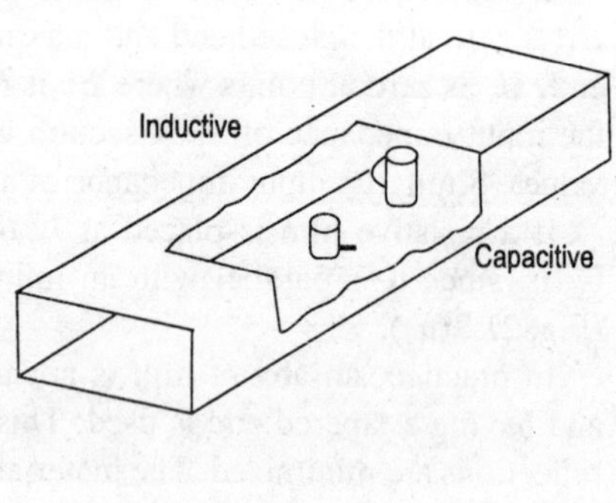

(*d*)

Fig. 21.7(*d*) *Reactive posts.*

- Inductive post
- Capacitive post

An iris physically forms an aperture within the guide (the gap between two portions of a diaphragm perpendicular to the broad face or perpendicular to the narrow face). It corrects mismatches inherent in the waveguide components, and its aperture size is determined by SWR measurement. Inductive irises have higher breakdown voltages.

Reactive posts perpendicular to broad face is capacitive load for smaller lengths. As the post's penetration increases, reactance becomes inductive.

Resonant apertures: A combination of inductive and capacitive irises exhibit reasonance at a particular frequency, as per reactance values (Fig. 21.7(*e*)). They find application in the construction of resonant windows (of glass or mica) by which a waveguide system can be sealed against moisture entry and at the same time its transmission properties remain unchanged. It may please be noted that the pressure window—the mica sheets inserted at the place where the waveguide run ends—have some capacitance. If these are replaced with perspex or plastic sheet, it may cause loading instead of resonance, i.e., it may not present the lowest impedance at operating frequency, and hence the receiver signal strength and AGC reading may fall.

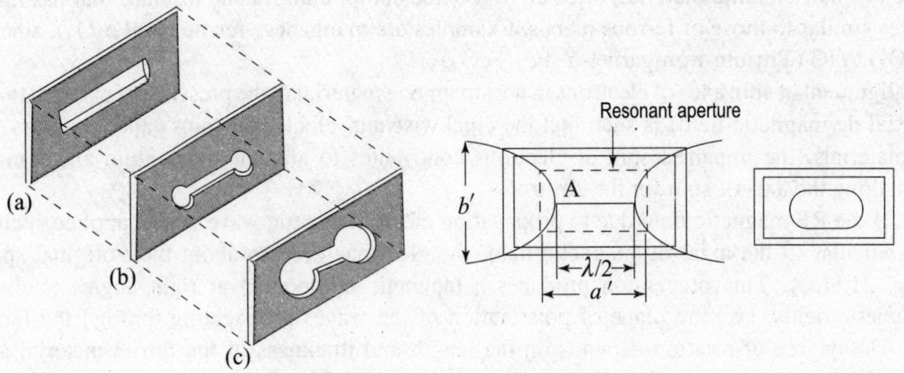

Fig. 21.7(*e*) *Resonant apertures.*

Filters: Consider a waveguide section correctly terminated at its far end having two irises inserted in it. There is a particular frequency at which reflections from the two irises will cancel. At this frequency, power transmission takes place along the guide without losses. The space between the irises is a reasonant cavity and the arrangement is a single cavity resonant fillter (Fig. 21.7(*f*)).

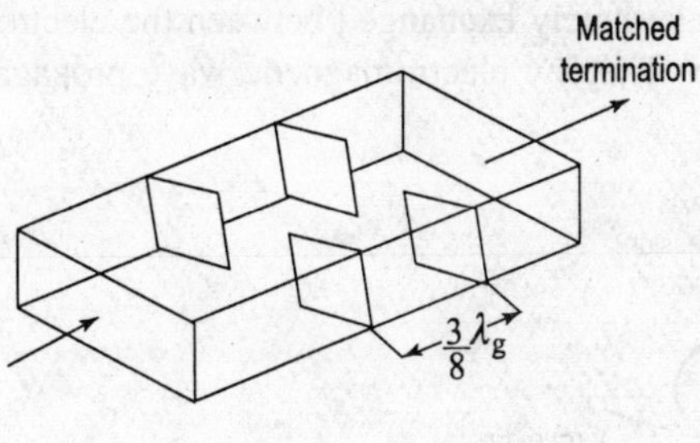

(*f*)

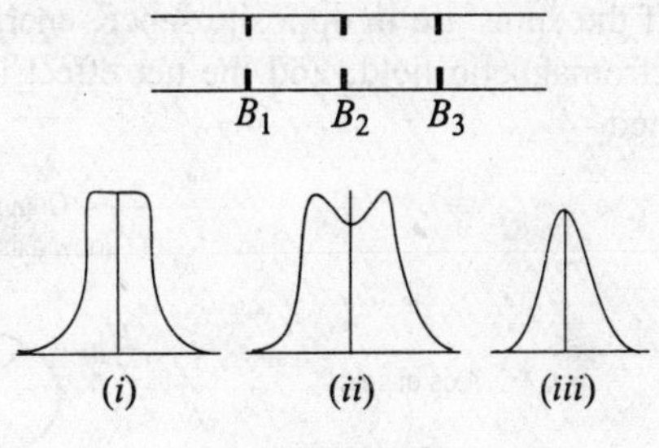

(*g*)

Fig. 21.7(*f*). *Single cavity resonant filter.*

Fig. 21.7(*g*) *Two-cavity filter and its response with (i) critical coupling, (ii) over-coupling, (iii) under-coupling.*

Using three or more irises, the filter can be designed to have a bandpass selectivity. The irises have susceptances B_1, B_2, B_3, and the arrangement consititutes a two-cavity resonant filter. It has the properties of two coupled tuned circuits. Critical coupling is obtained when $B_2 = B_1^2 = B_3^2 (= B_c$ the critical susceptance). If $B_2 > B_c$ it is over-coupling. If $B_2 < B_c$, it is under-coupling.

21.8 Ferrite Devices

The requirement to be satisfied often in microwave system design is to keep the coupling a one-way affair. There are devices other than hybrid juctions and directional couplers useful in such applications. These are the Isolators, Circulators, and Duplexers—commonly known as ferrite devices.

21.8.1 Ferrites

A ferrite is a non-metallic material, often an iron oxide compound. It is an insulator but has magnetic properties similar to those of ferrous metals. Examples are manganese ferrite ($MnFe_2O_3$), zinc ferrite ($ZnFe_2O_3$), YIG (Yitrium-iron-garnet-$Y_3Fe_2(FeO_4)_3$).

The alignment of spin axes of electrons in non-magnetic material in the presence of external magnetic field (axial dc magnetic field) is such that the clockwise/anti-clockwise spins cancel in pairs. But in ferrite materials, the impaired spin of electrons contributes to angular momentum and a magnetic moment along the axis of spin for the electrons.

Now, if the RF magnetic field due to propagating electromagnetic wave is also applied such that it is perpendicular to the axial dc magnetic field, the electrons precess about their original spin axis (see Fig. 21.8(*a*)). This precession produces a magnetic component at right angles to the other two magnetic fields, i.e., the plane of polarization of the waves propagating through the ferrites is rotated. The degree of rotation depends on the length and thickness of the ferrite material and the strength of axial dc magnetic field. The axial dc magnetic field must provide at least the saturation magnetization, the minimum value required to ensure that the axes of the spinning electrons are suitably aligned. (This in turn decides the lowest usable frequency of ferrites.) If the dc magnetic field is made stronger than this, the precessional frequency equals the frequency of EM waves being propagated. When this happens, 'gyromagnetic resonance interaction' (GMRI) takes place between the spinning electrons and the magnetic field of the propagating waves. If both have clockwise direction, energy is delivered to electrons, i.e., absorption of energy from magnetic field to EM waves takes place. If the spins are in opposite sence, energy is alternately exchanged between the electrons and the electromagnetic field, and the net effect is zero. Thereby electromagnetic wave propagation is unaffected.

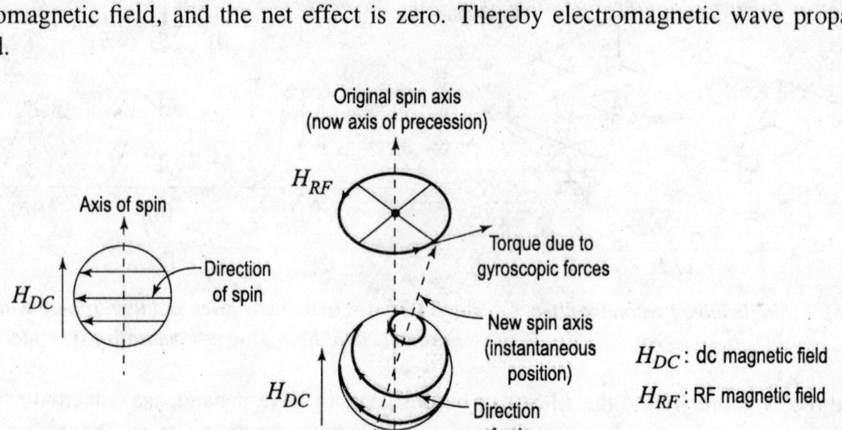

Fig. 21.8(*a*) *Spinning electrons and magnetic fields.*

The properties of rotating plane of polarization, gyromagnetic resonance interaction and electrical insulation give ferrites a unique place as non-reciprocal devices. The complete specification of ferrites are:

Saturation field	:	The minimum axial magnetic field to ensure alignment of axes of spinning electrons
Resonant frequency	:	Frequency at which GMRI takes place

Line width	:	Range of magnetic field strength over which absorption takes place.
Curie temperature	:	Temperature at which magnetic properties are lost
P_{max}	:	Maximum usable P
F_{max}	:	Maximum usable frequency

21.8.2 Isolator

An isolator allows propagation of electromagnetic wave in one direction by offering high attenuation for the reverse direction. Isolators are two types (*i*) Faraday rotation isolator, and (*ii*) resonant absorption isolator.

Faraday rotation isolator: It consists of a piece of circular waveguide carrying TE_{11} mode with transitions to rectangular guide with TE_{10} mode at both ends. The output end is twisted by 45° (Figs. 21.8(*b&c*)). A thin pencil of ferrite is located inside the circular guide, supported by polyfoam. The waveguide is surrounded by a permanent magnet.

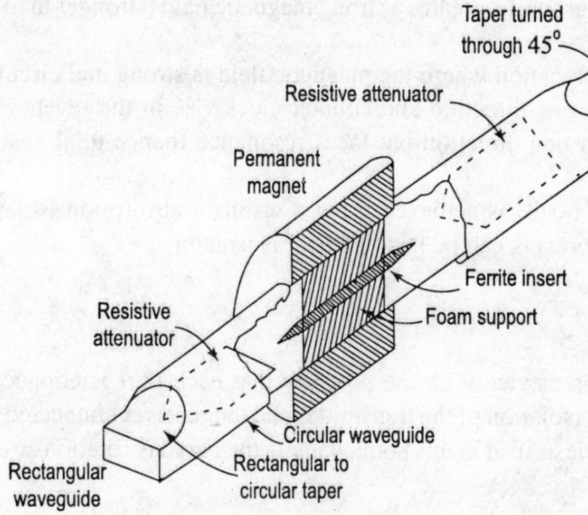

Fig. 21.8(*b*) *Cut-away view of Faraday isolator.*

Because of the dc magnetic field (below resonance), a wave propagating through an isolator has its plane of polarization rotated by the time it reaches the output end (by 45° in practical cases). It reaches its destiny through a suitable transition, and remains unaffected by resistive waves which are at 90° to the electric field. Only an insertion loss of 0.5 dB to 1 dB would be there.

A wave propagating in the reverse direction is also rotated clockwise, because the direction of Faraday rotation is dependent on the dc magnetic field only. When this wave reaches the input side, it is absorbed by the resistive wane. Moreover, it cannot propagate in the input rectangular guide because of its dimensions. The attenuation of reverse wave is of the order of 30 dB.

Faraday rotation isolators can handle peak powers up to 1 kW and have a VSWR of 1.1, and a bandwidth of 20% of the center frequency.

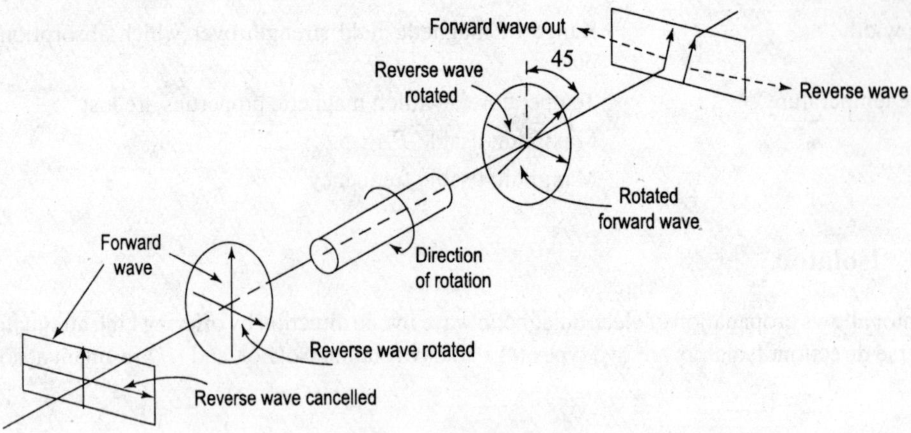

Fig. 21.8(*c*) *Faraday rotation isolator — operating principle.*

Resonant absorption isolator: This is used for high power/switched carrier applications. It consists of a piece of rectangular waveguide in TE_{10} mode with the ferrite material placed longitudinally inside. A permanent magnet placed around it creates a strong magnetic field (stronger than the Faraday rotation case).

The ferrite is placed at a location where the magnetic field is strong and circularly polarized. This polarization is clockwise in one direction and counter clockwise in the reverse direction. Thus, the propagation is unaffected in one direction but faces resonance (hence total absorption) in the other direction.

Isolation of the order of 60 dB is achieved using a resonant absorption isolator. Insertion loss is about 1 dB. 100 kW pulsed powers can be handled by this isolator.

21.8.3 Circulator

A circulator is a 3- or 4-port device with the property that each port is connected only to the next clockwise port. It is used for isolation of the transmitter and the receiver connected to the same antenna. Two types of circulators are described in this book, namely the Faraday rotation circulator and miniature Y circulator.

Faraday rotation circulator: It is a 4-port device. Power entering port 1 is converted to TE_{11} in the circular waveguide; passes port 3 unaffected because the electric field is not significantly cut; gets rotated through 45° by the ferrite; passes port 4 as in the cae of port 3; emerges from port 2. Similarly, port 2 is coupled to port 3; port 3 to port 4; port 4 to port 1 (Fig. 21.8(*d*)). Like the Faraday rotation isolator, this circulator is power limited.

Miniature Y circulator: This is 3-port circulator available in waveguide, coaxial and microstrip versions. A waveguide version used in a 7 GHz MW radio equipment is described here (see Fig. 21.8(*e*)).

With the magnet on one side of the ferrite only, a phase shift occurs for any signal fed to the circulator. If the three ports are arranged 120° apart, a clockwise shift with correct terminations will ensure that each signal is rotated so as to emerge from the next clockwise port, without being coupled to the remaining port.

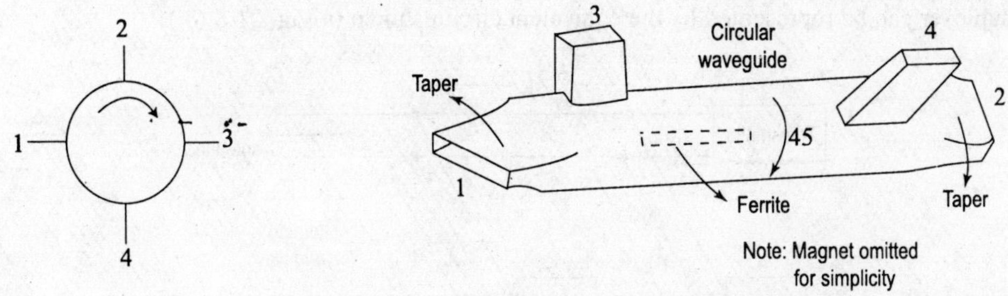

Fig. 21.8(*d*) *Faraday rotation circulator.*

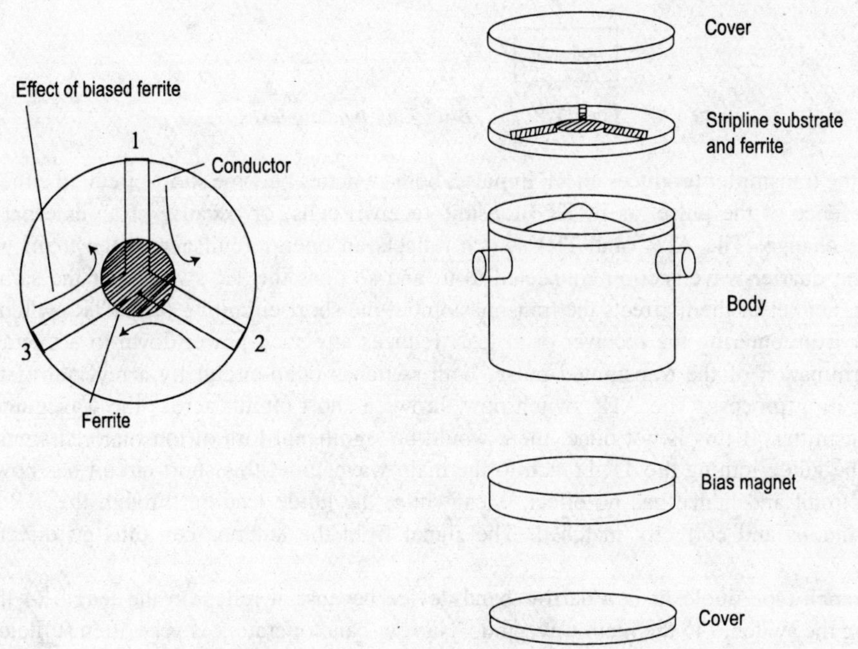

Fig. 21.8(*e*) *Miniature Y circulator.*

Size of Y circulator	12 mm (height) × 25 mm (diameter)
Insertion loss	0.5 dB
Isolation	30 dB

A 4-port circulator can be formed by joining two Y circulators.

21.8.4 Duplexer

A duplexer is a device used to facilitate the connection of a single antenna to both transmitter and receiver, with minimal interference between them. Its description is similar to a circulator, but in a duplexer the emphasis is on pulsed transmission, e.g. radar, and not CW transmission.

A duplexer can be represented by the equivalent circuit shown in Fig. 21.8.(f).

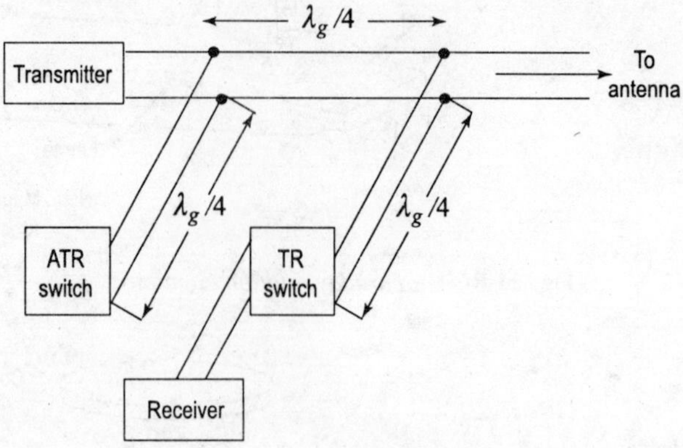

Fig. 21.8(f) *Branching-type duplexer.*

When the transmitter produces an RF impulse, both switches become short-circuited either because of the presence of the pulse, as in TR (transmit–receive) cells, or because of an external synchronized bias change. The ATR (anti-TR) switch reflects an open circuit across the main waveguide through the quarter-wave section connected to it, and so does the TR switch, for the same reason. Therefore, neither of them affects the transmission but the short-circuiting of the TR switch prevents RF power from entering the receiver or at least reduces any such power-down to a tolerable level. At the termination of the transmitted pulse, both switches open-circuit by a reversal of the initial short-circc :iting process. The ATR switch now throws a short circuit across the waveguide leading to the transmitter. If this is not done, there would be significant loss of the received signal. At the input to the guide joining the TR branch to the main waveguide, this short circuit has now become an open circuit and hence has no effect. Meanwhile, the guide leading through the TR switch is now continuous and correctly matched. The signal from the antenna can thus go directly to the receiver.

The branch-type duplexer is a narrow-band device because it relies on the length of the guides connecting the switches to the main waveguide. Narrow-band operation is very often sufficient, so the branch-type duplexer is quite commonly used.

Many a time, the words duplexer and circulator are interchanged in their usage. It should not be done so. Duplexer involves switching arrangement (TR/ATR switches) and is used in pulsed transmission systems like radar. In radar, the Tx frequency and echo frequency are the same. Normal isolation provided by the circulator between Tx and Rx links is inadequate. Hence the switching arrangement of a duplexer is needed.

21.9 Layout of Branching Circuit Components in DMR 770

In the branching circuit of DMR 770 digital MW radio equipment, a miniature Y circulator, an isolator, and a coaxial switch are used. The layout of components inside the equipment is illustrated in Fig. 21.9. The branching circuit is described in Chapter 22 in detail.

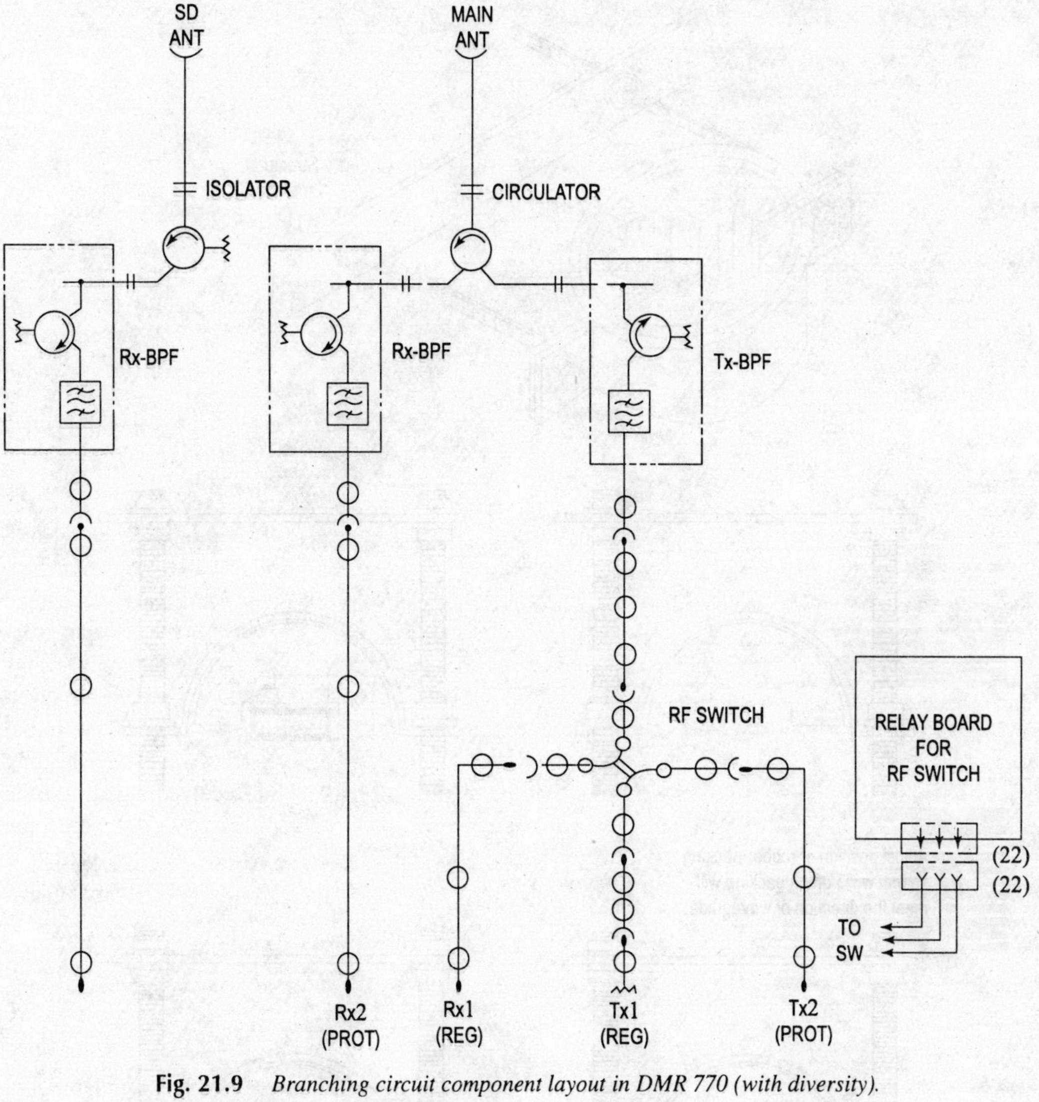

Fig. 21.9 *Branching circuit component layout in DMR 770 (with diversity).*

21.10 Waveguide Accessories

21.10.1 Waveguide clamps

Clamps are fixed 1 m apart along the waveguide ladder up the tower and also on the lead-out tray from the tower to the equipment room. The shape of waveguide clamp is different for rectangular and elliptical waveguides.

For rectangular waveguides, two C-shaped clamps are placed one over the other to form a circular groove inside through which the waveguides passes. The two C clamps are held together at the edges and with the ladder using bolts and nuts (see Fig. 21.10)

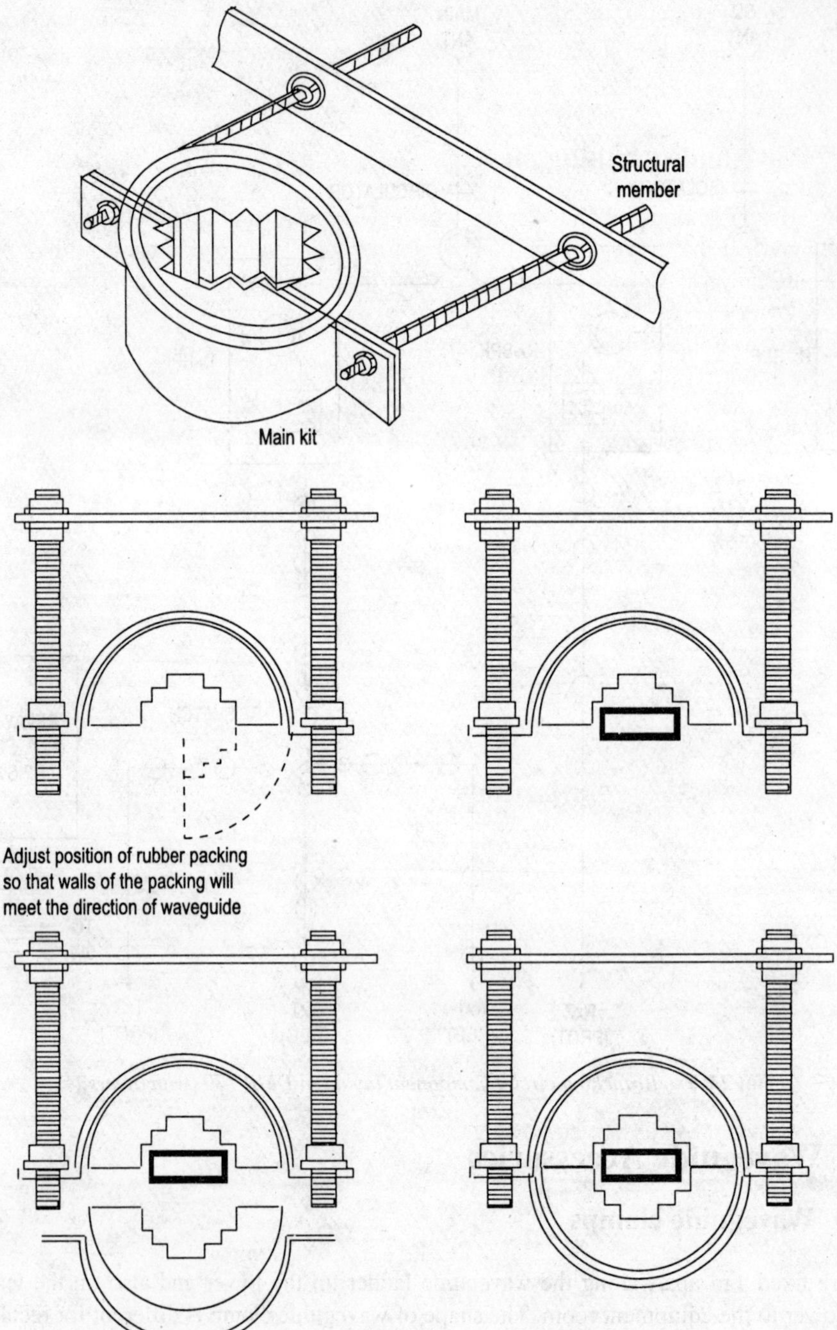

Fig. 21.10(*a*) *Waveguide clamp for rectangular waveguide.*

For elliptical waveguides, a C-shaped clamp is fitted to the ladder. The clamp has a steel tongue attachment which can be bent round the circumference of elliptical guide and closed at the edge with a nut and a bolt.

21.10.2 Waveguide earthing pit

A lightning arrestor is mounted on tower top which gets an earth connection through the tower body. Hence one may feel that waveguide earthing is not needed. The metal antenna at the tower top has waveguide connection at the back which runs down to the equipment. It is better to provide earthing for waveguide run (see Fig. 21.10(*b*)). This will safeguard the equipment in the event of electric continuity in lightning strike. Copper laces are put a round the guide at three or four places along the run and clamped. The clamps are connected to the tower body. (Tower body is earthed since tower legs are earthed.)

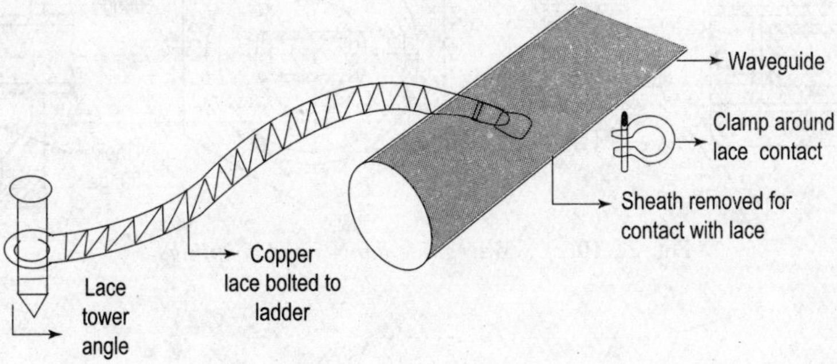

Fig. 21.10(*b*) *Waveguide earthing.*

21.10.3 Waveguide flanges and coupling

In a system employing waveguides and waveguide components, some form of coupling is required to join the separate pieces. First the flanges are soldered or brazed on the end of each waveguide piece/length and then the pieces are bolted together. It may be noted that

- after brazing, the ends to be coupled should be 'planed' and polished for reflection-free coupling,
- gaskets are fixed for air-tight coupling,
- flange holes at the ends of pieces being coupled are matching.

Coupling methods other than flange coupling are screwed ring type coupling and choke coupling (Fig. 21.10(*c*)).

21.10.4 Waveguide bending tools for elliptical waveguides

While hoisting the elliptical waveguide, the bends from equipment to the lead-out tray, the lead-out-tray to the waveguide ladder, and the waveguide ladder end to the feeding position are to be negotiated smoothly. The waveguide should never be bent with bare hands. Bending tools (E-bending tool and H-bending tool) as shown in Fig. 21.10(*d*) should be used.

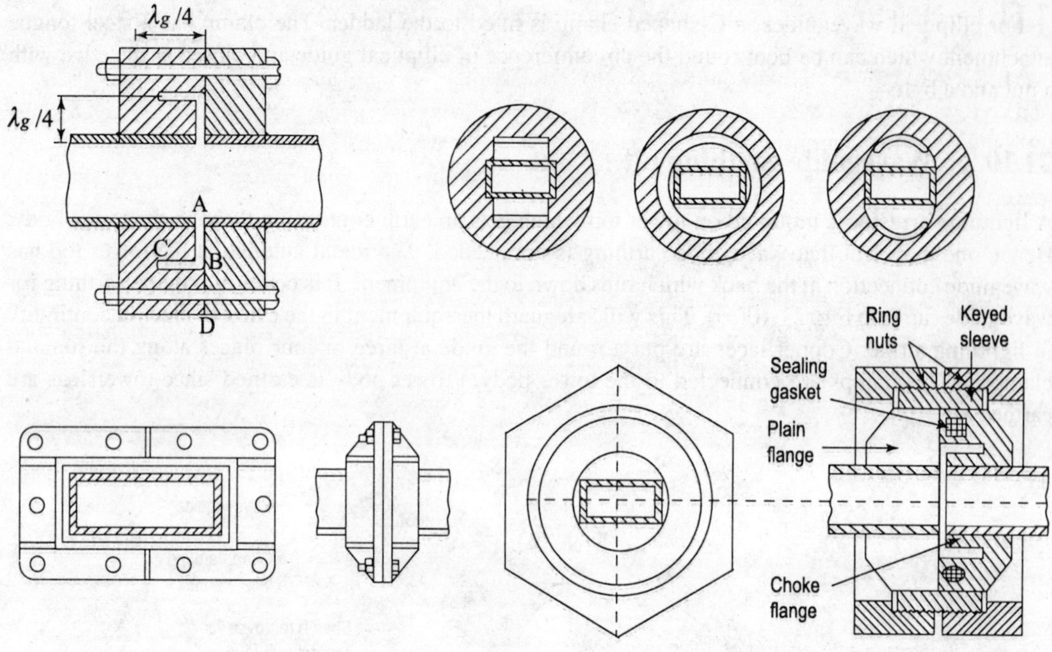

Fig. 21.10(*c*) *Waveguide flanges and couplings.*

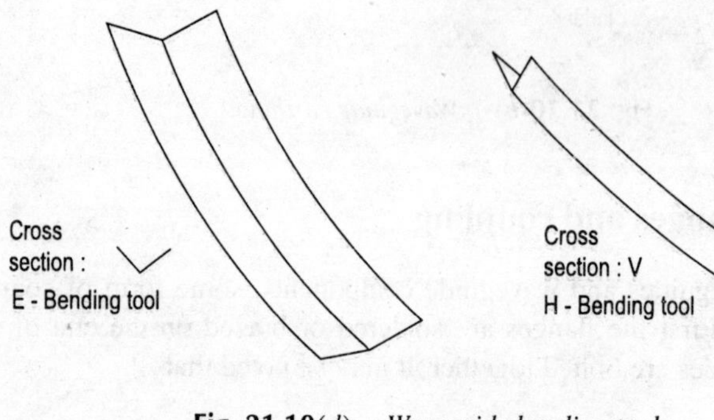

Fig. 21.10(*d*) *Waveguide bending tools.*

21.10.5 Wall feed-through

At the equipment room entry, the waveguide should be secured in the wall using feed-through. It comprises of a plant with a waveguide clamp at the centre to hold the waveguide firmly in position and without damage while entering through the wall.

21.11 Precautions while Hoisting Waveguide

The precautions to be taken while hoisting the waveguide, connecting various components, and using different accessories are briefly discussed here. Details are available in manuals and handouts accompanying individual products.

21.11.1 Hoisting waveguide–outdoor work

A) Check the waveguide running route for the complete fixing of ladders/supports.

B) Rectangular waveguides come in pieces of 3 m length and can be easily fixed using waveguide clamps. The focus is on elliptical waveguides which are used almost in all new installations. Elliptical waveguides comes in lengths of 150 m/200 m, wound on drums.

(a) Do not drop or lay down waveguide drum.
(b) Move the drum by rolling, it in the direction indicated on the drum.
(c) The waveguide, when received, is filled with nitrogen gas. Prior to hoisting the waveguide, check the N_2 gas by opening the gas-pressure measuring valve on the sealing cap attached to the starting end. If gas discharge is observed, a thorough examination is needed to ascertain the leakage/crush point on the waveguide coils.
(d) Never violate the prescribed bending radius.
(e) Do not remove sealing caps form both ends.
(f) Use bending tool to negotiate bends and curves (Fig. 21.11(*a*)).
(g) While unpacking, pull out bundles as in Fig. 21.11(*b*).
(h) Place manpower, roller, pulleys as in Fig. 2.11(*c*).

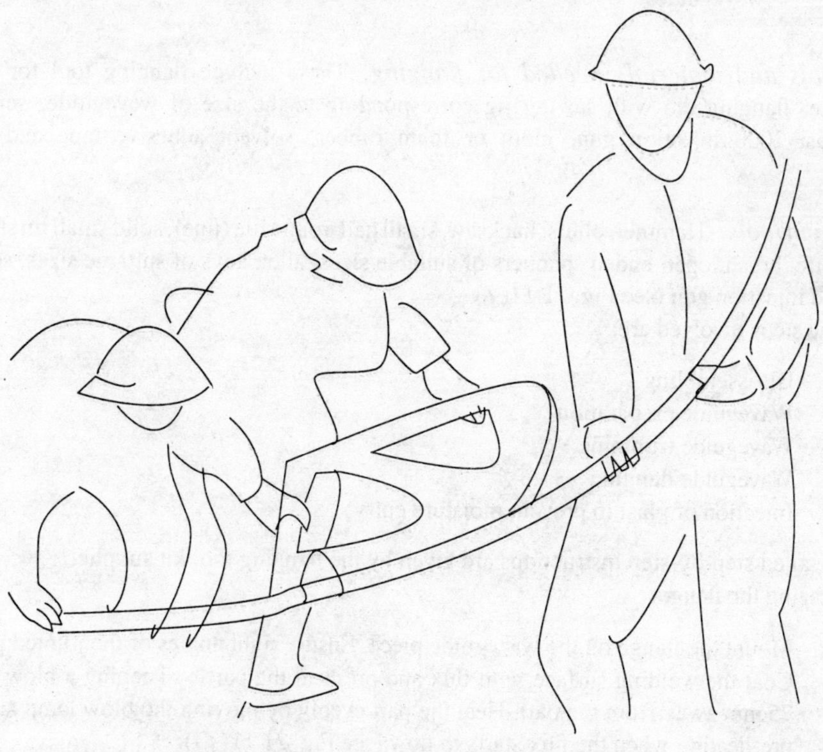

Fig. 21.11(*a*) *Waveguide bending method.*

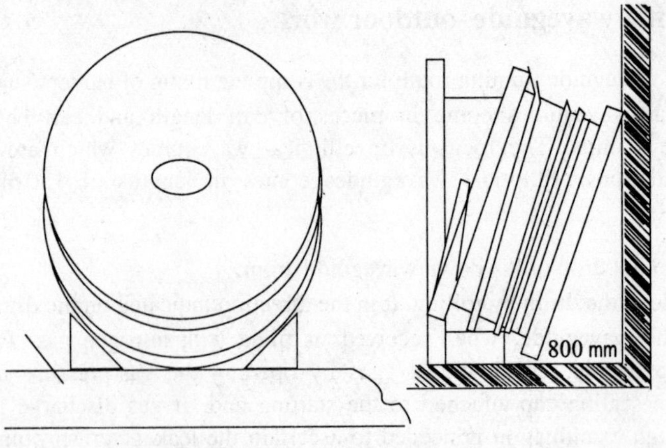

Fig. 21.11(*b*) *Removal of waveguide in bundle.*

(i) Prepare the length required for horizontal installation towards the antenna at tower top as in Fig. 21.11(*d*).

(j) Straightening of waveguide on leadout should be done as in Fig. 21.11(*e*).

(k) Mounting of terminations using flanging tool is described below for elliptical waveguides.

Tools and materials needed for flanging: These include flanging tool for waveguide size, flanging die with sawing jig corresponding to the size of waveguide, screw clamp, Plast 2000 injection gun, cloth or foam rubber, solvent adhesive tape and hand tool kit.

Hand tools: Hammer, pliers, hacksaw, small half-round file (fine), soild small tin shears, cable knife, brush, open-ended spanners of suitable sizes, allen keys of suitable sizes, screwdriver, and injection gun (see Fig. 21.11(*f*).
The steps involved are:

- Disassembling
- Waveguide preparation
- Waveguide trimming
- Waveguide flanging
- Injection of plast to prevent moisture entry

Detailed step-by-step instructions are given by the flanging tool kit supplier.

C) Brazing the flange

- Mount the flange on the waveguide piece. Ensure right angles of the jointed part.
- Coat the welding surface with flux and pre-heat the part by keeping a blow lamp 25 to 75 mm away from the part. Heat the part evenly by moving the blow lamp around. Stop pre-heating when the flux starts to flow (see Fig. 21.11(*f*)).

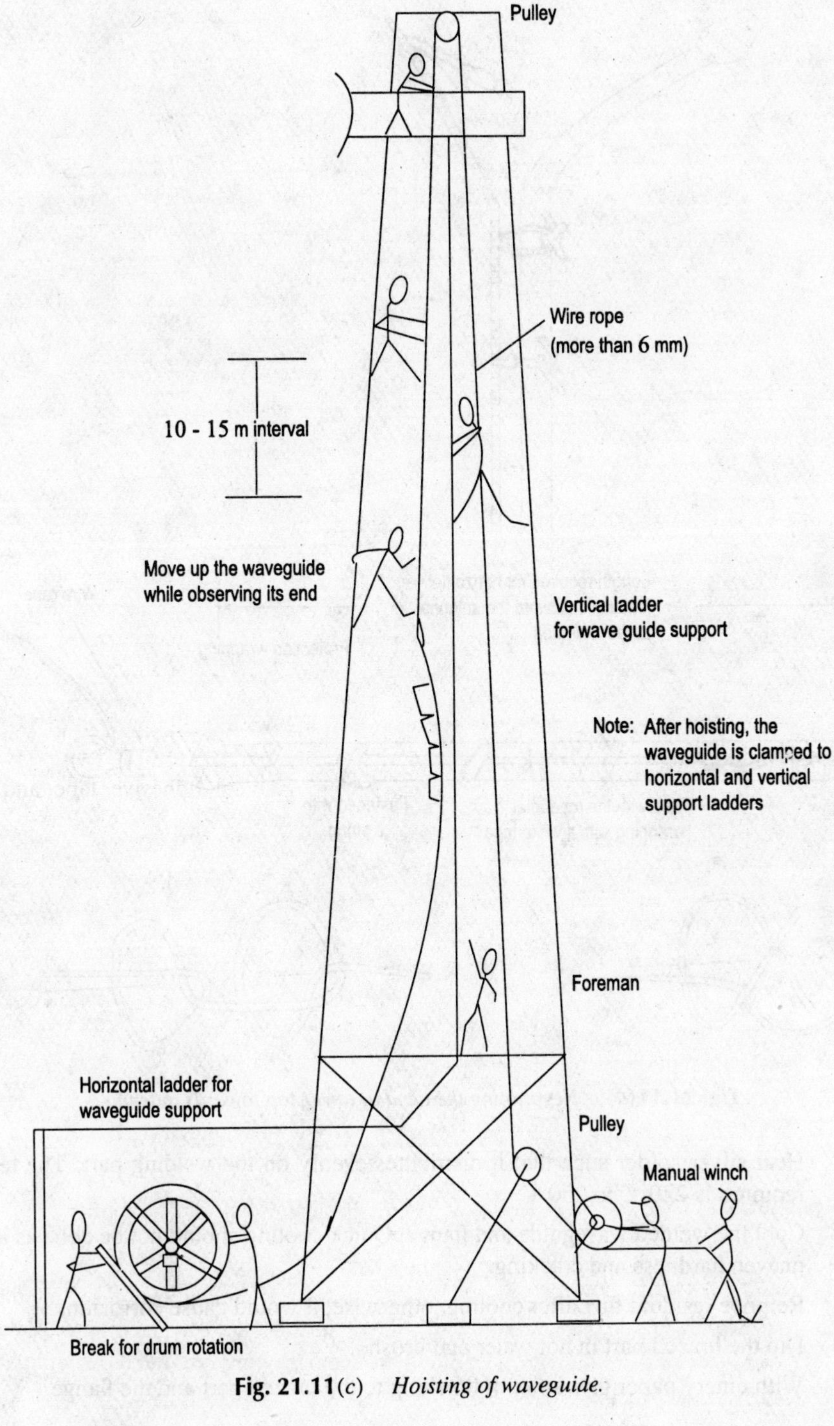

Fig. 21.11(*c*) *Hoisting of waveguide.*

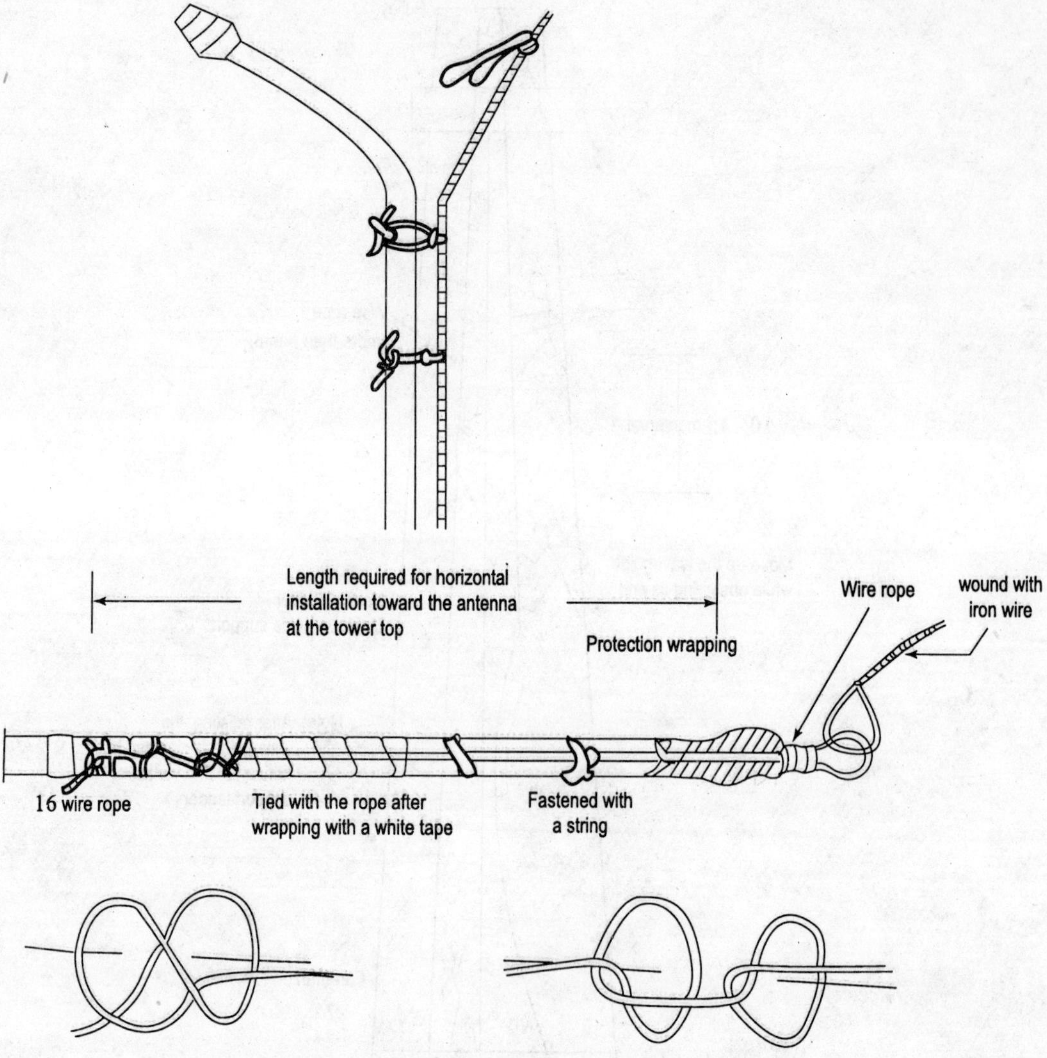

Fig. 21.11(*d*) *Negotiating the bend at tower top towards antenna.*

– Heat silver solder such that it distributes evenly on the welding part. The temperature required is 220°C to 250°C.
– Cool the welded waveguide and flange. Quick cooling should not be done as it results in uneven hardness and cracking.
– Remove residual flux after cooling, otherwise, it would cause corrosion.
– Dip the brazed part in hot water and brush.
– With emery paper give smooth finishing to the brazed part and the flange.

Fig. 21.11(*e*) *Straightening of waveguide.*

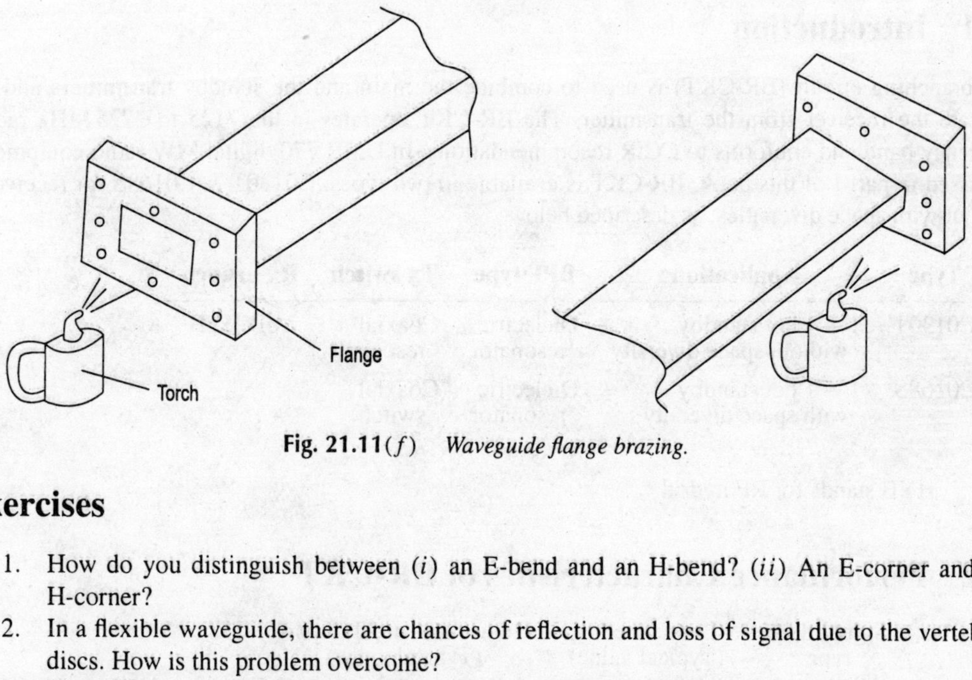

Fig. 21.11(*f*) *Waveguide flange brazing.*

Exercises

1. How do you distinguish between (*i*) an E-bend and an H-bend? (*ii*) An E-corner and an H-corner?
2. In a flexible waveguide, there are chances of reflection and loss of signal due to the vertebral discs. How is this problem overcome?
3. The MW tower is earthed for protection from lightning. Why is it necessary to earth the waveguide also?
4. Why is a ferrite named non-reciprocal device?
5. Distinguish between a circulator and a duplexer.

22

Branching Circuit in DMR 770 Digital MW Radio Equipment

22.1 Introduction

The branching circuit (BR-CKT) is used to combine the main and the standby transmitters and to separate the receiver from the transmitter. The BR-CKT operates in the 7125 to 7725 MHz radio frequency band and conforms to CCIR recommendations. In DMR 770 digital MW radio equipment discussed in part-I of this book, BR-CKT is available in two types: C0130T & C0168S for receivers without/with space diversities, as described below.

Type	Application	BPF type	Tx switch	Rx branch
C0130T	1 + 1 hot standby without space diversity	Dielectric resonator	Coaxial resonator	3 dB HYB
C0168S	1 + 1 hot standby with space diversity	Dielectric resonator	Coaxial switch	—

Note : HYB stands for RF hydrid.

22.2 Performance Characteristics of BR-CKT

	Signal flow type	Loss at center (Typical value)	Return loss at ANT port (Typical value)
C0130T	ANT-Rx1	Less than 6.6 dB	More than 21 dB
	ANT-Rx2	Less than 6.6 dB	
	Tx1-ANT	Less than 3.4 dB	
	Tx2-ANT	Less than 3.4 dB	
C0168S	ANT-Rx1	Less than 3.2 dB	More than 21 dB
	ANT-SD Rx	Less than 3.2 dB	
	Tx1-ANT	Less than 3.4 dB	
	Tx2-ANT	Less than 3.4 dB	

22.3 Principles of Operation

The BR-7G770B BR-CKT has two applications according to the applicable system requirements as described below.

22.3.1 1 + 1 BR-CKT for hot standby without space diversity (C0130T)

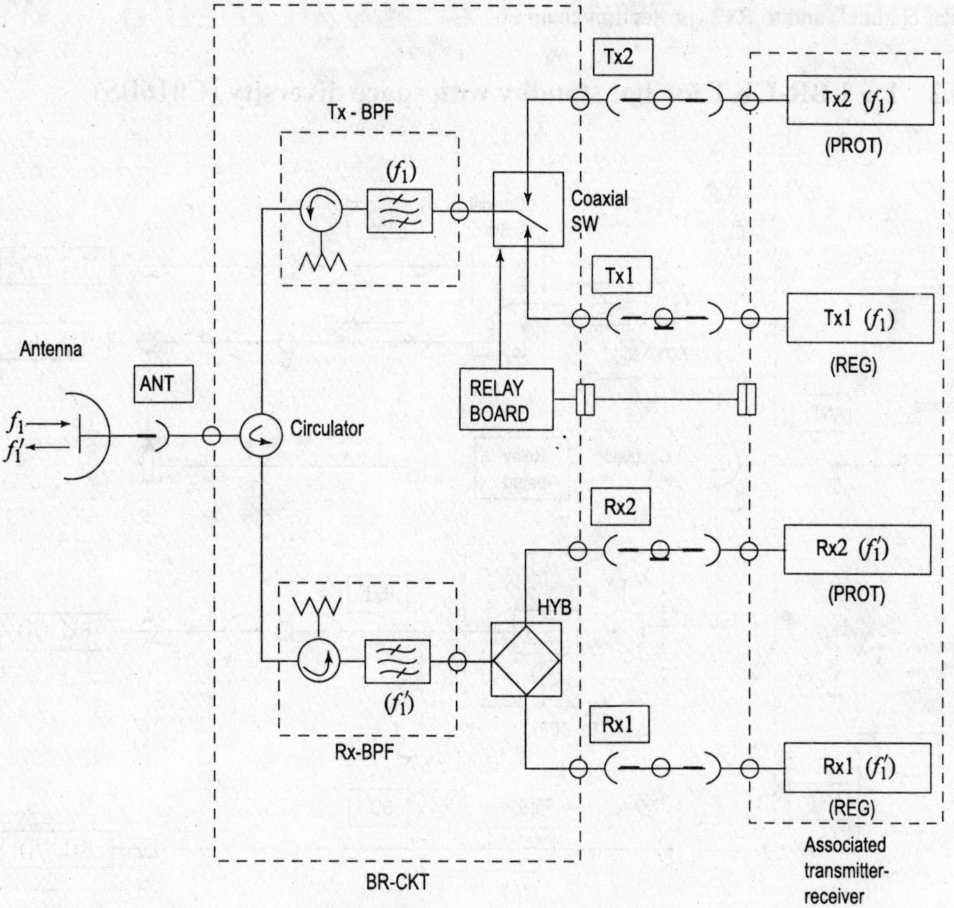

Fig. 22.1 1 + 1 *Hot standby without space diversity-BR-CKT block diagram.*

As shown in Fig. 22.1, the transmitting signals from transmitters Tx1 and Tx2 enter the coaxial switch. Normally, the coaxial switch selects a transmitting signal from Tx1 (regular channel) and terminates the signal from Tx2 (protection channel). A control signal (+9 V) from the TR equipment passes through relay contact on the relay board and is applied to the coaxial switch, and when the control signal reverses the polarity with relay contact absolutely, the coaxial switch selects a transmitting signal from Tx2. The transmitting signal is then sent to a bandpass filter (Tx-BPF) which is tuned to the respective transmitting frequency, and suppresses the undesired components of the signal. The signal is then passed to the circulator and sent to the antenna through the waveguide.

If the power supply (+5 V) for the relay board interrupts, the relay on the relay board is released and the coaxial switch selects the transmitting signal from Tx1.

The received signal from the antenna enters the Rx-BPF through a circulator and an isolator. The bandpass filter of the Rx-BPF, tuned to f_1', selects the desired input frequency and rejects the undesired input frequencies.

The received signal f_1', selected at the Rx-BPF is sent to the hybrid which delivers the signal to Rx1 (regular channel) and to Rx2 (protection channel).

22.3.2 1 + 1 BR-CKT for hot standby with space diversity (C016BS)

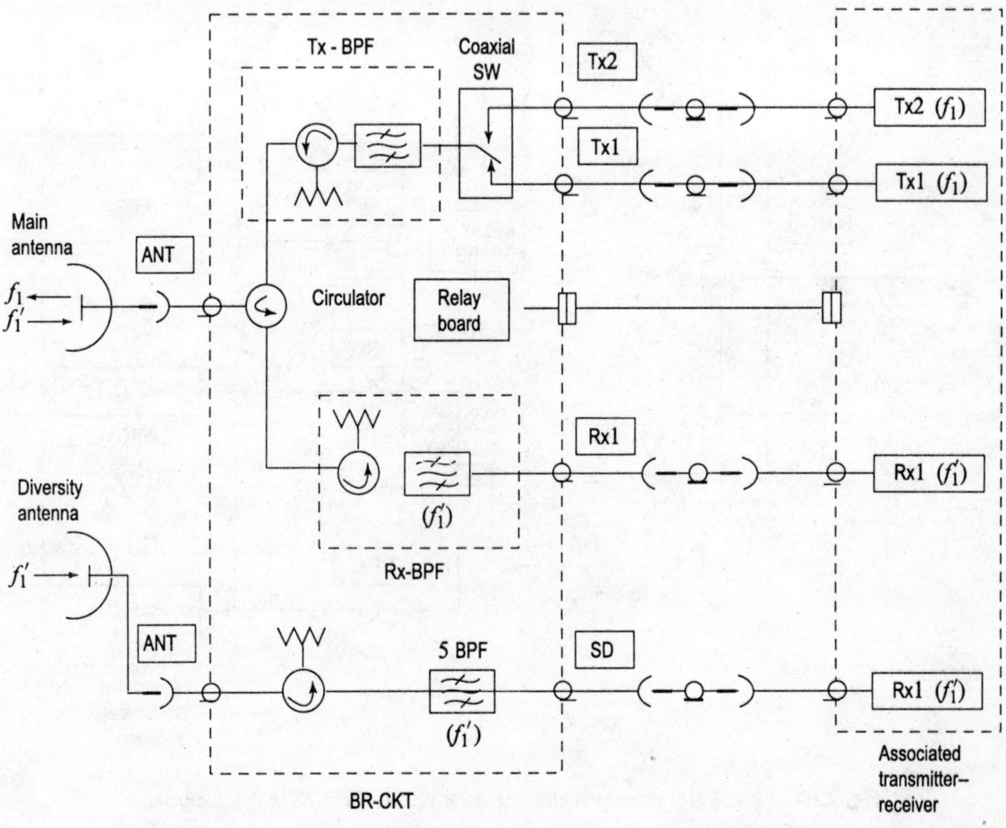

Fig. 22.2 1 + 1 *Hot standby with space diversity-BR-CKT block diagram.*

As shown in Fig. 22.2, the transmitting signals from transmitters Tx1 and Tx2 enter the coaxial switch. Normally, the coaxial switch selects a transmitting signal from Tx1 (regular channel) and terminates the signal from Tx2 (protection channel). A control signal (+9 V) from the transmitter equipment passes through the relay contact on the relay board and is applied to the coaxial switch, and when the control signal reverses the polarity absolutely with relay contact, the coaxial switch selects a transmitting signal from Tx2. The transmitting signal is sent to a bandpass filter (Tx-BPF) which is tuned to the respective transmitting frequency, and suppresses the undesired components of the signal. The signal is then passed to the circulator and sent to the antenna through the waveguide.

If the power supply (+5 V) for the relay board is interrupted, the relay on the relay board is released and the coaxial switch selects the transmitting signal from Tx1.

The received signal from the main antenna enters the Rx-BPF through a circulator and an isolator. The bandpass filter of the Rx-BPF, tuned to f_1', selects the desired input frequency and rejects the undesired input frequencies. The received signal f_1' selected at the Rx-BPF is sent to the main Rx.

The received signal from the SD antenna enters the 5-BPF through the isolator. The 5-BPF, tuned to f_1', selects the desired input frequency and rejects the undesired input frequencies. The signal passing through the 5-BPF is sent to SD-Rx.

Exercises

1. What is the function of a coaxial switch in a branching circuit?
2. In space-diversity arrangement, the diversity antenna is connected to the standby receiver alone. Why?
3. How many hybrids are used in 1 + 1 hot standby arrangement with or without space diversity?

23

Antenna Mounting Techniques

23.1 Introduction

7 GHz band is the 'work-horse' band for long-haul MW links. 3 m/3.3 m diameter antenna of 7 GHz band are hoisted on self-supporting as well as guyed towers for line-of-sight communication. Also, 4 ft fibre, 2 ft fibre, 1.2/1.8 m metal antennas are used on small towers (mostly on building terraces where the load is to be minimum) Mounting techniques for these different types of antennas are described in this chapter with the help of photographs of different types of installations.

23.2 Mounting of 3 m/3.3 m Metal Antenna

3 m or 3.3 m metal antenna has a dish diameter of 3 m or 3.3 m. When such a metallic body is hoisted on towers, the wind pressure due to high speed winds would be tremendous. The towers are designed to withstand the worst-case wind speed at the location of the microwave station. For example, in coastal areas, towers are designed to withstand 230 kmph of wind speed.

Microwave towers are of two types — guyed type and self-supporting type. towers are erected usually in low wind pressure areas or on building terraces. They have uniform cross-sections.

Self-supporting towers require strong foundation with 'tie-beams' between tower legs. These towers have a tapering structure.

On a guyed tower, a ring like structure (or part of ring) is fitted at the required height for hoisting the antenna. The antenna is mounted on this ring with a square frame.

On a self-supporting tower, the antenna is hoisted by 'pipe-mounting', i.e., a pipe is held in position on the tower face by two beams (top beam and bottom beam) which are fixed to the tower by brackets. The pipe has 4" dia and length suited to the height of antenna hoisting. A 'V' shaped frame attached at the back of antenna dish has groves suitable for the pipe, and is locked in position by 'U' bolts.

On self-supporting tower, if the antenna is to be hoisted on the tower leg as per the orientation requirements, brackets are fixed to the leg to hold the mounting pipe. Other arrangements are similar to the case of hoisting on tower face.

Two pipes of approximately 6 m length attached to the antenna in the rear are clamped to suitable tower members (angles) after the orientation to prevent subsequent tilt. This is known as anti-tilt arrangement.

A ring-like structure can also be used on a self-supporting tower to mount the antenna.

Thus we have four cases of mounting:

- Pipe mounting on tower face on self-supporting tower
- Pipe mounting on tower leg on self-supporting tower
- Ring mounting on self-supporting tower
- Ring mounting on guyed tower

Four sets of photographs (Plates 3-6) illustrate these four cases.

23.3 Mounting Fibre Antenna

2 sizes of fibre antenna are in use in the 7 GHz band.

- 4 ft. fibre antenna
- 2 ft. fibre antenna

The former can be mounted on pipe in a fashion similar to that followed for 3 m metal antenna. The antenna is light-weight, and four angles are needed at the back as a mesh to form an anti-tilt support.

The 2 ft. fibre antenna, used in surveys, comes with a mounting attachment. It also needs four angles at the back as anti-tilt support.

The radiating element of the fibre antenna is coupled to the waveguide through a coaxial feed. Hence there is an additional loss of 5 dB in the receive signal strength.

The 4 ft. fibre antenna, despite the coaxial feeder loss, gives good gain (just 10 dB to 15 dB less than the 3 m metal antenna) and hence can be used to probe the maximum Rx signal strength points effectively. Such probing is required if we are unable to orient the 3 m dia antenna to maximum Rx signal position. 2 ft. fibre antenna may not be effective in probing because of the narrow beam width and difficulties in exact positioning.

Exercises

1. How do we prevent the tilting of an antenna mounted on a tower top?
2. While orienting antennas on a MW hop, we achieve coarse adjustment by horizontal orientation and fine adjustment by vertical orientation. Why?
3. The waveguide connected to the antenna may break while turning the antenna for orientation. What are the precautions to be taken?
4. How do you retain the feed-horn position which is fixed at the focal point at the time of hoisting the antenna?

24

Waveguide Matching Techniques

24.1 Introduction

Waveguides and waveguide components like E and H bends are procured by organisations as per certain standard dimensions, in accordance with the equipment plan for the network. In due course of time, the following requirements may arise:

1. Installation of new equipment: The antenna ports of the new equipment may not have matching dimensions.
2. Provision of space diversity with a new matching port: The existing equipment may not have a diversity port.

Soluti ɔns for the above problems are presented in this chapter.

24.2 Changes in Waveguide Cross-section

The first problem is basically a problem due to changes in waveguide cross-section. Two solutions are available to tackle this problem:

- Waveguide transformer
- Trapezoidal coupling unit

24.2.1 Waveguide transformer

The wave impendance Z_w for a wave in a waveguide is the ratio of the transverse field components E_y/H_x. This simple expression cannot help us calculate Z_w since E_y and H_x are varying. If the field components E_y and H_x are written in terms of the maximum electric field amplitude $E_{\max}$ and the maximum amplitude of the current density $I_{\max}$ induced in the horizontal walls, they become

$$E_y = E_{\max} \sin \frac{\pi x}{a} \sin (\omega t - \beta z), \tag{24.1}$$

$$H_x = I_{\max} \sin \frac{\pi x}{a} \sin (\omega t - \beta z). \tag{24.2}$$

The expression for the mean power can then be written as

$$P = \frac{E_{max} I_{max} ab}{4}$$
$$= \frac{1}{2} \frac{b E_{max}}{\sqrt{2}} \cdot \frac{a I_{max}}{\sqrt{2}}. \tag{24.3}$$

Taking $b E_{max}/\sqrt{2}$ as representing an average potential difference V between the horizontal walls and $a I_{max}/\sqrt{2}$ as the average current I flowing along these walls, we obtain the usual power equation

$$P = \frac{1}{2} V I.$$

From this a characteristic impedance can be defined as

$$Z_0 = V/I$$
$$= \frac{b E_{max}}{a I_{max}} = \frac{b}{a} \left| \frac{E_y}{H_x} \right|$$
$$= \frac{b}{a} Z_w. \tag{24.4}$$

When there is no change of dimensions in the guide system, it does not matter whether Z_0 of Z_w is used in calculations so long as consistency is maintained, since the difference disappears when the impedances are normalized. When changes in dimensions are involved, however, the following examples show that the two definitions of impedance lead to different equivalent circuits, although to the same final result.

Please refer to Fig. 24.1(*a*) which illustrates a junction between two rectangular guides differing in their *b*-dimensions. These guides have the same wave impedance, but obviously cannot match each other since the change in dimensions at the junction must produce an abrupt change in field configuration with consequent reflection. It can be shown that if the waveguides are represented by two transmission lines of equal characteristic impedance in the equivalent circuit, then the junction can be represented by the transformer of turns ratio $\sqrt{b_2} : \sqrt{b_1}$ and a parallel capacitor as shown in Fig. 24.1(*b*). If,

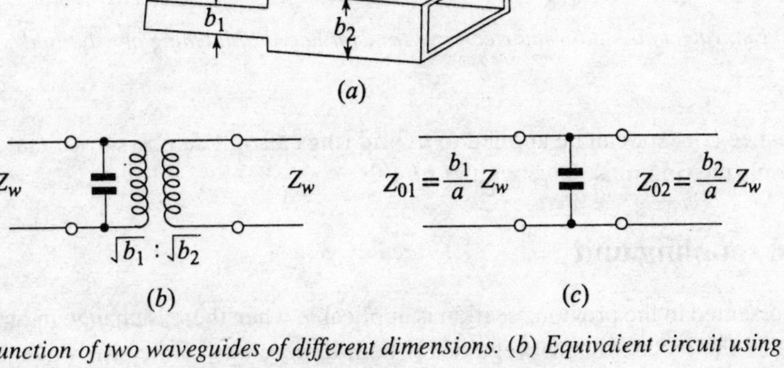

Fig. 24.1. *(a) Junction of two waveguides of different dimensions. (b) Equivalent circuit using wave impedance. (c) Equivalent circuit using characteristic impedance.*

however, the guide impedance is defined to be proportional to b, as in the characteristic impedance definition Z_0 of (24.4), then the two transmission lines of the equivalent circuit have characteristic impedances proportional to b_1 and b_2 and the transformer ratio can be shown to become unity. The effect of the junction can therefore be reduced to be simple shunt capacitor shown in Fig. 24.1(c). The use of Z_0 rather than Z_w is thus advantageous in simplifying the equivalent circuit. The value of the shunt capacitor is a complex function of the guide dimensions. Its derivation is beyond the scope of this book.

This equivalent circuit may now be used to design a waveguide quarter-wave transformer to match the two guides of Fig 24.1(a). Neglecting for the moment the shunt capacitor in the equivalent circuit, the characteristic impedance Z_0 of a quarter-wave transformer required to match the two transmission lines of characteristic impedances Z_{01} and Z_{02} is given by $Z_0 = \sqrt{Z_{01}Z_{02}}$. As Z_{01} and Z_{02} are proportional to b_1 and b_2 respectively, the b-dimension of the waveguide quarter-wave transformer is given by

$$b = \sqrt{b_1 b_2},$$

and the length of the transformer is $\lambda_g/4$. The effect of the capacitor at each of the two junctions can be tuned out by adding suitable inductors as shown in Fig. 24.2(a). In the waveguide these are provided in the form of an inductive iris at each step. Thus the matched junction takes the form shown in Fig. 24.2(b).

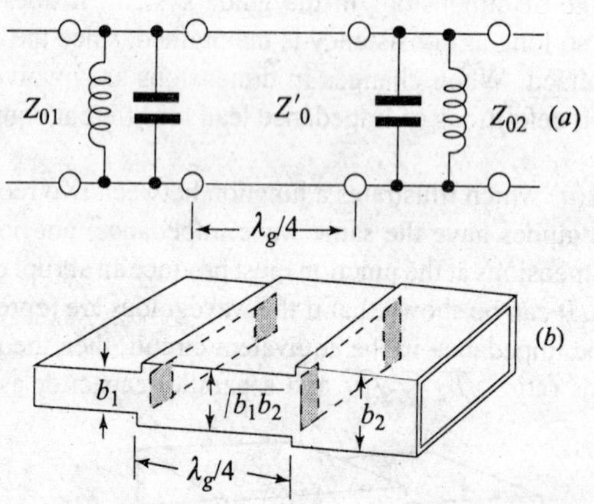

Fig. 24.2. *(a) Equivalent circuit for quarter-wave transformer. (b) Matching of waveguides by quarter-wave transformer.*

This impedance concept can be applied to hybrid rings also. It can be shown that for a hybrid with parallel junctions, the ring must have width of $b\sqrt{2}$.

Trapezoidal coupling unit

The solution presented in the previous section is applicable when there is change in only one dimension (either a or b) of the rectangular waveguide. If both dimensions of the two waveguide pieces to be joined are different, we have to use three matching transformers. For example, if the waveguide pieces

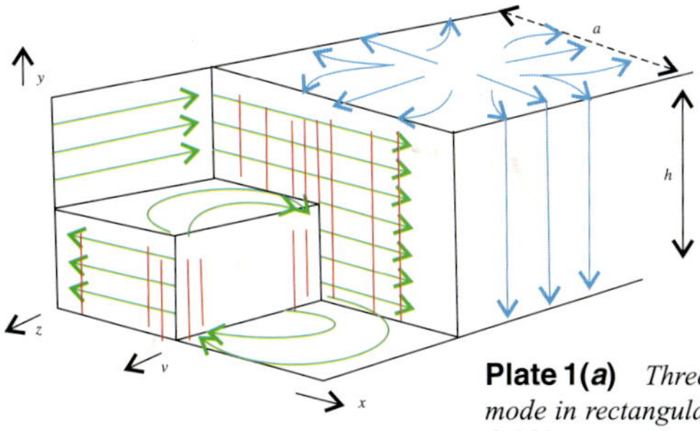

Plate 1(*a*) *Three-dimensional view of field configuration of H_{10} (or TE_{10}) mode in rectangular waveguide. Note: (i) Since it is H (or TE) wave, electric field lines are transverse to the direction of propagation and the magnetic field has a longitudinal component. (ii) Since it is H_{10} (or TE_{10}) mode, one row of magnetic field loops is present between the side walls. These are shown as a set of two half-loops (in green) in the cross sections. There is one maxima of electric field along x direction; hence, the red lines appear dense in the middle along x direction. (iii) The current lines in waveguide walls emanate from electric field maxima, i.e., from the middle in x direction.*

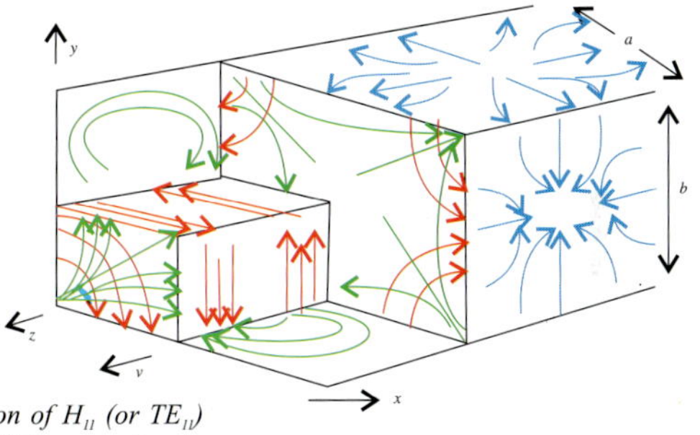

Plate 1 (*b*) *Three-dimensional view of field configuration of H_{11} (or TE_{11}) mode in rectangular waveguide. Note: (i) Since it is H (or TE) wave, electric field lines are transverse to the direction of propagation and the magnetic field has a longitudinal component. (ii) Since it is H_{11} (or TE_{11}) wave, one row of magnetic field loops are present between the side walls in x and y direction. These are shown as a set of two half-loops (in green) in the cross-sections along x and y directions. There is one maxima of electric field along x and y directions. Hence, the red lines appear dense in the middle along x and y directions. (iii) The current lines on waveguide walls emanate from the electric field maxima, i.e., the middle along x and y directions.*

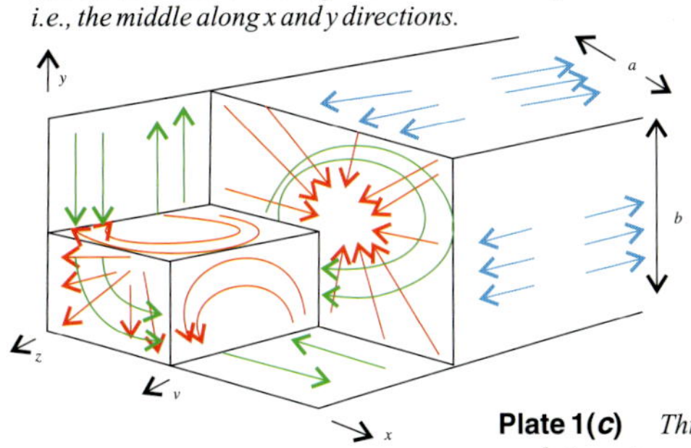

Plate 1(*c*) *Three-dimensional view of field configuration of E_{11} (or TM_{11}) mode in rectangular waveguide. Note: (i) Since it is E (or TM) wave, magnetic field lines are transverse to the direction of propagation and electric field has longitudinal component. (ii) Since it is E_{11} (or TM_{11}) mode, one row of electric field loops are present between the side walls along x and y directions. These are shown as set of two half-loops (in red) in the cross- sections along x and y. There is one maxima of electric field along x and y. Hence, the red lines appear dense in the middle along x and y. (iii) The current lines in the waveguide walls appear concentrated at the electric field maxima, i.e., in the middle along x and y direction.*

In all the above figures, electric field lines are shown in Red, magnetic field lines in Green, and current in waveguide walls in Blue. Direction of propagation is along vector 'v'

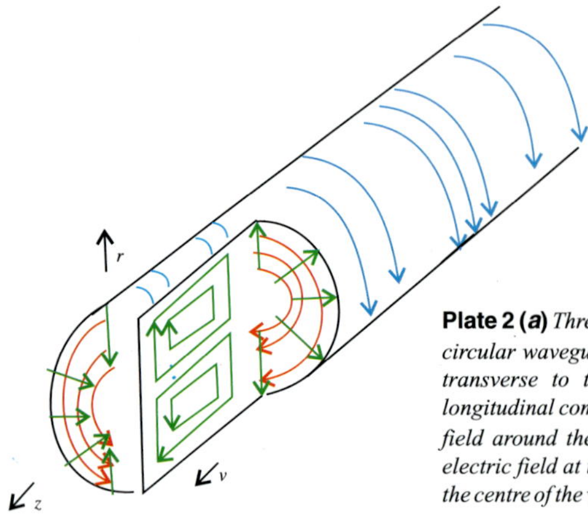

Plate 2 (*a*) *Three-dimensional view of field configuration of H_{01} (or TE_{01}) mode in circular waveguide. Note: (i) Since it is H (or TE) wave, electric field lines are transverse to the direction of propagation and the magnetic field has a longitudinal component. (ii) Since it is H_{01} mode, there is no variation of magnetic field around the circumference of the waveguide. (iii)There is one maxima of electric field at the center of the waveguide, hence, the red lines appear dense at the centre of the waveguide.*

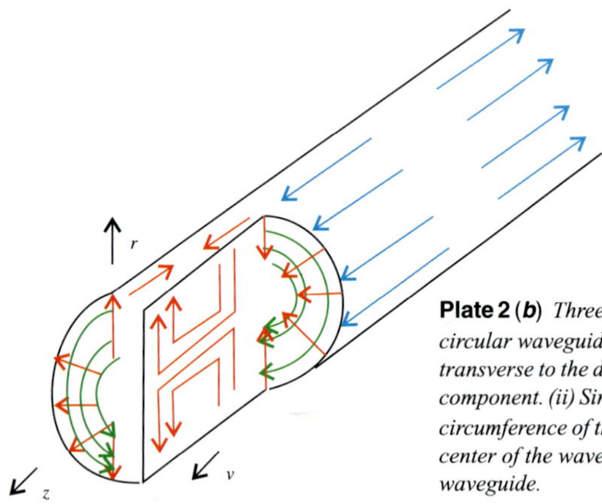

Plate 2 (*b*) *Three-dimensional view of field configuration of E_{01} (or TM_{01}) mode in circular waveguide. Note: (i) Since it is E (or TM) wave, magnetic field lines are transverse to the direction of propagation and the electric field has a longitudinal component. (ii) Since it is E_{01} mode, there is no variation of electric field around the circumference of the waveguide. (iii) There is one maxima of magnetic field at the center of the waveguide; hence, the green lines appear dense at the centre of the waveguide.*

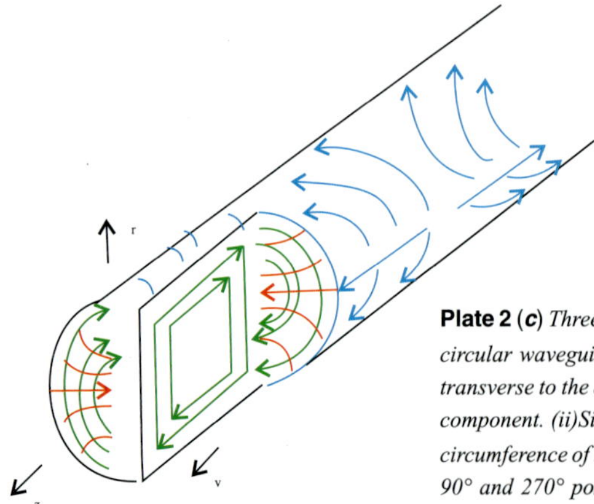

Plate 2 (*c*) *Three-dimensional view of field configuration of H_{11} (or TE_{11}) mode in circular waveguide. Note:(i) Since it is H (or TE) wave, electric field lines are transverse to the direction of propagation and the magnetic field has longitudinal component. (ii)Since it is H_{11} mode, there is variation of electric field around the circumference of the waveguide. It is zero at 0° and 180° position and maximum at 90° and 270° position. (iii) H_{11} is the fundamental mode as it has the longest critical wavelength.*

In all the above figures, electric field lines are shown in Red, magnetic field lines in Green, and current in waveguide walls in Blue. Direction of propagation is along vector 'v'

Plate 3*a* *Pipe-mounting of antenna on tower face on self-supporting tower (view from the back). The V frame attached to the antenna is mounted on 4" dia pipe held in position by U bolts. The top and bottom brackets extending from the tower face support the top beam and the bottom beam. The antenna mounting pipe (4" dia) is held in position between the top and the bottom beam. Anti-tilt pipes attached to V frame are also shown.*

Plate 3*b* *Pipe-mounting of antenna on tower face on self-supporting tower (view from the side). The V frame attached to the antenna is mounted on 4" dia pipe held in position by U bolts. The Mounting pipe is fixed to tower face (held in position rather) by the top beam and the bottom beam which are supported by brackets extending from tower face.*

Plate 3*c* *Pipe-mounting of antenna on tower face on self-supporting tower. Attachment of anti-tilt pipe).*

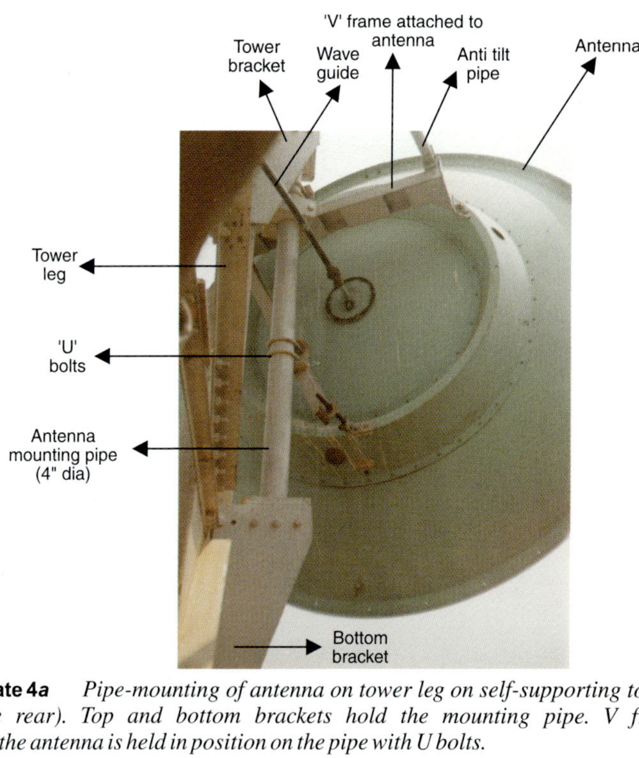

Plate 4*a* *Pipe-mounting of antenna on tower leg on self-supporting tower(view from the rear). Top and bottom brackets hold the mounting pipe. V frame attached to the antenna is held in position on the pipe with U bolts.*

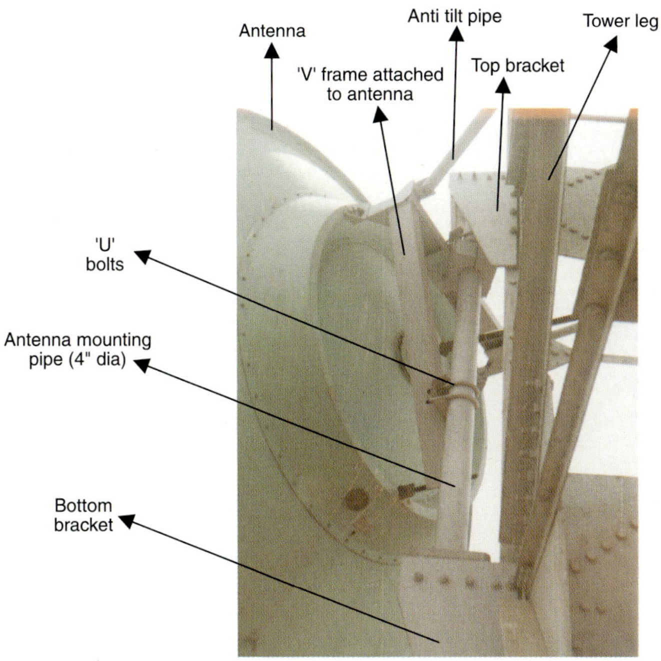

Plate 4*b* *Pipe-mounting of antenna on tower leg on self-supporting tower (view from the rear at an angle).*

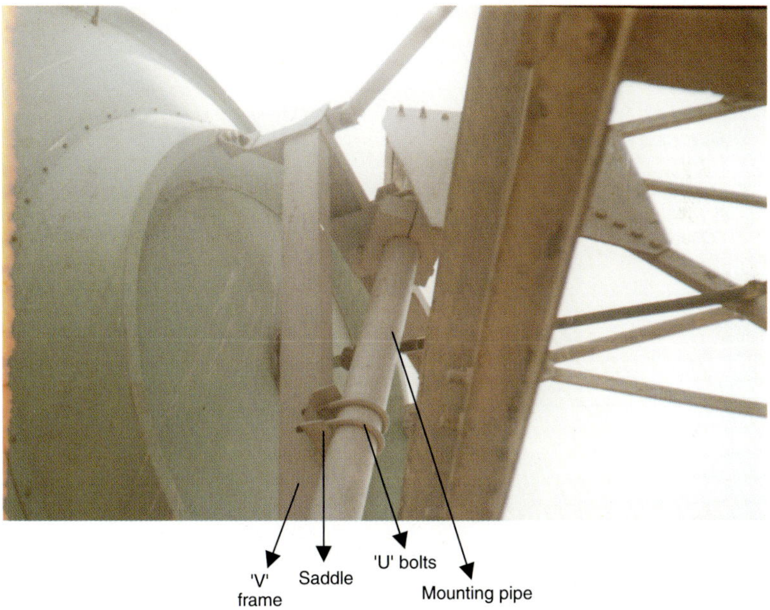

Plate 4*c* *Pipe-mounting of antenna on tower leg on self-supporting tower (view from the rear at an angle showing the seating of V frame on the pipe through saddle and U bolts).*

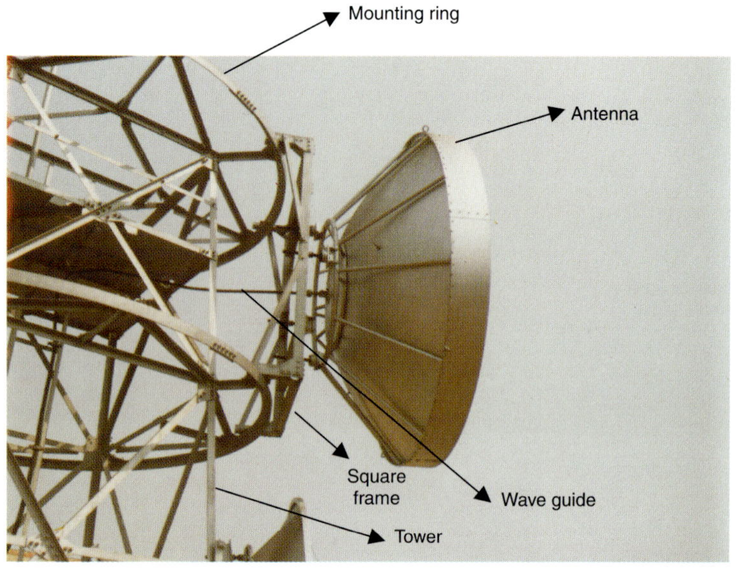

Plate 5*a* *Ring-mounting of antenna on self-supporting tower (view from a side).*

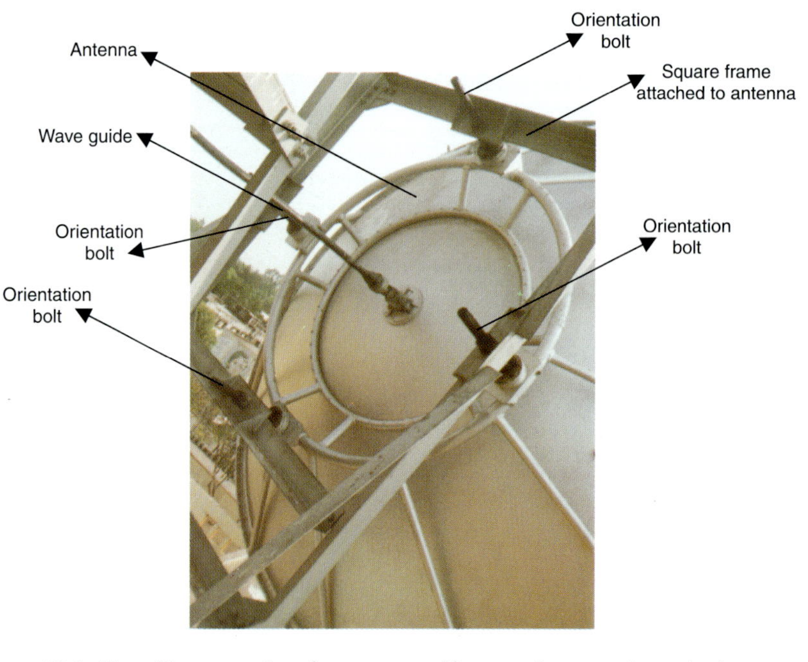

Plate 5*b* *Ring-mounting of antenna on self-supporting tower (rear view).*

Plate 5*c* *Ring-mounting of antenna on self-supporting tower (rear view at an angle). (1), (2), (3) and (4) are orientation bolts.*

Plate 6*a* *Ring-mounting of antenna on guyed tower (view from the ground).*

Plate 6*b* *Ring-mounting of antenna on self-supporting tower (top view).*

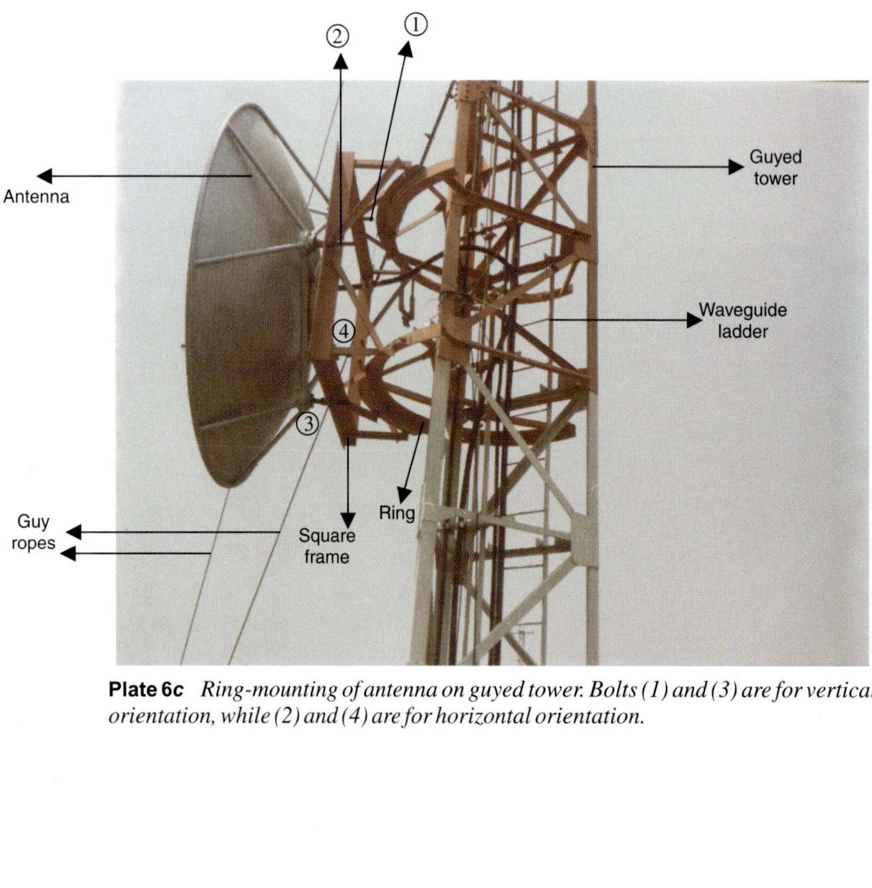

Plate 6*c* *Ring-mounting of antenna on guyed tower. Bolts (1) and (3) are for vertical orientation, while (2) and (4) are for horizontal orientation.*

have sizes $a_1 \times b_1$ and $a_2 \times b_2$, we have to interpose the 'matching waveguide transformers' in the sequence

$$(a_1 b_1) - (a_1, \sqrt{b_1 b_2}) - (a_1 b_2) - (\sqrt{a_1 a_2}, b_2) - (a_2 b_2).$$

A simpler solution is to use a trapezoidal coupling unit. As the same suggests, this component is in the shape of a trapezoid, with linear transition from cross-section $(a_1 \times b_1)$ to cross-section $(a_2 \times b_2)$. It is also known as 'gradual rectangular transition'.

24.3 Creation of Diversity Port

Sometimes, we may decide to provide space diversity on a hop where it was not planned for/existed earlier. Since all equipment have 1 + 1 standby, availability of a second receiver is not a problem. The second receiver can be made to function as a diversity receiver. But we require provision of diversity port on the equipment for connection to diversity antenna. Since the equipment is already installed and functioning, we have to place another equipment in its place, and send the previous one to the manufacturer to fit in the diversity port. This is very cumbersome and expensive procedure.

A very simple solution of negligible cost is to provide matching port with coaxial adaptor (see Fig. 24.3). It has a rectangular waveguide piece with coaxial feeder probe placed at $\lambda_g/4$ from the closed edge. The matching port fits the waveguide run (through transitions and bends). The coaxial feeder's other end is given to the 'RF-in' of the second receiver. The length of coaxial feeder should be kept bare minimum to reduce losses. Waveguide run and necessory supports in the interior of the equipment room should be suitably planned. It may please be noted that the matching unit length should be $\lambda_g/4$ for matched termination.

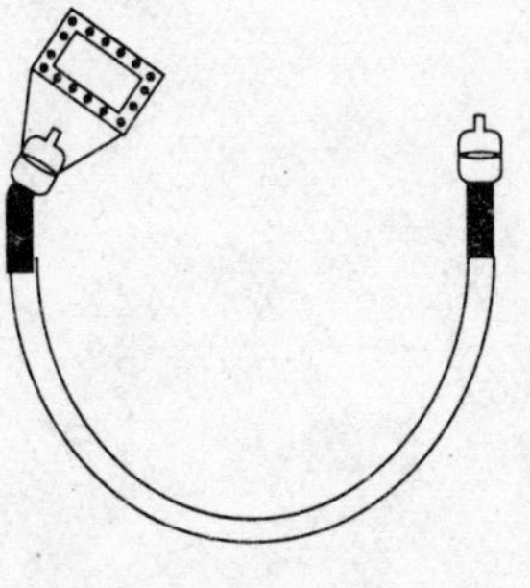

Fig. 24.3 *Matched termination with coaxial adaptor.*

This arrangement is also useful to investigate the problem of poor received signal strength for a hop without diversity. The main (regular) equipment will be working on the main port without causing interruption to the working MW link. A diversity port can be created as explained above and a 4 ft. fibre antenna can be provided on this port to act as spatial probe at different heights of the tower to ascertain the maximum receive signal position. Then the antenna and the waveguide can be hoisted for this maximum received signal position and the main (regular) equipment can be connected permanently.

Part - IV

25

Global System for Mobile Communications

25.1 Mobile Communications—An Introduction

In the years to come, most computers will be portable and mobile users would access networks through wireless systems. So, there are three dimensions to this futuristic scenario:

- Device portability
- Wireless access
- User mobility

With device portability, communication devices can be on the move, with or without a user. (Personal mobile phone is an example for portable device with user. The identification module of a mobile vehicle or unit being tracked by global positioning system is an example of a device without a user.) The network and the device shall have a mechanism to ensure communication while the device is moving. For example, in a mobile phone system the system itself 'hands-over' the device from one transmitter to other if the signal becomes too weak.

Wireless access describes the way of accessing a network or other users/devices without a wire, i.e., via transmission of electromagnetic waves in the allotted frequency band.

User mobility implies that a user has access to the same or similar telecommunication services at different places, i.e., the services follow the user. Examples of user mobility are call forwarding mechanisms, and roaming-supported desktops.

A communication device can thus fit into the following matrix:

	Wired	**Wireless**
Fixed	Desktop computers in office; conventional telephone instruments	Wireless LAN in exhibitions, buildings where wiring-up is inconvenient
Mobile	Laptops	Cellular telephone networks that support worldwide seamless roaming with the same phone, e.g. GSM

25.2 History of GSM

In the early eighties, analog cellular systems proliferated in Europe and USA. Each country developed its own system which was incompatible with the equipment and operation of other systems in other countries. This was a hindrance to achieving seamless roaming. Also, it posed problems of economies of scale due to fragmented market and technologies.

In 1982, the Conference of European Posts and Telegraphs (CEPT) formed a study group called 'Groupe Speciale Mobile' to study and develop a pan-European public land mobile system. The proposed system had to meet the following criteria:

- Good subjective speech quality
- Low terminal and service cost
- Support for international roaming
- Ability to support hand-held terminals
- Support for a range of new services and facilities
- Spectral efficiency
- ISDN compatibility

In 1989, the European Telecommunications Standards Institute (ETSI) took up the cause of GSM. GSM Phase-I specifications were published in 1990. Commercial services started mid 1991 and by 1993 there were 36 GSM networks in 22 countries with 25 more countries opting for GSM. By the beginning of 1994 there were 1.3 million subscribers worldwide for GSM. Thus the acronym GSM is more apt for 'Global System for Mobile Communications'.

The worldwide market figures for cellular networks are:

GSM :	40%,
Analog AMPS of US:	30%,
Japanese PDC:	15%,
TDMA or CDMA based and others:	15%.

In Europe GSM has 90% marketshare (over 60 million subscribers), whereas analog systems have 10% marketshare (less than 10 million subscribers). In USA and Canada, 80% of marketshare is for analog AMPS. The rest is split among TDMA, CDMA and GSM systems. This scenario is a consequence of the non-adoption of GSM standard in its early stages of development.

25.2.1 Different versions of GSM

GSM has three versions:

- GSM 900 (890–915 MHz for up-link and 935–960 MHz for down-link)
- GSM 1800 or DCS (digital cellular system) 1800 (1710–1785 MHz for up-link and 1805–1880 MHz for down-link)
- GSM 1900 or PCS (personal communication services) 1900 (1850–1910 MHz for up-link and 1930–1990 MHz for down-link)

25.3 GSM Services

GSM has three different categories of services—bearer, tele and supplementary.

Refer to Fig. 25.1 for a GSM services reference model. A mobile station (MS) is connected to the GSM public land mobile network (GSM PLMN) via the U_m (U interface of ISDN for mobile applications) interface. It is the infrastructure needed for GSM network. This network is connected to transit networks (ISDN or PSTN) and additional source/destination networks (optional) before another terminal TE is connected. Bearer services include all services that enable transparent transmission of data between the interface and the network. Bearer services are connection-oriented and circuit or packet switched. These services need only the lower three layers of the ISO/OSI reference model.

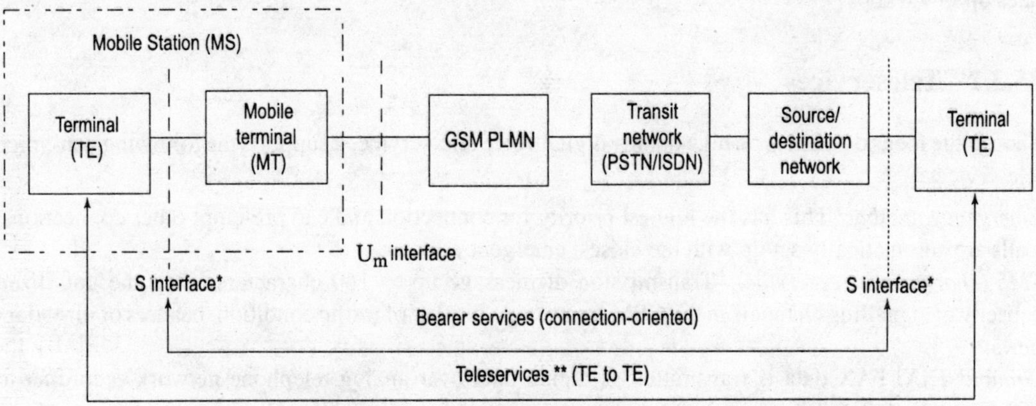

* U or R interface for a TE with ISDN capabilities as per ISDN reference model
** Supplementary services are enhancements over teleservices

Fig. 25.1 *GMS services reference model.*

Within the mobile station (MS), the mobile terminal (MT) performs all network specific tasks (TDMA, FDMA, speech coding, etc.) and offers an interface (interface S) for data transmission to terminal TE which can then be network-independent. Depending on the capabilities of TE, further interfaces may be required.

Teleservices are application-specific services and may need all the seven layers of the OSI reference model. These services are specified TE-to-TE, i.e., end-to-end.

Supplementary services are enhancements over standard telephony service. The services offered vary from one service provider to another.

25.3.1 Bearer services

Bearer services provided by GSM can be classified as 'transparent bearer services' and 'non-transparent bearer services'.

Transparent bearer services use the functions of the physical layer (layer-1) only for data transmission. Data transmission has a constant delay and throughput provided no errors occur. The transmission quality is improved with the use of forward error correcting codes (FEC). The redundancy built in the data stream due to the use of FEC codes helps in reconstructing data in case of transmission errors.

Depending on FEC, data rates of 2.4, 4.8 and 9.6 kbps are possible. However, recovery of lost data is not possible in case of shadowing or interruptions due to hand-over.

Non-transparent bearer services use protocols of layers 2 and 3 (data link and network layers) of OSI model for error correction and flow control. For a BER poorer than 1 in 1×10^{-7}, re-transmission is resorted to. The protocol is termed Radio Link Protocol (RLP), which is very much similar to High Level Data Link Control (HDLC) protocol.

Bearer services for inter-working with PSTN, ISDN and PSPDN (packet switched public data network, e.g. X.25) are specified through a set of transparent and non-transparent bearer services in GSM.

GSM bearer services provide for data transmission in synchronous and asynchronous modes at data rates up to 9.6 kbps.

25.3.2 Teleservices

Though the focus of GSM is on high quality digital voice tele-service, it supports the following non-voice teleservices:

Emergency number: This has the highest priority for connection and can pre-empt other connections. Calls are automatically set up with the closest emergency center.

SMS (short message service): Transmission of message up to 160 characters using the unutilized capacity of signalling channels in the GSM frame, e.g., display of traffic condition, headers of important news.

Group-3 FAX: FAX data is transmitted as digital data over analog telephone network according to ITS-T standards T.4 & T.30

25.3.3 Supplementary services

Supplementary services are enhancements over standard telephone services. Examples are call identification and call redirection. Further, 'closed user groups' and 'multi-party communications' are also possible in GSM.

25.4 System Architecture

25.4.1 Subsystems of GSM

GSM has a hierarchical architecture with many entities and interfaces. For easy reference, suitable acronyms are framed for these entities and interfaces. Figure 25.2 shows the simplified overview of GSM. GSM has three subsystems:

- Radio subsystem (RSS)
- Network and switching subsystem (NSS)
- Operation subsystem (OSS)

The cell phone or the mobile station (MS) carried by the subscriber is a part of the RSS.

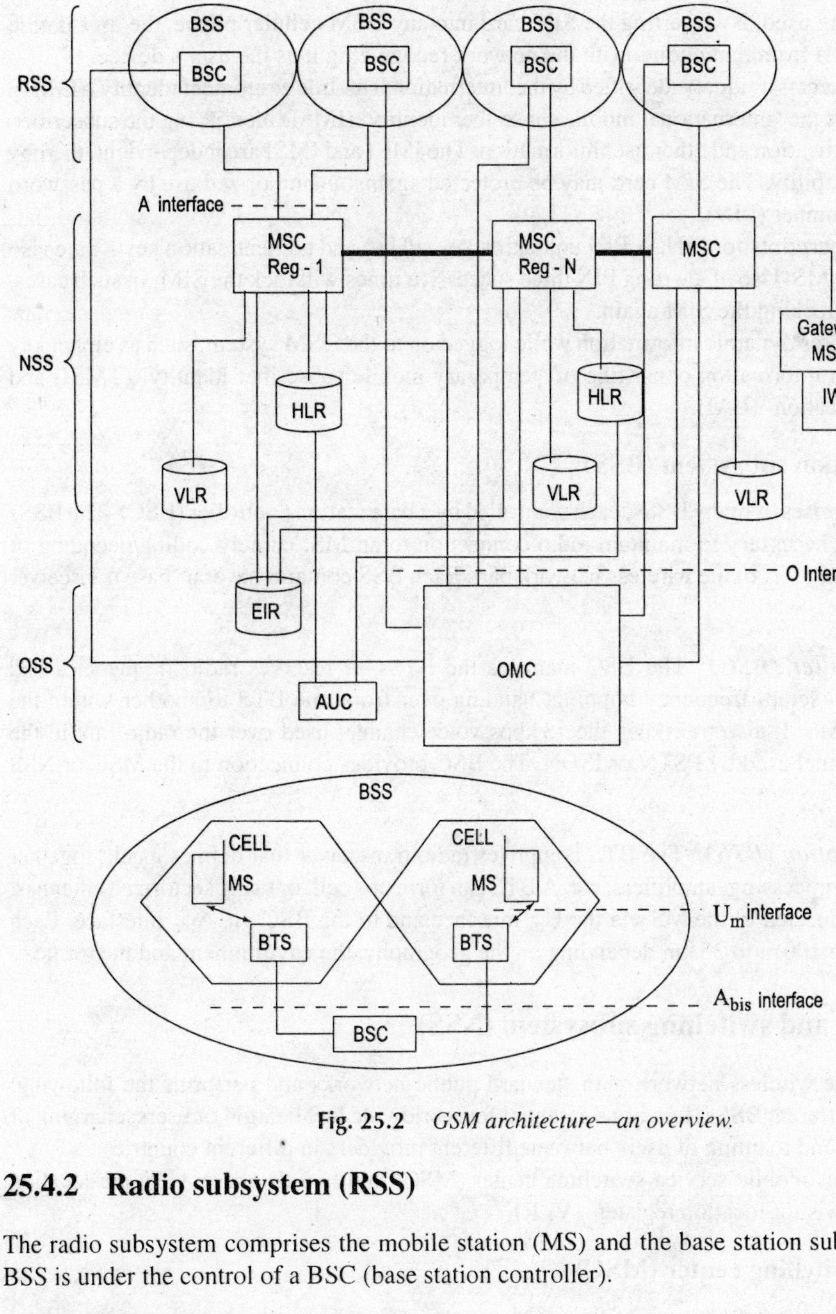

Fig. 25.2 *GSM architecture—an overview.*

25.4.2 Radio subsystem (RSS)

The radio subsystem comprises the mobile station (MS) and the base station subsystem (BSS); each BSS is under the control of a BSC (base station controller).

Mobile station (MS)

The MS comprises user equipment such as the radio transceiver, display and digital signal processors and a smart card called 'subscriber identity module (SIM)'. The SIM provides for personal mobility so that the user has access to all subscribed services irrespective of where the terminal is located and which terminal is being used. By inserting the SIM card into any GSM cellular phone, the user is able to receive or make calls from that phone, with the network recognizing it as the user's device.

The mobile equipment is uniquely identified by the 'international mobile equipment identity' (IMEI). The SIM card contains the 'international mobile subscriber identity' (IMSI) identifying the subscriber, a secret key for authentication and other user information. The IMEI and IMSI are independent, thereby providing personal mobility. The SIM card may be protected against unauthorized use by a password or personal identity number (PIN).

Some MSs use fingerprints for PIN. A PIN unlocking key (PUK) and authentication key Ki are also used. PIN unlocks the MS. Use of a wrong PIN three successive times will lock the SIM. In such cases, PUK is needed for unlocking the SIM again.

Further, the MS stores dynamic information while logged on to the GSM system, such as cipher key (Ki), and the location information consisting of 'temporary mobile subscriber identity' (TMSI) and 'location area identification' (LAI).

25.4.2.1 Base station subsystem (BSS)

A GSM network comprises of many BSSs, each controlled by a base station controller (BSC). The BSSs perform all functions necessary to maintain radio connection to an MS, namely coding/decoding of voice and rate adaption to/from the wireless network part. Each BSS contains several 'base transceiver stations' (BTS).

Base station controller (BSC): The BSC manages the BTSs. It reserves radio frequencies and handles radio channel setup, frequency hopping, handing over from one BTS to another within the BSS, and paging of MS. It also translates the 13 kbps voice channel used over the radio link to the standard 64 kbps channel used by PSTN or ISDN. The BSC provides connection to the MSC or NSS via the 'A' interface.

Base transceiver station (BTS): The BTS comprises radio transceiver that defines a cell, together with antennas, signal processing, amplifiers, etc. A BTS can form one cell, or using sectorized antennas, several cells. It is connected to the MS via the U_m interface and to the BSC via A_{bis} interface. Each cell can measure from 100 m to 35 km depending on the geography, the environment and the traffic.

25.4.3 Network and switching subsystem (NSS)

The NSS connects the wireless network with standard public networks and performs the following: hand-over between different BSSs, functions essential for worldwide localization of users, charging of services, accounting, and roaming of users between different providers in different countries.

The NSS comprises mobile service switching center (MSC) and two databases—'home location register' (HLR) and 'visitor location register' (VLR).

Mobile services switching center (MSC)

The MSCs form the backbone network of the GSM system. An MSC is a high performance digital ISDN switch and manages several BSCs of a geographical region. It uses standard signalling system No.7 (SS7) to handle all the signalling needed for connection setup, connection release and hand-over of connections to other MSCs. An MSC also performs functions necessary for multi-party calls, reverse charging, etc.

Home location register (HLR)

The HLR has static information relevant to the user such as 'mobile subscriber ISDN number' (MSISDN), subscribed services and authentication key (Ki). The user-specific information stored as a block has no duplication in HLR. Charging and accounting are also supported by HLR.

The current location area (LA) of an MS which is a dynamic information, is updated as soon as the MS leaves its current LA.

Visitor location register (VLR)

Information related to the MS users currently in an LA that is associated with the MSC is stored in the VLR. VLR thus is a dynamic database. User-relevant information for any MS coming into an LA that the VLR is responsible for, is copied from the HLR by the VLR.

The hierarchy of HLR and VLR is introduced to avoid frequent HLR updates and long distance signalling of user information.

25.4.4 Operation subsystem (OSS)

The OSS performs all functions necessary for network maintenance and operation. It processes network entities on its own and accesses other entities via SS7 signaling.

The OSS consists of OMC (operation and maintenance centre), AuC (authentication centre) and EIR (equipment identity register).

Operation and maintenance center (OMC)

The OMC monitors and controls all other network entities via the 'O' interface (SS7 with X.25). The management functions performed by OMC are:

- Traffic monitoring
- Status reports of network entities
- Subscriber and security management
- Accounting and billing

The OMC uses ITU-T's TMN (telecommunication management network) standard.

Authentication centre (AuC)

A separate AuC has been defined to protect user identity and data transmission in view of the vulnerability of radio networks to unauthorized access, and hence mobile stations. The AuC contains algorithms for authentication as well as keys for encryption. AuC can also be stored as part of HLR with restricted access.

Equipment identity register (EIR)

All device identification information registered with a GSM network, i.e., all IMEIs are stored in a database named EIR. Since the MS is mobile, anyone can use a stolen MS with a valid SIM card of his/her own. This is prevented by the EIR by maintaining a list of stolen MSs (black list or locked list). The EIR also contains a list of malfunctioning devices.

25.4.5 Interfaces in GSM

So far, several interfaces between different entities of GSM have been mentioned. These are now consolidated below for the convenience of the reader:

Between entities	Interface
MS and BTS	U_m
BTS and BSC	A_{bis}
BSC and MSC	A
OMC and MSC	O
OMC and BSC	O
MSC and PDN	SS7

The U_m interface is of special interest since all other interfaces are between fixed entities in GSM.

25.5 Digital Transmission in GSM

We have come to the most critical part of GSM—digital transmission. To understand the mechanism fully, refer to the sequence of speech coding, channel coding, modulation, multiplexing, etc. given in Fig. 2.1 of Chapter 2.

25.5.1 Speech coding

GSM is a digital system, hence analog speech signals are to be digitized. In digital transmission over microwave radio or optical fibre lines, the basic modulation is pulse code modulation (PCM) with channel data rate of 64 kbps. Several channels (speech signals or otherwise) in the PCM form are then multiplexed. 64 kbps is too high a rate to be feasible over a radio link in cellular networks. Also, the 64 kbps signal contains much redundancy although it is simple to implement.

The GSM group considered several voice coding algorithms on the basis of subjective speech quality and complexity. The best trade-off between processing delay on one hand, and cost and power consumption on the other was possible through 'regular pulse excited linear predictive decoder' (RPE-LPC) algorithm with long-term predictive loop included. The information from previous samples does not change quickly in the case of speech samples. Hence, the information from previous samples is useful in predicting the current sample. A linear combination of the previous samples with suitable coefficients plus an encoded form of the residual (i.e. the difference between the predicted and actual sample), represent the signal. Thus speech signal samples with 20 ms interval, each of which is encoded at 260 bits (i.e., a total bit rate of 13 kbps) would be sufficient for speech transmission in cellular radio.

25.5.2 Channel coding

The next step to speech coding is channel coding. GSM uses conventional encoding and block interleaving to achieve this protection.

The speech code of GSM system produces 260 bit block for every 20 ms. From subjective testing it was found that some bits of this block were more important for perceived speech quality than others. The bits can thus be divided into three classes:

Class Ia (50 bits)	-	Most sensitive to bit errors
Class Ib (132 bits)	-	Moderately sensitive to bit errors
Class II (78 bits)	-	Least sensitive to bit errors

Class Ia bits have a 3-bit cyclic redundancy code added for error detection. If an error is detected, the frame is judged incomprehensible and is discarded. It is replaced by a slightly attenuated version of the previous correctly received frame. These 53 bits, together with the 132 class Ib bits and a 4 bit tail sequence (a total of 189 bits) are input into a 1/2 rate convolution encoder of constraint length 4. Each input bit is encoded as two output bits based on a combination of previous 4 input bits. The convolution encoder thus outputs 378 bits which are unprotected. Thus, every 20 ms the speech sample is encoded as 456 bits, giving a bit rate of 22.8 kbps (456/20 ms = 22.8 kbps).

To further protect against burst errors common to radio interfaces, each sample is diagonally interleaved. The 456 bits output by the convolution encoder are divided into 8 blocks of 57 bits, and these blocks are transmitted in eight consecutive time-slot bursts.

In each time slot, 57 bit blocks of two different speech samples are put together as described below:

Tail bits	3 bits
Data field (first block)	57 bits
S flag (to indicate that the data field has user/network control data)	1 bit
Training sequence (to adapt parameters of the receiver to the current path propagation characteristics and to select the strongest signal in case of multi-path propagation)	26 bits
S flag	1 bit
Data field (first block)	57 bits
Tail bits	3 bits
Guard space (both sides put together in bit interval)	8.25 bits
Total	156.25 bits

Apart from normal burst, ETSI defines four more bursts:

Frequency correction burst: Allows the MS to correct the local oscillator to avoid interference with neighboring channels.

Synchronization burst: With an extended training sequence, the MS is synchronized with BTS.

Access burst: Allows initial connection setup between the MS and BTS.

Dummy burst: No data is available for a slot.

Data transmission over radio in GSM

Data transmission over radio line in GSM is to be tackled in two parts—radio spectrum utilization and multiplexing. In GSM this is achieved through a combination of FDMA and TDMA.

The International Telecommunication Union (ITU), which is responsible for the allocation of radio spectrum, allocated the 80–915 MHz band for up-link (mobile to base station) and the 935–960 MHz band for down-link (base station to mobile). This band is to be shared by as many users as possible. GSM uses a combination of TDMA and FDMA to achieve this. In the FDMA part, the total 25 MHz bandwidth is divided into 124 carrier frequencies of 200 kHz bandwidth. One or more carrier frequencies are assigned to each base station. Each of these carrier frequencies is then divided in time using the TDMA scheme, into eight time-slots. One time-slot is used by the mobile for transmission, and one for reception, thus enabling full duplex communicatione. It may be recalled that in digital MW radio, full duplex communication is achieved by using different carriers for data frame transmission /reception between two radio stations.

Each group of eight time-slots or the TDMA frame is transmitted every 4.615 ms so that each time-slot in the TDMA frame has a duration of $4.615/8 = 0.5777$ ms. As already discussed in Section 25.5.2, a burst of 156.25 bits is to be transmitted in each TDMA time-slot over a 0.577 ms period. Thus, the bit rate would be 270.833 kbps (156.25/0.5777 ms).

25.5.3 Logical channels in GSM

Physical versus logical channels

As seen in the previous two sections, the medium (transmission medium in the allotted frequency band) has 8×124 duplex channels in GSM. These are the 'physical channels'—each channel is a slot repeated every 4.615 ms.

Each physical channel has number of logical channels and the data field transmitted in each physical channel is structured as per the logical channel pattern. There can be logical channel that takes up every alternate slot; there can also be logical channel that takes up every fourth slot. Hence, a study of the logical channels of GSM is necessary.

Types of logical channels

Various types of logical channels are illustrated in Fig. 25.3. These are described in brief now.

Traffic channels(TCH): GSM uses a TCH to transmit user data in either full rate (TCH/F) with data rate of 22.8 kbps, or half-rate (TCH/H) with data rate of 11.4 kbps. Since voice codecs of GSM are of 13 kbps rate, the spare capacity of TCH/F is used for error correction.

Different coding schemes are available that make possible different data rates:

TCH/F 4 for 4.8 kbps; TCH/F 9.6 for 9.6 kbps; and TCH/F 14.4 for 14.4 kbps.

The number of error/erasure correction capacity varies with the data rate and the coding scheme used.

Control channels (CCH): As illustrated in Fig. 25.3, three groups of control channels are defined, each with a number of sub-channels so that the requirements of medium access, allocation of traffic channels and mobility management are fully addressed.

Broadcast control channel (BCCH): The BTS uses this channel to signal information to all MSs within a cell. Information transmitted in this channel are cell identifier , options available within the cell (for frequency hopping), frequencies available inside the cell and in neighboring cells.

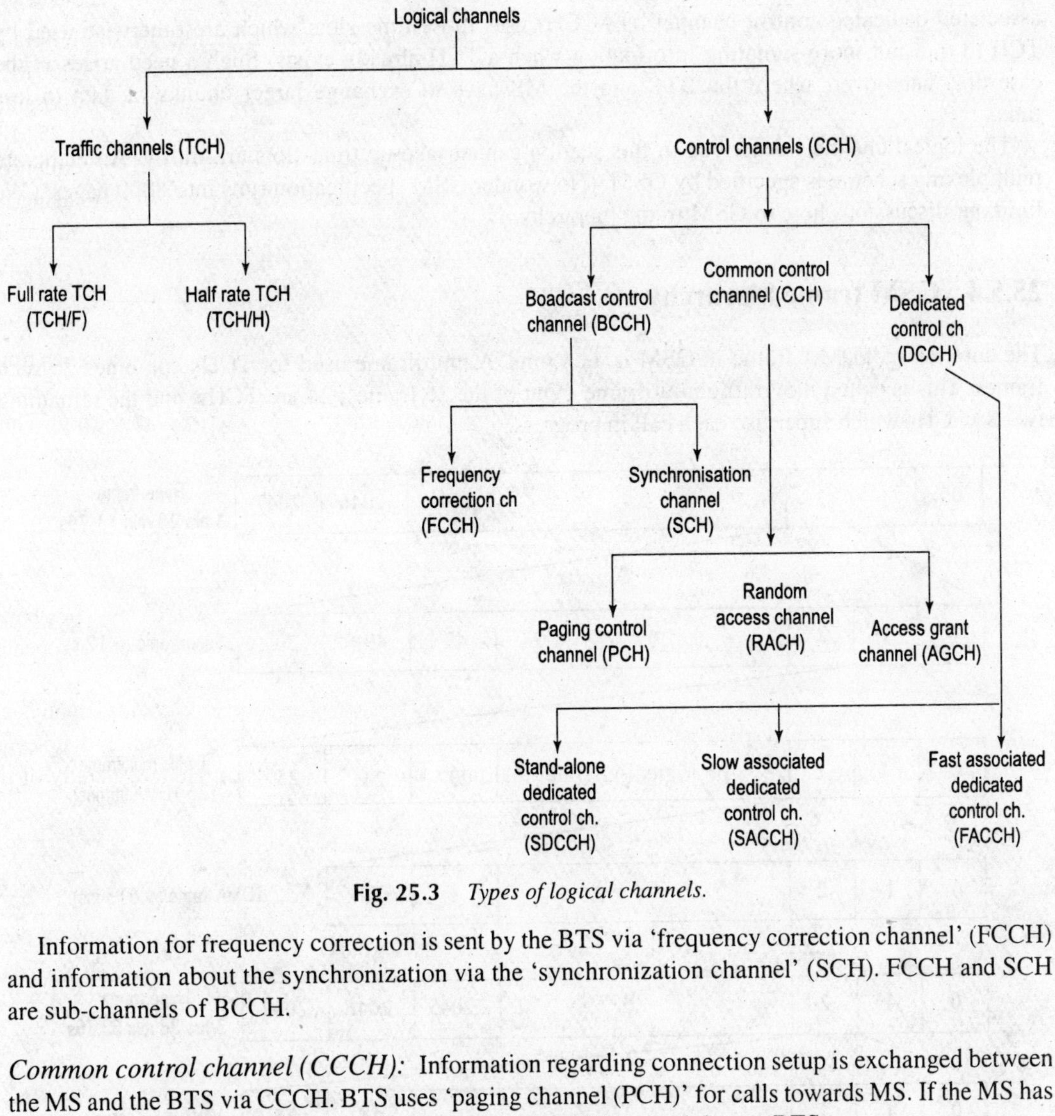

Fig. 25.3 *Types of logical channels.*

Information for frequency correction is sent by the BTS via 'frequency correction channel' (FCCH) and information about the synchronization via the 'synchronization channel' (SCH). FCCH and SCH are sub-channels of BCCH.

Common control channel (CCCH): Information regarding connection setup is exchanged between the MS and the BTS via CCCH. BTS uses 'paging channel (PCH)' for calls towards MS. If the MS has to set up a call, it uses a 'random access channel (RACH)' to send data to BTS.

All MSs within a cell can use RACH through multiple access. Since the possibility of collision exists, BTS uses an 'access grant channel' (AGCH) to signal to the MS that it can use a TCH or a SDCCH (stand-above dedicated control channel) for further connection setup.

Dedicated control channel (DCCH): BCCH and CCCH are unidirectional whereas DCCH is bi-directional. As long as an MS has not established a TCH with the BTS, it uses the 'stand-alone dedicated control channel' (SDCCH) with a low data rate (782 kbps) for signaling. This comprises authentication, registration and other data needed for setting up a TCH.

A 'slow associated dedicated control channel' (SACCH) is associated with each TCH and SDCCH to facilitate exchange of system information like channel quality, signal power level, etc. And a 'fast associated dedicated control channel' (FACCH) uses those time slots which are otherwise used by TCH to transmit more signaling information when a TCH already exists. Such a need arises in the case of a hand-over, where the BTS and the MS have to exchange larger chunks of data in less time.

The logical channels described in this section cannot take up time-slots arbitrarily. An elaborate multiplexing scheme is specified by GSM. (No wonder GSM specification runs into 8000 pages!) We limit our discussions here to GSM frame hierarchy.

25.5.4 GSM frame hierarchy

The duration of TDMA frame in GSM is 4.615 ms. A multiframe used for TCHs combines 26 such frames. This is called the 'traffic multiframe'. Out of the 26 frames, 24 are TCHs, and the remaining two SACCHs which supervise each call in progress.

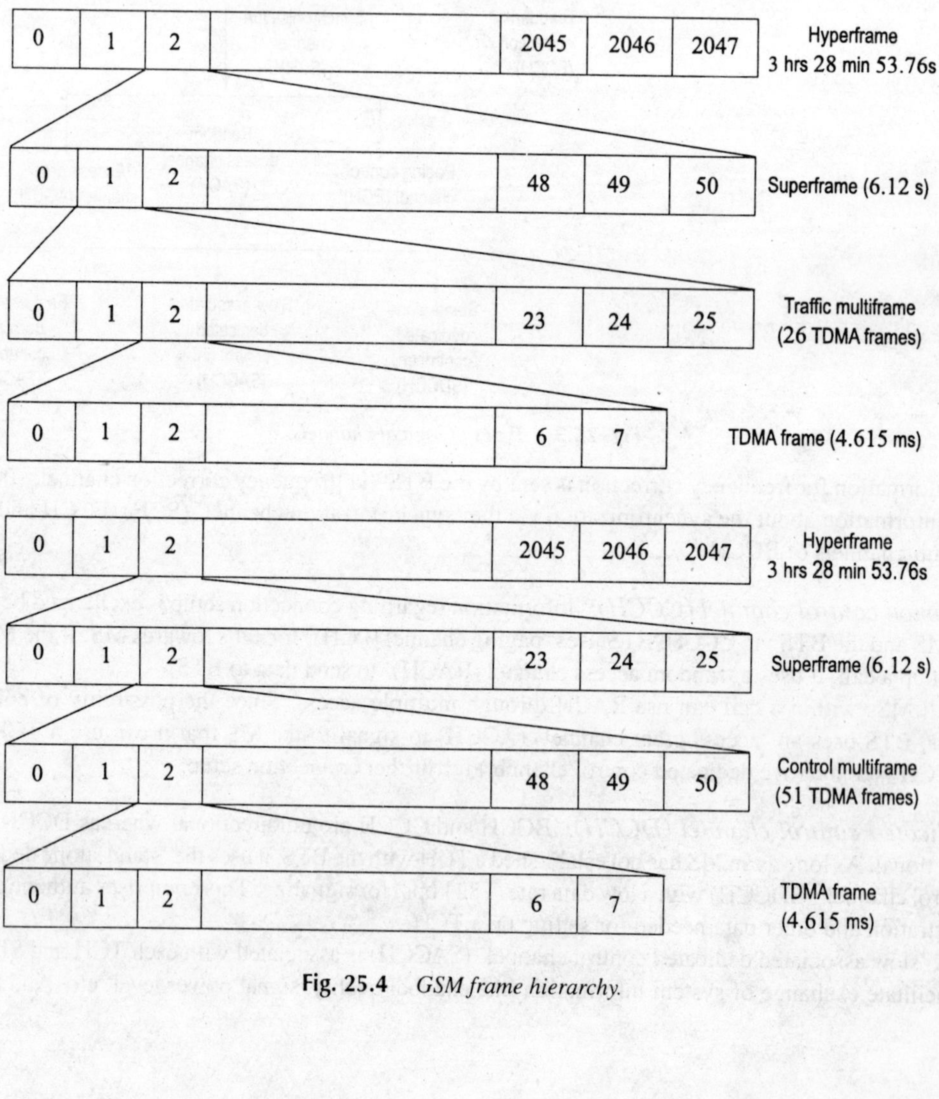

Fig. 25.4 *GSM frame hierarchy.*

The SACCH in frame-12 contains eight logical channels, one for each of the eight connections carried by the TCHs. The SACCH in frame-25 is not currently used, but will carry eight additional SACCH channels when half-rate traffic is implemented.

One FACCH also works by stealing slots from a TCH to transmit power control and hand-over signalling messages. The channel stealing is achieved by setting one of the control bits in the time-slot burst. As per the value of this designated bit, the frame control mechanism leaves the designated channel to be used for transmission of power control and hand-over signal messages.

The TDMA frames containing the signalling data for other logical channels are combined to form a 'control multiframe'. It has 51 TDMA frames of 235.4 ms duration.

The frame hierarchy continues, combining 51 'traffic multiframes' (each of 26 TDMA frames) or 26 'control multiframes' (each of which is 51 TDMA frames), to form a 'superframe' whose duration is 6.12 s. A 'hyperframe' consists of 2048 superframes which has a duration of 3.5 hours, and altogether 2715648 TDMA frames (see Fig 25.4). This hierarchical structure is needed for encryption . GSM counts each TDMA frame, and the frame number is input to the encryption algorithm. Since each time-slot in a frame is numbered (8 slots per frame), every time-slot is uniquely identified in GSM. Fig. 25.4 shows the GSM frame hierarchy.

25.5.5 Modulation

It may be recalled that each time-slot has a data burst at the bit rate of 270.833 kbps. This digital signal is modulated onto the analog carrier frequency which has a bandwidth of 200 kHz using 'Gaussian-filtered minimum shift keying' (GMSK). The Gaussian filter cuts down the large bandwidth requirement of conventional MSK. GMSK was selected over other modulation schemes as a compromise between spectral efficiency, complexity of the transmitter and number of spurious emissions. If the complexity is reduced to make the power consumption optimal, we would have handy mobiles. Spurious emissions outside the allotted bandwidth are to be controlled to allow for coexistence of GSM and the older analog systems.

25.6 GSM Protocols

25.6.1 Protocol architecture

Architecture of GSM protocols for signaling between various entities is shown in Fig. 25.5. The different layers are:

Layer 1	-	Physical layer
Layer 2	-	Link access control layer
Layer 3	-	Radio resource management layer
Layer 4	-	Mobility management layer
Layer 5	-	Call management layer

25.6.2 Physical Layer

Layer 1 or the physical layer handles the following functions:

- Channel coding

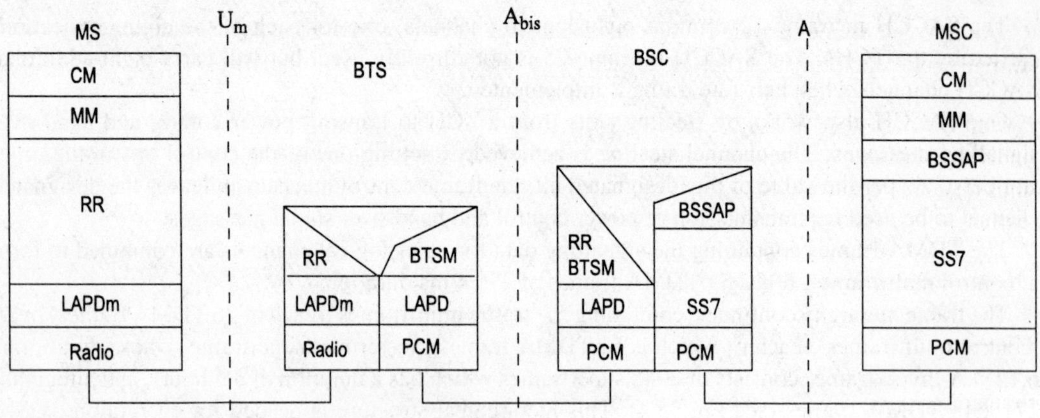

Fig. 25.5 *GSM protocol architecture.*

- Modulation
- Generation of time structure of frames and slots for TDM
- Synchronization of MSs with BTS.
- Measurement of channel quality
- Voice activated detection (VAD)

These functions are described in brief now.

Channel coding: Channel coding in GSM is for error correction rather than error detection alone as in the case with OSI reference model for computer networks. In the physical layer, GSM employs FEC (forward error correction) coding scheme. FEC adds redundancy to data, and the number of redundant bits depends on the error /erasure correction capability desired. Different coding schemes are used for different logical channels. It may please be recalled that voice codecs give a data rate of 13 kbps, whereas the data burst rate of I channel is 22.8 kbps. Hence there is scope for redundancy which is utilized for FEC.

Modulation: The physical layer at U_m interface uses GMSK for digital modulation and performs encryption/decryption of data, i.e., encryption is not needed end-to-end but only between the MS and the BTS.

Generation of time structure of slots and frames for TDMA: The BTS generates the time structure of frames and slots for TDMA. The crux of the problem lies here—the round trip time (RTT) is different for different MSs. The MSs close to BTS have short RTTs whereas MSs at the far end of the cell have RTTs of around 0.23 ms.

To cater for all MSs, i.e., if we design time structure with uniformity, too much of guard space (in time domain) would be wasted. The duration of 0.23 ms consumes 40% of the 0.577 ms time frame

Hence, this particular function of Layer 1 is designed such that the BTS sends the current RTT to the MS, which then adjusts its access time so that all bursts reach the BTS within their time limits. This mechanism requires guard space of only 30.5 μs, i.e., 5% of the frame time.

Adjustment of access time is controlled using the 'timing advance' variable, where, a burst can be shifted up to 63 bit times earlier, with each bit having a duration of 3.69 μs ($63 \times 3.69\,\mu s = 0.23$ ms).

Synchronization of MS with BTS: Synchronization of the MS with the BTS involves correction of individuai path delays between the MS and the BTS. All the MSs within a cell use the same BTS and thus must be synchronized to the BTS. The BTS generates the time structure of the frames as described earlier, which is implemented by all MSs.

Measurement of channel quality: This is done on down-link, i.e., 935–960 MHz band (BTS to MS transmission).

Voice activity detection (VAD): This is a special function implemented by the physical layer. VAD transmits voice data only when there is voice signal. During periods of silence (e.g. the conversation between philosophers), the physical layer generates a comfort noise to create the feeling of connection to the user at the other end, but no actual transmission takes place.

Layer-2 functions

Layer-2 of GSM, the LAPDm, is the mobile version of 'link access procedure for the D-channel' (LAPD) of ISDN systems which is a derivative of HDLC. LAPDm is simpler than LAPD since it does not need synchronization or error detection functions. These functions have been relegated to the physical layer in GSM.

Another subtlety of GSM is the absence of buffering between layers 1 and 2. So, LAPDm has to obey the frame structures, resource patterns, etc., defined for the U_m interface. Other services provided by LAPDm are segmentation and re-assembly of data and acknowledged/unacknowledged data transfer.

Layer-3 functions

Layer-3 functions namely resource management functions are implemented partly in the BSC. The functions in the BTS are supported by the BSC via BTS management . The main functions of RR (radio resource management) are setup, maintenance and release of radio channels . RR can directly access Layer-1 for radio information.

Layer-4 functions

Mobility management functions which pertain to Layer-4 support the following:

- Registration
- Authentication
- Identification
- Location updating
- Provision of TMSI

TMSI is the temporary mobile subscriber identity number which replaces IMSI (international mobile subscriber identity). TMSI hides the real identity of user over air. It is valid in the current location of VLR.

Layer-5 functions

Layer-5 functions or the call management functions are:

- Call control service (CCS)
- Short message service (SMS)
- Supplementary service (SS)
- Tone transmission service (TTS)

CCS provides point-to-point connection between two terminals and is used by higher layers for call establishment, call clearing and change of call parameters.

SMS provides message transfer using the control channels SDCCH and SACCH (if no signaling data is sent).

SS offers services like user identification, call redirection, closed user groups, and multi-party communication.

TTS facilitates transmission of in-band tones using DTMF scheme over the GSM network. These tones are used for remote control of answering machines; entry of PINs for electronic banking; entry of PNRs/train numbers on the interactive voice response system (IVRS) for reservation/arrival/departure information; dialing in traditional analog telephone system, etc. These tones cannot be sent directly over the voice codec of a GSM MS as the codec would distort the tones. These are transferred as signals and then converted into tones in the fixed network part of the GSM system.

25.6.3 Protocols across A_{bis} interface and A interface

The physical layer across A_{bis} interface, i.e. between BTS and BSC, uses the PCM system. Here, GSM allows for sub-multiplexing of four 16 kbps channels into a single 64 kbps channel. Since 16 kbps is higher than the voice codec data rate of 13 kbps, there will not be any problem.

The physical layer across A interface, i.e. between BSC and MSC, uses PCM on leased 2 Mbps lines. LAPD is used for layer-2 at A_{bis}, together with BTSM for MTS management.

Signalling between MSC and BSC (Layer-2 protocol across A interface) follows SS7 (see Appendix A13). SS7 facilitates transfer of call management information between MSCs, HLR, VLRs, AuC, FIR and OMC.

25.7 Localization of Users in GSM

25.7.1 Concept of roaming

In GSM system, localization of user worldwide is feasible, i.e., the current location of a user is known to the system (MS is logged on) and the same phone number is valid worldwide. The HLR–VLR pair makes this feasible. HLR contains information about the current location, and the VLR currently responsible for the MS informs the HLR about location changes. As soon as the MS moves into the range of a new VLR, the HLR sends all the user data needed to the new VLR.

Thus, 'roaming' means changing VLRs to ensure uninterrupted service. Roaming can be of three types:

Internal roaming:	Within the network of one service provider
National roaming:	Between two providers in a country
International roaming:	Between different providers in different countries

Numbers used in locating MS

The following numbers are used in locating an MS:

1. *Mobile station international ISDN number (MSISDN):* MSISDN follows ITU-T standard E.164. It consists of the country code (CC); national destination code (NDC) which is the address of the network provider, and JLR; and subscriber number (SN).
2. *International mobile subscriber identity (IMSI):* GSM uses the IMSI for internal unique identification of a subscriber. IMSI has mobile country code (MCC); mobile network code (MNC); and mobile subscriber identification number (MSIN).
3. *Temporary mobile subscriber identity (TMSI):* To hide IMSI over air interface, a 4- byte long TMSI is used for local subscriber identification. It is selected by the current VLR and is valid within the location area of the VLR. VLR can also change TMSI periodically for additional protection.
4. *Mobile station roaming number (MSRN):* MSRN hides the identity and location of the subscriber. It is generated by VLR on request by the MSC and stored in the VLR. It contains the visitor country code (VCC), visitor national destination code (VNDC) which is the identification of the current MSC, together with the subscriber number.

25.7.2 Mobile terminated call

A mobile terminated call (MTC) results when either the calling station is outside of GSM, or another MS calls an MS.

The steps involved in connecting the called MS are (refer Fig. 25.6):

Step 1 User dials the phone number of the GSM subscriber.

Step 2 PSTN forwards the call setup to the gateway MSC (GMSC) as it identifies the destination code.

Step 3 GMSC identifies the HLR for the subscriber which is coded in the phone number and signals call setup to HLR.

Step 4 The HLR checks whether the number exists and whether the user has subscribed to the requested services and requests an MSRN from the current VLR.

Step 5 MSRN is received by HLR.

Step 6 HLR determines the MSC responsible for the MS and forwards the information to the GMSC.

Step 7 GMSC forwards the call setup request to the MSC indicated.

Step 8 MSC requests the current status of the MS from the VLR.

Step 9 The VLR sends the status information as either (*i*) MS available, or (*ii*) not available

Step 10 If MS is available, the MSC initiates paging in all the cells it is responsible for.

Step 11 BTSs of all BSSs transmit this paging signal to the MS.

Step 12 and 13 MS answers; VLR performs security checks.

14 Security check is carried out, and if violation is detected, connection is denied.

Step 15 to 17 VLR signals to the MSC to set up connection to the MS.

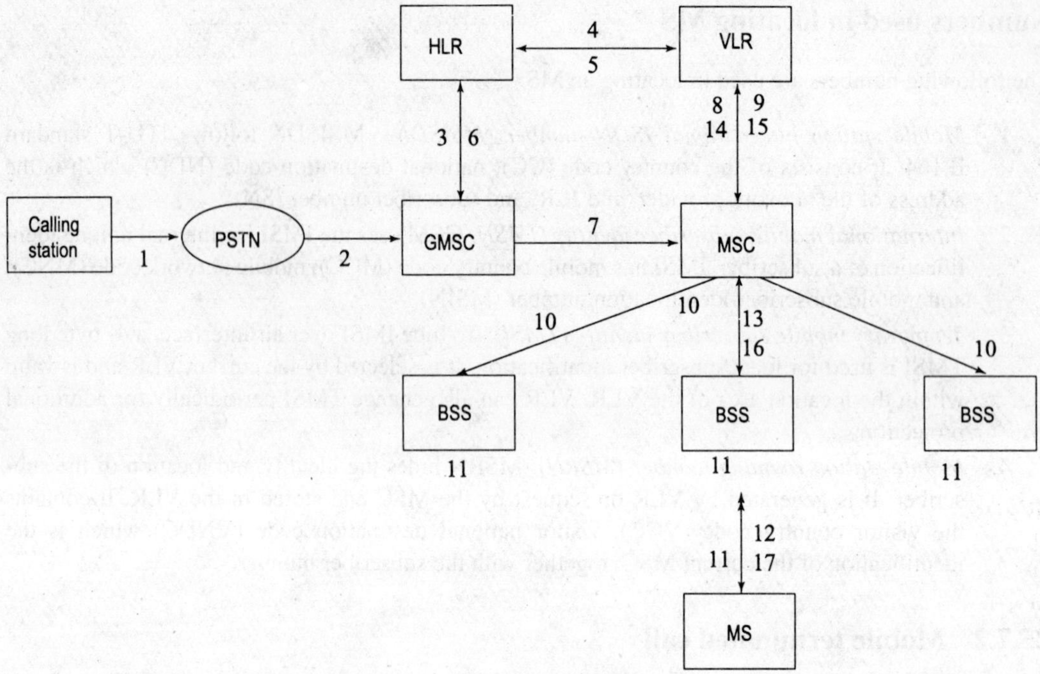

Fig. 25.6 *Steps in mobile terminated call.*

25.7.3 Mobile originated call (MOC)

The steps involved in setting up the call when a mobile originates it are given below (refer Fig. 25.7):

Step 1	MS transmits request for a new connection
Step 2	BSS forwards this request to MSC.
Step 3 & 4	MSC checks if with VLR the user is allowed to set up a call under the requested service.
Step 5 to 8	Checks the availability of resources through the GSM network into the PSTN.
Step 9 to 10	If resources are available, a connection is set up.

The messages exchanged between MS and BTS during connection setup—either MTC or an MOC—are shown in Fig. 25.8.

25.8 Security Services in GSM

25.8.1 General features of GSM security

The confidential information stored in a SIM card is utilized by GSM in the security features it offers.

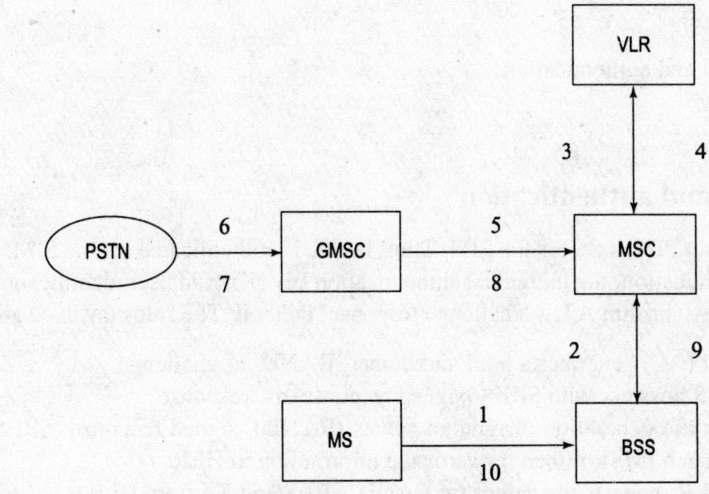

Fig. 25.7 *Steps in MOC.*

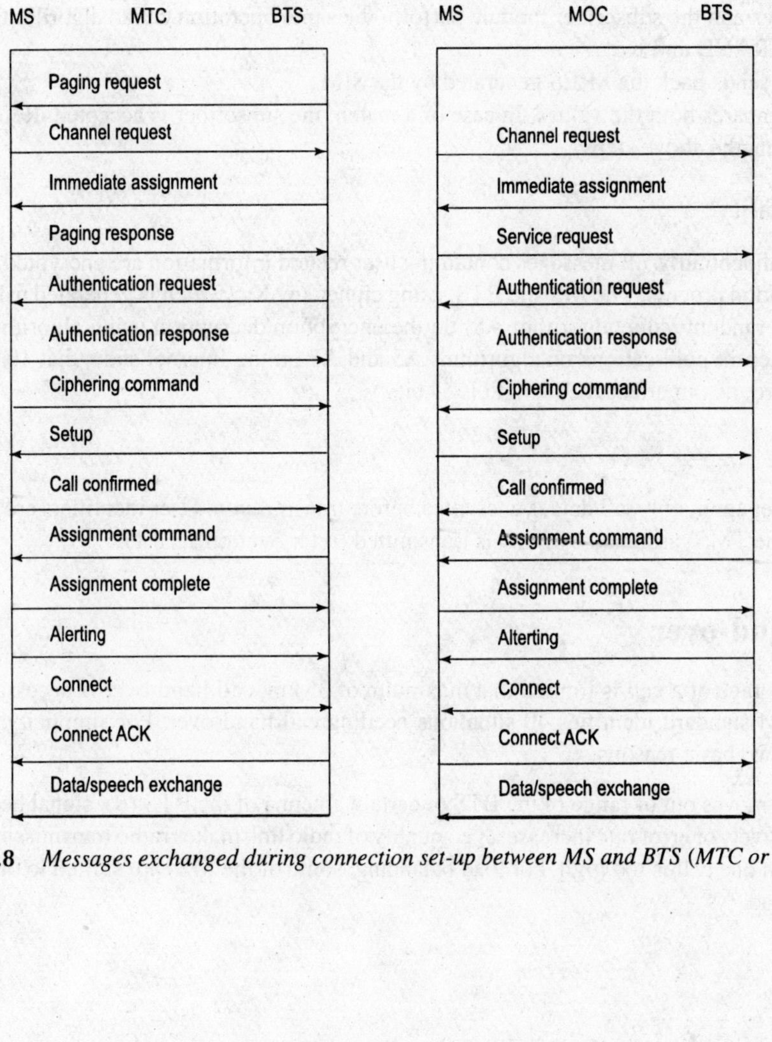

Fig. 25.8 *Messages exchanged during connection set-up between MS and BTS (MTC or MOC).*

These features are:

- Access control and authentication
- Confidentiality
- Anonymity

Access control and authentication

The user uses a secret PIN to access the SIM. Thus, he/she is authenticated for the SIM. In the process of subscriber authentication, the individual authentication key (Ki) and user identification IMSI stored in SIM are used in Algorithm A3, a 'challenge response' method. The steps involved are:

- Access control (AC) generates a random number 'RAND' as challenge
- SIM within MS answers with SRES (signed response) as 'response'
- AuC performs the generation of random values (RAND), signed response (SRES), and cipher keys (Kc) for each IMSI; it then forwards the information to HLR
- The current VLR requests the values for RAND, SRES and Kc from HLR
- For authentication, VLR sends the random value RAND to the SIM
- Both sides and the subscriber module perform the same operation called algorithmic calculation A3 with RAND and Kc
- The MS sends back the SRES generated by the SIM
- VLR compares both the values. In case of a match, the subscriber is accepted. Refer to Fig 25.9 illustrating the above steps.

Confidentiality

To ensure confidentiality, all messages containing user related information are encrypted. This follows the authentication process. The MS and BTS, using cipher key Kc (which is generated using individual key Kc and a random value algorithm A8) do the encryption/decryption using algorithm A5. (Kc is 64 bit long. Recent publications on algorithms A3 and A8 on the Internet show that 10 bits out of 64 are always zero, i.e., in effect Kc's length is 54 bits.)

Anonymity

To provide user anonymity, all data is encrypted before transmission. User identifiers are not used over air. Instead, the TMSI assigned by VLR is transmitted (refer Section 25.8.1).

25.9 Hand-over

Since the coverage of a cell is limited to a maximum of 35 km, call hand-over is necessary in cellular systems. GSM standard identifies 40 situations needing call hand-over. For simplicity, these can be attributed to two basic reasons:

- The MS moves out of range of the BTS or certain antenna of the BTS (Rx signal becomes poorer progressively or error rate increases, i.e., quality of radio link makes radio transmission infeasible).
- Traffic in one cell is too high. For load balancing, some of the MSs are shifted to other cells with lower load.

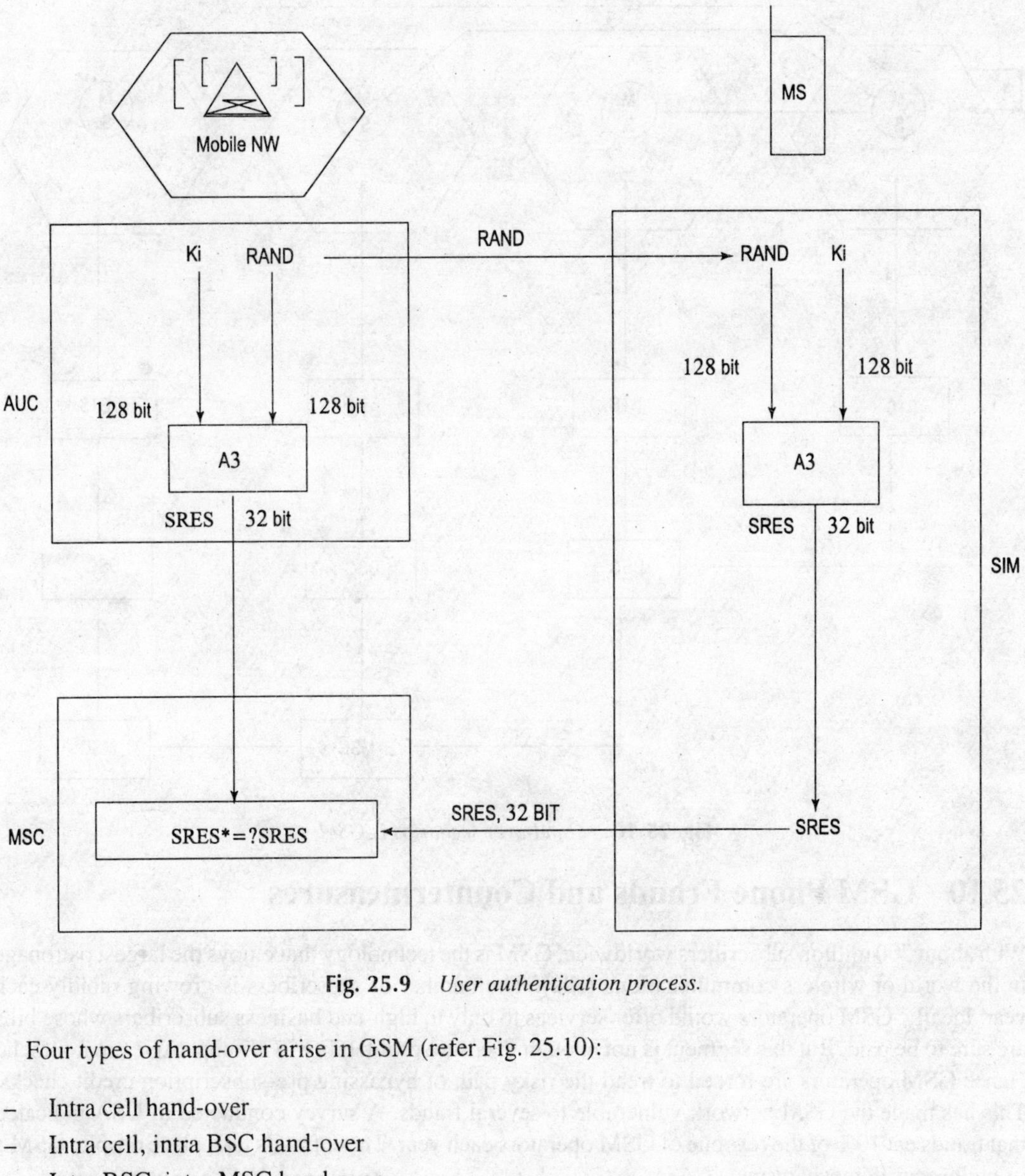

Fig. 25.9 *User authentication process.*

Four types of hand-over arise in GSM (refer Fig. 25.10):

- Intra cell hand-over
- Intra cell, intra BSC hand-over
- Inter BSC, intra MSC hand-over
- Inter MSC hand-over

When the MS moves away from one BTS to another, the hand-over decision does not depend on the actual value of the received signal level but on the average signal level. The BSC collects all signal values (BER, Rx signal level up-link and down-link) from the BTS and the MS to calculate the average values. These are compared to threshold values (hysteresis is in-built to avoid to-and-fro hand-over).

As an example, the message flow in inter BSC, intra MSC hand-over is shown in Fig. 25.11.

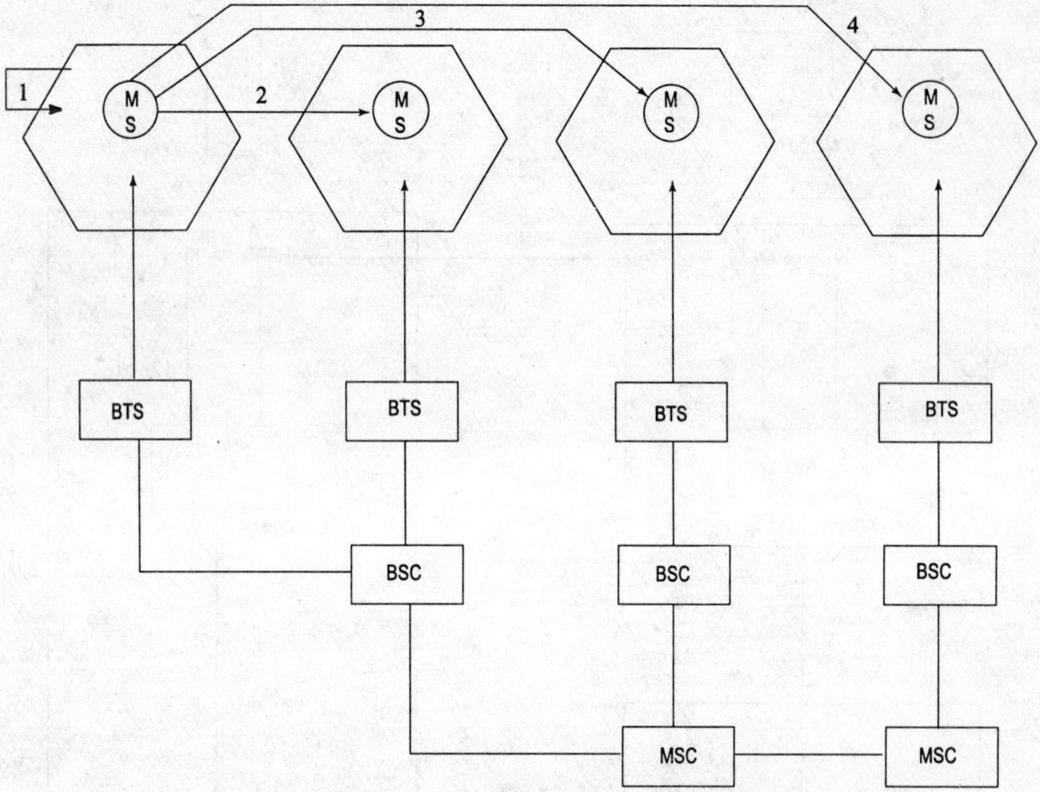

Fig. 25.10 *Hand-over scenarios is GSM.*

25.10 GSM Phone Frauds and Countermeasures

With about 200 million subscribers worldwide, GSM is the technology that enjoys the largest patronage in the world of wireless communications today. The number of subscribers is growing rapidly each year. Ideally, GSM operators would offer services to only to high-end business subscribers whose bills are sure to be paid. But this segment is not enough to ensure profitability in a competitive environment. Hence GSM operators are forced to tread the risky path of bypassing pre-subscription credit checks. This has made the GSM network vulnerable to several frauds. A survey conducted in USA indicated that frauds eat 10% of the revenue of GSM operators each year. These frauds take place despite GSM's in-built security mechanisms.

The types of frauds committed on GSM network/operators are:

- SIM cloning
- Subscription fraud
- Call-sell operations
- Black market in stolen handsets
- Frauds based on billing delays
- Premium-service scam

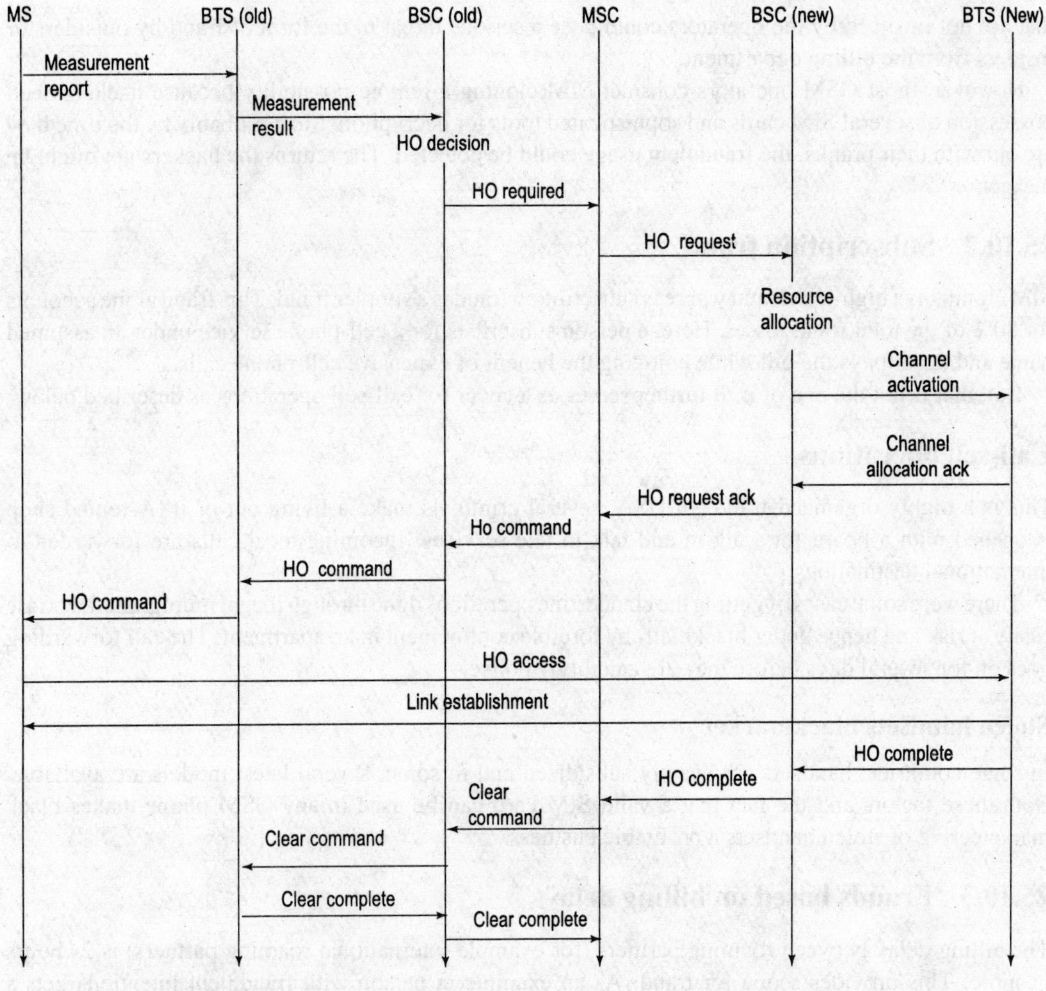

Fig. 25.11 *Message flow in inter BSC, Intra MSC hand-over.*

– SMS (short message service) scam
– Parallel call forwarding
– Publicity of perceived weakness

25.10.1 SIM cloning

SIM (subscriber identification module) is the smart card issued to each GSM subscriber which when inserted into a GSM phone, enables that phone to make calls. The bill is directed to the subscriber. The cell-phone identity and authentication are encrypted on the SIM card.

In April 1998, University of California, Berkely, announced that a group of researchers at their university have broken the encryption code using decryption algorithm and several hours of machine time. This has given rise to speculations of SIM cloning. The appearance of cloned SIMs tied to accounts that are not invoiced by the operators could pose a serious threat in the form of fraud by outsiders or insiders from the billing department.

However, most GSM operators consider SIM cloning a remote possibility because hackers need possession of several SIM cards and sophisticated tools for decryption. Most probably by the time they are out with their pranks, the fraudulent usage could be detected. The returns the hackers get might be meager.

25.10.2 Subscription fraud

SIM cloning is a high-tech fraud whereas subscription fraud is a simpler fraud. This fraud alone accounts for 80% of the total fraud losses. Here, a person subscribes for a cell-phone service under an assumed name and never pays the bill while enjoying the benefit of expensive cell-phone calls.

But, that is not the end of it. It further serves as a cover for call-sell operations as described below.

Call-sell operations

This is a highly organized fraud. In USA, several criminals make a living out of it. A rented shop is opened with a board for walk-in and talk-in tele services. Incoming local calls are forwarded to international destinations.

There were some cases of getting the clandestine operations done through illegal immigrants (who are easily at risk and hence under blackmail) by forcible confinement in an apartment. The call forwarding goes on for several days before they are caught by police.

Stolen handsets blackmarket

In some countries, handsets are heavily subsidized and in some, several latest models are available. Both these factors and the fact that a valid SIM card can be used in any GSM phone makes black marketeering of stolen handsets a profitable business.

25.10.3 Frauds based on billing delays

The billing delay between roaming partners (for example international roaming partners) is 24 hours or more. This provides scope for fraud. As an example, a person with fraudulent intentions gets a GSM phone with international roaming enabled (usually through subscription fraud) and takes it out of the country and uses it for several expensive toll calls. Until the bill is received at the subscriber's home network, the fraud cannot be detected. Often, some well organized criminals go to the extent of exchanging SIM cards with friends in different countries to carry out the frauds on a wide scale.

The only check against this fraud is to monitor roamers within their home networks and inform the home network (by e-mail or FAX) when the roamer exceeds a certain daily limit of calls. This notification depends on the roaming agreement between the operator and the user. The minimum period is generally 24 hours, which is enough for committing fraud worth lakhs of rupees.

25.10.4 Premium-rate service scam

A GSM network operator pays the owner of a premium-rate service a percentage of the anticipated revenue whether or not callers to the premium rate service actually pay their bills. This service is exploited by criminals to make money. They can procure a bank of phones, subscribe to a premium rate service and generate revenue by calling the premium number repeatedly and not paying the bills.

Further, if the GSM phones are taken outside the country, calls to the premium number can be initiated via international roaming that continue throughout the 24-hour (or more) time-slot. By the time the billing data is generated and the fraud noticed, the damage is already done.

Very long premium-rate-service calls are often caught at the billing stage since they invite suspicion. However, some of the new sophisticated phones provide for programming telephone calls of different random duration. Identification of frauds in such cases is difficult.

25.10.5 SMS scam

In SMS (short message service) scams, a large number of handsets are polled anonymously over the Internet and the owners are asked to call a premium-rate-service number to claim a prize. Many people follow the advice.

25.10.6 Parallel call forwarding

GSM phones can be programmed to forward an incoming call to a pre-defined phone number. This feature is exploited by criminals by using the facility of 'parallel-call forwarding'. In this fraudulent scheme, international or high-rate calls are sold for an attractive price. A purchaser makes his call to the GSM phone which then forwards the call to its faraway destination. This frees the GSM phone to receive the next purchaser's call.

Thus, numerous calls can be in progress simultaneously. Again, the criminal brains ride on the carpet of international roaming and execute fraud in an un-monitored time-slot (24 hours or more).

25.10.7 Publicity of perceived weakness

Any perceived weakness of the network is many a time given wide publicity by the media, especially on the Internet. This often works out to escalate the losses incurred by the network due to widespread exploitation of this weakness for fraudulent purposes.

25.10.8 How to defend against frauds

Defense against frauds begins with the monitoring of GSM call setup information and subscriber activity. The monitoring falls in two domains:

1. Check for suspicious calling patterns such as single long duration calls, high volume calls in a specified period, and calls made to high risk destinations.
2. Check for high number of calls made while roaming, frequent forwarding, and use of international call forwarding.

Any excesses in these two domains should generate a real-time alert, prompting the operators to take action.

A system can be designed to capture ongoing maintenance messages that the GSM phones transmit to the HLR for updating the location information. By tracking the geographic velocity change, suspicious features can be spotted, viz.,

- handset traveled too fast,
- handset traveled too far,
- handset is in two places (SIM cloning).

The operators can readily intervene by blocking the authentication from HLR.

Fraud detection systems like Hewlett Packard's HPACCESS7 are available in the market. The returns are great for a little investment in this direction.

Exercises

1. In GSM 900, 124 channels, each 200 kHz wide, are used for FDMA whereas in GSM 1800, 374 channels are used. Will there be any change in the TDMA frame, superframe and hyperframe of a GSM 1800 compared to GSM 900?
2. What is the purpose of using TMSI in GSM for local subscriber identification when IMSI is already available?
3. Why do we use two separate algorithm (A8 and A5) in the encryption process?
4. Compared to the TCHs offered, standard GSM can provide higher data rate (33.8 kbps) when considering the air interface. What lowers the data rates available to a user?

The topics in this chapter are based on ETSI (European Telecommunications Standard Institute) and GSM recommendations listed below.

References

[1] ETSI(1991a), Bearer services supported by a GSM PLMN, ETSI-GSM Rec. 02.02.

[2] ETSI(1991b), General description of a GSM PLMN, ETSI-GSM Rec. 01.02.

[3] ETSI(1991c), Subscriber identity modules, Functional characteristics, ETSI-GSM Rec. 02.17.

[4] ETSI(1993a), Multiplexing and multiple access on the radio path, ETSI-GSM Rec. 02.17.

[5] ETSI(1993b), MS-BSS data link layer-General aspects, ETSI-GSM Rec. 05.02.

[6] ETSI(1993c), MS-BSS data link layer specifications, ETSI-GSM Rec. 04.06.

[7] P.V.Sreekanth, GSM phone frauds, *Telecommunications*, (Journal of DOT/BSNL), p.51–53, Oct 2000.

[8] http://www.gsmworld.com/

26

New Data Services in Digital Cellular Networks

26.1 Introduction

The most successful standard for mobile communications based on digital cellular network is the GSM, as already discussed in the previous chapter. However, there are other standards that have carved niches for themselves by addressing the needs of specific user groups. Prominent among them which have come to stay with the 'big brother' GSM are

- DECT (digital enhanced cordless telephony),
- TETRA (terrestrial trunked radio),
- UMTS (universal mobile telecommunication system).

The mobile communications domain, based on the degree of customization and traffic density, can be categorized into three segments:

- Low degree of customization and high traffic density
- Medium degree of customization and medium traffic density
- High degree of customization and low traffic density

Cordless telephony comes in the first segment where the traffic density is very high, for example, in multi-storied premises of offices or industrial campuses where the number of instruments is high, contributing to high density of traffic; customization requirements are however minimum. The emerging mobile communication standard for such a situation is the DECT (digital enhanced cordless telephony). GSM occupies the second segment which is of course the segment with the highest market potential. The third segment pertains to certain professional user groups like police, military and transport, which are closed groups requiring highly customized mobile communication solutions; also the traffic density is low. The emerging standard for this segment is TETRA (terrestrial trunked radio).

Based on cell size and user densities, mobile communication domain can be categorized into three segments:

- Cell size in hundreds of meters, dense user density
- Cell size in kilometers, medium to high user density
- Cell size in tens of kilometers, low user density

UMTS (universal mobile telecommunication system) addresses the needs of the first segment. GSM is for the second segment with its high market potential, and TETRA is for the third segment. In fact, the spectrum allocation has been decided on these factors. UMTS operates in the GHz range, GSM in 900 MHz band and TETRA in lower hundreds of MHz. In this chapter, we will be studying DECT, and UMTS. Chapter 27 is devoted to TETRA.

Before we proceed further, let us look at systems which offer better data services than GSM but do not occupy a position as prominent as the ones mentioned above. These are:

- HSCSD (high speed circuit switched data)
- GPRS (general packet radio system)

GSM was developed focussing on voice transmission. The tremendous growth of data communication compared to voice communication was not anticipated. A data rate of 9.6 kbps was big enough for GSM planners. To meet the data transfer requirements associated with web-browsing, file download, e-mail exchange with attachments via cellular networks, two approaches have been tried. The first approach is to combine several GSM channels. This is a connection-oriented approach and the system is called HSCSD (high speed circuit switched data). The second approach, which is a packet-oriented approach, is to think in terms of packets and internet. This system is called GPRS (general packet radio system).

A brief description of HSCSD and GPRS is presented next.

26.2 HSCSD

In Section 20.6, we had seen that GSM uses TCH (traffic channels) to transmit user data. The data rates are 4.8 kbps (TCH/F 4.8) or 9.6 kbps (TCH/F 9.6) or 14.4 kbps (TCH/F 14.4). In HSCSD, higher data rates are achieved by aggregating several TCHs. An MS requests one or more TCHs from the GSM network, i.e., it requests allocation of several TDMA slots within a TDMA frame. The 'air interface user rate' (AIUR), i.e. the data rate made available to the user for transmission, depends on the number of TDMA slots aggregated. It may be noted that TDMA slot allocation can be asymmetrical, i.e., more slots can be allocated on the down-link than on the up-link (since users usually download more data than they upload).

To implement HSCSD, software upgradation is required in both MS and MSC since both should be able to split a traffic stream into several streams using a separate TCH for each, and to combine these streams again.

An MS can use all eight slots within a TDMA frame to achieve an AIUR of, say 8 TCH/F14.4 channels ($8 \times 14.4 = 115.2$ kbps). But, in this configuration the MS is required to send and receive at the same time. In GSM up-link and down-link, slots are always shifted for three slots. Therefore, in HSCSD, the AIUR available is 57.6 kbps (duplex) using four slots in the up-link and four in the down-link.

If we consider TCH/F 4.8, only one 4.8 kbps slot is available for HSCSD, while using TCH /F 14.4, it can be (*i*) three 4.8 kbps, one 14.4 kbps, or two 14.4 kbps combined to achieve 28.8 kbps, (*ii*) three 14.4 kbps combined to achieve 43.2 kbps, or (*iii*) four 14.4 kbps combined to achieve 57.6 kbps. The permitted combinations of traffic channels and allocated slots for non-transparent services are listed below.

AIUR	TCH/F4.8	TCH/F9.6	TCH/F14.4
4.8 kbps	1	–	–
9.6 kbps	2	1	–
14.4 kbps	3	–	1
19.2 kbps	4	2	–
28.8 kbps	–	3	2
38.4 kbps	–	4	–
43.2 kbps	–	–	3
57.6 kbps	–	–	4

Though HSCSD appears attractive due to the simple solution it offers for data traffic, there are the following disadvantages:

(*i*) It uses the connection-oriented mechanisms of GSM which are not efficient for computer data traffic which is typically bursty. Downloading a larger file keeps all channels busy, whereas web browsing would leave the channels idle most of the time.

(*ii*) How the channels are allocated is reflected directly in the service costs, as, once the channels have been reserved, other users cannot use them even though they may be idle.

(*iii*) For *n* channels HSCSD requires *n* times signalling during hand-over, connection setup and release. Since each channel is treated separately, this implies that the probability of blocking or service degradation increases during hand-over as the BSC checks for resources on *n* channels and not just one.

Thus, HSCSD is only an interim solution for higher bandwidth constant traffic, and not the answer for the requirements of higher rate infrequent traffic or lower rate frequent traffic.

26.3 GPRS

26.3.1 Introduction

From the discussion on HSCSD, it is clear that a fully packet-oriented system can provide a better solution for data communication. Packet mode transfer suits applications that exhibit traffic patterns such as frequent transmission of small volumes (e.g. typical web requests) or infrequent transmissions of small or medium volumes (e.g. typical web responses). GPRS (general packet radio service) has been designed with the following features to meet the above requirements:

(*i*) Compared to existing data transfer services, GPRS uses existing network resources more efficiently for packet mode applications.

(*ii*) It provides a selection of QoS (quality of service) parameters for the service requesters.

(*iii*) GPRS allows for broadcast, multicast, and unicast services.

(*iv*) Network providers can charge on volume and not connection time (GSM and HSCSD charges are based on connection time).

GPRS requires additional network elements (that are inclusive of software and hardware) to incorporate the above features. These elements called the GSNs (GPRS support nodes), are integrated into standard GSM architecture. The details are presented in Section 26.3.4.

The main concepts of GPRS, as stated by ETSI (European Telecommunications Standards Institute) in the document 'GPRS Service Description-Stage 2 : EN 301 344 V6.3.1' are:

- For new GPRS radio channels, the GSM system can allocate between one and eight time-slots within a TDMA frame. Time-slots are not allocated in a fixed, pre-determined manner but on demand. All time-slots can be shared by active users, up- and down- link are allocated separately. Allocation of the slots is based on current load and operator preferences. Depending on the coding, a transfer rate of up to 150 kbps is possible, e.g., allocating all time-slots using the coding for 14.4 kbps traffic channels results in a 115.2 kbps channel.
- GPRS system is independent of channel characteristics and the type of channel (traditional GSM traffic or control channel), and does not limit the maximum data rate (only the GSM transport system limits the rate). All GPRS services can be used parallel to conventional services.

26.3.2 Phases in GPRS implementation

GPRS implementation is in 2 phases:

Phase-I: GPRS offers a point-to-point (PTP) packet transfer service. In this again, two versions are possible:

- PTP connection-oriented network service (PTP-CONS), which includes the ability of GPRS to maintain a virtual circuit upon a change of cell within the GSM network. This type of service corresponds to X.25, the typical circuit-switched packed-oriented transfer protocol available worldwide.
- PTP connectionless network service (PTP-CLNS), which supports applications that are based on the Internet Protocol (IP).

Phase-II: Multicasting, called point-to-multi-point (PTM) service.

26.3.3 QoS profile for GPRS

Users of GPRS can specify a QoS (quality of service) profile. This profile determines the following parameters:

- Service precedence (high, normal, low)
- Reliability class of the transmission
- Delay class of the transmission
- User data throughput

GPRS adaptively allocates radio resources to fulfil these user specifications.

Table 26.1 shows the three reliability classes together with the maximum probabilities for a lost service data unit (SDU), a duplicated SDU, an SDU out of the original sequence, and the probability of delivering a corrupt SDU to a higher layer.

Reliability class 1: This is used for very error-sensitive applications that cannot perform error corrections themselves.

Reliability class 2: This is for applications that exhibit greater error tolerance.

Table 26. 1 *Reliability classes in GPRS*

Reliability class	Lost SDU probability	Duplicate SDU probability	Out of sequence SDU probability	Corrupt SDU probability
1	10^{-9}	10^{-9}	10^{-9}	10^{-9}
2	10^{-4}	10^{-5}	10^{-5}	10^{-6}
3	10^{-2}	10^{-5}	10^{-5}	10^{-2}

Table 26. 2 *Delay classes in GPRS*

Delay	SDU size 128 byte		SDU size 1024 byte	
class	Mean	95 percentile	Mean	95 percentile
1	< 0.5 s	< 1.5 s	< 2 s	< 7 s
2	< 5 s	< 25 s	< 15 s	< 75 s
3	< 50 s	< 250 s	< 75 s	< 375 s
4		Unspecified		

Reliability class 3: This is for error-insensitive applications or applications that can handle error corrections themselves.

Delay within a GPRS network is incurred due to channel access delay and transfer delays in the fixed and wireless part of the GPRS network. The delay introduced by external fixed networks is out of scope of GPRS network. However, GPRS does not produce additional delays since there is no buffering of packets as in the store-and-forward networks. GPRS tries to forward packets as fast as possible. Table 26.2 shows the specified maximum mean and 95 percentile delay values for packet sizes of 128 bytes and 1 k byte.

It can be seen that the delays are orders of magnitude higher than that for fixed network, irrespective of class. This should be given careful consideration when implementing higher layer protocols such as TCP on top of GPRS networks.

Finally, GPRS includes several security services such as authentication, access control, user identity confidentiality, and user information confidentiality. Even a completely anonymous service is possible, for example the service for road toll systems that only charges a user via the MS, independent of the user's identity.

26.3.4 New network elements in GPRS

As discussed in Section 26.3.1, the GPRS architecture introduces new network elements which are called GPRS support nodes (GSN). All GSNs are integrated into the standard GSM architecture, and many new interfaces have been defined (Fig. 26.1). The GSNs are of two types:

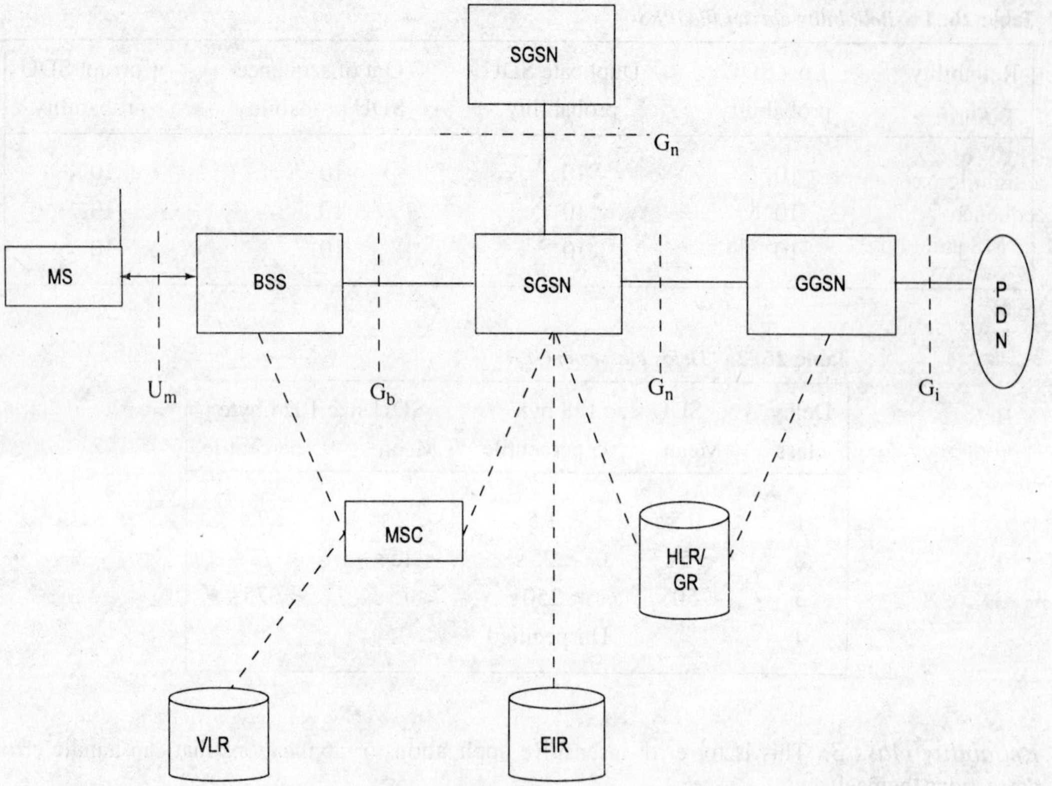

Fig. 26.1 *GPRS architecture.*

- GGSN (gateway GPRS support node)
- SGSN (serving GPRS support node)

The gateway GPRS support node (GGSN) is the inter-working unit between the GPRS network and the external packet data networks (PDN). This node contains routing information for GPRS users, performs address conversion, and tunnels data to a user via encapsulation. The GGSN is connected to external networks (e.g. IP or X.25) via the G_i interface and transfers packets to the SGSN via *n* ZIP-based GPRS backbone network (G_n interface).

The serving GPRS support node (SGSN) supports the MS via the G_b interface. The SGSN requests user addresses from the GPRS register (GR), keeps track of the individual MS's location, collects billing information and performs several security functions such as access control. The SGSN is connected to a BSC via frame relay and is basically on the same hierarchy level as an MSC. The GR, which is typically a part of the HLR, stores all GPRS-relevant data.

26.3.5 Packet transmission in GPRS

Figure 26.1 shows the transmission of packet data from a PDN via the GGSN and SGSN directly to the BSS and finally to the MS. The MSC, which is responsible for data transport in the traditional circuit-switched GSM, is only used for signalling in the GPRS scenario. Additional interfaces to further network elements and other PLMNs (public land mobile networks) can be found in ETSI Jocument 'GPRS Service Description-Stage 2 EN 301 344 V 6.3.1'.

Before sending any data over the GPRS network, an MS must attach to it (i.e., get associated with the network), following the procedures of the mobility management. The attachment procedure includes assigning a temporal identifier called a temporary logical link identity (TLLI), and a ciphering key sequence number (CKSN) for data encryption. For each MS, a GPRS context is set up and stored in the MS and in the corresponding SGSN. This context comprises:

- the status of the MS (which can be ready, idle, or standby),
- the CKSN, a flag indicating if compression is used, and
- the routing data (TLLI, routing area (RA), cell identifier, packet data channel (PDCH), identifier).

Besides attaching and detaching, mobility management also comprises functions for authentication, location management, and ciphering. In GRPS, ciphering involves a lot of transactions between MS and SGSN and its scope is more than that of GSM.

26.3.6 Protocol architecture of GPRS

Figure 26.2 shows the protocol architecture of the transmission plane (the information plane) for GPRS. Architectures for the signaling planes (control plane) can be found in ETSI document 'GPRS Service Description- Stage 2 : EN 301 113 V 6.3.1'.

(*i*) All data within the GPRS backbone, i.e. between the GSNs, is transferred using the GPRS tunnelling protocol (GTP).

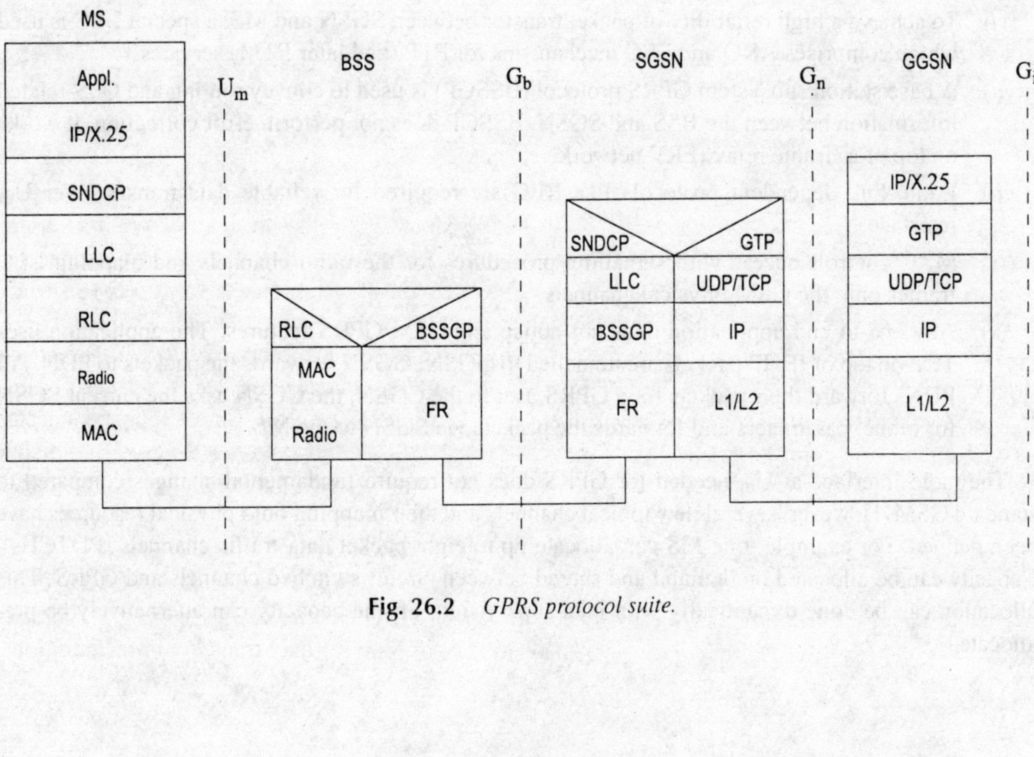

Fig. 26.2 *GPRS protocol suite.*

(*ii*) GTP can use two different transport protocols, either the reliable TCP (needed for reliable transfer of X.25 packets) or the non-reliable UDP (user datagram protocol) used for IP packets.

Note: UDP is a TCP/IP transport layer protocol that supports unreliable, connectionless communication between hosts on a TCP/IP network. It is a connectionless protocol that does not guarantee delivery of data packets between hosts. It differs from its companion transport layer protocol, the TCP (transmission control protocol), which is a connection-oriented protocol for reliable packet delivery. UDP offers only "best effort delivery" service and is used for both one-to-one and one-to-many communications where small amounts of data are exchanged between hosts. UDP is used by applications and services that do not require acknowledgements. It can transmit only small portions of data at a time because it is not capable of segmenting and re-assembling frames and does not implement sequence numbers.

UDP is used for services that perform broadcasts. These broadcasts can be directed to one of the following:

- All hosts on the network
- Hosts on a specific network
- Hosts on a specific sub-net

(*iii*) The network protocol for the GPRS backbone is IP (using any lower layers).

(*iv*) In order to adapt to the different characteristics of the underlying networks, the sub-network dependent convergence protocol (SNDCP) is used between an SGSN and the MS.

(*v*) On top of SNDCP and GTP, user packet data is tunnelled from the MS to the GGSN and vice versa.

(*vi*) To achieve a high reliability of packet transfer between SGSN and MS, a special LLC is used which comprises ARQ and FEC mechanisms for PTP (and later PTM) services.

(*vii*) A base station subsystem GPRS protocol (BSSGP) is used to convey routing and QoS-related information between the BSS and SGSN. BSSGP does not perform error correction. It works on top of a 'frame relay (FR)' network.

(*viii*) Radio-link dependent protocols like RLC are required for reliable data transfer over U_m interface.

(*ix*) MAC controls access with signalling procedures for the radio channels and mapping LLC frames onto the GSM physical channels

(*x*) Any end-to-end application does not notice any GSM/GPRS features. The application uses TCP on top of IP; IP packets are tunnelled to GGSN; GGSN forwards the packets to PDN. All PDNs forward their packets for a GPRS user to the CGSN, the CGSN asks the current SGSN for tunnel parameters and forwards the packets via SGSN to the MS.

The radio interface at U_m needed for GPRS does not require fundamental changes compared to standard GSM. However, several new logical channels and their mapping onto physical resources have been defined. For example, one MS can allocate up to eight packet data traffic channels (PDTCHs). Capacity can be allocated on demand and shared between circuit-switched channels and GPRS. This allocation can be done dynamically with load supervision, or, the capacity can alternatively be pre-allocated.

26.4 DECT

26.4.1 Introduction

The capabilities of analog cordless telephone systems are limited to the following:

- They allow roaming over utmost few hundreds of meters
- If the range is to be increased the transmission power has to be increased which brings in newer problems like interference, licensing
- They are not suitable for data transmission
- Channel capacity is low
- Security is ensured to a very limited extent

Hence, a new 'digital enhanced cordless telecommunications (DECT)' system has been specified by ETSI in 1999 (DECT Forum, 1999), as replacements for older analog cordless phone systems. The reader might wonder why again another standard when GSM exists. The following points dispel such doubts:

- Cordless telephony fits that segment of mobile communication where there is low degree of customization and high traffic density; cell sizes are in hundreds of meters and user density very high. GSM, on the other hand, with its medium to high degree of customization, is designed for outdoor use with a cell diameter of up to 70 km. The range of DECT is limited to about 300 m from the base station (only around 50 m are feasible inside buildings, depending on the thickness of the walls). Due to this limited range, with additional multiplexing techniques, DECT can offer its service to some 10,000 people within one sq.km. This is a typical scenario in a big city, say the zonal headquarters of the Indian Railways, or the State Government secretariats in India where hundreds, sometimes even a few thousands of employees are located in skyscrapers or closely clustered buildings.
- Access points to the PSTN can be established within railway stations, large government buildings and hospitals, offering a much cheaper telephone service than the GSM system.
- DECT solves the 'last mile' problems of new network operators. Using a 'small range' local loop, new companies can offer their service without installing their own lines in the streets.
- DECT systems offer different inter-working units, e.g. with GSM, ISDN, or data networks. Currently, more than 20 million DECT units are in use, so that the system holds more than 30 per cent of all wireless local loop (WLL) installations.
- DECT also uses base stations, but these base stations together with the mobile station are priced 100 times less than a GSM base station. DECT can also handle hand-over. (It may please be noted that DECT was not designed to work at a higher speeds like that of mobiles, i.e. like GSM systems up to 250 kmph.)

Thus the development of a new digital cordless system with enhanced capabilities is fully justified.

26.4.2 Data transmission in DECT

DECT works in the frequency range of 1880–1990 MHz offering 120 full duplex channels. Time division duplex (TDD) is applied using 10 ms frames. The frequency range is subdivided into 10 carrier frequencies using FDMA, each frame being divided into 24 slots using TDMA. For the TDD mechanism, 12 slots are used as up-link, 12 slots as down-link. The digital modulation scheme used is GMSK — each station has an average transmission power of only 10 mW with a maximum of 250 mW.

26.4.3 System architecture of DECT

Depending on the actual use, DECT has different physical implementation variants. Different DECT entities can be integrated into one physical unit, entities can be distributed, replicated etc., However, all implementations are based on the same logical reference model of the system architecture as shown in Figure 26.3. A global network connects the local communication structure to the outside world

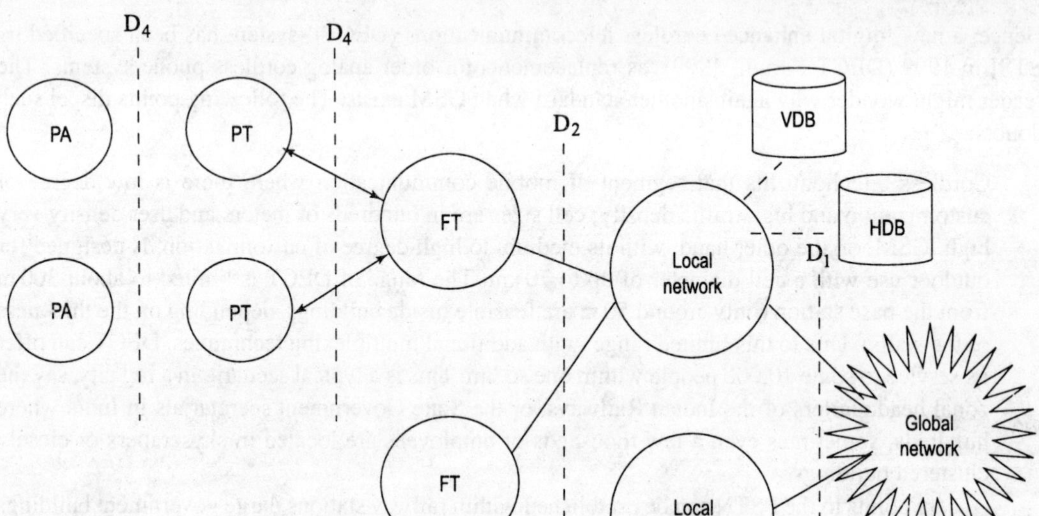

Fig. 26.3 *DECT architecture.*

and offers its services via interface D1. Global networks could be integrated services digital networks (ISDN), public switched telephone networks (PSTN), or public land mobile networks (PLMN), e.g. GSM, or packet switched public data network (PSPDN). The services offered by these networks include transportation of data on one hand and the translation of addresses and routing of data between the local networks on the other.

Local networks in DECT offer services like call forwarding, and address translation. These networks can be PBXs or IEEE 802.X LANs.

The database comprises of 'home database' (HDB) and 'visitor database' (VDB) to support mobility similar to HLR and VLR in GSM.

The DECT core network consists of 'fixed radio terminals' (FTs) and 'portable radio terminals' (PTs). FT covers layers 1 to 3 on the fixed network side, whereas RT covers the same layers on the mobile network side.

26.4.4 Protocol architecture in DECT

DECT protocol suite can be well understood by viewing it at 'control plane (C-plane)' and 'user plane (U-plane)'. In C-plane, we have an additional layer compared to OSI so that the user data from layer-2 can be forwarded directly to the U-plane. The management layer covers all the layers of the U- and C- planes. (Fig. 26.4 depicts the ETSI specified protocol suite for DECT. It has, for the C-plane the physical, data link and network layers, and for the U-plane just the physical and the data link layers.

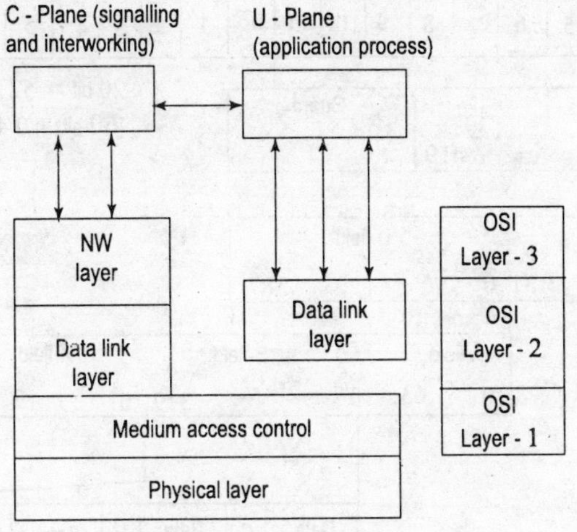

Fig. 26.4 *DECT protocol suite.*

Physical layer

It performs the functions of modulation/demodulation, incoming signal detection, synchronization, etc. It guarantees certain throughput. It assigns a channel for data transmission, on request by MAC layer.

The DECT frame is shown in Fig. 26.5. Following are its features:

- Each frame has 10 ms, duration
- It has TDMA structure with 12 slots for down-link and 12 for up-link in the basic connection mode.
- If the mobile receives data in slot 's', it sends data in slot 's + 12'.
- Each slot has 0.4167 ms duration, and has several different physical packets.
- In each slot, 420 bits are for data inclusive of the 32-bit synchronization pattern. The rest of 52 μs is for guard space.
- The 388 bits of actual data field (420 − 32 = 388) have three fields:
    - A field for network control
    - B field for user data
    - X field for transmission quality parameters
- Data pertaining to network control is transmitted at a data rate of 6.4 kbps
- User data requires additional error correction mechanisms depending on the mode of transmission (protected/unprotected) and simplex/duplex communication. Four combinations arise here:

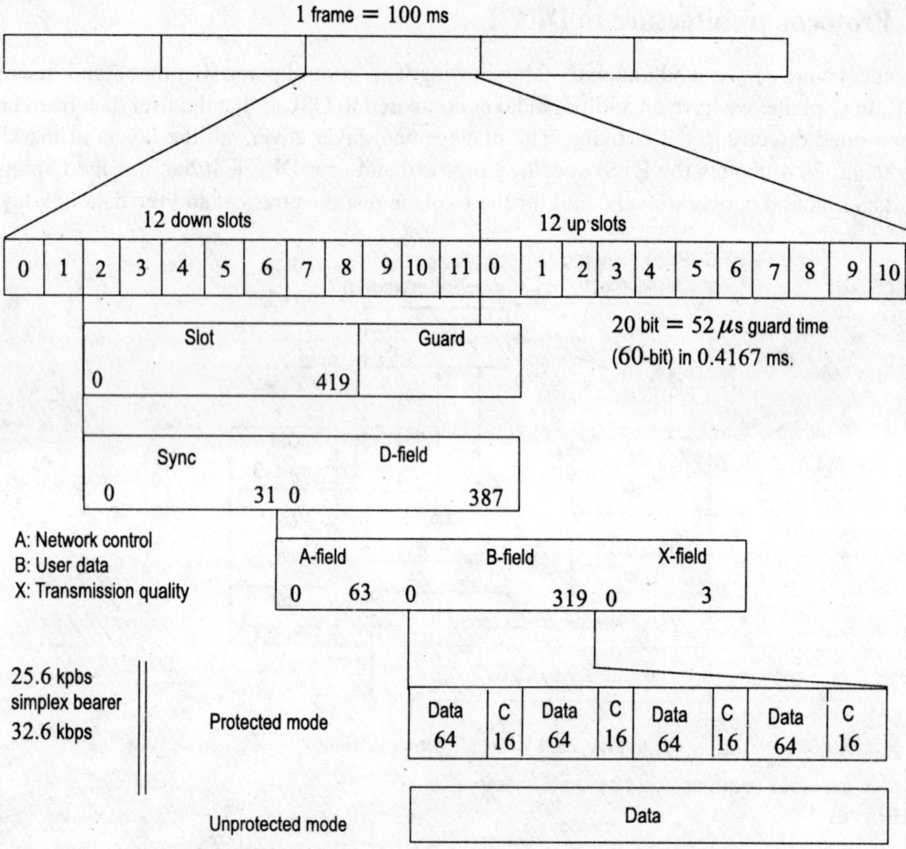

Fig. 26.5 *DECT frame structure.*

- Simplex bearer provides 32 kbps data rate in unprotected mode
- Simplex bearer provides 64 kbps with 16 bit CRC for every 64 bit block (i.e., the effective data rate is 25.6 kbps) in protected mode
- Duplex bearer combines two simplex bearers
- The double duplex bearer offers 80 kbps full-duplex

Medium access control layer

The medium access control (MAC) layer establishes, maintains, and releases channels for higher layers by activating and deactivating physical channels. MAC multiplexes several logical channels onto physical channels. Logical channels exist for signalling network control, user data transmission, paging, or sending broadcast messages. Additional services offered include segmentation/re-assembly of packets and error control/error correction.

Data link control layer

This layer is for creating and maintaining reliable connections between the mobile terminal and the base station. Two services have been defined for the C-plane:

- A connectionless broadcast service for paging (called Lb)
- A point-to-point protocol similar to LAPD in ISDN, but adapted to the underlying MAC (called LAPC+Lc).

The data services offered across this layer are:

- A transparent unprotected service (basically a null service)
- A forward error correction service
- Rate adaptation services
- Services for future enhancements

If services like ISDN data transfer at 64 kbps are used, DECT tries to transfer 64 kbps. However, in case of errors, DECT raises the transfer rate to 72 kbps, and includes FEC and a buffer for up to eight blocks to perform ARQ. This buffer then introduces an additional delay of up to 80 ms.

Network layer

The network layer of DECT is similar to those in ISDN and GSM and only exists for the C-plane. The services provided by this layer are:

- Resource management (request, check, reserve, control, and release resources) at the fixed station (connection to the fixed network on wireless connection) and the mobile terminal (wireless connection)
- Mobility management (MM) which involves identity management, authentication, and management of the location databases
- Call control (CC) which handles connection setup, release, and negotiation
- Message services (connection-oriented message service (COMS) as well as connectionless message service (CLMS)) which transfer data to and from the inter-working unit that connects the DECT system with the outside world

26.4.5 Future of DECT

DECT has its berth ensured in the III-generation mobile communication systems. Its frequency band is part of IMT-2000. ITU confirmed in 1999 that DECT fulfils the requirements of IMT-2000 and also adopted DECT as a member of IMT-2000 family. At present, over 45 million terminals are in use, and the number would go up to 200 million terminals in the next 4 years. 110 countries have accepted DECT as their standard for digital cordless systems.

26.5 UMTS

26.5.1 The third generations mobile communication system

We have seen how different mobile communication systems were developed to suit different needs like indoor use, for vehicles, pedestrians and even satellites. Perceived chronologically, there are three generations of mobile communication systems:

- First generation systems are the analog cellular phone systems like advanced mobile phone service (AMPS)

- Second generation systems are digital cellular systems like digital advanced mobile phone service (D-AMPS), time division multiple access (TDMA) systems, code division multiple access (CDMA) systems and GSM

Though the II-generation systems had solved the problems of analog cellular systems, there was the problem of fundamental inoperability between various types of digital cellular technologies. This has prompted ITU to take the initiative for creating a global standard in the new millenium that permits terminal and user mobility. The ITU wanted to usher in the III-generation (3G) mobile communication system supporting 'universal personal telecommunication' (UPT) and 'wireless data network'. The program was named as 'future public land mobile telecommunication system' (FPLMTS) and later became popular as IMT-2000. The 2000 does not stand for year 2000 but stands for data rates of 2000 kbps or 2 Mbps.

The ITU made a request for proposals for radio transmission technologies (RTT) for the International Mobile Telecommunications (IMT) 2000 program. To achieve the goal of UPT, ITU has created several recommendations for systems such as

- network architectures for FPLMTS (M.817),
- requirements for the radio interface(s) for FPLMTS (M. 1034), and
- framework for services supported by FPLMTS (M.816).

The WRC (Wireless Regulatory Commission) in 1992 identified 1885–2025 and 2110–2200 MHz as the frequency bands that should be available worldwide for the new IMT-2000 systems (recommendation ITU-R.M. 1036).

26.5.2 IMT-2000 set of standards

The European Telecommunications Standards Institute (ETSI), which was also toying with ideas of universal system for mobile communications, came-up with a proposal for IMT-2000 known as Universal Mobile Telecommunications System (UMTS). The specific proposal for the radio interface, i.e., radio transmission technology (RTT), is the UMTS terrestrial radio access (UTRA). UMTS is proposed by ETSI as an evolution from the second generation GSM system to the third generation.

In this way, many solutions have been proposed for a smooth transition from different systems to IMT-2000 in order to be able to save money by extending the current system rather than introducing a new one. Finally, ITU proclaimed the following standards, known as IMT-2000 set of standards:

- CDMA 2000 : A proposed hardware upgrade to the existing CDMA cellular phone systems used in USA and some parts of Asia
- W-CDMA : A proposed upgrade to GSM networks that provide alternative interface to CDMA 2000 for upgraded CDMA systems in Europe and Japan
- UMTS : A standard for third generation wireless networks proposed by ETSI
- Universal wireless communication (UWC-136) A standard for third generation wireless network proposed by the Telecommunications Industry Association (TIA)

The upgrade for CDMA bearers is simple and straight which moves them from the current CDMA1 (CDMA development group's standard that suports data transmission at 16 kbps) to CDMA 2000.

The upgrade for TDMA bearers has the following options:

- TDMA to GPRS to UMTS
- GSM to EDGE to UMTS

The enhancement of GSM towards UMTS is via EDGE (enhanced data rates for global evolution). EDGE uses enhanced modulation schemes and other techniques for data rates of up to 384 kbps using the same 200 kHz wide carrier and the same frequencies as GSM (i.e., a data rate of 48 kbps per time-slot is available). EDGE can be introduced incrementally, offering some channels with EDGE enhancement that can switch between EDGE and GSM.

Besides enhancing data rates, new additions to GSM like CAMEL (customized application for mobile enhanced logic) introduce intelligent network support. This system supports, for example, the creation of a virtual home environment (VHE) for visiting subscribers. The GSM Memorandum of Understanding (1999) provides many proposals for covering QoS aspects, roaming, services, billing, accounting, radio aspects, core networks, access networks, terminal requirements, security, application domains, operation and maintenance, and several migration aspects.

26.5.3 Data rates and services provided by UMTS

The data rates provided by UMTs depend on radio transmission technology (RTT). As already mentioned, ETSI's specific proposal for RTT concerning UMTS is termed UTRA (UMTS terrestrial radio access). Its key features are:

- Minimum data rates of 144 kbps for rural outdoor access (with the goal of 384 kbps) at a maximum speed of 500 km/h
- For suburban outdoor use a minimum of 384 kbps with the goal of 512 kbps at 120 kmph
- For indoor or city use with relatively short ranges, upto 2 Mbps are required at 10 kmph (walking)

Coming to services, UMTS should provide several bearer services, real-time and non real-time services, circuit and packet switched transmission, and many different data rates. Hand-over should be possible between UMTS cells, and also between UMTS and GSM or satellite networks.

The system should be compatible with GSM, ATM, IP, and ISDN-based networks. In order to reflect the asymmetric bandwidth needs of typical users, UMTS should provide a variable division of up-link and down-link data rates.

As the global UMTS approach is rather ambitious, a more realistic alternative for the initial stages world be UMTS cells in cities providing a subset of services.

26.5.4 UMTS basic architecture

Several companies and interest groups had submitted proposals for UTRA of which ETSI selected two:

- For the paired band (using FDD (frequency domain duplex) as duplex mechanism) — W-CDMA (wide-band CDMA)
- For the unpaired band (using TDD as duplex mechanism) — TD-CDMA (time division CDMA)

The paired band is typically used for public mobile network providers (wide area like GSM), while the unpaired band is used for local and indoor communication (like DECT). These are discussed in detail in the next two sections. Let us turn our attention to the architecture of UMTS.

Figure 26.6 shows a simplified UMTS architecture which applies to both UTRA solutions. The UTRA network (UTRAN) handles cell level mobility and comprises several radio network subsystems (RNS). The functions of the RNS include radio channel ciphering and deciphering, hand-over control, radio resource management, etc. The UTRAN is connected to the user equipment (UE) via the radio interface U_u (which is similar to the U_m interface in GSM). UTRAN communicates with the core network (CN) via the I_u interface (which is similar to the A interface in GSM). The CN contains functions for inter-system hand-over, gateways to other networks (fixed or wireless), etc. The functions of I_u have been defined but not yet standardized.

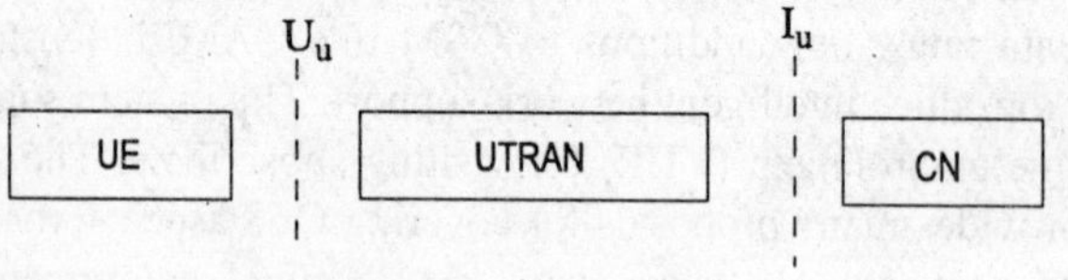

Fig. 26.6 *UMTS architecture (simplified).*

To separate signaling and user data, UMTS introduces a C-plane (control) and U-plane (user data). The architecture includes the following three lower layers:

- Layer-1 at the radio interface comprises functions for FEC, signal level measurements, multiplexing and de-multiplexing, modulation, spreading, synchronization, and power control.
- Layer-2 is responsible for MAC connections, medium access control, priority handling, contention resolution, and the scheduling of data packets.
- Layer-3 performs all control functions for connection setup.

26.5.5 UTRA FDD mode

Radio link aspects of UTRA FDD mode

The FDD mode for UTRA uses wideband CDMA (W-CDMA) with direct sequence spreading. As implied by FDD, up-link and down-link use different frequencies, and hence the name paired band. A mobile station up-links using a carrier between 1920 MHz and 1980 MHz. The base station uses 2110–2170 MHz for down-link. This mode roughly provides 250 channels for user traffic, e.g., voice channels. Each physical channel corresponds to a carrier frequency, a certain spreading code (DS), and a relative phase (up-link only), and carries up to 2 Mbps data.

Physical and logical channels in UTRA FDD Mode

Similar to GSM, several physical and logical channels have been defined:

- On the up-link, user data (from layer two and higher) is transported over the up-link dedicated physical data channel (up-link DPDCH).
- Layer-1 control data (e.g. for power control) is transported via the up-link dedicated physical control channel (up-link DPCCH).
- Both channels are transmitted in parallel, DPCCH on the Q branch of the signal (quadrature component of carrier), DPDCH on the I branch of the signal (in-phase component of the carrier). For both channels, different spreading factors are possible.

- The physical random access channel (PRACH) carries control data from a mobile station for random access. It uses slotted Aloha with 1.25 ms slots.
- A common broadcast channel carries information to indicate which slots are available for random access within a cell.
- On the down-link, the down-link dedicated physical channel (down-link DPCH) carries user data and control data from Layer-1 in a time multiplexed fashion.
- UTRA provides for additional common physical channels for radio control.

Frame Structure for data transmission in UTRA FDD mode

Figure 26.7 shows the frame structure of UTRA in FDD mode and the selected channels. A superframe with a duration of 720 ms consists of 72 frames. A frame contains 16 slots and has a duration of 10 ms. The example shows three different channels that can use a slot. The up-link DPCCH contains a pilot to support channel estimation, transmit power control (TPC), and the optional transport format identifier (TFI). The up-link DPDCH, which is transmitted at the same time as the up-link DPCCH, contains user data. The down-link DPCH contains the same fields (i.e., pilot, TPC, TFI, and data), but in a time-multiplexed version in one slot.

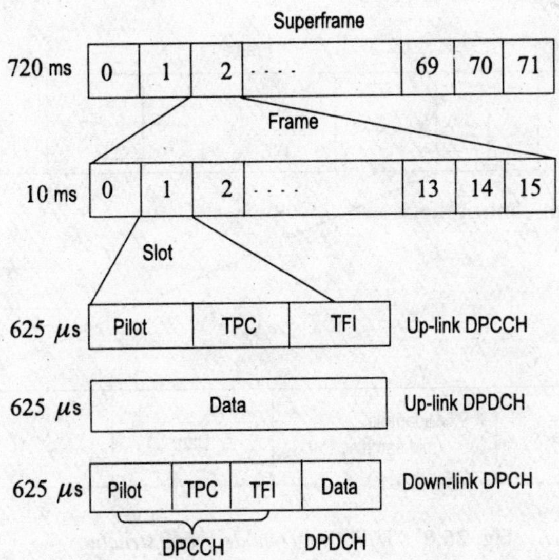

Fig. 26.7 *Frame structure of UTRA FDD mode.*

The bit rate used in this scheme is 4.096 Mbps and can be extended to 8.192 Mbps and 16.384 Mbps for future applications. For modulation of data bits, QPSK is used. Due to the use of CDMA, soft hand-over is possible between different base stations. A hand-over between different sectors within the same base station is termed 'softer hand-over'.

Furthermore, the CDMA scheme allows for localization of mobile stations with a maximum accuracy of approximately 18 m. However, CDMA needs a more complex power control scheme with 1,600 power control cycles per second. Sender and receiver operate continuously during a call.

26.5.6 UTRA TDD mode

Radio link aspects of UTRA TDD mode

The TDD mode of UTRA uses wideband TDMA / CDMA for medium access, and in this mode up-link and down-link use the same frequency (hence the name unpaired band). This mode also offers 2 Mbps for approximately 120 orthogonal channels for user traffic, which is roughly half the capacity of the FDD mode, but the TDD scheme only requires half the bandwidth.

Channels in UTRA TDD mode

Physical channels are represented by a slow and spreading code (direct sequence) in TDD mode. Typically, the system can transmit eight bursts from different channels within a single slot.

Frame structure of UTRA TDD mode

Figure 26.8 shows the frame structure and 'burst structure' (denoting a traffic burst within a slot) of UTRA TDD mode frame. Each frame has 16 time-slots and each slot has traffic burst. This burst contains (*a*) user data, (*b*) a 'mid sample' for channel estimation, and (*c*) a guard period (GP) to avoid interference between different slots (23.4 μs).

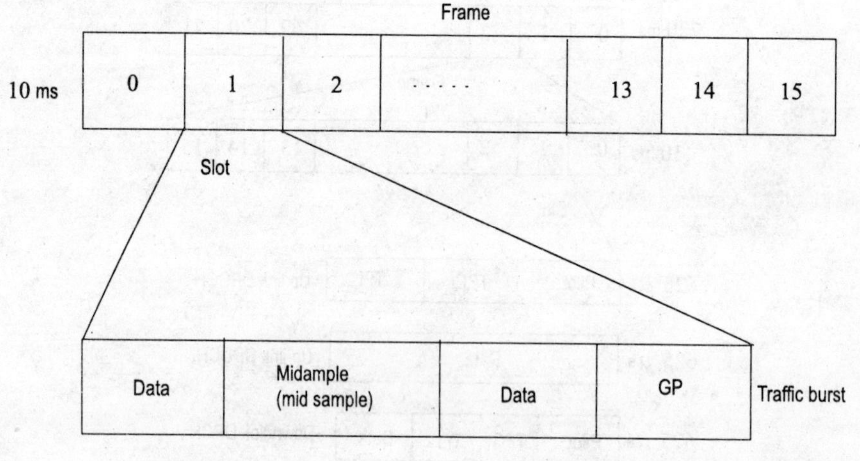

Fig. 26.8 *UTRA TDD mode frame structure.*

Each slot has a duration of 625 μs and contains 2,560 bits. This implies a data rate of 4.096 Mbps (2560/625 μs). Since 16 slots make a frame, the duration of a frame is 10 ms which is same as the frame duration in FDD mode.

The TDD frame can be symmetrical or asymmetrical, i.e., the frame can contain the same number of up-link and down-link slots or any arbitrary combination so that different data rate requirements are met with in an optimum way. The system can even change the spreading factor as a function of the desired data rate.

The TDD frame can have either one switching point from up-link to down-link or several switching points. However, at least one slot must be allocated for the up-link and down-link respectively.

The modulation scheme of TDD mode is QPSK. The power control can be slower compared to FDD (only 100 to 800 power control cycles per second are needed). Sender and receiver only operate in allocated slots, i.e., they do not operate continuously.

Due to the time division multiplexing scheme, tight synchronization is required within a small area (several cells) to ensure successful hand-over. The TDD mode allows hand-over for GSM-like situations only.

Due to similar frame structure and same frame duration of 10 ms, hand-over between FDD and TDD is feasible. So, both FDD and TDD schemes can coexist. Furthermore, the multiframe structure specified for the UTRA modes allows for a seamless hand-over to GSM systems.

Exercises

1. What are the problems in data transmission over a cellular network designed for voice transmission?
2. What are the specific advantages of (*a*) DECT, (*b*) TETRA and (*c*) UMTS?

References

[1] ETSI (1997a), High speed circuit switched data(HSCSD), Stage 1, ETSI-GSM Rec. 02.34, V5.2.1.

[2] ETSI (1998a), General packet radio service (GPRS), Requirements specification of GPRS, ETSI spec. TR 101 186.

[3] ETSI (1998a), General packet radio service (GPRS), Service description, Stage-2, ETSI spec. EN 301 113.

[4] ETSI (1998b), General packet radio service (GRPS), Service description, Stage-2, ETSI spec. EN 301 344.

[5] ETSI (1998d), General packet radio service (GPRS), Overall description of the GPRS radio interface, Stage-2, ETSI spec. TS 101 350.

[6] ETSI (1998e), High speed circuit switched data (HSCSD), Stage-2, ETSI spec. TS 101 038.

[7] ETSI (1998f), Universal mobile telecommunication system (UMTS), Concept groups for definition of UMTS and UTRA, ETSI spec. TR 101 397.

[8] ETSI (1998g,h), UMTS-High level requirements relevant for the definition of the UMTS terrestrial radio access (UTRA) concept, ETSI spec. TR 101 398.

[9] ETSI (1998i), Concept groups for the definition of the UMTS terrestrial radio access (UTRA), ETSI spec. TR 101 397.

[10] ETSI (1998k), Digital enhanced cordless telecommunications (DECT), Common interface (CI), ETSI spec. EN 300 175.

[11] ETSI (1998j), DECT – Generic access profile (GAP), ETSI spec. EN 300 444.

[12] ETSI (1998n), The ETSI UMTS terrestrial radio access (UTRA) ITU-R radio transmission Technologies (RTT).

[13] http://www.umts-forum/com/

[14] http://www.tetramou.com/

[15] http://www.itu.int./imt/

27

Terrestrial Trunked Radio (TETRA)

27.1 Introduction

In Chapter 2, we saw that TETRA (terrestrial trunked radio) is for the 'high degree of customization–low traffic density' segment of mobile communications. TETRA has its origins in PMR (private mobile radio), the doyen of mobile radio communications.

On hearing the term PMR, the reader might think that this is yet again a new mobile communication system of recent times. The reader might find it difficult to believe that the first commercial PMR system was deployed in the UK in 1947!

Yes, it is true! PMR was developed during second world war for military applications but was licensed for the first ever commercial application in 1947 to Camtax, a taxi company in Cambridge, UK. The equipment were made by Pye Telecommunications (now Simoco International). PMR was based on conventional operation, i.e. dedicated communication channels to user groups and manual selection of channels from the available set.

Cellular telephony arrived on the scene in 1980s with the advent of trunking technology with features of automated channel selection and frequency sharing among users. Mobile stations had become handy and the entire system was spectrally efficient leading to the storming of markets by cellular telephony, the best known and the most widely accepted one among them being GSM.

However, professional user groups like military, police, transport sector, etc. were reluctant to jump on to the GSM bandwagon for the following reasons:

- Slow call set-up (few seconds in GSM)
- Risk of loss of terminal autonomy, i.e., loss of communication in areas not covered by the network or loss when the system failed
- GSM offers reasonable confidentiality over radio path with its encryption schemes, but cannot guarantee strict confidentiality to professional user groups in a system commonly shared with the public
- For SOS messages, direct terminal-to-terminal communication is preferred instead of going via trunking network

Basically, the above user groups were looking for highly customized solution for their mobile communication requirements. TETRA with the following features is the solution for them.

- It offers the benefits of automatic operation and frequency efficiency of trunking and also retains the terminal autonomy of conventional PMR

- It offers a novel multi-mode capability by combining trunking and direct modes of working on single terminal
- It also provides inter-working between direct mode and trunking system

Thus, TETRA provides the best of both the worlds namely,

- spectral efficiency, calls on demand of trunking system,
- terminal autonomy, direct terminal-to-terminal mode of conventional system.

In this chapter, we will study the architecture of TETRA, the direct mode of TETRA, provision for data transmission in TETRA, security features of TETRA, and the distinctive features of TETRA with respect to GSM.

27.2 TETRA Architecture

We have a proven architecture for GSM. Then why not modify the cell diameter and some other elements in GSM to suit TETRA? This is not as straightforward a process as it might seem. The architecture of mobile communication system is determined by the traffic profiles.

27.2.1 Traffic profiles of GSM and TETRA

Traffic profiles of GSM and TETRA are quite different. With a typical call duration of two minutes, GSM traffic profiles show that 85 per cent of all mobile calls are made via PSTN, 20 per cent of which are PSTN initiated and 80 per cent mobile initiated. A further 5 per cent of all calls are from mobile to mobile between different regions, and only 10 per cent of all calls are mobile to mobile calls within the same region.

In contrast, in a private mobile radio system (to be implemented as TETRA) a typical call duration is only 20 seconds. Ninety per cent of all calls are made within the same region and out of these 60 per cent are initiated by a control dispatcher. A further 5 per cent of all calls are between two regions and only 5 per cent via PSTN, of which less than 1 per cent are initiated from PSTN.

Another striking difference is that while GSM calls are point-to-point (individual calls), TETRA calls are a mixture of point-to-point and point-to-multipoint (group calls), with a predominance of group calls.

27.2.2 TETRA architecture

To suit the above mentioned traffic profiles of GSM, the system transmission architecture is,

- highly hierarchical,
- optimised for centralised delivery, and
- call is validated on each transaction at the regional and central level.

The consequences of such an architecture are long call setup times, high transmission costs, and low inherent resilience, as single system failure can cause major disruption in services. This in turn demands sophisticated duplication and back-up, with consequential effects on system costs.

By contrast, TETRA (i.e., a PMR system to be implemented as TETRA) which is optimised for predominance of local traffic, needs the following features to be incorporated in its transmission architecture:

- Distributed intelligence
- Optimisation for local delivery
- Call validation to be performed per session.

This requires the base site equipment of TETRA to be more intelligent than that of GSM—it shall be capable of setting up a call between two users without the involvement of the regional centre; it shall have the functionality and interfaces required for a single site system.

Considering the above architectural differences arising out of diverse traffic profiles and also the facts that (*i*) cellular frequency plan is the only one available, (*ii*) the cell sizes in GSM are relatively smaller than in PMR, (*iii*) a single GSM channel is 200 kHz wide, (*iv*) number of users in PMR for a region is usually below 10,000, it can be concluded that GSM architecture cannot be 'scaled down' economically to suit TETRA. We need a distributed architecture for GSM, instead.

The consequences of such an achitecture are low transmission costs, fast call setup times and inherent system resilience. Please refer to Fig. 27.1 showing a typical TETRA network. The TETRA network comprises of the following entities:

- Base site equipment (BS)
- Mobile station (MS)
- Switches
- Switching and management infrastructure (SwMI)
- Standard interfaces

The standard interfaces to (discussed in detail in Section 27.4) are marked 1 to 5 in Fig. 27.1. The gateway, i.e. TETRA-to-public network-interface is depicted as '6.'

- 1 V+D air interface ('a' : trunked mode, 'b' : direct mode)
- 2 Inter system interface (ISI)
- 3 Peripheral equipment interface (PEI)
- 4 Line station interface (LSI)
- 5 Network management interface (NMI)
- 6 Gateway to public networks

These network entities are discussed in Section 27.5.

27.2.3 Base site equipment (BS)

The distributed architecture of TETRA requires the base site equipment to be more intelligent than its counterpart in GSM. It should be capable of locally setting up a call between two users on a site without referral to the regional node. Furthermore, such architecture is highly scaleable. The base site equipment represents the simplest variant of the TETRA infrastructure, and given the functionality and the necessary interfaces, it contains all that is required for a complete single site system.

27.2.4 Mobile station (MS)

TETRA MS is similar to GSM MS but with capabilities of terminal-to-terminal contact.

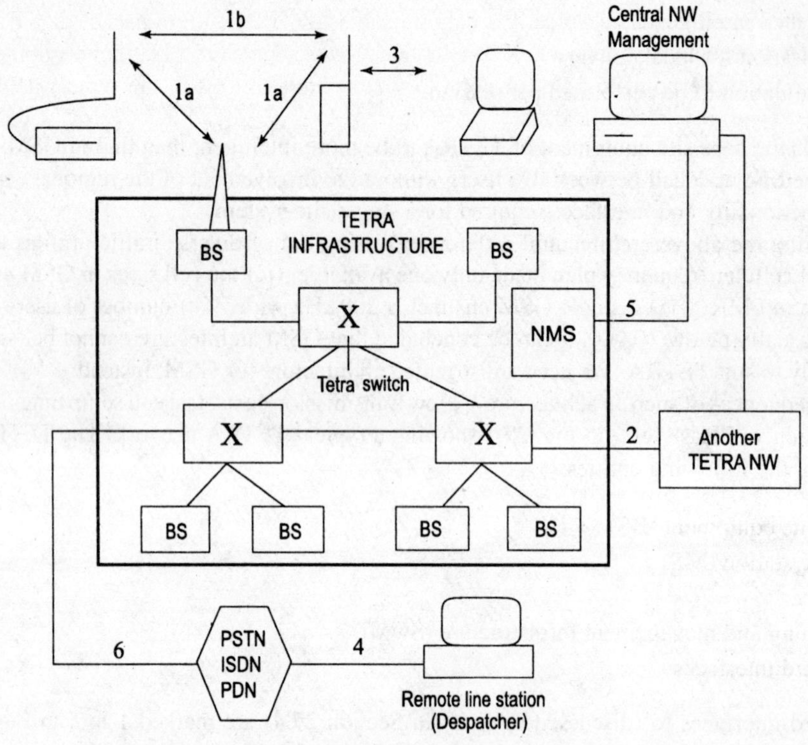

Fig. 27.1 *TETRA architecture.*

27.2.5 TETRA switches

The network switch for a TETRA system needs to be different in several aspects from its GSM counterpart. Because of the need for faster call setup, the call switching and management is devolved to the lowest possible point in the network, i.e., the switch. In addition to the distributed system architecture, distributed databases are needed in making the network an entity of loosely coupled autonomous parts.

Furthermore, TETRA switch, unlike its GSM counterpart, is a parallel non-blocking core switch with support for point-to-multipoint (group) calls, so that when necessary, it can provide immediate access to all communications channels.

27.2.6 SwMI

The mobile station connects to the SwMI (switching and management infrastructure) via the radio interface U_m. The SwMI contains the user databases (HDB, VDB), the base station and the interface to PSTN, ISDN or PDN. The implementation is simpler since there is no hand-over involved.

27.3 TETRA Services

Basically, TETRA offers two types of services viz., (*i*) voice + data (V+D) service, (*ii*) packet data optimized (PDO) service.

V+D offers circuit-switched voice and data transmission. PDO offers packet data transmission, either connection-oriented to connect to X.25 or connectionless for ISO CLNS (connectionless network service). CLNS can be point-to-point or point-to-multi-point. In CLNS, the typical delay for short messages (128 byte) is less than 100 ms.

V+D mode comprises unicast and broadcast connections, and group communications within a certain protected group. However, delays for short messages can be up to 500 ms or higher, depending on the priority.

The PDO standard was completed by the Working Group-4(WG4) set-up by ETSI and was published by ETSI in May 1996. PDO is basically for narrowband operation. Later, the same WG4 have worked on a broadband version called DAWS (digital advanced wireless service) which is in the early stages of standardization. Hence the emphasis in this chapter is on V+D and PDO.

In V+D again, there are two modes of operation:

- Trunked mode (TM)
- Direct mode (DM)

Trunked mode is the main mode of operation of the V+D TETRA network. In this mode, a call is handled by the switching and management infrastructure (SwMI), referred to as 'infrastructure' henceforth. The infrastructure makes use of the trunking gain plus queuing and routing techniques to handle a larger number of subscribers than a non-trunked network.

The direct mode air interface is designed for direct communication between mobile terminals within or outside the infrastructure coverage without using the infrastructure. Direct mode is discussed in detail in Section 27.5.

27.4 Data Transmission in TETRA

27.4.1 Radio frequency and bandwidth

Several frequencies have been specified for TETRA which uses FDD (e.g. 380–390 MHz uplink/390–400 MHz down-link, 410–420 MHz uplink/420–430 MHz down-link). Each channel has a bandwidth of 25 kHz and can carry 36 kbps. Modulation is DQPSK. While V+D uses up to four TDMA voice or data channels per carrier, PDO performs statistical multiplexing.

27.4.2 TETRA data frame

Figure 27.2 shows the typical TDMA frame structure of TETRA. Each frame consists of four slots (thus four channels in the V+D service per carrier), with a frame duration of 56.67 ms. Each slot carries 510 bits within 14.17 ms, i.e., 36 kbps. 16 frames, together with the control frame (CF) form a multiframe, and finally, a hyperframe contains 60 multiframes. In order to avoid sending and receiving at the same time, TETRA shifts the up-link for a period of two slots compared to the down-link.

Similar to GSM, TETRA offers traffic channels (TCH) and control channels (CCH). Typical TCHs are TCH/S for voice transmission, and TCH/7.2, TCH/4.8, and TCH/2.4 for data transmission (depending on the FEC mechanisms required). However, in contrast to GSM, TETRA offers additional services like group call, acknowledged group call, broadcast call, and discreet listening.

Emergency services need a sub-second group-call setup in harsh environments which possibly lack all infrastructure. These features are currently not available in GSM or other typical mobile telephone networks.

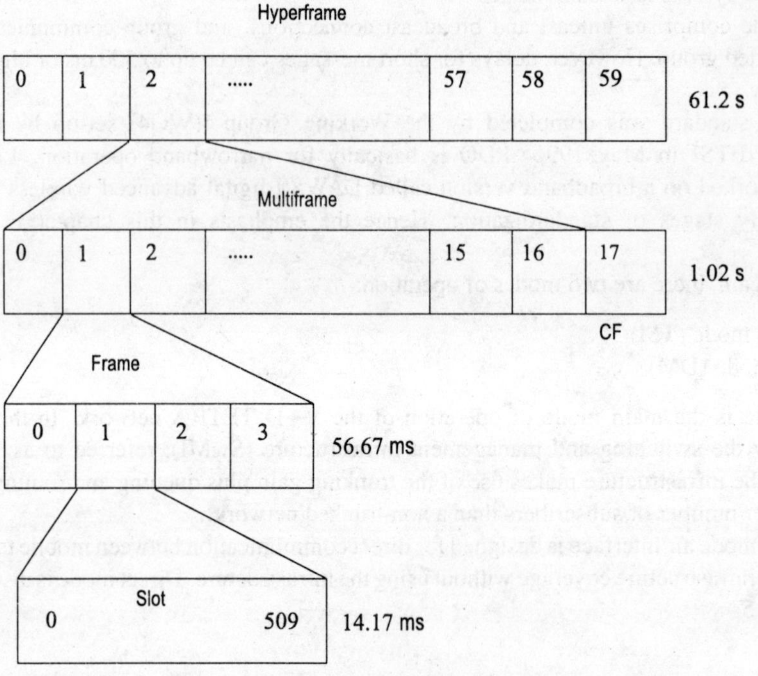

Fig. 27.2 *TETRA-TDMA frame.*

27.5 Standard Interfaces and Gateways

In Section 27.2, five types of TETRA standard interfaces have also been mentioned as the Gateway. In this section, we will study them along with the services for which these are meant. The discussion on each interface is divided into three parts:

- Features of the interface
- Services supported by the interface
- Protocols for the interface

27.5.1 TETRA voice plus data air interface

The V+D standard air interface has the following two components (refer to Fig. 27.3) related to two distinct modes of operation, defined earlier in Section 27.3, namely

(*a*) trunked mode (TM), and
(*b*) direct mode (DM).

Let us discuss the TM and DM air interfaces.

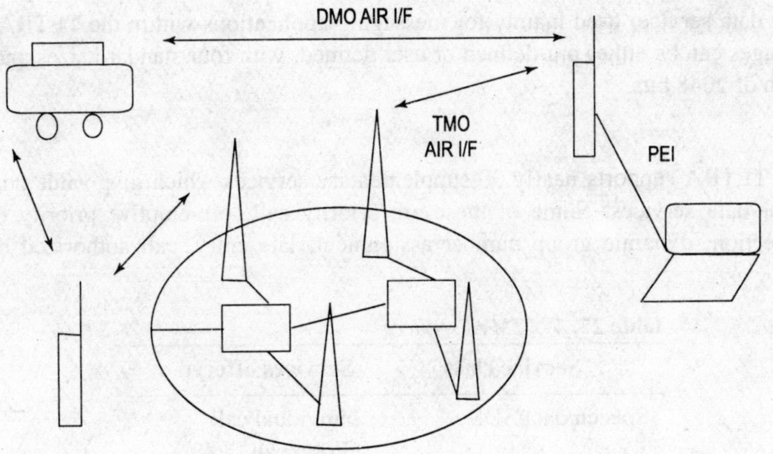

Fig. 27.3 *Air interface in V+D mode.*

(a) TM air interface

Features of TM AI: The TM air interface aims at providing bandwidth on demand as well as supporting multimedia services. It is achieved in the following manner. TETRA air interface uses TDMA technique to allow sharing of the same radio frequency carrier by multiple channels. In the TM air interface, 4 channels are transmitted on each frequency (TETRA carrier), thereby facilitating

(*i*) simultaneous voice and data operation by the same terminal, with each service on a different time-slot,

(*ii*) easier duplex operation with portable (hand-held) device by switching rapidly between transmit and receive time-slots (TDMA duplex), and

(*iii*) multi-slot working where time-slots 1 to 4 can be aggregated to increase the data transmission speed.

In addition, TETRA offers a more spectrally efficient system compared to the existing mobile communication systems because of the use of low speed codec and the RF features chosen for the air interface.

TM AI services: Table 27.1 summarizes the main services and features offered over trunked mode air interface (TM AI) standard. All speech and data calls can be point-to-point or point-to-multipoint. The speech calls can be performed either in clear or encrypted form. The data calls can be group calls (half-duplex) or a broadcast calls (simplex).

An extensive range of data services listed below are supported on TETRA air interface.

1. Circuit mode data service, using one to four time-slots in aggregate
2. Packet mode data service (based on X.25 protocol)
   - Connection-oriented packet data service which requires setting up of a virtual circuit
   - Connectionless packet data service in which individual packets are routed to the destination through the system
   - A new packet data service based on the Internet Protocol (IP)

3. Short data service, used mainly for messaging applications within the TETRA system. The messages can be either pre-defined or user defined, with four standard sizes and a maximum length of 2048 bits.

Furthermore, TETRA supports nearly 30 supplementary services which give value additions to the basic voice or data services. Some of these are priority call, pre-emptive priority or emergency call, area selection, dynamic group number assignment, late entry, call authorized by dispatcher, etc.

Table 27. 1 *TM AI services*

Service class	Services offered
Speech/data/SDS	Individual call Group call Broadcast call
Speech only	Half duplex Full duplex Clear/encrypted
Packet mode data	Connection-oriented Connectionless (X.25) IP packet data
Short data service	16 or 32 or 64 kbps packet 2 Mbps frame

Call control, mobility management and security management are the three vital tasks performed by the infrastructure(SwMI) over air interface. Call control is responsible for setting up and clearing up calls for trunked mode operation between mobile and stationary users connected to the TETRA infrastructure. It is responsible for assigning system resources required by each call. Mobility management caters to user mobility in the network and features registration, cell re-selection, group attachment, etc. Security services include authentication and air interface encryption. The TETRA standard also provides the capability to accept end-to-end encryption. A full length treatment on security management in TETRA is deferred to Section 27.7.

Data transmission rate over the V+D TM AI can be selected on demand from 2.4 to 28.8 kbps, depending on the number of time-slots used in aggregate and the type of error protection selected. There are three levels of error protection resulting in different net channel speeds as shown in Table 27.2.

TM AI protocols: The complete protocol stack for the two ends of the trunked mode air interafce consists of the three layers shown in Fig. 27.4. Standard service access points (SAPs) at the user end (top) of layer-3 ensure standard access to the AI facilities throughout the network.

Table 27. 2 *Data transmission rates over TM V+D air interface*

	Bandwidth on demand			
No. of time-slots	1	2	3	4
No protection	7.2	14.4	21.6	28.8
Low protection	4.8	9.6	14.4	19.2
High protection	2.4	4.8	7.2	9.6

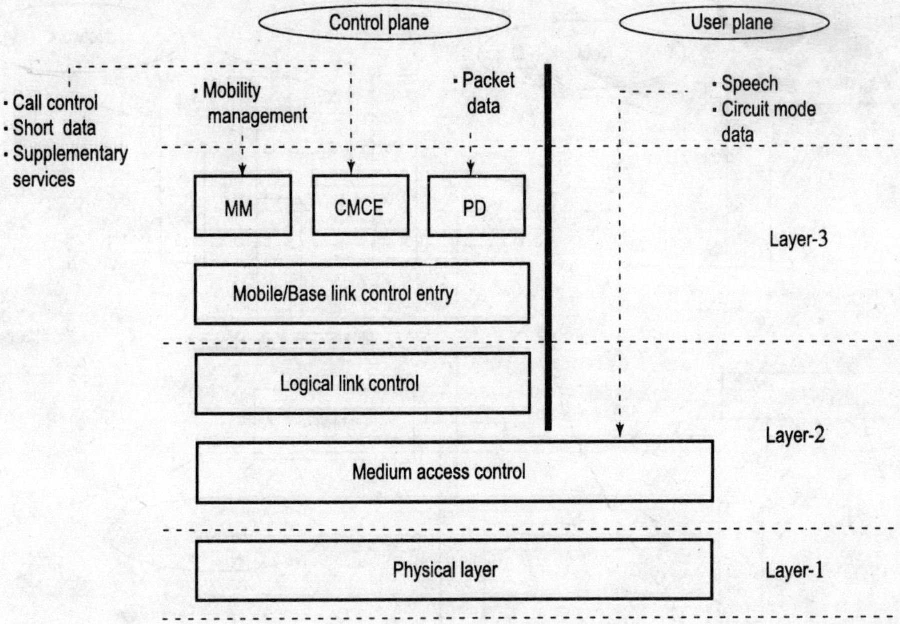

Fig. 27.4 *TM AI protocol stack.*

(b) DM air interface

DM AI features: The second part of V+D air interface devised for the direct mode operation has features which suit various direct mode scenarios. The various DM operational scenarios are depicted in Fig. 27.5.

Scenario 1: Individual call—Call setup between two mobiles without involvement of infrastructure.

Scenario 2: Direct mode group call—Calls setup between number of mobiles without the involvement of the infrastructure.

Scenario 3: Dual watch capability—In this, the mobile user operating in DM can monitor the TM control channel, or if operating in TM, the user can monitor a direct mode channel.

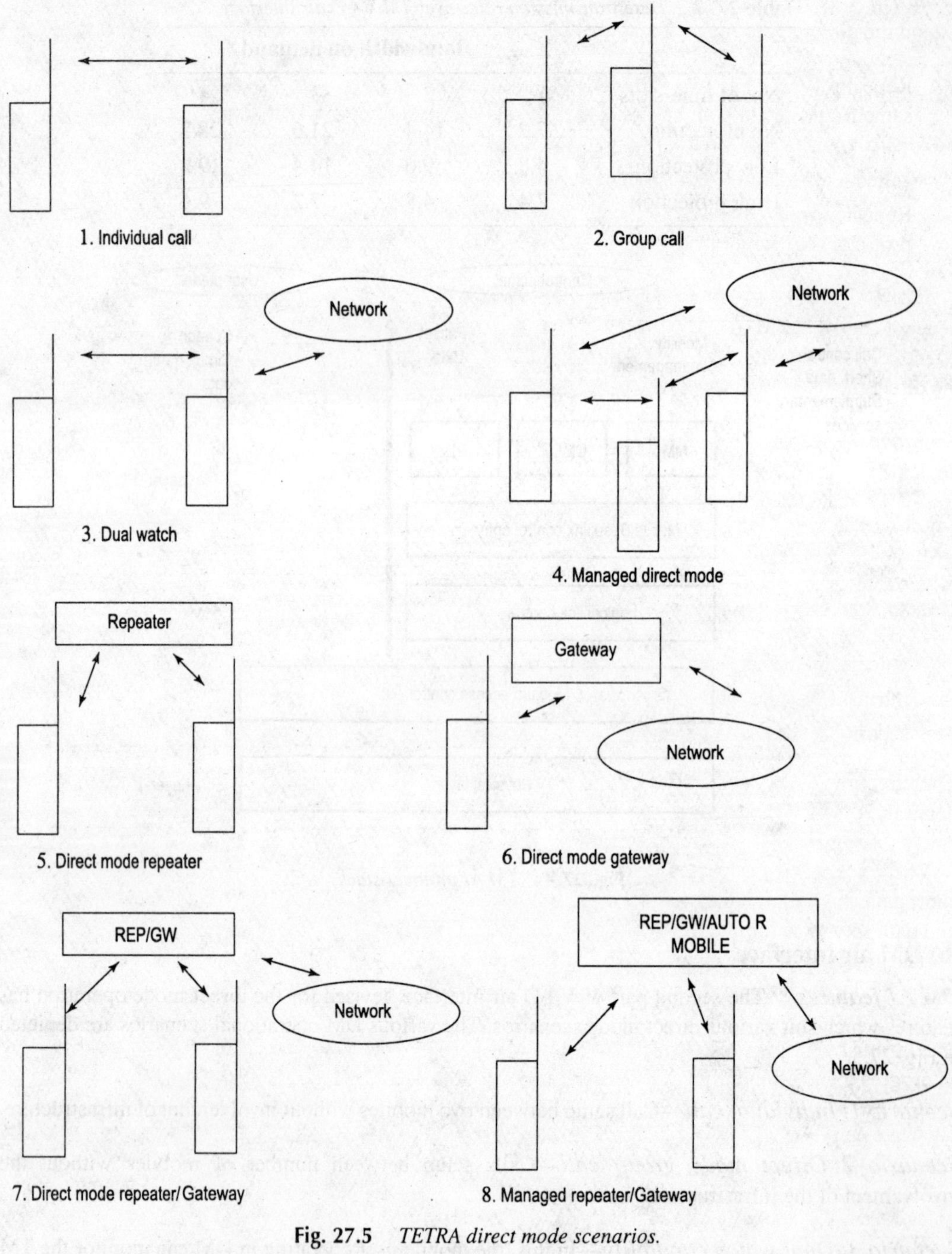

Fig. 27.5 *TETRA direct mode scenarios.*

Scenario 4: Managed DM—In this scenario, the DM users need authorization, normally provided by an authorizing mobile station which is in direct communication with the infrastructure.

Scenario 5: Direct mode repeater—The repeater is purely a direct mode device, used essentially to extend the direct mode range. Two types of repeaters have been standardized for the DM operation:

- Repeater type 1 has two variants-1a and 1b, both handling a single call only. Type 1a uses the same frequency in both up-link and down-link directions as a time division regenerative repeater. Type 1b uses two different frequencies in the up-link and the down-link directions. This type is called a regenerative transponder. It performs the same regeneration process as type 1a.
- Repeater type 2 handles two calls simultaneously. This is also a regenerative repeater but the protocol is different from type 1 repeater because the type 2 repeater supports two carriers.

Scenario 6: Direct mode gateway—The gateway is used to connect a DM user to the infrastructure, hence it acts as a DM to TM protocol converter. Thus, a gateway positioned at fringes of the network effectively extends the coverage, although the services offered by the gateway to the subscribers are the direct mode type services.

Scenario 7: Repeater/Gateway—In this, the repeater and gateway are implemented as a combined device.

Scenario 8: Managed repeater/Gateway—It is the combination of repeater and gateway in managed DM.

DM AI services: The services offered across the direct mode air interface are

- voice and data services,
- intrinsic services, and
- security services.

The direct mode services are fewer and a subset of the TM AI services. These are covered in greater detail in Section 27.6 where a full-length treatment of direct mode is presented. For the time being, the above three groups of services are briefly explained.

The voice service is half-duplex. The data services are single slot, limited to 7.2 kbps capacity. The short data and circuit mode data services are the only data services supported, and simultaneous voice and data service is limited to voice plus SDS data.

The intrinsic services in the TETRA DM operation play the same role as the supplementary services in the TETRA TM operation. However intrinsic services are only three in total, as against the 30 supplementary services of TM. They differ from their TM counterparts in their setting-up procedure. The intrinsic services are activated during the call set-up process and cannot be invoked when a call is in progress. The three intrinsic services are:

- late entry,
- pre-emptive call or emergency call, and
- talking party identification.

Security services in the DM operation are identical to their TM operation counterpart. However, in the DM operation, addressing is required not only for the called and calling parties, but also for repeaters and gateways.

DM AI protocols: The protocol stack in DMO is a simplified version of the TMO as shown in Fig. 27.6 for the case of a basic DM call. Layer-3 is called call control layer in the DMO stack and is responsible for controlling the call, providing the DM intrinsic services and supporting the passage of short data messages in the control plane. The user plane is responsible for transporting circuit mode speech or data information without addressing. All signalling and addressing takes place in the control plane.

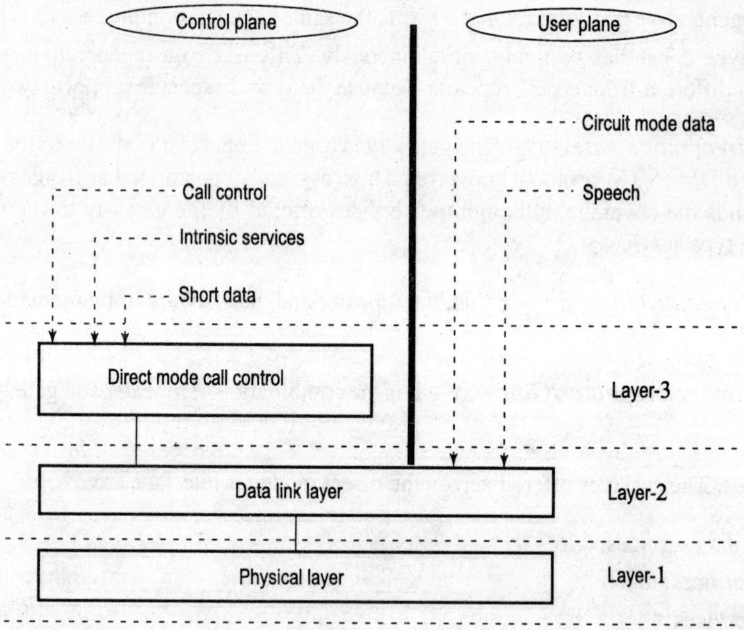

Fig. 27.6 *Direct mode protocol stack.*

27.5.2 TETRA inter system interface (ISI)

ISI features

Inter system interface (ISI) is designed to allow a number of *separate* TETRA networks to be interconnected to form a unified network. This interconnection is achieved in two ways:

Tiered approach: By interconnecting a number of local networks via their switching nodes, a single, say a regional TETRA network run by a single call control entity and network management sub-system, is formed first. Then a number of regional networks are be interconnected to form a national network. This requires planning for the overall network at the outset, and procuring of equipment from a single manufacturer.

Phased approach: Some operators opt for a phased implementation plan that may include procuring the infrastructure for different regions and/or customers from different suppliers.

Whatever the approach, ISI plays a key role in providing unified TETRA services across the whole network. These services will be available not only for inter-working between users in the two networks but also for roamers moving from one network to the other. The available services across the ISI will be based on a common denomination of all services offered separately in the two networks.

Services supported on ISI

The services available across the ISI include:

- Individual call
- Group call
- Mobility management
- Short data service
- Supplementary services
- Security

These services are established between service access points (SAPs) in the adjoining TETRA networks. The services are termed ANFs (additional network features), and as far as ISI is concerned they are considered as part of an application layer residing above the ISI lower layers which are subject of ISI standardization.

As an example of the services provided across the ISI, consider the group call configuration shown in Fig. 27.7. This is one of many possible group call configurations. In this example, the following takes place:

1. A user in the top left network requests a group call.
2. The top right network is in charge of setting-up the group call, linking/unlinking of groups in different TETRA networks, controlling and clearing the call.
3. Users in all three networks can then be participating in the call.

Let us now turn our attention to features of mobility management (MM) within the interconnected TETRA networks, i.e., across ISI. These features are:

- migration (individual and group),
- de-registration (individual),
- group attachment/detachment,
- database fault recovery, and
- OTAR for key management.

These features are explained below.

Migration: The use of the ISI allows migration from one TETRA system to another with the migrating user having access to TETRA services offered in the visited systems. If the migration occurs during an established call, then the migration could support the call restoration function.

De-registration: The migrating user has to register in the visited TETRA system. De-registration of the migrating user is another ISI feature which enables the home network of the migrating subscriber to be informed of this by the visited network.

Group attachment/Detachment: Another MM feature called group attachment enables the migrated subscribers to participate in a group call belonging to their home system.

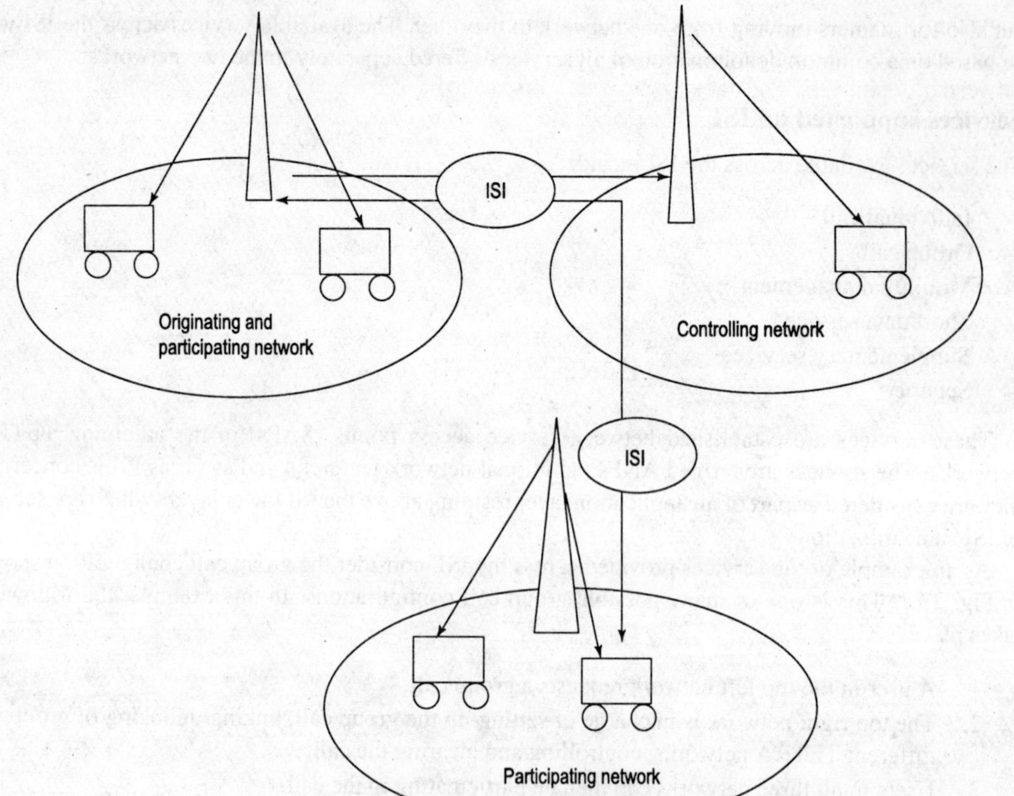

Fig. 27.7 *Group call configuration across inter system interface.*

Database fault recovery: The database fault recovery service is to correct inconsistencies in the back-up MM databases used for managing the migration. The authentication service enables a visited network to authenticate the migrating subscriber.

OTAR key management: The 'over the air re-keying (OTAR)' service enables the assignment of static cipher keys to the migrated individual subscriber.

ISI protocols

Nodes in various TETRA systems interconnected by ISIs form a private integrated service network or PISN. The lower layers of the ISI protocol stack (Fig. 27.8) are based on private signaling system Number 1, PSS1, also known as the 'Q signaling' originally developed by the European Computer Manufacturers Association, ECMA. This signalling system is based on the ISDN standard using (*i*) circuit mode bearer services, and (*ii*) DSS1 link layer signalling.

The generic functional protocol defined originally to support supplementary services in the ISDN is used on top of the PSS1 to communicate with additional network features or the ISI services. This is done via the remote operations service element (ROSE) that directly handles each ISI service.

At the ISI physical layer, each TETRA channel is of 8 kbps rate, thus maintaining TETRA voice coding across the ISI. These channels are multiplexed into 64 kbps bearers as the unit of interconnection between different ISIs. The network designers are free to use a 2 Mbps bearer and to multiplex as many ISI 64 kbps time-slots as deemed necessary.

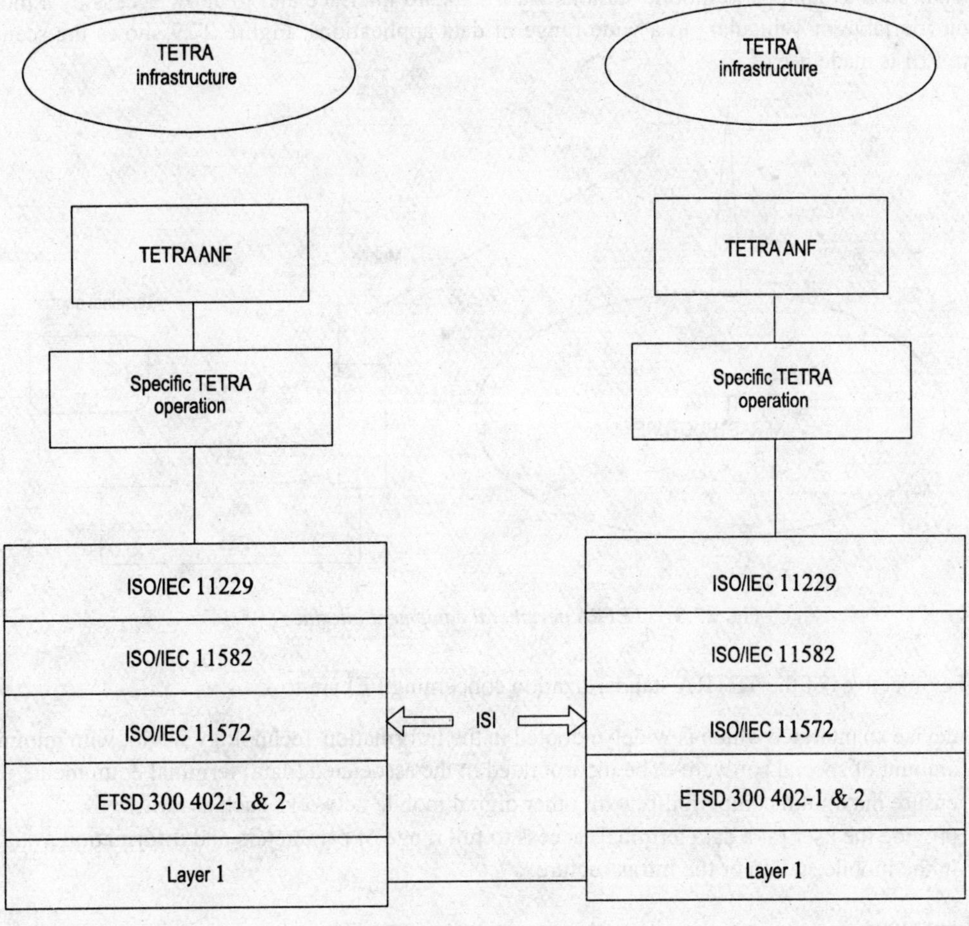

Fig. 27.8 *ISI protocol stack (Private signalling system 1 (PSS1) used in ISI standard).*

To summarize:

- ISI is a very useful tool in staged implementation of a medium to large TETRA network
- It facilitates multi-sourcing of the network from different suppliers
- It guarantees certain minimum innter-working between different TETRA networks
- It allows full interoperability between separate TETRA networks if such networks share the same frequency band.

No wonder then that the ISI standardization is very extensive resulting in documentation over 1000 pages!

27.5.3 TETRA peripheral equipment interface (PEI)

Features of TETRA PEI

The peripheral equipment interface or PEI was introduced by TETRA to enable the connection of data terminals such as laptops to mobile stations via a standard interface and to allow access by a mobile station (portable or vehicular) to a wide range of data applications. Figure 27.9 shows the scenario where PEI is made use of.

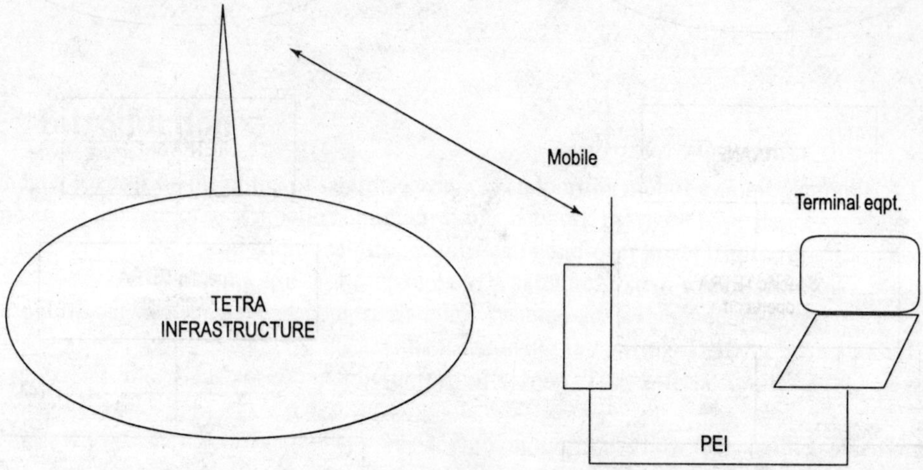

Fig. 27.9 *TETRA peripheral equipment interface (PEI).*

The objectives of the TETRA standardization concerning PEI are to

- devise an interface which is widely adopted in the Information Technology world, with minimum amount of special software to be incorporated in the associated (data) terminal equipment,
- ensure maximum compatibility with other digital mobile networks such as GSM,
- provide the user (at a data terminal) access to full range of parameters and information available in the mobile station or the infrastructure.

PEI services

All TETRA V+D data services can be carried out via the terminal equipment connected to the mobile terminal using PEI. PEI services also include setting up and control of a voice call, but not voice communication itself.

Three service access categories are defined in the PEI standard:

- *AT commands:* Used to allow access to circuit data services, short data services and information in the mobile terminal.
- *PPP and packet data:* Used to allow access to packet mode services via the PEI.
- *PPP and radio remote control:* Uses the TETRA network protocol type 1 (TNP1) to allow call control, short data service and mobility management functions to be accessed from a terminal equipment such as a PC.

PEI protocols

The protocol structure of the PEI is shown in Fig. 27.10. As seen from the protocol stack, the PPP platform is shared by TNP1 and IP protocols. The latter supports layer-3 TCP and UDP protocols used for packet data services. IP versions 4 and 6 will be supported.

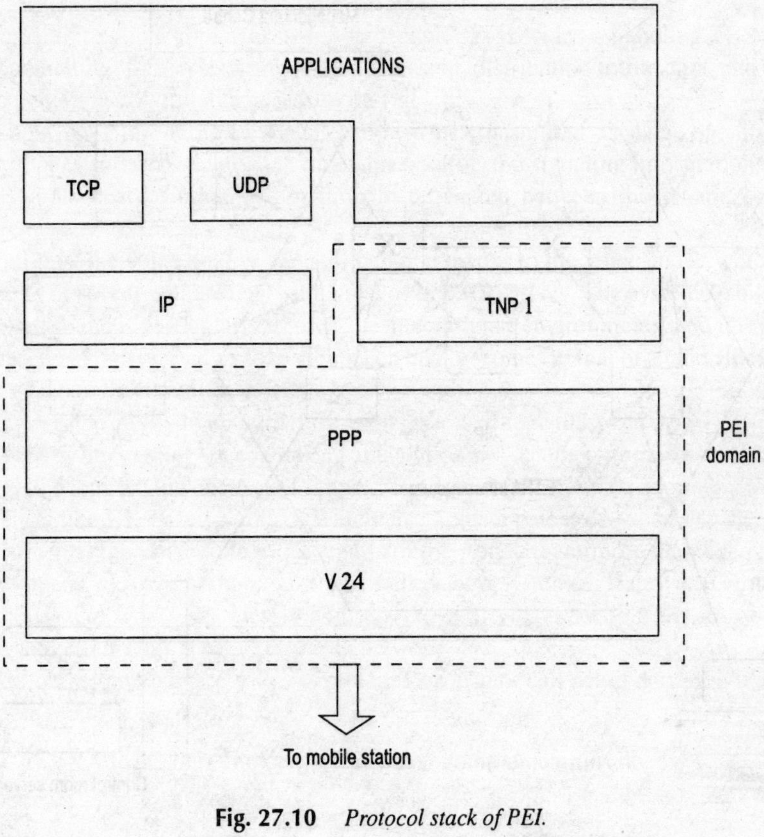

Fig. 27.10 *Protocol stack of PEI.*

The PEI physical layer is based on ITU-T V.24/V.28 with speeds normally below 20 kbps but with possibility of reaching up to 64 kbps.

Internet access, multimedia and other applications

IP packet data service and SDS are two powerful TETRA bearer services that will allow a wide variety of applications to be used in conjunction with the TETRA network (Fig. 27.11). Use of open standard APIs has resulted in flexibility, future-proof network and also multi-vendor interoperability. The data application possibilities for use in a TETRA network is shown in Fig. 27.12.

A good example of these open standard APIs is the wireless application protocol (WAP), a de-facto industry standard currently being developed by the WAP community.

The WAP architecture defines a framework to meet the challenges of advanced services, differentiation and fast/flexible service creation in wireless networks. WAP defines a protocol stack with its

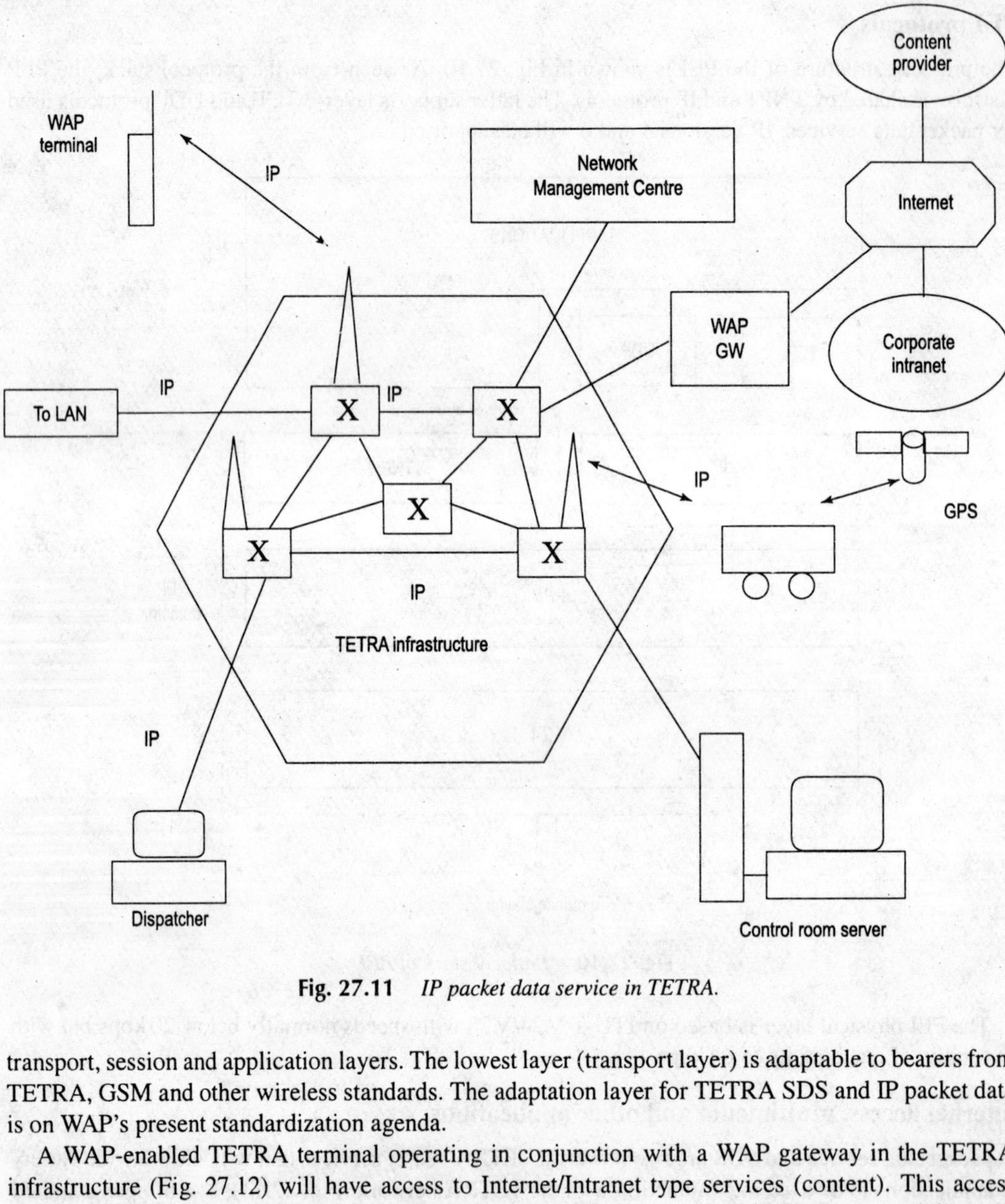

Fig. 27.11 *IP packet data service in TETRA.*

transport, session and application layers. The lowest layer (transport layer) is adaptable to bearers from TETRA, GSM and other wireless standards. The adaptation layer for TETRA SDS and IP packet data is on WAP's present standardization agenda.

A WAP-enabled TETRA terminal operating in conjunction with a WAP gateway in the TETRA infrastructure (Fig. 27.12) will have access to Internet/Intranet type services (content). This access can be from the WAP terminal or a data terminal connected to it via the PEI. This capability will allow the use of internet-based/multimedia services in the TETRA network as and when they become available.

The gateway will do all the necessary conversions from WWW features to WAP layers. For example, WAP terminals will use wireless mark up language (WML) instead of HTML. WAP gateway provides this conversion.

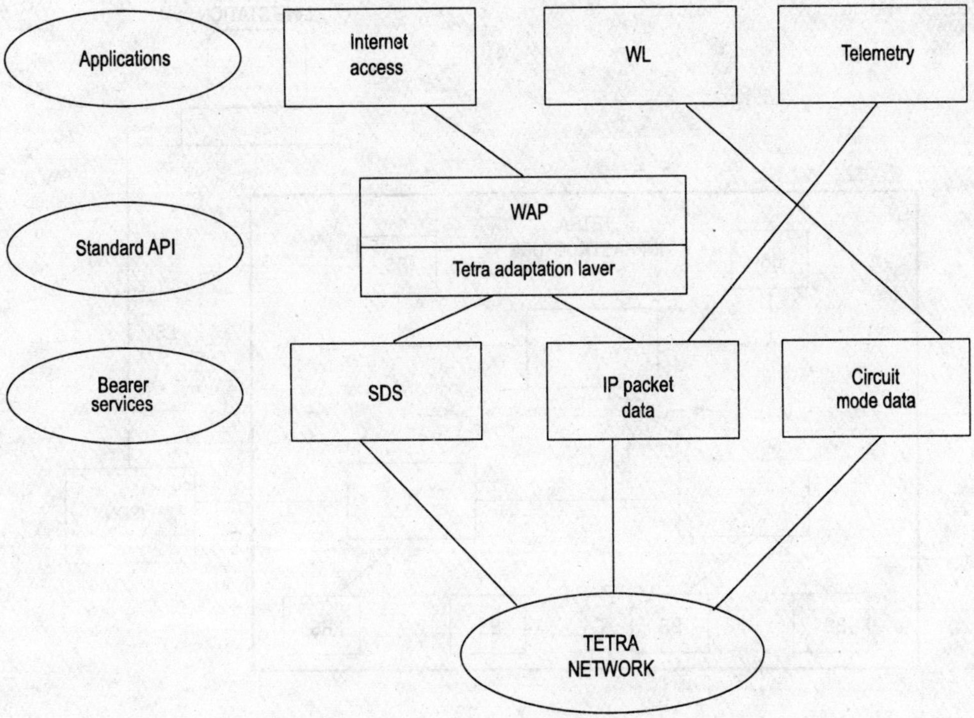

Fig. 27.12 *Data application possibilities in TETRA.*

27.5.4 TETRA line station interface (LSI)

TETRA LSI features

This interface was included in the TETRA standardization program from the outset with a view of providing the same services to a mobile user as that of a line-connected user. However, the specification is yet to be issued. Meanwhile, several TETRA manufacturers have developed proprietary interfaces for static subscribers, mainly for use in dispatching stations. (Refer to Fig. 27.13 showing TETRA line station interface).

The LSI is based on transporting each TETRA channel as 8 kbps payload with TETRA coding intact across the interface. This necessitates the use of a bank of TETRA codecs at the line station.

The same approach could be applied to end-to-end encryption service with encryption devices positioned at the line station.

Services supported across TETRA LSI

All TETRA V+D services, with the exception of a couple of supplementary services, will be available across the LSI. The services are:

- voice,
- circuit mode data,

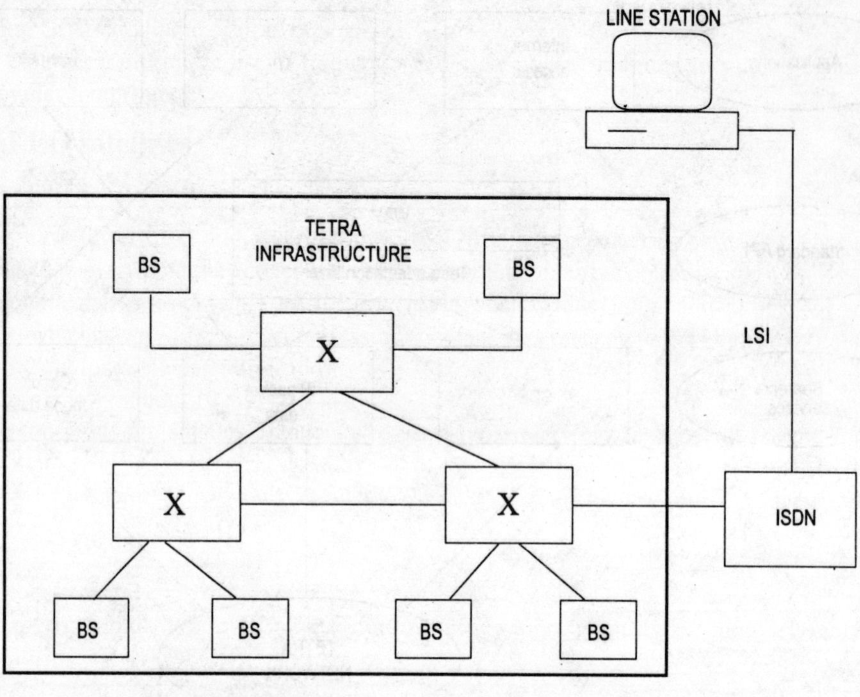

Fig. 27.13 *TETRA line station interface.*

- packet mode data,
- short service data,
- supplementary services (except access priority, ambience listening),
- control functions, viz., mobility, registration, authentication, and group attachment/detachment.

TETRA LSI protocols

The physical layer in the LSI is based on a 64 kbps link. This link could be a leased line or a B channel of an ISDN connection, at either basic rate or primary rate, depending on the application. Furthermore, the connectivity can be direct or through an intervening network such as ISDN as shown in Fig. 27.14.

27.5.5 Central network management (CNM) interface

Features of TETRA CNM interface

The CNM interface was aimed at providing an external network coordination and management entity for at least two separate TETRA networks interconnected by the ISI. The internal network management system (known as the local network management or LNM) and its interconnections within each TETRA network are not part of this standardization. However, it includes the interconnection of LNM and CNM, and is meant for providing guidance to TETRA network designers. This standardization is based on the ITU-T TMN methodology. It deals only with high level functions carried out from a central entity.

The CNMI has its own authorization and authentication processes, and could optionally include encryption.

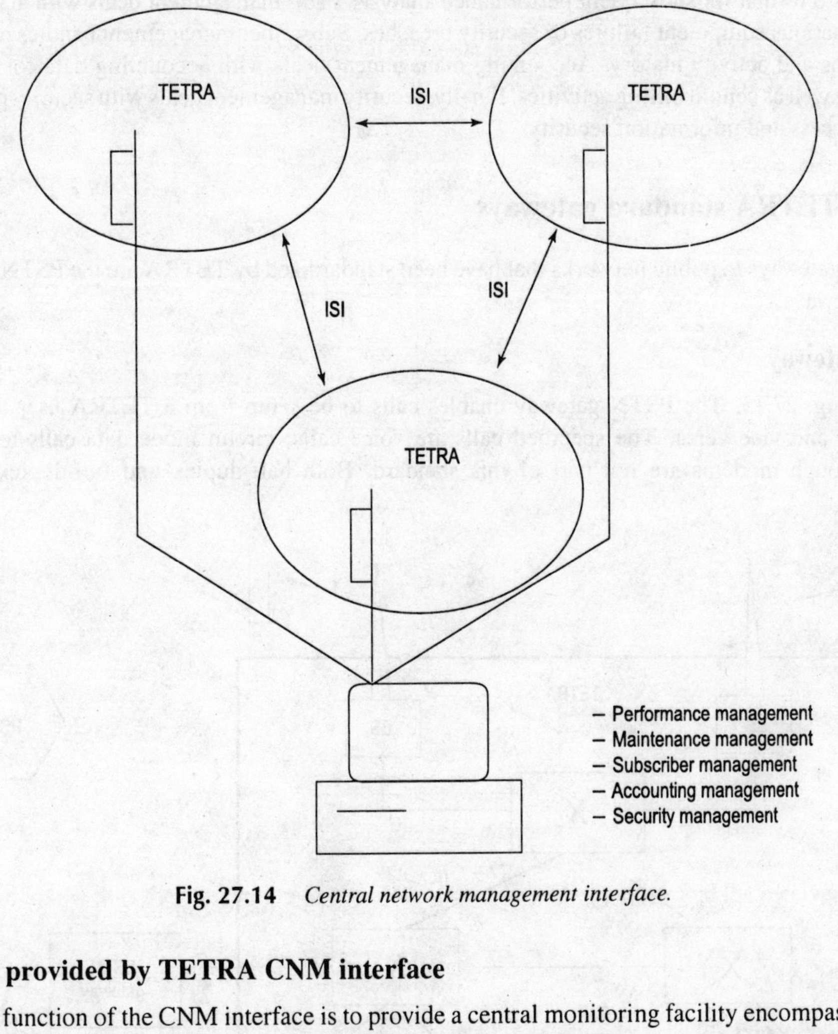

Fig. 27.14 *Central network management interface.*

Services provided by TETRA CNM interface

The main function of the CNM interface is to provide a central monitoring facility encompassing more than one TETRA system (Fig. 27.14). No real-time control of system resources or databases is intended. In particular, the status of ISIs and inter-system links are made available at the central facility.

The network management functions performed at the CNM facility are more limited in scope compared with those carried out at the non-standardized local network management entities. Some examples of these network management functions are:

- performance management,
- fault management,
- subscriber management,
- accounting management, and
- security management.

Performance management deals with the transfer of performance data transfer from LNM to CNM in a standard format for subsequent performance analysis. Fault management deals with alarm trigger related to serious equipment failures or security breaches. Subscriber management handles registration permissions and activity history. Accounting management deals with accounting data for wide area roamers as well as central billing activities. Finally, security management deals with security procedures both for access and information security.

27.5.6 TETRA standard gateways

The three gateways to public networks that have been standardized by TETRA are the PSTN, PDN and ISDN gateways.

PSTN gateway

Refer to Fig. 27.15. The PSTN gateway enables calls to be setup from a TETRA user to a PSTN subscriber and vice versa. The specified calls are voice calls; circuit mode data calls to analogue PSTN through modems are not part of this standard. Both half-duplex and full-duplex calls are included.

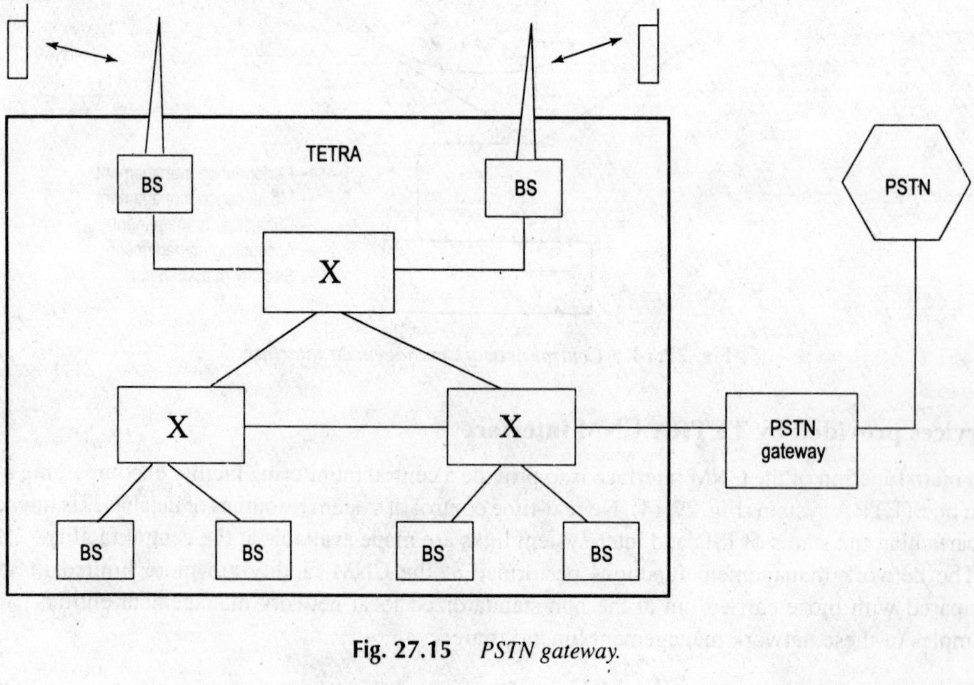

Fig. 27.15 *PSTN gateway.*

From the PSTN gateway and the TETRA network, users are seen as PSTN subscribers. Incoming calls from the PSTN use a two-step DTMF (dual tone multiple frequency) dialing. The interface has to carry out functions such as TETRA voice coding/decoding, echo cancellation, numbering conversion, provision of comfort tones and so on. This standard interface is already published by ETSI.

PDN gateway

Refer Fig. 27.16. The gateway to the public data network is simply an X.25 gateway to allow TETRA packet mode services a wider access to services and databases available on the public network. Looking at this gateway from the external network, the TETRA network appears as a data network node (working on X.75 protocol) or a data terminal (working on X.25 protocol. The PDN gateway is capable of mapping the internal TETRA addresses to standard X.25 addresses for use in the external network.

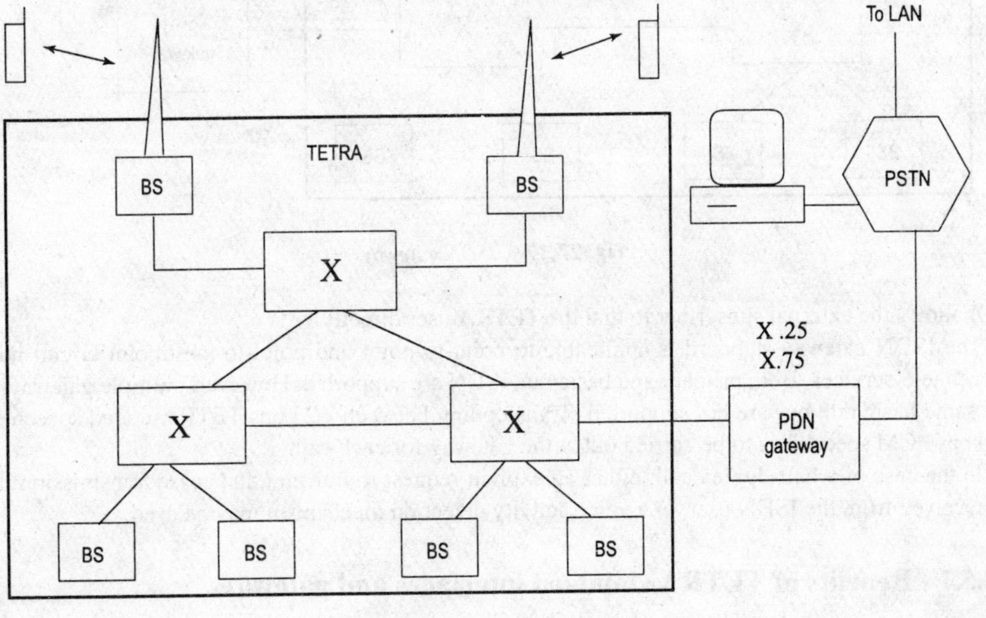

Fig. 27.16 *PDN gateway.*

ISDN gateway

The ISDN gateway (Fig. 27.17) enables calls to be setup from an ISDN subscriber to a TETRA user and vice versa. During incoming calls the ISDN gateway allows ISDN signalling information to be passed from the external network to the TETRA infrastructure in a form compatible with TETRA call control procedures. Conversely, for an outgoing call TETRA signalling information will be passed to the ISDN network in accordance with the ISDN DSS1 call control procedures.

In an incoming call, the called TETRA user can be addressed in either of the following ways:

– ISDN direct dial in (DDI) supplementary service
– ISDN sub-addressing (SUB) supplementary service
– Two stage dialing

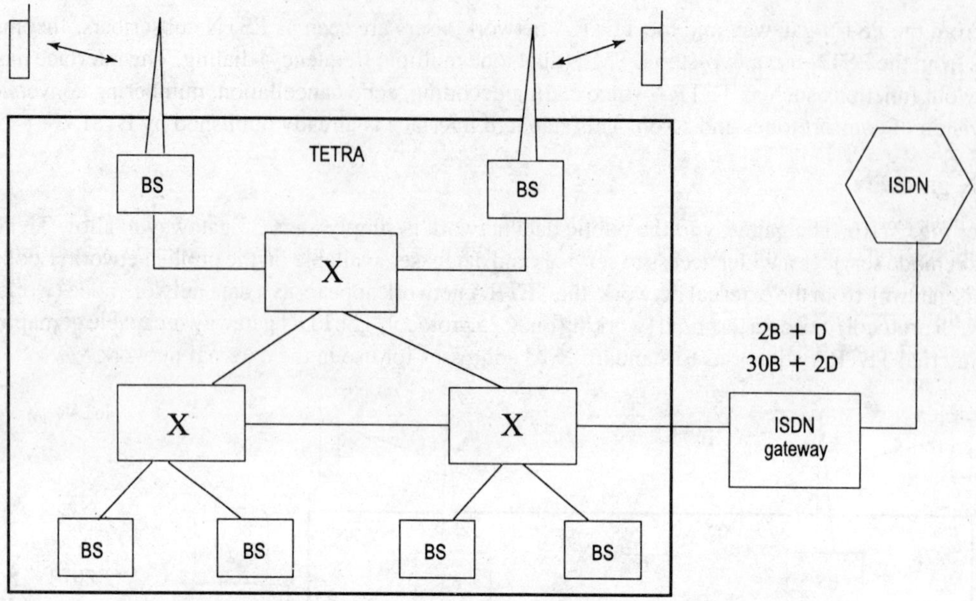

Fig. 27.17 *ISDN gateway.*

DDI allows the external subscriber to dial the TETRA user directly.

The ISDN gateway standard is applicable to point-to-point and point-to-multipoint circuit mode speech tele-services. Both primary and basic rate ISDN are supported. However, multiple gateways on the same basic rate bus are not supported. Trans-coding between 7.2 kbps TETRA-coded speech and 64 kbps PCM speech has to be carried out at the gateway for each call.

In the case of a half-duplex call, either an explicit request to transmit and cease transmission may be received from the ISDN user, or a voice activity detection mechanism may be used.

27.5.7 Benefits of TETRA standard interfaces and gateways

TETRA standard and its standardized interfaces are attractive because of its features like low cost, multiple sourcing, inter-operability and future upgradability.

In terms of services and facilities, the use of TETRA V+D standard interfaces brings about the following advantages:

- Spectral efficiency
- Two types of air interface namely trunked mode and direct mode
- The capability of interconnecting different TETRA systems as unified system
- Possibility of inter-working via standard gateways with fixed public network
- Support for a range of security features across all interfaces

However, one major benefit that makes TETRA an applications-rich network and ensures future upgradability and long operating life cycle is the open standards of the TETRA data communications platform. This platform with its three data communication bearer types and 28.8 kbps digital pipe will go a long way towards supporting new services including internet type multimedia services, even into the III-generation mobile communications era.

27.6 TETRA Direct Mode

We have already seen how TETRA offers the benefits of automatic operation and frequency efficiency of trunking, combined with the terminal autonomy benefit of conventional PMR. The most important feature of TETRA is that it offers a novel multi-mode capability by combining these two modes—trunking and direct (conventional)—into a single terminal equipment. TETRA also provides a standardized way of inter-working between the direct mode and the trunking system. For the professional users, all this means unrivalled flexibility when choosing the most appropriate mode of response to an increasingly complex operational needs of today.

In this section, a brief overview of 'direct mode' is presented, followed by description of services and facilities that are part of the standard. The benefits they offer to the users and some of the typical applications for professional users are also discussed.

27.6.1 What is direct mode?

In its basic form, direct mode represents direct communication between two or more TETRA DM terminals/mobile stations (DM-MS) without the use of trunking network infrastructure (TETRA SwMI). The PSTN gateway enables calls to be setup from a TETRA user to a PSTN subscriber and vice versa. The specified calls are voice calls. Circuit mode data calls to analogue PSTN through modems are not part of this standard. Both half-duplex and full-duplex calls are included. Various direct mode scenarios have already been presented earlier, and depicted in Fig. 27.5. Let us quickly recapitulate.

The simplest form of DM is the two-way communication between two or more MS terminals, back-to-back. Through the use of direct mode repeater (DM-REP), it is possible to further extend the area coverage over which these terminals can communicate.

Direct mode also includes a mode of communication between DM terminals/mobile station and trunking TETRA terminal. This could be done using either an optional facility— the so-called 'dual watch'— or by deploying a special type of equipment called the DM gateway.

In the dual watch mode (DM-DW), the TETRA MS is a dual-mode terminal, which can be in either mode, periodically monitoring (scanning) one or both systems to which calls are directed.

Direct mode gateway (DM-GATE) on the other hand, makes it possible for a TETRA MS to make and receive calls from other terminals, which could either be in trunking or direct mode system, irrespective of which TETRA mode it is in.

In managed DM, DM users need authorization, usually provided by an authorizing mobile station which is in direct communication with the infrastructure.

27.6.2 Direct mode equipment and interfaces

Direct mode mobile station (DM-MS)

The basic reference model for DMO applies to a simple point-to-point or point-to-multipoint communication between DM-MSs using the DM air interface U_d. Figure 27.18 shows the most basic configuration and is the most simple reference to define the range of services which are to be supported by DMO.

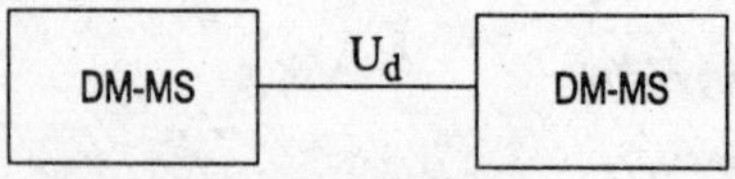

Fig. 27.18 *DM-MS connected to DM-MS.*

Dual Watch mobile station (DW-MS)

This reference model (Fig. 27.19) applies to a DM-MS, a multi-mode equipment capable of dual watch operation.

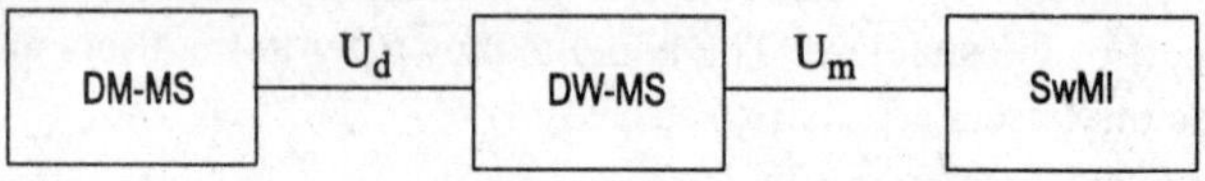

Fig. 27.19 *Dual watch reference model.*

The DW-MS can be in any one of three following states:

- Idle in both modes and periodically monitoring both the trunking mode control channel and a selected DM frequency
- Communicating with another DM-MS and monitoring the trunking mode control channel without interrupting direct mode operation
- Communicating with the TETRA switching and management infrastructure (SwMI) in trunking mode and periodically monitoring a selected DM frequency without interrupting trunking mode operation

Direct mode repeater (DM-REP)

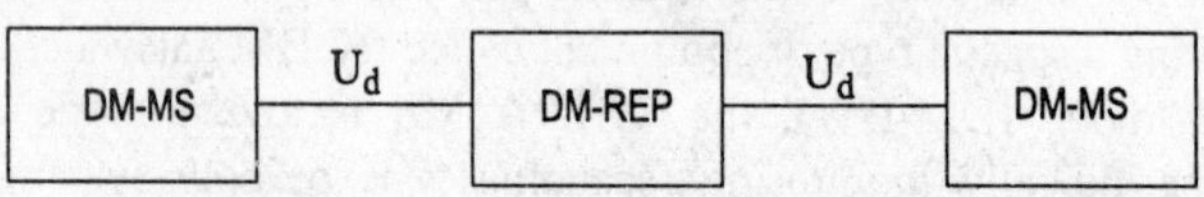

Fig. 27.20 *DM-REP reference model.*

The DM-REP (see Fig. 27.20) receives information from a transmitting mobile on an up-link time-slot and re-transmits this information to another mobile or group of mobiles on a down-link time-slot. The DM-REP specified within the TETRA standard is regenerative, i.e., it decodes and re-encodes the received speech and signalling bursts which it receives (one slot's worth at a time) to improve the overall link performance. The standard defines three different types of repeater, (refer to Section 27.5.1), namely

- single frequency single call repeater (Rep-Ia),
- two frequency single call repeater (Rep-Ib),
- two frequency two call repeater (Rep-II).

Direct mode gateway (DM-GATE)

Fig. 27.21 *Gateway into a TETRA SwMI.*

The DM-GATE provides both-way interconnectivity between DM and the TETRA trunking network (see Fig. 27.21). Communication received from trunking mobile is retransmitted to the mobiles on the DM side and vice versa.

27.6.3 Direct mode services and facilities

A summary of the services supported within TETRA DMO is listed in Table 27.3.

Table 27. 3 *Voice and data (trunking) services supported in DMO*

	Teleservice	Bearer service
TETRA speech	Individual call (point-to-point) Group call (point-to-multipoint)	N/A
Circuit mode unprotected data	N/A	7.2 kbps (point-to-point) or (point-to-multipoint)
Circuit mode low protected data (one of three different interleaving schemes can be demanded).	N/A	4.8 kbps (point-to-point) 4.8 kbps (point-to-multipoint)
Circuit mode high protected data (one of three different interleaving schemes can be demanded).	N/A	2.4 kbps (point-to-point) 2.4 kbps (point-to-multipoint)
Short data service - type 1		16 bits user defined data
Short data service - type 2		32 bits user defined data
Short data service - type 3		64 bits user defined data
Short data service - type 4		2048 bits user defined data
Status messages		16 bits

Speech and data services may, as an option, have end-to-end encryption and/or air interface encryption. Let us now discuss DM teleservices and bearer services listed in Table 27.3 in detail.

Direct mode teleservices

TETRA DM voice teleservices support transmission of speech utilizing a TETRA-specific voice codec as defined in ETSI's ETS 300 392-5 standard. They use standard TETRA speech and channel coding as defined in ETSI's ETS 300 396-2 standard. Optional encryption is also available. TETRA DM voice teleservices support speech on a point-to-point (individual call) and point-to-multipoint (group call) circuit mode basis. The voice mode of operation is always simplex.

Individual call: An individual call is a point-to-point communication between one calling party and the called party. It may only be set up between two MSs which have selected the same DM channel. An individual MS has a pre-defined ITSI (individual TETRA subscriber indentity) number by which it is addressed. The mode of operation is simplex.

Group call: A group call is a two way point-to-multipoint communication between a calling party and one or more called parties. It may only be set up between MSs which have selected the same DM channel. The members of a group have one common pre-defined number which is called their group number GTSI (Group TETRA subscriber identity) and by which they are addressed. The U_d air interface uses the same addressing scheme for group numbers as defined in ETSI standard ETS 300 392-1. The air interface may support more than one group on a DM channel although only one group is able to communicate using the channel at any given time.

In addition, there may be an 'open' or 'common' group number (which includes all users) to allow calls to be made to all users who have selected the same DM channel, providing for open channel operation. Only one number (group number GTSI) is sent on the air interface and no acknowledgement is expected. The primary objective is to have a fast call setup. The mode of operation is simplex.

Direct mode bearer services

A circuit mode bearer connection is a point-to-point or point-to-multipoint data communication between one calling MS and one or more called MSs. It may only be set up between MSs which have selected the same DM channel. The mode of operation is simplex.

Three types of circuit mode bearer services are offered within TETRA DM depending on whether or not the data is protected or unprotected, and depending on the level of protection provided. The difference between the protected and unprotected bearer services is that the protected bearer service provides error protection for the user data being transmitted. The result, as far as the user is concerned, is a more reliable and robust channel at the expense of a reduction in the net user data rate.

Circuit mode unprotected bearer services: Circuit mode unprotected bearer services support circuit mode data on a point-to-point (individual call), and point-to-multipoint (group call) basis. Data throughput at the user interface is 7.2 kbps.

Circuit mode protected bearer services: Circuit mode protected bearer services support data on a point-to-point and point-to-multipoint basis. Six protected bearer services are defined in TETRA DM offering two different levels of protection against bit error by using forward error protection in the transmitted bit stream. Error protection offer data throughput at the user interface at 4.8 kbps or 2.4 kbps with error protection rates of approximately 2/3 or 1/3 respectively. In order to provide further protection against errors, interleaving at depths 1, 4 or 8 may be applied along with the two levels of error protection, resulting in the six service options.

Short data service (SDS): The TETRA DM SDS is similar to the SDS offered within the TETRA trunking standard. Both point-to-point and point-to-multi point services are supported in DM SDS. Point-to-point SDS offers optional acknowledgement whereas the point-to-multi point service is unacknowledged.

The SDS is essentially a message service which can be optimized to be quick, enabling the user to exchange a short user-defined message or a short pre-defined message such as an emergency message. The message is sent within signalling capacity and may be sent or received in parallel with an ongoing speech or data call. The SDS may be used for applications such as automatic vehicle location, status or OTAR, etc.

The SDS in DM supports up to the same limited number of data bits as the SDS in trunking mode (2048 bits max.) with the content being user-defined or pre-defined. DMO SDS may be carried on an individual call or group call basis across the DMO–air interface. It may also be carried via a DM-REP and may be directed into or received from the trunking system via a DM-GATE.

A point-to-point short data message is sent from one originating MS to one receiving MS using the currently selected DM channel. The receiving MS is addressed by its ITSI in the usual manner. The receiving MS acknowledges receipt of the message if an acknowledgement has been requested; the originating MS may retry a number of times if the expected acknowledgement is not received.

A point-to-multi point short data message is sent from one originating MS to a group of one or more receiving MSs using the currently selected DM channel. The group is addressed by its GTSI in the normal manner. There is no acknowledgement from the receiving MSs in this case, but the originating MS may retransmit the message a number of times for reliability.

The following SDS/status message functional capabilities are supported in DMO terminals:

- Ability to send/receive a user-defined message of limited length
- Ability to send/receive a pre-defined status message, e.g. emergency message
- Ability to address a single party by using the ITSI address of the target message receiver
- Ability to receive the ITSI address of the message sender and relay it to the user application
- Ability to address multi-parties using the GTSI of the target message receivers
- Ability to apply priority to the message
- Ability to receive the priority applied to the message and to relay it to the user application

Intrinsic services

Intrinsic services are offered in association with all voice and data services described previously, and form an integral part of the signalling on the U_d interface. As already mentioned, there are three types of intrinsic services, namely, DM late entry, emergency call and transmitting party number identification.

DM late entry: This intrinsic service allows an MS when it accesses an active DM channel to enter an ongoing call if that call is addressed to a group of which it is a member. An MS may access a DM channel in a number of ways, for example,

- the user switches on the MS and selects that channel;
- the user switches from another DM channel to that channel;
- the MS returns to DM coverage after a period of lost coverage;
- the user switches the MS from trunking mode to DM and selects that channel (applies to DW-MS and DU-MS).

Pre-emptive call or emergency call: The DM air interface supports emergency calling. A DM-MS initiating an emergency call out of coverage of the system may use a DM channel, and if necessary, pre-empt any lower priority communication using that channel.

Transmitting party number identification: This intrinsic service provides the ability for the receiving MS to receive the number of the currently transmitting MS. This requires that at the beginning of each call transaction the transmitting MS transmits an individual subscriber number. This allows the receiving MS(s) to display the identity of the transmitting MS.

A user option allows details of the originating ITSI to be concealed. This option allows a transmitting MS to withhold its identity, a feature which may be required for certain authorised users.

OTAR (over the air rekeying)

DM supports OTAR by means of a specially designated MS. The OTAR MS may generate and distribute SCKs (static cipher keys) for each MS using a sealing key distributed from an authentication centre in a secure manner that is specific to each DM-MS (derived from the secret key K associated with each ITSI). A multi-pass protocol for key request and key delivery shall be defined with only implicit authentication being required (as the static cipher key can only be recovered by the holder of K).

Any DM-MS may act as a store-and-forward station for a sealed static cipher key (SSCK).

27.6.4 Security features

TETRA DMO contains mechanisms for the security of control signalling, user speech and data at the air interface. Security mechanisms (see Section 27.7 for a detailed treatment of the topic) included in DM are:

- authentication,
- confidentiality,
- key management including OTAR mechanism, and
- secure enabling/disabling of terminals.

TETRA DMO can also support end-to-end encryption using a synchronous stream cipher to provide high level protection for user traffic.

Authentication

Authentication is concerned with establishing the identity or authenticity of the equipment (DM-MS or DM-GW) involved in the communication transaction. It covers all three types of operations of the equipment as explained below.

Mobile to mobile operation: Since static cipher keys are used (which are generated, controlled and distributed through the DMO system security management) in encrypting the communication, mobile to mobile operation in the same DMO net receives implicit authentication.

Dual watch operation: In dual-watch mode a DM-MS is a valid member of the TETRA trunking network and authenticates to that network using the procedures for trunking operation.

Gateway mode operation: In this mode, the calls are multi-hop as they are established through a gateway following a multi-pass call setup procedure. For secure calls the gateway authenticates itself to the TETRA trunking network. There are two modes of operation when using a gateway from DMO to TETRA trunking, where,

- DM-MS can act as a full member of the TETRA trunking network, or
- DM-MS can use a gateway as an agent to access the TETRA trunking network.

Confidentiality

Confidentiality in DM is concerned with the protection of data/communication on the radio link between a DM-MS and either a single DM-MS or a group of DM-MSs. This is achieved by two types of encryption, viz.,

- air interface encryption,
- end-to-end encryption,

both of which are explained below, followed by a note on key management in these schemes.

Air interface (AI) encryption: In AI encryption, the output of an encryption algorithm KSS (key stream sequence) implemented in a key stream generator (KSG) is combined with the contents of the messages to be transmitted across the air interface to generate a 'synchronous stream cipher.' Synchronization applies only to the current call. The key stream bits are modulo-2 added (XORed) with plain text bits in data, speech and control channels to obtain encrypted cipher text bits, with the exception of the MAC header bits and fill bits. The ciphering process is shown in Fig. 27.22.

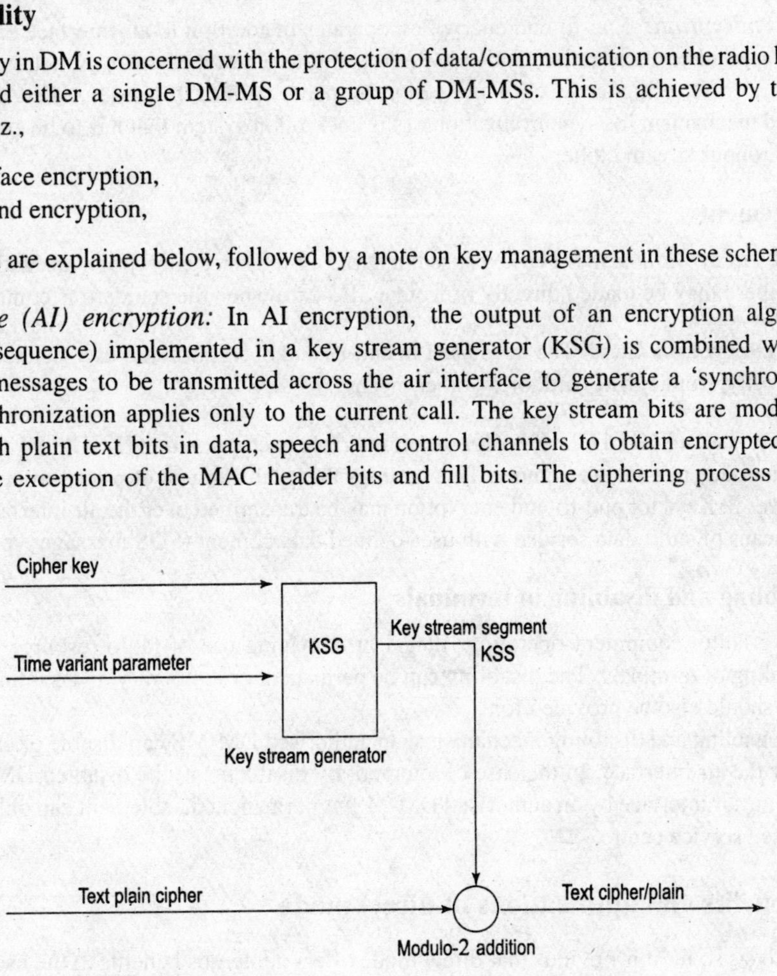

Fig. 27.22 *Speech and control information encryption.*

The KSG has two inputs—a time variant parameter (TVP) and a cipher key. The KSG produces a sequence of key stream bits referred to as a key stream segment (KSS).

Cipher Key: For direct mode, only one type of cipher key is defined, i.e. the static cipher key (SCK). The SCK can be considered to be a binary vector of 80 bits. Up to 32 SCKs can be stored for use in direct mode.

The time variant parameter (TVP): To prevent 'spoofing,' i.e. the same segment of keystream being used more than once, a time variant parameter (TVP) is included in the keystream generation. The initial value of the TVP can be randomly chosen by the call master. The TVP on messages from master to slave is independent of the TVP on messages from slave to master. To provide some degree of protection against the possibility of messages being recorded and replayed later, the initial TVP may contain a 'time of the day' element. The TVP is incremented in every time-slot.

End-to-end encryption: End-to-end encryption operates in addition to air interface encryption, but can only be applied to user traffic, and not control traffic. The mechanism is not standardized as it is intended to offer a higher level of protection by being specific to a user. There is however, a recommended mechanism for synchronization of the encryption system that has to be employed when using a synchronous stream cipher.

Key management

Air interface encryption keys: The keys used in DMO air interface encryption are static cipher keys (SCKs), and these may be loaded directly or from a SIM card when the standard is completed.

End-to-end encryption keys: The keys used in end-to-end encryption are user defined. These may be loaded directly, from a SIM card, or by other means.

Over the air re-keying (OTAR): Keys for the air interface encryption unit (KSG) may be transmitted over the air interface in a secure manner. This requires the establishment of a peer-to-peer messaging service at Layer-3. Keys for end-to-end encryption may be transmitted over the air interface in a secure manner by means of short data service with user-defined data content (SDS message type-4).

Secure enabling and disabling of terminals

Situations like faulty equipment operation, illegal or damaging use of radio resource by user, etc., warrant disabling of terminals. The disabling can be permanent or temporary. If disabling is provided for, enabling should also be provided for.

In secure enabling and disabling mechanisms, an authorised DM-MS can disable or enable another DM-MS over the air interface. In the case of a temporary disablement, the disabled DM-MS may be enabled over the air interface by an authorised DM-MS. A permanent disablement can only be reversed at an authorised service centre.

27.6.5 Benefits and applications of direct mode

From the analysis so far it is obvious that direct mode offers numerous benefits to the users. The main ones include

- DM operation outside the coverage of trunking infrastructure,
- DM operation in the areas of poor coverage inside the trunking infrastructure,
- extending trunking operation into areas of poor coverage,

– DM fall-back operation when the trunking system is inoperative, and
– DM operation of special nature inside the trunking infrastructure.

With each of these DM benefits there are some useful applications for each of the main user groups as detailed below.

Operation outside the coverage of trunking infrastructure: Consider an example of a derailment taking place on a railway network at a point outside the coverage of the trunking system. Each of the emergency services can retain not only its communication at the site of the accident by the use of direct mode, but can also maintain the interoperability between various services active at the site with careful planning of the allocated channels.

Another application of this kind of operation is for the network operators, especially during the early period of the network roll-out.

Operation in the areas of poor coverage inside trunking infrastructure: In an urban environment, the application may encompass underground offices, metro station platforms, car parks, shopping plazas, tunnels, subways, or any coverage "hole" within the system, which may come to light only in an emergency. In rural setting, such areas include caves, potholes, deep river valleys, and gorges.

Extending trunking operation to areas of poor coverage: Through DM-gateway, TETRA offers the possibility of extending the coverage of trunking system into areas of poor coverage by relaying messages originated in the trunking system to direct mode terminals and vice versa. Such an application may be that of the fire control centre talking to its fire crew deep inside a burning skyscraper via a fire tender positioned at the bottom of the building.

Fall-back when trunking system is inoperative: When trunking system becomes inoperative due to say a major system fault, an act of sabotage, or traffic overload, we can fall back on DM operation.

Operation of special nature, inside trunking infrastructure: In the working of convoys deployed for the protection of highly dangerous and/or valuable goods and substances and also VIPs (basically these are local operations where secrecy is vital), operating via trunked infrastructure may not be desirable. Also, in a railway network where intense activity is on at a derailment site, local conversations amongst the members of the service crew and executives need not clog-up a 'thin' trunking network available on the telecom cable/OFC along the railway line. In such situations, DM is extremely useful.

To conclude, direct mode options extend information exchange beyond the fringes of the trunked network.

27.7 Security in TETRA

In all the mobile systems, security is essentially an integral aspect. The GSM experience in this regard is worth mentioning — the inclusion of authentication of the mobile terminal by the network stopped the massive fraud that was occurring in the previous generation analog mobile systems. Further, GSM provides very reasonable confidentiality over the radio path using an encryption that cannot easily be broken in practice. TETRA has built on the GSM security and added features which are relevant for professional mobile radio users. These include end-to-end encryption, encryption for closed user groups and secure enabling and disabling of mobile terminals. We will now focus on TETRA security and describe in detail the TETRA security functions.

When describing TETRA security functions, it is important to make a distinction between different categories of functions and their specific application. Otherwise, we may tend to mix the issues while doing an analysis, and make the implementation difficult in an operational environment. In TETRA the following categories or classes of security functions exist:

- Security mechanisms
- Security management features
- Standard cryptographic algorithms
- Lawful interception mechanisms

Figure 27.23 depicts the basic relations between different security functions.

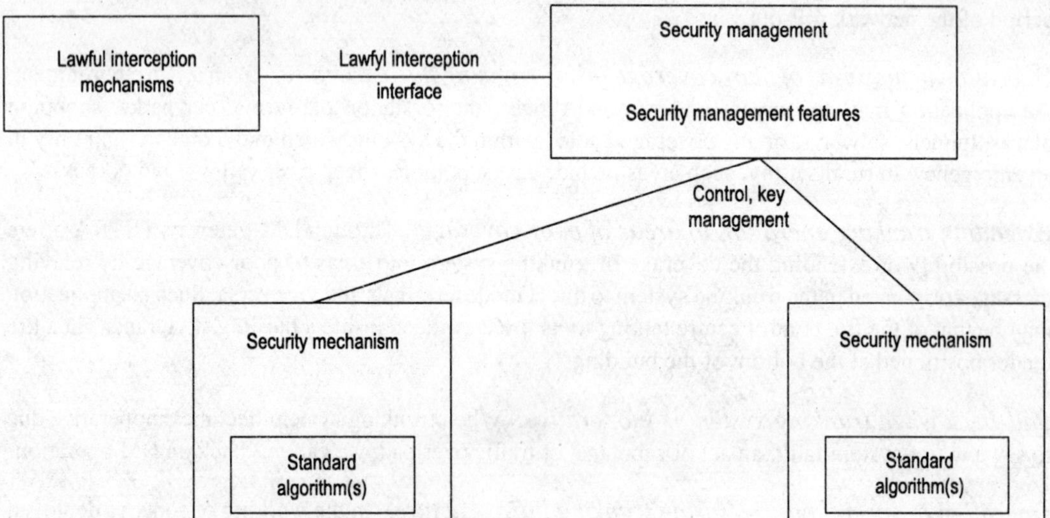

Fig. 27.23 *Relationship between security functions.*

Security mechanisms: These are independent, self-contained functions that aim to achieve a specific security objective such as confidentiality of information or authentication of mobile terminals. Security mechanisms are the main organs of a security system.

Security management features: These are functions that are used to control, manage and operate individual security mechanisms. They form the heart of the security and should guarantee that the security features are integrated into a consistent security system. Furthermore, they are used to realize interoperability of the security mechanisms over different networks. Key management is the most essential security management function.

Standard cryptographic algorithms: These are standardized system-specific mathematical functions that are used normally in combination with parameters called cryptographic keys to provide an adequate level of security for the security mechanisms and the security management features. Standardized cryptographic algorithms are offered in TETRA to support interoperability between different TETRA systems.

Lawful interception mechanisms: These are functions that are used within communication systems to provide the lawfully required access to information and communication, with the aim to fulfil national regulatory requirements. It is essential that such functions do not undermine the regular security of the system. Therefore, these functions should be controlled through security management.

27.7.1 Security mechanisms

TETRA standard has four types of security mechanisms in it. They are

- mutual authentication over the air interface,
- encryption,
- anonymity, and
- secure enabling and disabling of terminals,

which are discussed one by one next.

Mutual authentication over the air interface

In TETRA, as in most other secure systems, authentication is the firm basis for overall security. The TETRA standard supports mutual authentication of a mobile station (MS) on one hand and the network (which in TETRA terminology is referred to as switching and management infrastructure (SwMI)) on the other. This makes it possible for a TETRA system to control the access to it and for an MS to check if a network can be trusted. The mutual authentication provided in TETRA is useful for the following purposes:

- Control access of the MS to the network and its services
- Ensure that MSs are connected to the legitimate TETRA system
- Derive a unique session encryption key which is the derived cipher key (DCK) linked to authentication, and establish other security parameters. (The DCK is a unique encryption key used to encrypt the information exchanged on the link between the network and the MS.)
- Create a secure distribution channel for sensitive information such as other encryption keys
- Control the disabling and enabling of an MS/SIM in a secure way
- Ensure correct billing if the system supports public access also

A mutual authentication security mechanism is available for voice and data (V+D) and packet data optimized (PDO) mode. In the direct mode operation (DMO), an explicit authentication mechanism is not available; in this case the use of static cipher keys provide implicit mutual authentication.

The use of several authentication algorithms, both standard and proprietary, is supported (more on this later).

Mutual authentication is done on the basis of an authentication key K, which is unique for every MS or SIM. The key 'K' is stored in the MS/SIM and in the network. Normally a specific network element is used to store the authentication keys. This is called the authentication centre (AUC). Mutual authentication using an authentication centre is illustrated in Fig. 27.24.

Encryption

The reader might wonder why 'end-to-end encryption' at all when mutual authentication is available on the air interface. There is a subtle difference between the scope of these two. The air interface security is intended to secure the connection between the MSs and the network. However, in TETRA

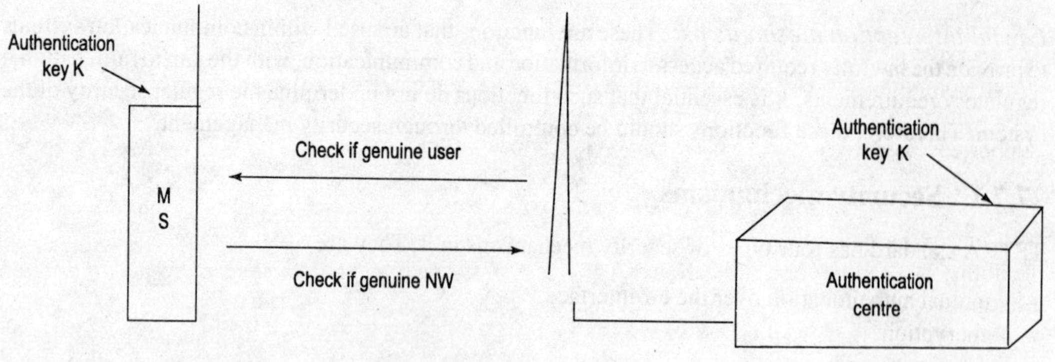

Fig. 27.24 *Mutual authentication.*

systems needing a very high level of security, additional security is required to protect information transmitted from one MS to another not only over the air interface but also within the network. In this case, end-to-end security provides an efficient solution.

The difference between the scope of air interface security and end-to-end security is illustrated in Fig. 27.25.

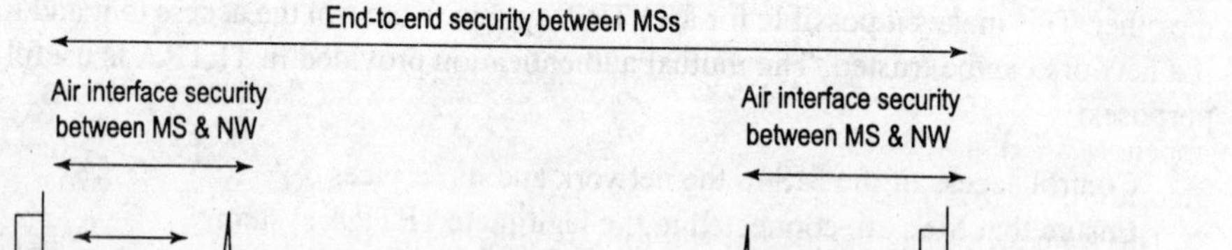

Fig. 27.25 *Air interface security versus end-to-end security.*

Air interface encryption: Using a variety of keys, the user and the signalling information can be encrypted over the air interface between the MS and the SwMI, both for individual and group communications. The air interface encryption mechanism is available for V+D as well as direct mode operation. Use of several encryption algorithms, both standard and proprietary, is supported.

End-to-end encryption: The TETRA end-to-end service can be realised in any number of ways. This means that a user may easily tailor an end-to-end encryption system to his own requirements. This flexibility is essential for a standard like TETRA that will be implemented in many forms for different user groups.

Public safety organisations will have specific (high) national security requirements for their implementation of end-to-end encryption which will be different from the requirements of military user groups which have even greater security requirements. All such organisations need to be able to specify an end-to-end encryption system according to their own requirements. It can also be expected that commercial user groups will have a need for secure end-to-end encryption systems.

The TETRA MoU end-to-end encryption framework

Many users, though having particular requirements over the cryptography used, may not wish to specify the rest of the end-to-end system (including key management). TETRA forum published the TETRA MoU SFPG recommendations which fully specify all that is required for an end-to-end service other than the detail of cryptographic algorithms. These are treated as black-box functions. However, in order to provide a complete solution for the general user, these recommendations conclude with an appendix showing how these cryptographic functions can be realised using the well known IDEA algorithm.

The framework has been designed to be adaptable to a range of security policies, with flexibility being achieved through a number of simple operational choices.

Anonymity

Anonymity can be achieved by the SwIM assigning temporary individual or group identities, and then encrypting these identities over the air interface. It is possible to make this encryption dynamic, i.e., an identity encrypted in a different way on different occasions. Anonymity is available for voice and data as well as direct mode operation.

Secure enabling and disabling of terminals

TETRA supports different options for a direct secure disabling or enabling of

- the MS equipment, based on the terminal equipment identity (TEI),
- the MS subscription, based on the individual TETRA subscriber identity (ITSI), or,
- both the MS equipment and the MS subscription.

If the TEI is disabled the MS cannot be used anymore even if another ITSI (which can be stored in a detachable module such as a SIM) is inserted in the MS. If the ITSI is disabled, an MS can still be used in combination with another (enabled) ITSI. The disabled ITSI cannot be used in any other MS. In addition, the disabling can be either temporary (which leaves the possibility to enable again) or permanent (which is irreversible). This results in the following nine states:

TEI	ITSI
Enabled	Enabled
Enabled	Temporarily disabled
Enabled	Permanently disabled
Temporarily disabled	Enabled
Temporarily disabled	Temporarily disabled
Temporarily disabled	Permanently disabled
Permanently disabled	Enabled
Permanently disabled	Temporarily disabled
Permanently disabled	Permanently disabled

The state diagram (Fig. 27.26) extracted from the TETRA security standard (ETS 300 392-7, TETRA-V+D, Part-7, Security Version 2.1.1) describes these functions and states.

In systems demanding high security, disabling and enabling should take place only after mutual authentication has been performed. Otherwise, the feature (especially the disabling) can be used to attack the system. TETRA standard provides the option to disable and enable without mutual authentication taking place first. User groups which do not need high level of security allow it so as to save costs, but user groups which require high level of security go in for mutual authentication before enabling/disabling terminals.

27.7.2 Security management features

By integrating security functions in TETRA, the security risks are 'condensed,' that is, they are concentrated to specific elements in the system. These are to be controlled, which is achieved by the 'security management features'. Security management features ensure the following:

- Security mechanisms are used in the proper way
- Different mechanisms are integrated in an appropriate way to achieve an overall secure system
- Interoperability between different (TETRA) systems is achieved in secure way

The form in which the security is condensed is normally through keys. A key is a piece of secret information that is used, often in combination with cryptographic algorithms, to provide the actual security of a security mechanism. The security management is responsible for dealing with the keys in a secure way. Though security management is partly an issue of implementation, TETRA has laid down certain specifications for security management features. In addition, the TETRA MoU SFPG has produced recommendations on 'key' management. In TETRA key management, *functionality* and *flexibility* are the key words. A large number of features have been integrated to support key management. A summary of these is provided below.

Authentication key

The authentication key K is used for mutual authentication between an MS and the SwMI. There are three possible methods for generating K which are outlined below.

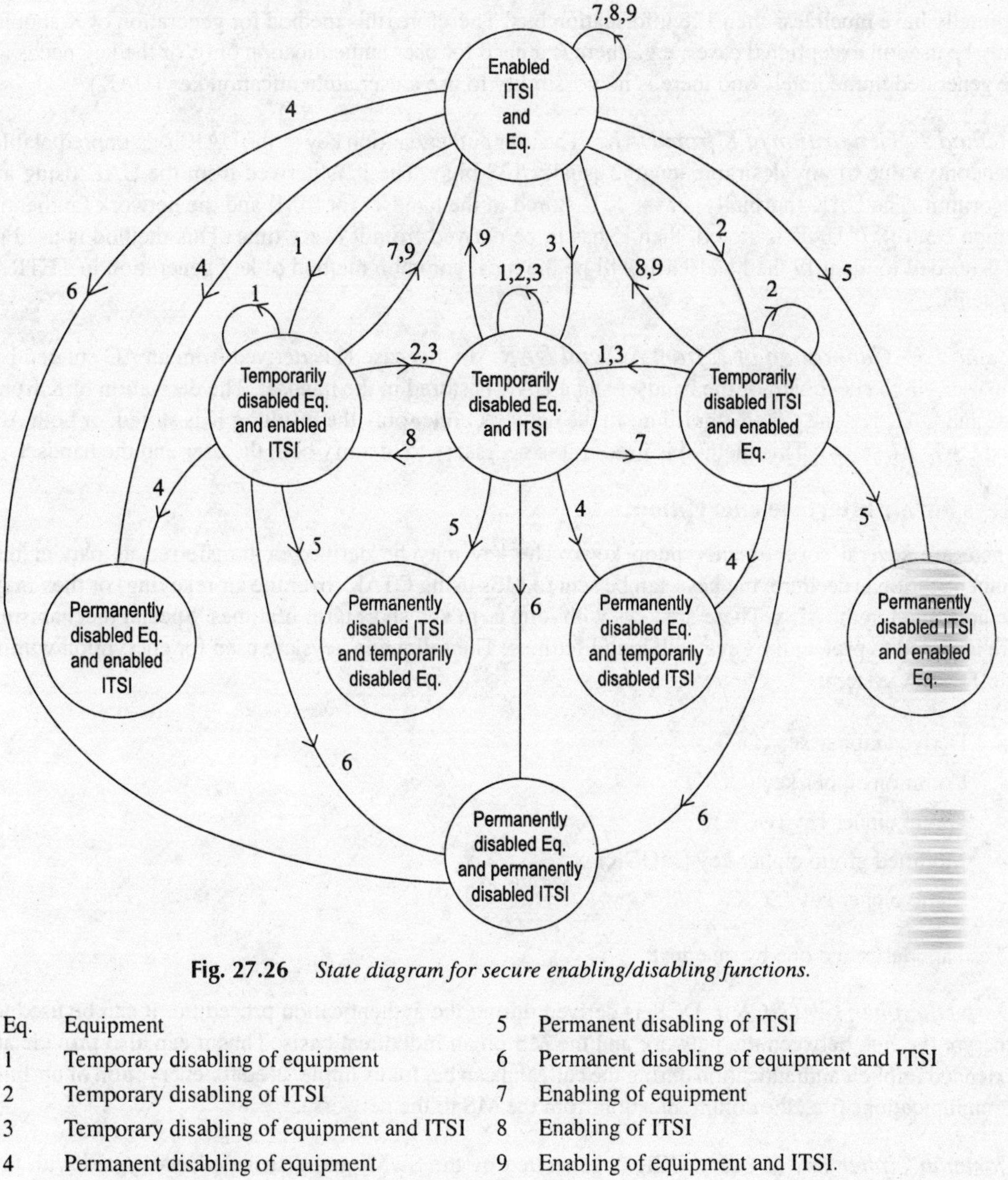

Fig. 27.26 *State diagram for secure enabling/disabling functions.*

Eq.	Equipment	5	Permanent disabling of ITSI
1	Temporary disabling of equipment	6	Permanent disabling of equipment and ITSI
2	Temporary disabling of ITSI	7	Enabling of equipment
3	Temporary disabling of equipment and ITSI	8	Enabling of ITSI
4	Permanent disabling of equipment	9	Enabling of equipment and ITSI.

Method 1 - Generation of K from an authentication code (AC): In this case the user types in an authentication code via the keyboard of the handset. The digits of the AC are represented as a string of bits. This bit string is translated using an algorithm to key K.

The AC is normally not stored in the handset. In the network (authentication centre), either the K or AC is stored. In the latter case K is derived from the AC every time it is needed. This method is used if it is required to identify the user of a handset, but not the handset. It should be noted that the AC would normally have much less then 128 information bits. Therefore, this method for generation of K should only be used in exceptional cases, e.g., there is a need for user authentication only, or the key needs to be generated immediately and there is no possibility to use a user authentication key (UAK).

Method 2 - Generation of K from UAK: The user authentication key or the UAK is an unpredictable (random) value of any desirable length (usually 128 bits). The K is derived from the UAK using an algorithm. The UAK (normally) or the K is stored in the handset (or SIM) and the network (authentication centre). If UAK is stored, then K has to be derived from it every time. This method is used if it is needed to identify the handset. It will be the most common method of key generation in TETRA systems.

Method 3 - Generation of K from AC and UAK: In this case K is derived from an AC entered by the user via the keyboard of the handset and a UAK is stored in the handset. The derivation of K from AC and UAK is done via an algorithm. In the network either only the resulting K is stored, or both AC and UAK are stored. This method is used if it is necessary to identify both the user and the handset.

Keys for air interface encryption

There are several sorts of encryption keys. The key may be derived or transferred as part of the authentication procedure, the keys can be sent to MSs using OTAR (over the air rekeying) or they may be preloaded in the MSs. There are keys with long term and short term lifetimes. Special mechanisms are included to protect the keys with long lifetimes. The following keys are used for encryption within the TETRA system:

- Derived cipher key (DCK)
- Common cipher key (CCK)
- Group cipher key (GCK)
- Modified group cipher key (MGCK)
- Static cipher key (SCK)

These are described one by one next.

Derived cipher key (DCK): DCK is derived during the authentication procedure. It can be used to encrypt the link between the network and the MS on an individual basis. Thus it can also provide an extended implicit authentication during the call, and can be, for example, used for encryption of up-link communications (i.e., the communication from the MS to the network).

Common cipher key (CCK): CCK is generated by the SwMI and distributed, encrypted with the DCK to mobile stations. It is efficient to use this key for encryption of messages that are directed to a certain location area (LA). A location area is a geographical area where a network and a number of MSs having logical connections are operational, for example, public safety organization of a city, administative departments, etc. In practice, the CCK can be used to set up a group call with all MSs that are at the moment in a certain area, independent of the specific closed user groups that these MSs are part of.

When the CCK is distributed to an MS over the air interface using OTAR, it is encrypted with the DCK of that particular MS.

Group cipher key (GCK) and modified group cipher key (MGCK): GCK is linked to a specific closed user group. It is generated by the SwMI and distributed to the MSs of a group (e.g., similar to the CCK, on a smart card, or using OTAR). Within a location area the GCK is always used in a modified form. It is encrypted by the CCK to obtain the modified group cipher key (MGCK). If an MS is in this location area, the MGCK is used to encrypt the closed user group messages for this MS.

When the GCK is distributed to an MS over the air interface using OTAR, it is encrypted with a session encryption key derived from the authentication key for this MS.

Static cipher key (SCK): SCK is a predetermined key which can be used without prior authentication. It is 'static' in the sense that it is not changed until replaced (for e.g., by an authentication exchange). TETRA supports the use of up to 32 SCK's. They can be distributed similarly to the GCKs. Their use is largely implementation-dependent but they can be used for e.g., encryption in direct mode operation (where they may also provide explicit authentication), and in certain TETRA systems for encryption for group and individual communications.

When an SCK is distributed to an MS over the air interface using OTAR, it is encrypted with a session encryption key derived from the authentication key for this MS.

OTAR

As indicated above, there is a possibility for distributing or updating CCKs, GCKs and SCKs using OTAR mechanism. This mechanism makes it possible to send air interface encryption keys from the SwMI over the air directly to an MS in a secure way, and can be applied as long as an authentication key K is available for the MS. The OTAR messages to an MS are encrypted using session encryption keys that are derived from the authentication key for this MS.

The OTAR mechanism can be used for both individual MSs and groups of MSs. A mechanism similar to OTAR is also available for the management of end-to-end encryption keys.

Transfer of authentication information between networks

If a TETRA MS roams to another TETRA network other than its home network, the visited TETRA network will need to obtain authentication information from the home network of the MS in order to be able to perform mutual authentication and generate and/or distribute encryption keys. The transfer of authentication information in networks is in principle supported in three ways:

- Transfer the authentication key K to the visited network
- Transfer certain information that can be used for one single authentication procedure
- Transfer only once a session authentication key for an MS (which can be used for repeated authentications) to a visited network without revealing the original authentication key of the MS.

The first method is simple (very fast) but not advisable for security reasons. The second method is basically the same method as applied in GSM, and can be implemented in a very secure way. However, it might cause too much overhead to transfer this information on a regular basis in TETRA systems. The third alternative is the best one—this option combines security and efficiency. Refer to Fig. 27.27 illustrating this method.

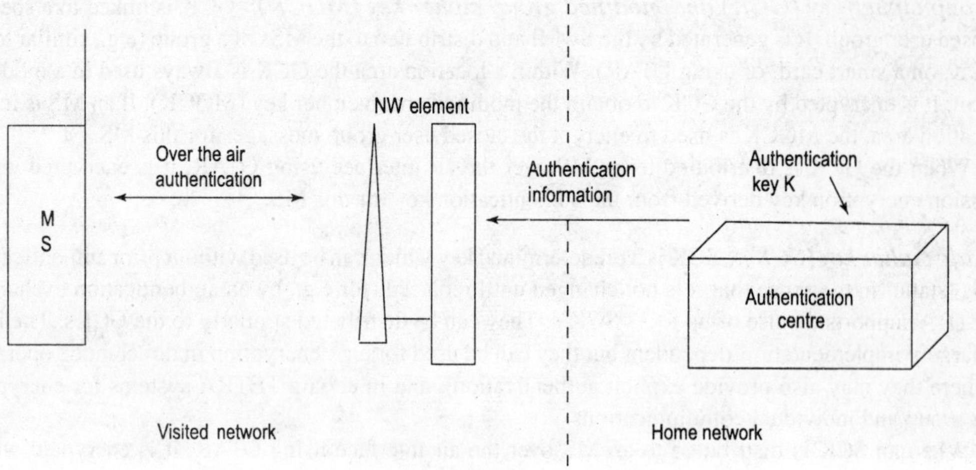

Fig. 27.27 *Authentication in a visited network (without disclosing the authentication).*

27.7.3 Standard TETRA cryptographic algorithms

TETRA standard offers a number of standard cryptographic algorithms which all have their own specific purposes. This section explains the purpose and use of these standard algorithms.

Air interface encryption algorithms

TETRA users can specify their own air interface encryption algorithm. However, for reasons of easy interoperability in multi-vendor systems, a number of air interface encryption algorithms have been specified as part of the TETRA standard. Several requirements have been taken into account when specifying these standard algorithms. The most important of these are the need for diversity and export control regulations.

Need for diversity: The reader is already aware that there will be a wide range of TETRA networks and applications. Not all users want to 'share' their standard encryption algorithms with all other TETRA users. For example, public safety organizations require their own standard air interface encryption algorithm.

Export control regulation: Equipment that includes encryption algorithms is likely to be subject to specific export controls in addition to any other functional controls. The encryption related controls are slowly being relaxed. Such controls are country-specific, but around 33 developed countries have evolved a common policy and derived their national controls from it. This policy is published under the banner of the Wassenaar Arrangement (details can be found at *http://www.wassenaar.org*). Controls on cryptography fall under Category 5, Part 2.

Four standard encryption algorithms are currently available for use in TETRA systems. These have been developed by ETSI's Security Algorithm Group of Experts (SAGE) for two different criteria. These are explained below.

TEA2 and TEA3 — Restricted export algorithms: These algorithms are controlled items under the 1998 Wassenaar Arrangement rules. The algorithms have been primarily designed for use by public safety organizations. The former algorithm (TEA2) has been assigned for use by public safety organizations associated with the European Schengen organization in European countries.

TEA1 and TEA4 — Readily exportable algorithms: TEA1 was an early design. TEA4 reflects the more relaxed controls of the 1998 Wassenaar Arrangement. Standard TETRA encryption algorithms are available to TETRA users and manufacturers. They are distributed by a custodian. In case of the TEA1, TEA3 and TEA4, the custodian is ETSI. TEA2 is distributed by the Dutch Police IT organization.

Air interface authentication and key management algorithms: TETRA users can specify their own air interface authentication and key management algorithm. Again, for easy interoperability in multi-vendor systems, a set of standard air interface authentication and key management algorithms have also been specified as part of the TETRA standard.

The requirements on diversity and export control regulations do not exist in the case of authentication and key management algorithms. Therefore, only a single set of standard air interface authentication and key management algorithms has been specified. This algorithm set is called the TAA1. Its specification is distributed by its custodian, ETSI.

End-to-end encryption algorithms

The end-to-end standard is written around four blackbox cryptographic functions designated E1 to E4. Those users with the necessary expertise may define how these are realised using algorithm(s) of their own choice. The only constraint is that the algorithm(s) have to fit within the broad parameters of functions E1 to E4.

For those users who are content to follow a public standard, the recommendations include an appendix which shows how these cryptographic functions can be realised using the IDEA algorithm. So, the body of the recommendation, together with the appendix, forms the complete specification for a standard TETRA end-to-end encrypted voice service. On this basis, the TETRA MoU has established a licence agreement with ASCOM (the owners of IDEA) covering the use of IDEA in this context.

Details of the licensing agreement and how to make an application are available with the TETRA forum's SFPG secretariat.

27.7.4 Lawful interception mechanisms

Public as well as private telecommunication network operators have to provide lawful interception facilities to the responsible national authorities. Since a standardized solution is much more cost efficient than proprietary implementations on a case-by-case basis, support for such lawful interception has been provided within the TETRA standard. A subgroup of the TETRA security group has standardized a Lawful Interception Interface (refer ETSI standard EN301 040 : TETRA Security – Lawful Interface V2.0 1998) to support the mechanisms for lawful interception. The detailed implementation of this interface might differ on a country-to-country basis.

27.8 TETRA versus GSM

While explaining TETRA system, reference to GSM has been made at several places for the purpose of similarity or differentiation. In this section, all such similarities/differences are put together at one place for the benefit of the reader.

27.8.1 Comparision between GSM and TETRA traffic profiles

GSM traffic profiles show that 85 per cent of all mobile calls are made via the PSTN, 20 per cent of which are PSTN initiated and 80 per cent mobile initiated. A further 5 per cent of all calls are from mobile to mobile between different regions, and only 10 per cent of all calls are mobile to mobile calls within the same region.

The traffic profile of a private mobile radio system (to be implemented as TETRA) shows a typical call duration of only 20 seconds. Furthermore, 90 per cent of all calls are made within the same region, 60 per cent of which are initiated by a control dispatcher. A further 5 per cent of all calls are between two regions and only 5 per cent are via the PSTN, of which less than 1 per cent are initiated from the PSTN.

While GSM calls are point-to-point (individual calls), TETRA calls are a mixture of point-to-point and point-to-multipoint (group calls), with a predominance of group calls.

27.8.2 Comparison between TETRA and GSM voice services

Along with telephony-style full duplex voice services and supplementary services, PMR (private mobile radio) users like TETRA users need the 'despatcher' style of working. This includes call types suitable for group-style communication, like group call, acknowledged group call and broadcast call and supplementary services such as dynamic group number assignment (DGNA), late entry (LE) and discreet listening/monitor.

The GSM standard was largely based on the ISDN standard which does not have group and broadcast calls, monitor and other despatcher facilities. The fact that it takes several seconds to establish a call, which then normally lasts for several minutes, is not an issue for GSM users. In contrast, TETRA users such as the emergency services, require sub-second group-call call setup as they handle incidents where calls are typically of very short duration and speed of communication is vital.

TETRA being private mobile radio system, has to work in noisy environments to a background of sirens, traffic, loudspeaker announcements, multiple speakers, etc. The TETRA codec was designed to provide good quality speech in these harsh environments and has a voice quality superior to GSM in conditions of poor signal.

27.9 Data Services and Multimedia in TETRA and GSM

Private mobile radio users are the most demanding mobile users of data services. Their main data applications include status report, job despatching from the control centre and – for users in the field — access to corporate information databases. For these applications, due to the intrinsic limitations of analogue systems, TETRA has to complement its voice systems with secondary, data-only technologies, with consequential implications on cost and complexity.

The new requirements include transmission of drawings, photographs and even full motion video, to be used, for example, for sending back pictures from an incident to the control centre or transmitting video from CCTV surveillance cameras. TETRA can support all this—TETRA, in addition to providing status and short data services of up to 256 characters, provides a user (on a dynamic basis) with up to four TDMA slots in a single communication channel — effectively 'bandwidth on demand.' This applies to packet-mode and circuit-mode data, with data rates of up to 28 kbps. When combining this with the rapid advances in data compression technique, currently being finalised in two international fora (MPEG-4 of the Moving Picture Expert Group and SG15 of the ITU), the quest for mobile multimedia is rapidly becoming a reality.

In comparison, GSM offers 160 characters of SMS (short message service) and circuit-mode data only (no packet data) at less than half the rate of 9.6 kbps.

27.9.1 TETRA's direct mode

One powerful operational mode of TETRA not provided by GSM is the so-called direct mode (DMO), which equips radio handsets to communicate directly with each other without using the network infrastructure. There are several additional DMO features like repeater operation, dual watch (ability to work in either direct or trunked mode and the Gateway), relaying communication between direct mode and trunked mode in real time, etc. The DMO functionality comprises voice, individual, broadcast and group call (including all-informed) services, circuit data and short data service (SDS) — both clear or encrypted.

27.9.2 TETRA security versus GSM security

GSM offers the option of over-the-air encryption, although most GSM users do not require it. TETRA, in contrast, offers comprehensive encryption capabilities which can be tailored to match the varied security requirement of private mobile radio users.

Authentication procedures check that a mobile station requesting the service is a bona fide one, and similarly, the infrastructure trying to remotely disable or 'kill' a terminal is legitimate. The air interface can be encrypted using two standard algorithms to meet the security needs of public safety organizations and business users.

For users who require confidentiality throughout the whole network, proprietary end-to-end encryption can be added. Standard features include encryption of speech, signalling and data, support of late entry, OTAR, and protection against spoofing (replay).

27.9.3 Architectural differences

TETRA is the most frequency-efficient standard for mobile communication, equivalent to 6.25 kHz per channel, voice or data. Its architecture is fully scaleable, from a large high-capacity nationwide system to a low-capacity single site system with up to four channels and only 25 kHz of spectrum. In contrast, GSM frequency allocations are in multiples of 200 kHz for eight channels, equivalent to 25 kHz per channel, and the system design has been optimized for high-capacity, multi-site nationwide systems for 10,000 subscribers upwards.

The system architecture also had to meet the needs of PMR users who presently have their own analogue system but will share a common TETRA network. It allows different users to have different functional profiles — a different view of the common physical network (i.e., the base stations, backbone and switches), with full control of their own 'virtual network.'

Such virtual networks offer the benefits of trunking through a shared common resource. They can also provide system performance equivalent to current private conventional (analog) systems, by combining various TETRA services and facilities (viz., open channel, access priority, pre-emptive priority), and an equivalent level of control.

27.9.4 GSM extensions to provide TETRA like facilities

There is now a move towards developing additional, TETRA-like facilities for GSM, pushed by users attracted by the competitive price of GSM equipment. These are known as Phase 2 + +, ASCI and GSM-R. The major additions under consideration are group and broadcast calling, fast call setup and pre-emptive priority; however there are still uncertainties about the technical and commercial feasibility of these extensions, since they are addressing a comparatively small market, while requiring significant modifications to the relatively narrow technological platform of GSM.

27.10 TETRA and the Railways

Railway organizations in Europe and USA, though using PMR, were weary of the shortcomings of this analog system, particularly in terms of continuous geographic coverage in urban areas, security, transmission quality of communication in trains running at high speeds, and data handling. As already seen in this chapter, these shortcomings of analog systems were foremost among the issues felt and addressed by ETSI when formulating the TETRA standard. Also under consideration was the belief that in the interests of keeping the overall system costs low, end-users would have a choice of suppliers for both terminals and network infrastructure. A common open standard is the best way forward.

As rolling stock and track management in Railways have become more sophisticated and subject to even greater pressure for efficient working, the limitations of analog radio have forced rail and track operators to adopt a plethora of communications systems such as PMR for train radio; GSM for executives and administrators; pagers for operating and running staff; track-side fixed lines/cables for communication between maintenance gangs and control office; and even for emergencies; VHF radios between driver and guard/station staff, etc. Railways in developing countries are trying out all types of systems that they can access, while the mobile communications industry is busy developing solutions for next-generation systems. In Europe, the railways has been quick to jump on the bandwagon of digital wireless communications, and in concert with the likes of Ericsson, Alcatel, Matra and Siemens, is developing both country-specific and pan-European signalling protocols. The network platform being considered by many is a modified version of GSM, currently used around the world for public digital cellular communications. Much work has been done in this area and much investment made in R&D, but TETRA has a very strong case for its adoption on the ground of its being ubiquitous rather than application-specific communications platform in the rail industry. Let us see the strong points of TETRA from the point of view of Railways.

1) TETRA operates in the 450 MHz frequency range, around half of that of the 800 MHz of GSM; it has the inherent range advantage in that the transmitters/receivers (base stations) can be located much further apart (up to 100 km at present) while offering the same geographical coverage. This implies significant savings in infrastructure costs.
2) TETRA can also offer raw data transmission rates far above those of GSM. It has been proved in trials and in fully functional pilot networks that TETRA will deliver high-quality, full duplex

voice transmission from a radio handset no bigger than a mobile phone, while simultaneously transmitting or receiving data at 28.8 kbps.

3) Information in visual, textual and graphical forms can be shared by many or few users depending on a given situation, without putting any limitations on voice traffic.
4) Data can be downloaded to and displayed on a variety of terminals, from handsets to mobile data pads (similar to hand-held computers).
5) The range of possible applications for the technology is wide — handling TBS (transmission-based signalling) traffic, securing track possession, distributing traffic data, and carrying voice traffic including controllable access to PSTN.
6) Just about anything audio or visual can also be transmitted as data, opening the way for video images and stills displaying timetables, traffic and signaling information, route changes and a whole host of other information to be shared between those, and only those who require it.
7) With all this, the voice quality can still be good thanks to the voice codec which provides for standards of speech hitherto only dreamed of in PMR circles. The system's ability to filter out unwanted sound is particularly impressive in high-noise environments, and it is probably more appropriate to measure how close the speech quality comes to that of a land-line call than dwell on its distance from traditional analog mobile radio.
8) TETRA has the ability to operate from vehicles travelling with speed up to 450 km/hr and cope with widely-fluctuating speeds during transmission—whether in the open or while moving through the longest and deepest of tunnels. Rigorous testing and development by mobile radio specialists like UK-based Simoco (which also holds a patent on voice enhancement in noisy environments) has ensured that these parts of the standard are more than adequately covered. The company recently completed trials on the metro system in Hanover (Germany), proving that the system works seamlessly as users move at great speeds from above the ground to below, and all through the subterranean network.
9) In addition to the above, are the advantages of instant call setup and a wide range of user-definable features.

With no doubt, TETRA is the best choice for Railways' communications needs.

27.11 TETRA and the Future

Building early confidence in a new technology is usually something of an uphill task for those keen to promote it, but although the first full network installations are only just beginning, confidence among potential buyers, apart from the vendors, already seems high. At the end of year 2000, the UK's Home Office specified TETRA as the standard for the country's police and emergency services. It awarded its largest ever contract, said to be worth around £1.5 billion over three years, to a consortium led by British Telecom. A trial system has been set up for appraisement. The West Midlands Ambulance Trust, who's area includes England's second biggest city Birmingham, decided not to wait for the official trial and bought a complete TETRA mobile radio network from Simoco in February 2001.

Although TETRA is primarily a European standard, telecommunications industry watchers are confidently predicting that it will become a global standard in much the same way as GSM has with public cellular telephony market. Indeed, signatories to the Shengen Treaty, which provides for among other things cross-border policing and includes most European countries, are expected to adopt TETRA as their platform for police and public safety mobile communications. Organizations that have been holding back on system upgrades till an open digital standard emerged, are also likely to adopt TETRA systems.

According to a report from market analysts The Strategies Group of USA, the total number of mobile radio users is expected to reach around 11 million in Europe alone by 2002, resulting in cumulative subscriber equipment sales of around US$ 4.1 billion. Significantly, it adds that from 1997 to 2002, US$ 850 million of that total value will be attributable to the sale of TETRA subscriber units. It is rare indeed for a new technology to carve out such a significant slice of the market in its early years of implementation, however the enthusiasm with which European governments and commercial enterprises alike have greeted TETRA standard suggests that the predicted market projections will have to be only revised upwards.

Following in the footsteps of the success of GSM, TETRA is making rapid advances in the Asia Pacific region. In China, where analogue trunk systems are used by the Emergency Services, the spectral efficiency and powerful feature set of TETRA are very attractive. At a recent forum organized by the Chinese Ministry of Electronics Industry, TETRA was presented to a forum of nearly hundred representatives from the national offices of major bureaus, research organizations and equipment suppliers.

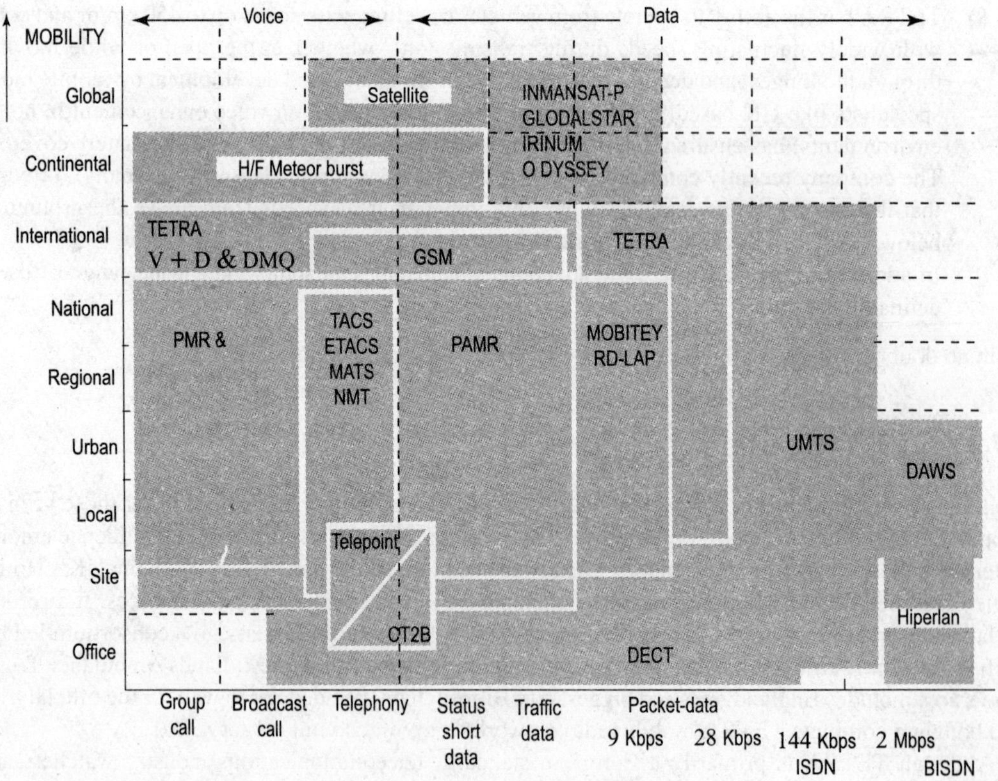

Fig. 27.28 *Positioning of TETRA w.r.t. other mobile communication systems.*

There are also opportunities for TETRA in Thailand, where a tender has recently been issued for a digital PAMR system.

Significant Australian interest in TETRA is evident from the formation of the Australian branch of the TETRA MoU earlier this year. A number of police forces in the region are likely to upgrade to TETRA from their old PMR systems.

Figure 27.28 shows the positioning of TETRA w.r.t. other mobile communication systems. On the x-axis, the information (data) rate is indicated, while on the y-axis, mobility is indicated.

References

[1] ETSI (1998l) Terrestrial Trunked Radio (TETRA), Voice plus Data (V + D), ETSI spec. ETS 300 392.

[2] ETSI (1998m) TETRA – Packet DATA Optimized (PDO), ETSI spec. ETS 300 393.

[3] ETS 300 396 – 1 to 7, Radio equipment and system (RES) for TETRA direct mode, Part 1 to 7:

Part – 1 : General network design

Part – 2 : Direct MS-MS air interface – Radio aspects

Part – 3 : Direct MS-MS air interface protocol

Part – 4 : Repeater air interface

Part – 5 : Gateway air interface

Part – 6 : Security

Part – 7 : Type-2 repeater

[4] Ranko Pinter, TETRA Direct Mode, Papers presented at the introductory workshop on TETRA 2000.

[5] TETRA MoU SFPG Recommendations:

01 Key distribution

02 End-to-end encryption

06 Management of static cipher keys in DMO

[6] G. Roelofsen, TETRA security – Information security Technical Report, Vol. 5, No. 3 pp. 44—54, 2000.

28

Satellite System for Mobile Communication

28.1 Introduction

This chapter focuses on the utility of satellite system in supporting mobile communications. A satellite system can be thought of as a mobile communications network in which base stations are also on the move in addition to MSs, and additional instances of call hand-over arise. But for this complexity, we have the advantage of making use of the features of a satellite system which score over a terrestrial mobile phone network on several counts. Satellite systems are independent of varying population densities, and they offer global coverage without incurring the wiring costs for base stations. Their coverage extends into areas not covered by terrestrial system. They offer a simple and quick solution for a global network.

We start with the basics of satellite systems and go on to discuss how the satellite systems support mobile communications.

28.2 Basics

Satellites move in circular or elliptical orbits around the earth. Satellites in circular orbit maintain a constant radial distance from earth in accordance with the law

$$F_g = mg(R/r)^2, \quad \text{and} \quad F_c = mr\omega^2,$$

where

F_g = attractive force of due to Earth's gravity,
F_c = centrifugal force working on satellite,
m = mass of the satellite,
R = radius of Earth,
r = distance of the satellite from the centre of Earth, i.e., the radius of the satellite orbit,
g = acceleration due to gravity, and
ω = angular velocity of the satellite (equals $2\pi f$, where f is the frequency of rotation).

From $F_g = F_c$, it can be derived that

$$r = \left[(gR^2)/(2\pi f)^2\right]^{1/3}.$$

When the period of rotation equals 24 hours, r equals 35,786 km. A satellite with orbiting period of 24 hrs and placed above the equator (i.e. in equatorial plane) is called a geostationary satellite.

28.3 Some Important Definitions

Inclination angle: It is angle between equatorial plane and the plane described by the satellite's orbit (for geostationary satellites, inclination angle is 0 degrees).

Elevation angle C: It is the angle between the center of satellite beam and the plane tangential to earth's surface.

Footprint: It is the area on earth where the signals of the satellite can be received.

Perigee: If the satellite does not have a circular orbit, the point closest to the earth in its orbit is called perigee. Figure 28.1, and Fig. 28.2 illustrate the above parameters.

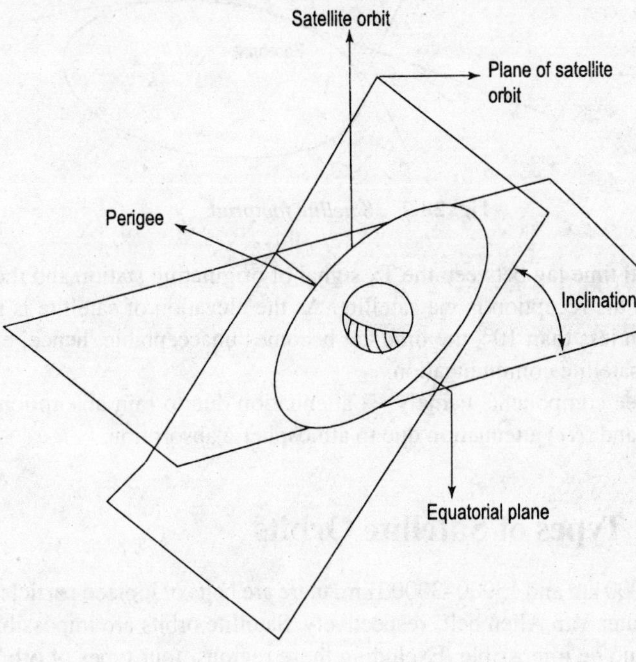

Fig. 28.1 *Satellite orbit parameters.*

Propagation loss: It is the attenuation of signal power between the satellite and the receiver on earth. It is given by the expression.

$$\ell = \left(\frac{4\pi d f_c}{c}\right)^2,$$

where

f_c = carrier frequency,

c = velocity of light, and

d = distance between the satellite and the receiver on earth.

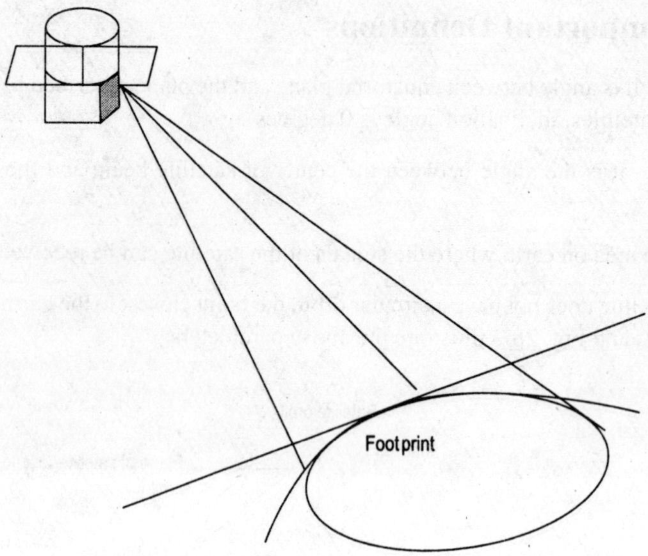

Fig. 28.2 *Satellite footprint.*

Latency: It is the total time lag between the Tx signal of originating station and the Rx signal of the receiving station when the reception is via satellite. As the elevation of satellite is reduced, time lag increases. For elevation less than 10°, the time lag becomes unacceptable, hence, elevation less than 10° is not suitable for satellite communication.

Attenuation has three components namely (*i*) attenuation due to rain absorption, (*ii*) attenuation due to fog absorption, and (*iii*) attenuation due to atmospheric absorption.

28.4 Different Types of Satellite Orbits

At altitudes of 2000–5000 km and 15000–30000 km, there are belts of ionized particles known as 'inner Van Allen belt' and 'outer Van Allen belt' respectively. Satellite orbits are impossible in these ranges as communication would be impossible. Excluding these regions, four types of orbits as listed below are possible for satellites (Fig. 28.3).

Geostationary or geosynchronous earth orbit: Satellites in this orbit are known as GEO satellites. The radius of the orbit is about 36000 km, i.e., beyond outer Van Allen Belt.

Medium earth orbit: Satellites in this orbit are known as MEO satellites. The radius of this orbit is between 5000 km and 12000 km, i.e., above inner Van Allen belt and below outer Van Allen belt.

Low earth orbit: Satellites in this orbit are known as LEO satellites. The radius of the orbit is 500–1500 km, i.e., below inner Van Allen belt.

Highly elliptical orbit: Satellites in this orbit are known as HEO satellites. Their orbits are non-circular (elliptical) but within outer Van Allen belt, they may cross inner Van Allen belt.

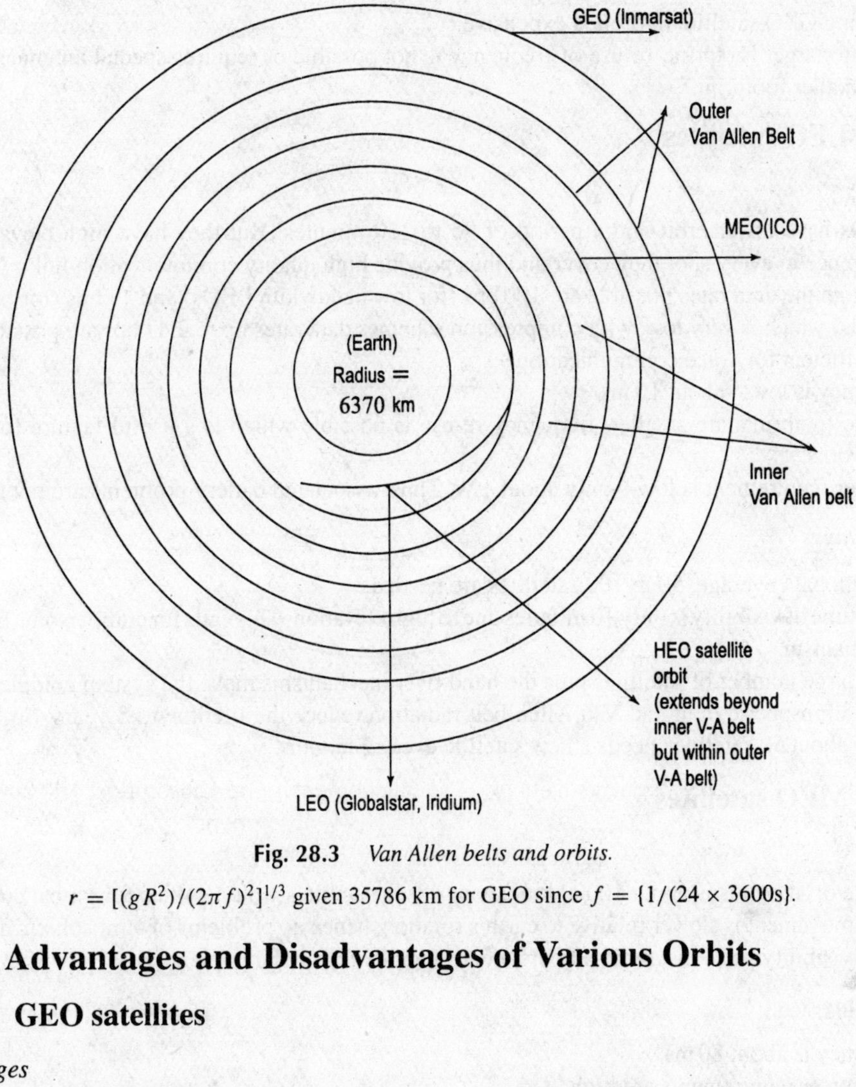

Fig. 28.3 *Van Allen belts and orbits.*

$r = [(gR^2)/(2\pi f)^2]^{1/3}$ given 35786 km for GEO since $f = \{1/(24 \times 3600\text{s})\}$.

28.5 Advantages and Disadvantages of Various Orbits

28.5.1 GEO satellites

Advantages

- With three GEO satellites, complete coverage of earth is possible.
- Antenna positions are fixed and no adjustments are required.
- High lifetime for satellites (about 15 years).
- No hand-over needed due to large footprint.
- Since relative movement with earth is zero, no Doppler shift is exhibited by these satellites.

Disadvantages

- Due to low elevation of GEO satellites, the northern and the southern regions of the globe have reception problems. To overcome this, larger antennas are needed which increase the payload.
- Tx power needed is 10 W which brings in associated battery problems.
- High latency of 0.25 s would be a problem for voice and data communications.

- Placing GEO satellite in orbit is expensive.
- Due to larger footprint, re-use of frequency is not possible or requires special antennas to focus on smaller footprint.

28.5.2 LEO satellites

Advantages

- LEOs have lower orbit and a period of 95 to 120 minutes. But they have high elevation with reference to every spot they cover and thus provide high quality communication link.
- Though the data rates possible are 100 bps (for low bandwidth LEOs) and 1 kbps (for broadband LEOs) which is very low, with compression schemes data rates up to 2.4 kbps are possible which is sufficient for voice communication.
- Latency is low—about 10 ms.
- Since footprints are smaller, frequency re-use is possible which is a useful feature for cellular networks.
- Power requirement is low—only about 1 W. Thus, associated battery problems are not there.

Disadvantages

- For global coverage, 50 to 100 satellites are needed.
- The time of visibility is only 10 minutes due to high elevation. This calls for cumbersome hand-over mechanism.
- The large number of satellites plus the hand-over mechanisms make the system complex.
- The atmospheric drag and Van Allen belt radiation reduce the lifetime to 8 years. So, a system with about 50 satellites needs a new satellite every 2 months.

28.5.3 MEO satellites

Advantages

- Since orbit radius is about 10000 km, about dozen satellites would suffice for global coverage
- The movement is slower relative to earth's rotation, hence no problems of atmospheric drag arise. The visibility period is about 6 hours, hence hand-over is not complex.

Disadvantages

- Latency is about 80 ms.
- Tx power requirements are high
- Special antennas are needed for appropriate footprints

28.5.4 HEO satellites

Advantages

- The orbits are selected to have perigee over large cities, hence it provides better quality of communication in cities.

Disadvantages

- The orbits cross inner Van Allen belt, hence black-out is possible.
- Orbits are not circular. The design is complex.

28.6 Satellite System for Global Mobile Communications

28.6.1 Communication links

A satellite system supporting global mobile communication is shown in Fig. 28.4. Three links are identified in this figure:

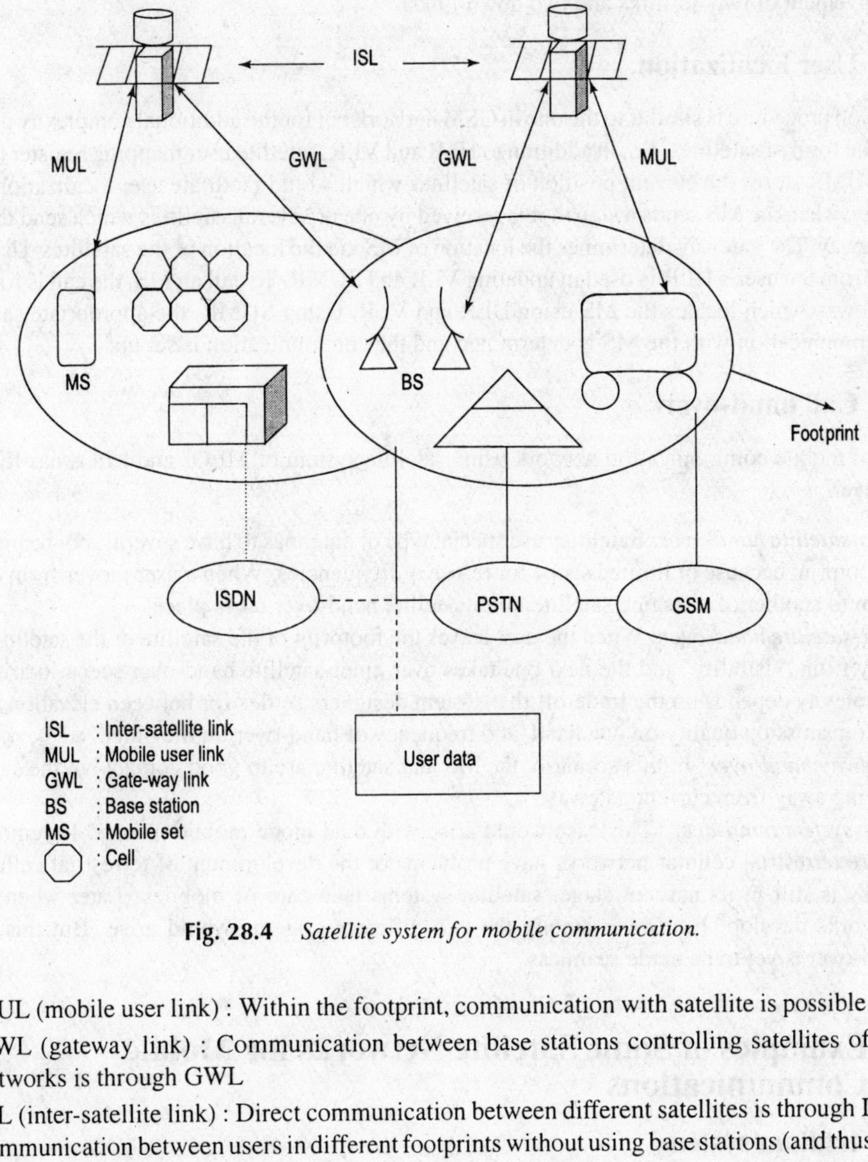

Fig. 28.4 *Satellite system for mobile communication.*

- MUL (mobile user link) : Within the footprint, communication with satellite is possible via MUL
- GWL (gateway link) : Communication between base stations controlling satellites of different networks is through GWL
- ISL (inter-satellite link) : Direct communication between different satellites is through ISL. Also, communication between users in different footprints without using base stations (and thus reducing latency) is possible via ISL.

28.6.2 Call routing

Routing in fixed segment is similar to the process in terrestrial mobile communication network. For the network portion in space, i.e., routing between satellites, the two possibilities are

- through ISL if the network supports ISLs (low data latency),
- traffic relayed to earth, routed there and relayed back to satellite (high data latency) due to involvement of two up-links and two down-links.

28.6.3 User localization

Localization procedure is similar to the one in GSM network but for the additional complexity of mobile BTSs in the form of satellites. So, in addition to HLR and VLR, satellite user mapping register (SUMR) is used. SUMR stores the current position of satellites which would facilitate user localization.

Initially, when the MS sends a signal, it is received by one or several satellites which send the report the to gateway. The gateway determines the location of MS via the location of the satellites. User's data obtained from the user's HLR is used in updating VLR and SUMR. To call an MS, the call is forwarded to the gateway which locates the MS using HLR and VLR. Using SUMR, the appropriate satellite to set up communication with the MS is determined and the communication is set up.

28.6.4 Call hand-over

The global mobile communication network using satellite system of MEOs and LEOs has four types of hand-over.

- *Intra-satellite hand-over:* Satellites use special type of antennas to have several spot beams within its footprint because of limited scope for re-use of frequencies. When a user moves from one spot beam to another of the same satellite, intra-satellite hand-over takes place.
- *Inter-satellite hand-over:* When the user leaves the footprint of the satellite or the satellite moves away from 'visibility' and the next one takes over, inter-satellite hand-over scenario arises. The complexity depends on the trade-off that system designers settles for between elevation angle or the transmission quality on one hand, and frequency of hand-over on the other.
- *Gateway hand-over:* In this scenario, the MS and satellite are in good contact but the satellite is moving away from current gateway.
- *Inter-system hand-over:* This case would arise with dual mode mobile phones. In remote areas where terrestrial cellular networks have problems or the development of terrestrial cellular networks is still in its nascent stage, satellite systems take care of mobiles. Later when cellular networks develop, hand-over among the two different systems would arise. But this type of hand-over is yet to be made seamless.

28.7 Examples of Some Satellite Networks for Mobile Communications

28.7.1 Iridium system

This is the first LEO system commissioned to cover the entire world. It was envisaged to have 77 satellites (hence the name Iridium, whose atomic number is 77), but finally had 66 working plus 6 spare. Its orbit altitude is 780 km, and the weight of each satellite is 700 kg. The minimum elevation is 5°.

The system has ISLs; MSs operate in the 1.6138–1.6256 GHz band in FDMA/TDMA scheme with TDD. Feed links to satellites use 29.1–29.3 GHz for up-link and 19.4–19.6 GHz for down-link. ISLs use 23.18–23.38 GHz. The number of channels supported is 4000, at bit rate of 2.4 kbps. The lifetime of each satellite is 6 years on an average.

The latest news is that the Iridium project is 'out of orbit' (due to high replacement costs of satellites!).

28.7.2 Globalstar system

Though the name is Globalstar, this system does not provide global coverage. But the bandwidth granted to customers is high (a data rate of 9.6 kbps is allowed). The number of channels supported is 2700. This is also a LEO system.

There are no ISLs, hence hand-over is indirect. The orbit has an altitude of 1414 km ±70° latitude. Minimum elevation is 20°. The number of satellites in the system is 48 (plus 4 spare), each weighing 450 kg. The lifetime of each satellite is 7.5 years. The multiplexing scheme followed is CDMA.

In this system, MS to satellite link is on 1.610–1.6265 GHz band and satellite to MS link is on 2.4835–2.5 GHz band. Gateway to satellite links are on 5.091–5.250 GHz band, and satellite to gateway links are on 6.875–7.055 GHz band.

28.7.3 Teledesic system

This is an ambitious LEO system planned to provide worldwide access not only for mobile phones but also for internet via satellite. Teledesic is designed to offer 64 Mbps down links and 2 Mbps up-links/64 Mbps up-links with special terminals. Receivers have roof-mounted equipment with connection to local networks within the building.

The number of satellites to be used in this system is 288 (in 12 planes with 24 satellites per plane). The target for commissioning the system is August, 2003. Lifetime of each satellite would be 10 years. Due to higher bandwidth planned, higher bands of

28.6 – 29.1 GHz for up-link
18.8 – 19.3 GHz for down-link

are used.

Elevation planned is 40° since at 18 GHz and 28 GHz rain attenuation would be high and should be avoided.

Teledesic would have several ISLs and would use TDMA/FDMA multiplexing schemes. The number of channels offered would be 2500.

28.7.4 Intermediate circular orbit (ICO) system

This is a MEO system with 10 satellites plus 2 spares with an orbit of 10390 km. The data rate possible is 4.8 kbps and number of channels is 4500. The multiplexing scheme is FDMD/TDMA. The system provides for ISLs.

Each satellite weighs 2600 kg and works at an elevation of 20° and can cover 30% of earth. The lifetime of a satellite is 12 years.

MS-to-satellite link is on 2 GHz band and satellite-to-MS link is on 2.2 GHz. Base station to satellite link is on 5.2 GHz, and satellite to base station link is on 7 GHz band.

28.8 Present Status

Satellites functioning as backbone for telecom networks are being replaced by optical fiber cables crossing oceans. However, satellite systems for mobile communications have carved a niche for themselves by their coverage of remote areas/low population areas where it is too expensive to install base stations, and also in areas where mobile communications are yet to be adopted in a big way. Thus, satellite system serve the purpose of extending mobile phone networks into areas without coverage.

Exercises

1. How do inclination and elevation determine the use of a satellite?
2. Elucidate the functions of a satellite as an interworking unit in a communication network.
3. What problems can you perceive with a mobile phone linked to a satellite system in (*a*) a skyscraper (*b*) an office in a congested road of populous city?
4. Why is OFC being preferred to satellite system?

References

[1] http://www.teledesic.com/.
Teledesic Corporation, Kirkland, WA, USA

[2] http://www.ico.com/
ICO Global Communications, London, UK

[3] http://www.iridium.com/
Iridium LLC, Washington DC, USA

[4] Miller, B., Satellites free the mobile domain, *IEEE Spectrum*, March 1998.

29

Synchronous Digital Hierarchy (SDH)

29.1 Introduction

Digital multiplexing hierarchy of plesiochronous bit-streams has been described in Chapter 3 up to 5th order multiplexing, i.e., up to 565 Mbps data frame. Fig. 29.1 illustrates PDH (plesiochronous digital hierarchy) in a different manner.

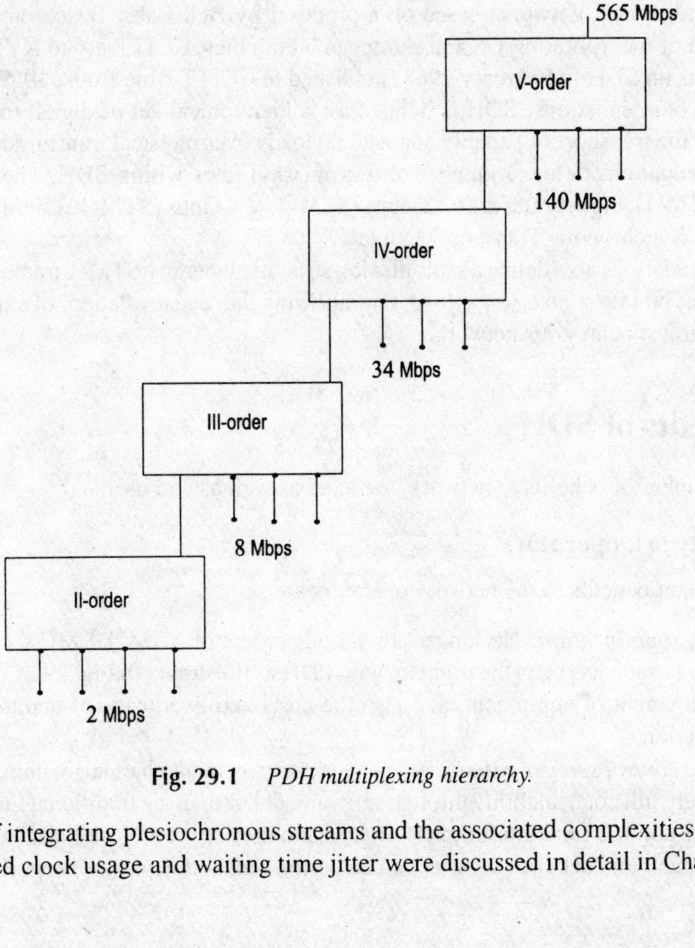

Fig. 29.1 *PDH multiplexing hierarchy.*

The process of integrating plesiochronous streams and the associated complexities like justification process, perforated clock usage and waiting time jitter were discussed in detail in Chapters 12 to 15.

As end users become more and more aggressive in their demands for flexibility, bandwidth, and also sophisticated network management, communications system designers have to transcend to a higher strata from PDH. With higher bandwidth being no longer a technological limitation thanks to optical fibre, the solution provided by communication system engineers is SDH or synchronous digital hierarchy.

SDH defines a structure which enables the encapsulation of plesiochronous signals to form standard digital frames that are transmitted over the network in which all transmission (includes reception) equipment are synchronized to an overall network clock.

29.2 Evolution of SDH

The evolution of SDH has in fact been triggered by the inability of PDH to extract individual circuits from high capacity systems without resorting to demultiplexing, down the hierarchical level. The evolution has been propelled by the promise of higher bandwidth by optical fibre, the demands for flexibility in transmission networking by end users, and the sophistication required in such network management.

In 1984, pioneering work towards formulation of standards for synchronous transmission began in USA. The subcommittee TIXI of ANSI (American National Standard Institute) began developing SONET (standard optical network) based on a proposal by Bell Labs. Based on the SONET work and also the idea of incorporating US and European hierarchies, CCITT group XVIII brought out the recommendations on SDH in February 1988 (published in CCITT Blue Book, 1989).

In CCITT recommendations, SDH is defined as 'a hierarchical set of digital transport structures, standardized for the transport of suitably adapted payloads over physical transmission networks'.

The recommendations define a number of transmission rates within SDH. The first is 155 Mbps (referred to as STM-1); higher rates are 622 Mbps (STM-4), 2.4 Gbps (STM-16) and 10 Gbps (STM-64). STM stands for 'Synchronous Transport Module'.

The recommendations also define a multiplexing structure whereby STM-1 frame can accommodate a number of lower bit rate signals as payload, thus allowing the 'encapsulation' of existing PDH signals to be carried over a synchronous network.

29.3 Benefits of SDH

SDH offers a number of benefits to network operators as well as end users.

Benefits to network operators

The three important benefits to the network operators are :

1. *Network simplification:* No longer are various orders of MUX-DEMUX needed to extract a tributary (which has been the problem with PDH as illustrated in Fig. 29.2). This would reduce the deployment of equipment, and also the costs and overheads of maintenance and power consumption.
2. *Software control sophistication:* The network management channels within SDH frame structure ensure full controllability through software. Other than the traditional functions like alarms and event control, functions like performance monitoring, configuration management, resource management and network security can be performed, and is part of network management.

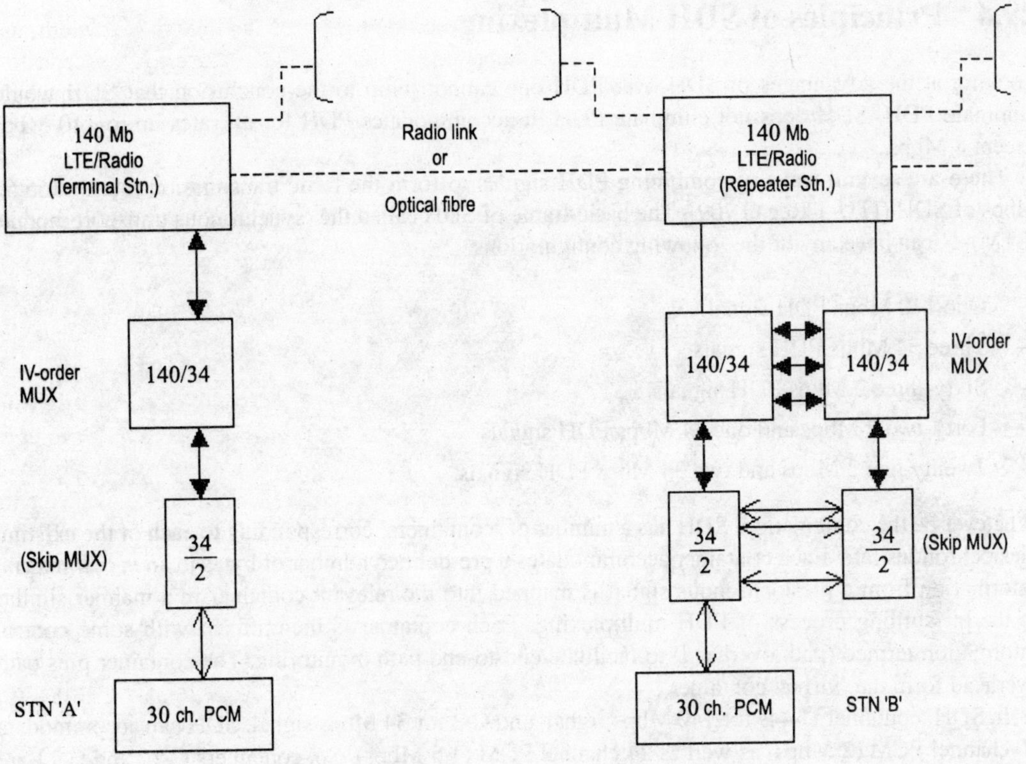

Fig. 29.2 *Problem of MUX/DEMUX operations down the hierarchy in PDH.*

(Note : Station A is terminal station and station B is repeater. At station B, to extract one 2 Mb tributary, we have to proceed down the hierarchy 140/134 and 34/2, duly making through the unwanted 34 Mb and 2Mb bit-streams.)

3. *Survivability:* Using a self-healing ring architecture, the network will be automatically reconfigured and the traffic instantly re-routed in case of a fault over a section, and thus the repair of a faulty section is facilitated.

Benefits to end-users

The three important benefits of PDH to the end user are :

1. *Bandwidth on demand:* Users anywhere in the network can subscribe at short notice for services requiring high bandwidth (e.g. dial-up video conference).
2. *Future-proof networking:* SDH is an ideal platform for a host of services ranging from POTS (plain old telephone service) to ISDN; from mobile radio to WAN; from video conferencing to digital video broadcasting.
3. *Standardization:* Because of the standards defined for fiber-to-fiber interfaces at physical level, transmission equipment from different manufacturers can easily get inter-worked. Users who lease SDH bandwidth and provide their own services can derive considerable benefits from this feature.

29.4 Principles of SDH Multiplexing

Looking at the advantages of SDH over PDH one cannot jump to the conclusion that SDH would eliminate PDH. SDH does not eliminate PDH, it accommodates PDH for all rates up to 140 Mbps except 8 Mbps.

There are several ways of combining PDH signals to form the basic transmission rate of 155.52 Mbps of SDH (ITU-T Rec.G 709). The basic frame of SDH called the 'synchronous transport module (STM)-1' can have any of the following configurations:

- One 140 Mbps PDH signal
- Three 34 Mbps PDH signals
- Sixty-three 2 Mbps PDH signals
- Forty-two 2 Mbps and one 34 Mbps PDH signals
- Twenty-one 2 Mbps and two 34 Mbps PDH signals.

Whatever be the combination, SDH has a number of 'containers' corresponding to each of the existing plesiochronous rate. Each container accommodates a pre-defined number of bytes in an $m \times n$ matrix. Information from a plesiochronous signal is mapped into the relevant container in a manner similar to the bit stuffing process of PDH multiplexing. Each container is then affixed with some control information termed 'path overhead' to facilitate end-to-end path monitoring. The container plus path overhead form the 'virtual container'.

In SDH, container C-4 is for 140 Mbps signal, and C-3 for 34 Mbps signal. SDH can accommodate 30 channel PCM (2 Mbps) as well as 24 channel PCM (1.5 Mbps). So, containers C-12 and C-11 are for 2 Mbps and 1.5 Mbps signals.

Corresponding to the containers, we have virtual containers VC-1, VC-3, and VC-4.

Though in the synchronous network of SDH all the equipment are synchronized to a master clock—the time delay associated with transmission link not being constant—we have problems of 'disturbance' of the locations of VC within the STM frame. This problem is overcome by associating a pointer with reference to the beginning of the frame with each VC to indicate the relative position of VC within the STM frame.

Filling up of containers C-12, C-3, C4 and forming of VC-1, VC-3, VC-4 are discussed in detail in Sections 29.10, 29.11 and 29.12 respectively. The process of filling up containers forming VCs is done in a nested way to first fill up the STM frame's payload area and then 'section overhead' bytes are added (for O & M facilities, protection switching, monitoring, etc.). The detailed structure of STM frame is presented in Section 29.5.

The basic rate of 155.52 Mbps is for STM-1 frame. Higher rates of transmission are achieved by byte-interleaved multiplexing to have

STM-4 (four STM-1s) –622.08 Mbps;

STM-16 (sixteen STM-1s or four STM-4s) –2488.32 Mbps (2Gbps);

STM-64 (sixty-four STM-1s or four STM-16s) - 9953.28 Mbps (10Gbps).

As can be seen in this interleaving, there is no addition of overhead bits. The higher rates are exact multiples of lower rates.

Note: To exploit the higher bandwidth provided by fibre optical transmission lines, the work on synchronous multyplexing hierarchy began in the form of SONET (synchronous optical network) in North America and SDH (synchronous digital hierarchy) in Europe. This resulted in a series of ANSI standards for SONET and ITU-T recommendatations for SDH. Although there are some implementation differences between SONET and SDH, all SONET specifications conform to SDH recommendations. In SONET, the basic frame in STS-1. If has 90 columns by 9 rows of bytes and is of 125 μs duration. Hence, the transmission rate of basic SONET frame is

$$\frac{(90 \times 9 \text{ bytes}) \times (8 \text{ bits per byte})}{125\ \mu s \text{ per frame}},$$

which works out to 51.834 Mbps.

STS-N signal has N times the basic data rate (i.e., $N \times 51.84$ Mbps). The basic SDH data frame (i.e. STM-1) is equivalent to STS-3 frame, i.e., $3 \times 51.84 = 155.52$ Mbps. The commonly used SONET and SDH transmission rates and their equivalences are shown below.

SONET level	SONET frame	Line rate (Mbps)	SDH equivalent
OC-1	STS-1	51.84	–
OC-3	STS-3	155.52	STM-1
OC-12	STS-12	622.08	STM-4
OC-24	STS-24	1244.16	STM-8
OC-48	STS-48	2488.32	STM-16
OC-96	STS-96	4976.64	STM-32
OC-192	STS-192	9953.28	STM-64

29.5 STM Frame Structure

The basic STM frame is a matrix of 9 rows and 270 columns with each column accommodating one byte. The transmission is row by row, starting from left column and moving to right (please refer to Fig. 29.3). Columns 9 to 270 form the payload. Columns 1 to 9 are for overhead bytes.

The frame repetition rate is 125 μs, which implies that a byte in the payload represents 64 kbps channels. The bit rate of 155.52 Mbps for basic STM frame can be arrived at as shown below.

$$\text{Number of rows in a frame} = 9$$

$$\text{Number of columns} = 270$$

$$\text{Number of bites/frame} = 9 \times 270$$

$$\text{Number of bites/frame} = 9 \times 270 \times 8$$

A frame repeats at 125 μs, i.e., 8 kbps.

$$\text{Therefore, bit rate of frame} = \text{number of bits frame} \times \text{frame repetition rate.}$$

$$= 9 \times 270 \times 8 \times 8 \text{ kbps} = 155.52 \text{ Mbps.}$$

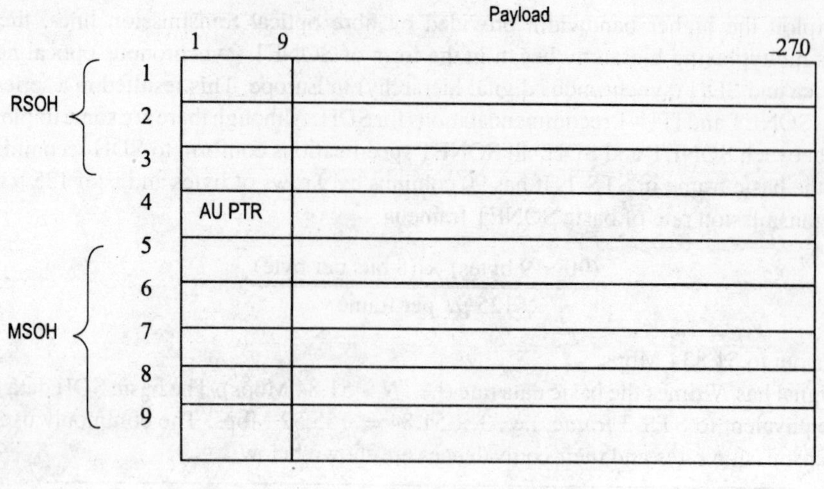

Fig. 29.3 *STM frame structure.*

In each row, columns 1 to 9, i.e. 9 bytes, are for overhead bytes called 'section overhead' (SOH) bytes. A further classification of SOH is made into 'regenerator section overhead' (RSOH) bytes and multiplex section overhead (MSOH) bytes. The fourth row is designated for administrative unit (AU) pointer. An SOH bytes map is shown in Fig. 29.4. The purpose of each byte is detailed below the figure.

The payload of STM frame is the transport capacity of the frame (i.e. excluding overhead bytes). Thus, columns 10 to 270 constitute the payload of STM frame.

29.6 Network Elements in SDH

Synchronous multiplexers used in SDH accept a wide range of tributaries and offer a number of possible output data rates. The multiplexers of SDH perform line terminator's functions also. The regeneration at repeaters is similar to PDH. However, to perform functions like cross-connections, 'digital cross-connects' are needed.

The following types of equipment exist in SDH network:

- Terminal multiplexers (TM)
- Add-drop multiplexers (ADM)
- Digital cross connects (DXC)
- Regenerators(RGNR)
- Network management system (NMS)

29.6.1 Terminal multiplexer (TM)

TM combines plesiochronous and synchronous input signals into higher bit rate STM-N signal (refer to Fig. 29.5). All the currently available plesiochronous bit rates (except 8 Mbps) are accepted by TM.

A1	A1	A1	A2	A2	A2	J0	X	X
B1			E1			F1	X	X
D1			D2			D3		
B2	B2	B2	K1			K2		
D4			B5			D6		
D7			D8			D9		
D10			D11			D12		
S1					M1	E2		

A1, A2	Frame alignment	J0	Trace identifier
B1, B2	Parity bytes for error monitoring	K1, K2	Automatic protection switching (APS channel)
D1, D2, D3	Data communication channel (DCC)	M1	Transmission error ACK
D4 to D12	Network management	S1	Clock quality monitor
E1, E2	Order wire channel		Media dependent bytes
F1	Maintenance		

Fig. 29.4 *SOH byte map of STM frame.*

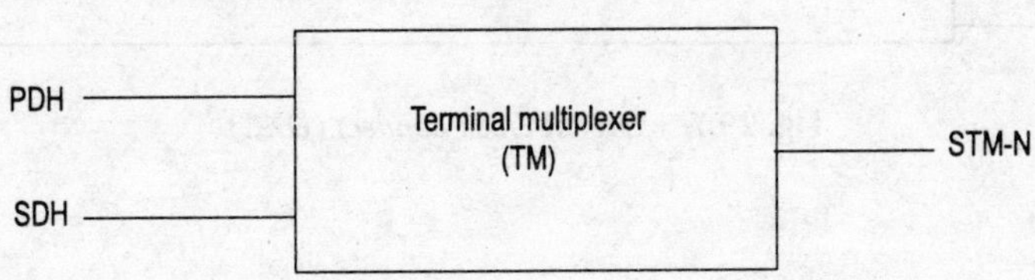

Fig. 29.5 *Terminal multiplexer.*

29.6.2 Add–Drop multiplexer (ADM)

Plesiochronous and lower bit rate synchronous signals can be extracted from or inserted into high speed SDH bit streams by means of ADMs. The ADMs enable setting-up of ring structures to provide back-up path switch-over. ADMs are similar to drop–insert MUXes of PDH (refer to Fig. 29.6).

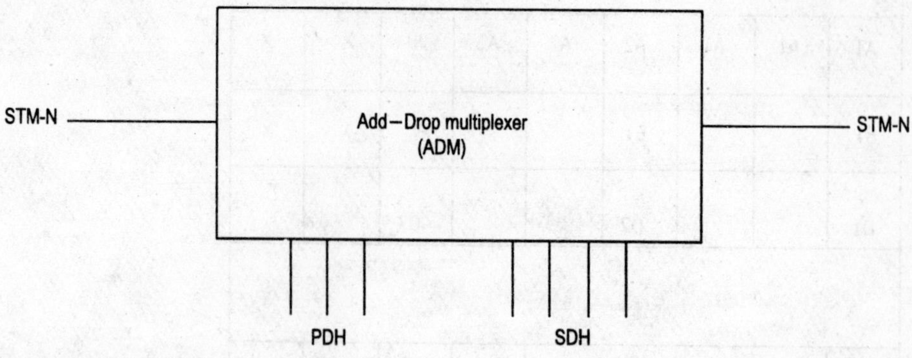

Fig. 29.6 *Add–drop multiplexer.*

29.6.3 Digital cross connects (DXCs)

Sometimes, semi-permanent interconnections are needed between different channels to perform routing down to VC level. Digital cross connects (DXCs) are used to provide mapping of PDH tributary signals into virtual containers (VCs) and switching of various containers up to VC-4, inclusive of VC-4 (Fig. 29.7).

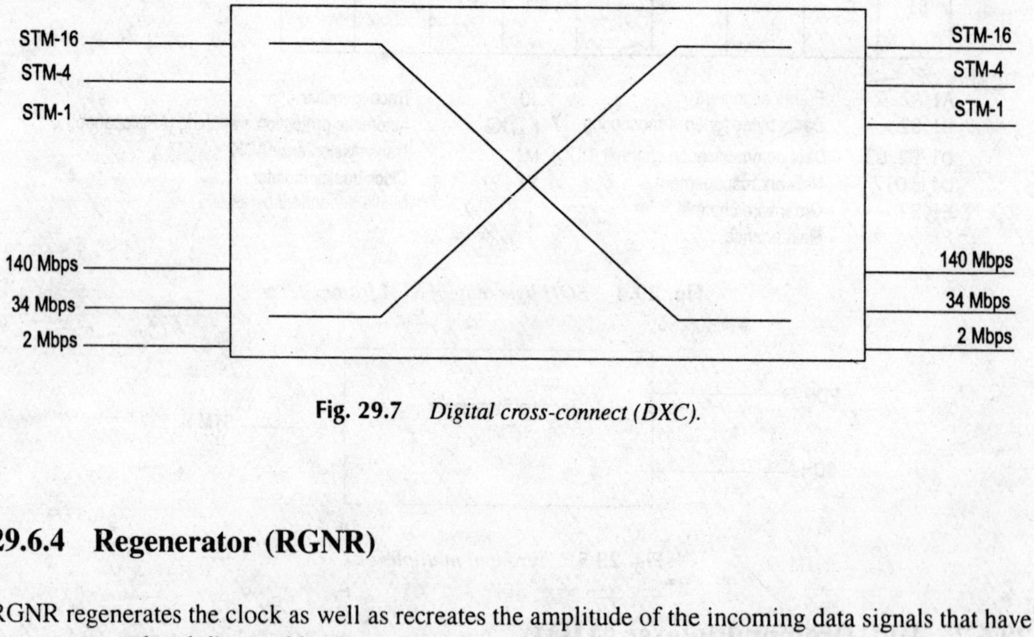

Fig. 29.7 *Digital cross-connect (DXC).*

29.6.4 Regenerator (RGNR)

RGNR regenerates the clock as well as recreates the amplitude of the incoming data signals that have been attenuated and distorted by dispersion. The clock is derived by RGNR from the incoming data stream. RGNR also receives messages by extracting various 64 kbps channels (for example E1, F2, service channels of RSOH). RGNR can output messages on these channels. RGNR is represented as shown in Fig. 29.8.

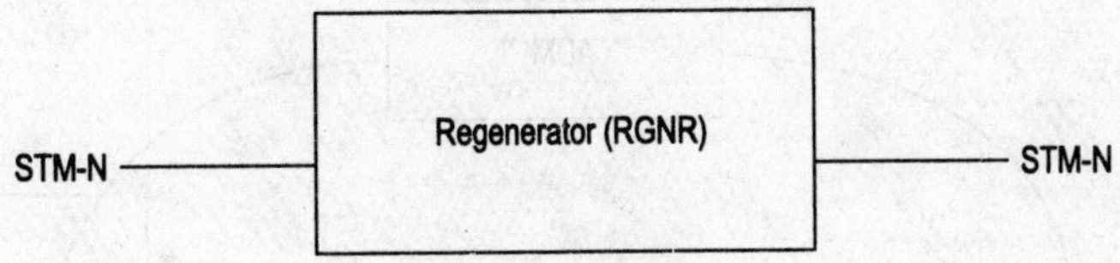

Fig. 29.8 *Regenerator.*

29.6.5 Network management system (NMS)

A computer system with software for controlling various equipment is a prominent entity in an SDH network. It is termed network management system (NMS).

29.7 SDH Network Topology

Commonly used topologies for SDH network are point-to-point topology; point-to-multipoint topology; ring topology with same type of elements; and ring topology with different types of elements.

In point-to-point topology, two TMs are connected as shown in Fig. 29.9. In point-to-multipoint topology, two TMs are connected via ADM or DXC to provide 'drop-insert' facility at ADM location (refer to Fig. 29.10). A ring using only ADMs is shown in Fig. 29.11. In the event of single failure on ring (break in medium), 'self-healing' is possible.

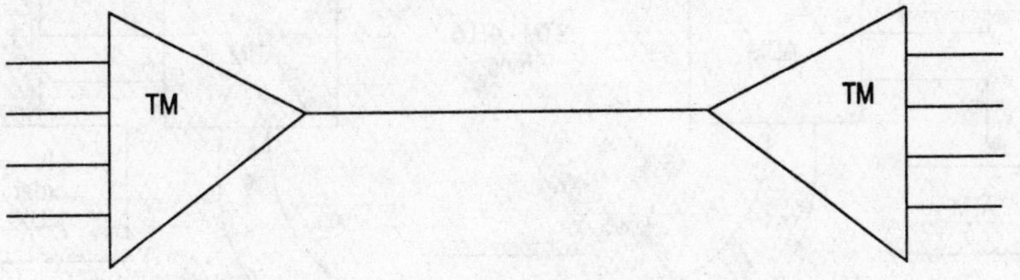

Fig. 29.9 *Point-to-point topology.*

Fig. 29.10 *Point-to-multi-point topology.*

A ring with different elements is shown in Fig. 29.12.

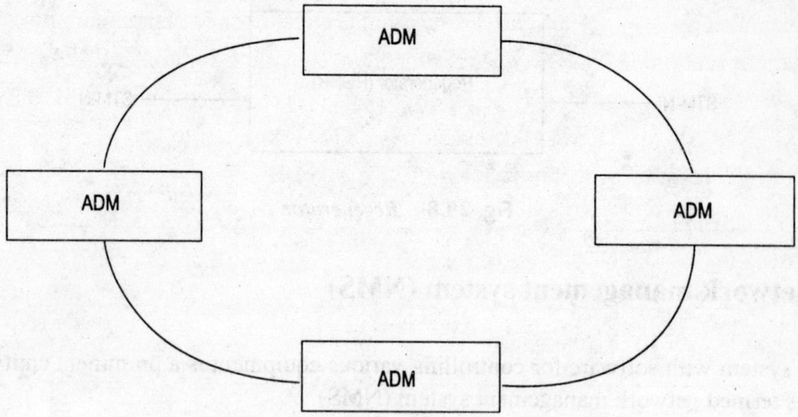

Fig. 29.11 *Ring topology (with ADMs).*

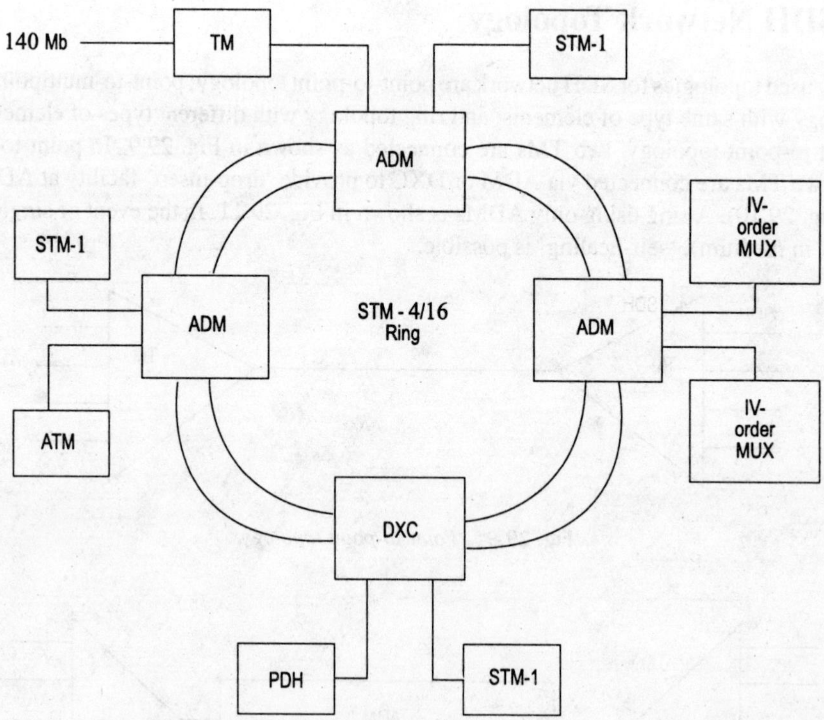

Fig. 29.12 *STM-4 and STM-16 ring with different elements.*

29.8 Layer Model of SDH

Similar to ISO's OSI model for computer network, we can present SDH in layer model as shown in Fig. 29.13. In this model also, peer-to-peer connection is envisaged. The different layers are:

The physical layer :	Represents TX medium (optical fiber or digital radio link)
The regenerator section:	Path between regenerators signalling is through RSOH bytes.
The multiplex section :	Path between multiplexers processing is through MSOH bytes.
VC-3/VC-4 layer :	Payload connection via STH frame end-to-end

Further, the VC layer provide interface with PSTN/ATM/IP.

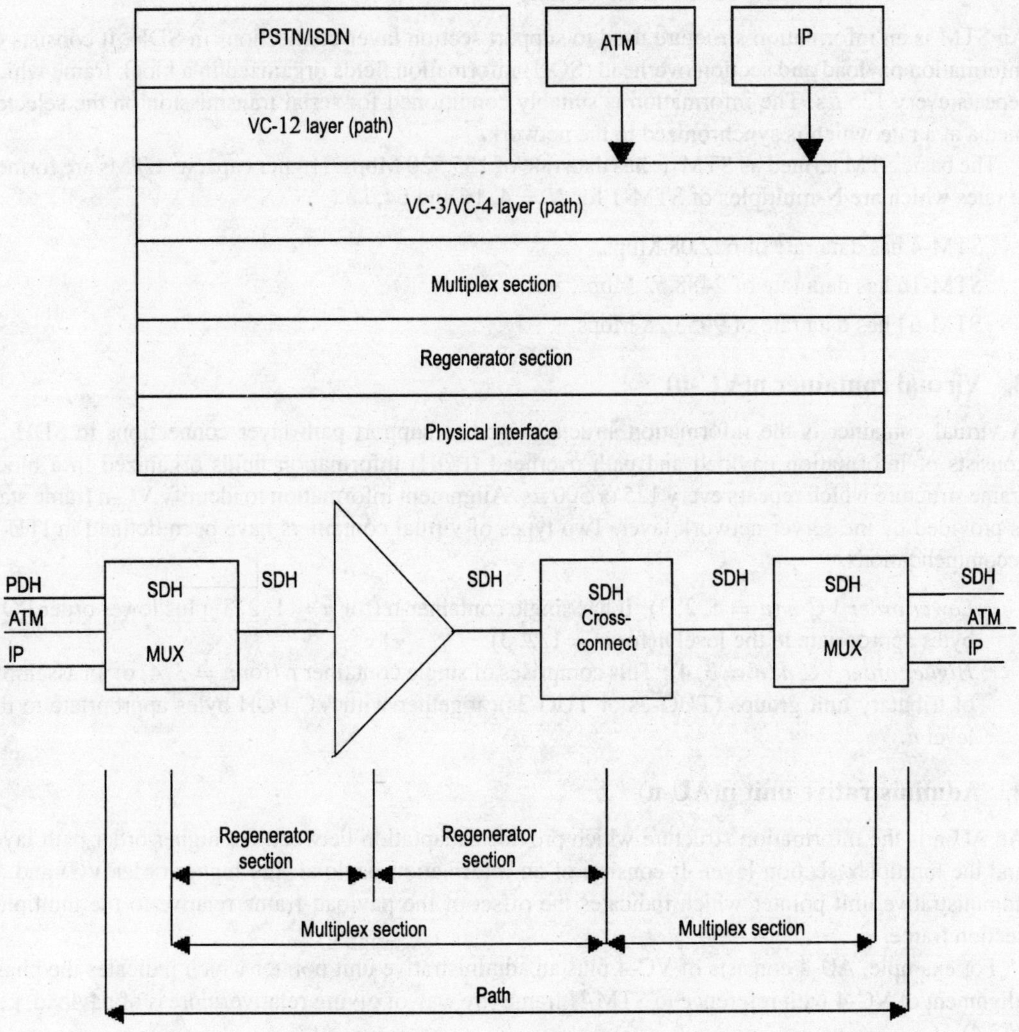

Fig. 29.13 *Layer model of SDH network.*

29.9 Important Terms associated with SDH

1. Synchronous digital hierarchy (SDH)

SDH is a hierarchical set of digital transport structures standardized for the transport of suitably adopted payloads over physical transmission networks.

2. Synchronous transport module (STM)

An STM is an information structure used to support section layer connections in SDH. It consists of information payload and section overhead (SOH) information fields organized in a block frame which repeats every 125 μs. The information is suitably conditioned for serial transmission on the selected media at a rate which is synchronized to the network.

The basic STM termed as STM-1, has data rate of 155.520 Mbps. Higher capacity STMs are formed at rates which are N-multiples of STM-1 for $N = 4$, 16 and 64, i.e.,

STM-4 has data rate of 622.08 Mbps,

STM-16 has data rate of 2488.52 Mbps, and

STM-64 has data rate of 9953.28 Mbps.

3. Virtual container n(VC-n)

A virtual container is the information structure used to support path layer connections in SDH. It consists of information payload and path overhead (POH) information fields organized in a block frame structure which repeats every 125 or 500 μs. Alignment information to identify VC-n frame start is provided by the server network layer. Two types of virtual containers have been defined in ITU-T recommendations.

- *Lower order VC-n* ($n = 1, 2, 3$)*:* It has single container-n (for $n = 1, 2, 3$) plus lower order POH bytes appropriate to the level n(for $n = 1, 2, 3$)
- *Higher order VC-n* ($n = 3, 4$)*:* This comprises of single container n (for $n = 3, 4$) or an assembly of tributary unit groups (TUG-2s or TUG-3s), together with VC POH bytes appropriate to the level n.

4. Administrative unit n(AU-n)

An AU-n is the information structure which provides adaptation between the higher order path layer and the multiplex section layer. It consists of an information payload (the higher order VC) and an administrative unit pointer which indicates the offset of the payload frame relative to the multiplex section frame.

For example, AU-4 consists of VC-4 plus an administrative unit pointer which indicates the phase alignment of VC-4 with reference to STM-N frame (by way of giving relative address of payload, i.e., VC-4).

One may get the doubt—we know the relative address of payload frame from the start of multiplex frame via the AU pointer, but how to locate the pointer? The answer is simple — AU pointer's location is fixed with respect to STM-N frame. So, we can access the pointer and then the payload area.

We can have more than one AU in an STM frame occupying fixed, defined positions. These are termed 'administrative unit group' (AUG).

5. Tributary unit n(TU-n)

A tributary unit is an information structure which provides adaptation between the lower order path layer and the higher order path layer. It consists of an information payload (the lower order virtual container) and a tributary unit pointer which indicates the offset of the payload block relative to the beginning of higher order virtual container frame.

The TU-n (for $n = 1, 2, 3$) consists of a VC-n together with a tributary unit pointer. A set of two or more tributary units, occupying fixed, defined positions in a higher order VC-n payload is termed a tributary unit group (TUG). The purpose of TUGs is to make feasible mixed capacity payloads made up of different size TUs so that the transport network has increased flexibility.

A TUG-2 consists of a homogeneous assembly of TUG-1s or a TU-2.

A TUG-3 consists of a homogeneous assembly of TUG-2s or a TU-3.

6. Container n (n = 1 to 4)

A container is the information structure which forms the network synchronous information payload for a virtual container. For each VC, there is a corresponding container.

7. Network node interface (NNI)

The interface between two network nodes for interconnection is termed 'network node interface' or (NNI).

8. Pointer

As already stated while discussing VC-n, pointers provide the frame offset of a VC with respect to the start of STM frame in which it is supported.

9. Concatenation

When multiple VCs are associated with one another in such a way that the resultant formation can be used as a single container, the process is called 'concatenation'. Across a concatenated container, bit sequence integrity is maintained.

10. SDH mapping

It is the process of adapting PDH tributaries into virtual containers at the boundaries of SDH network.

11. SDH multiplexing

It is a process in which multiple lower order path layer signals are adapted into a higher order path, or multiple higher order path layer signals are adapted into a multiplex section.

12. SDH aligning

It is the process in which the frame offset information is incorporated into the tributary unit (TU) or the administrative unit (AU) while adapting to the frame reference of the supporting layer.

29.10 Synchronous Multiplexing of 140 Mbps Signal

Conversion of one 140 Mbps signal into an STM-1 signal is less cumbersome than converting combination of 34 Mbps signals and 2 Mbps signals into an STM-1 signal. Hence, the case of 140 Mbps is examined first. The steps involved are :

- formation of container C-4,
- formation of virtual container VC-4,
- formation of administrative unit AU-4, and
- formation of STM-1 frame by addition of SOH.

Formation of container C-4

As already seen in Section 29.4, the payload area of STM-1 frame has 261 columns by 9 rows, after the exclusion of overhead bytes. Leaving one more byte in each row, we have 260 bytes in each row.

Consider one such row having 260 bytes (refer is Fig. 29.14). These 260 bytes are divided into 20 blocks of 13 bytes each. For the sake of convenience, the 260 bytes are shown spread over four rows but in fact are from the same row.

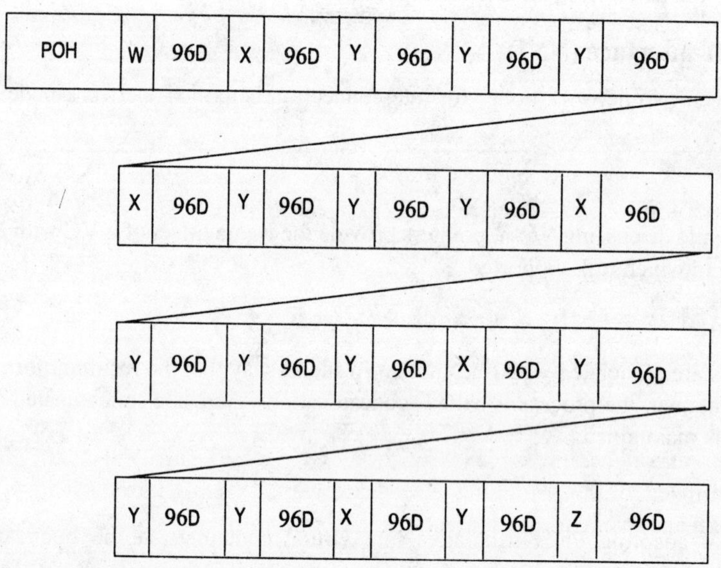

W : DDDDDDDD X : CRRRRROO Y : RRRRRRRR Z : DDDDDDSR
D : Data bit O : Overhead bit R : Fixed stuff bit
C : Justification control bit S : Justification opportunity bit

Fig. 29.14 *Formation of C-4 (byte map of C-4).*

The first byte of each block is designated as W or X or Y or Z byte and contains 'D' bits (data) or 'R' bits (fixed stuff information) or 'O' bits (overhead) or 'S' bits (justification opportunity) or 'C' bits (justification control bit).

The remaining 12 bytes of each block contain data bits, i.e. $12 \times 8 = 96$ D bits. The individual bits in W/X/Y/Z bytes are shown below :

W : DDDDDDDD X : CRRRRROO
Y : RRRRRRRR Z : DDDDDDSR

The purpose of having W/X/Y/Z bytes is to ensure that one S bit and five C bits are there in each row. Out of five C bits in a row, if majority are 0 (zero), S bit will be data bit.

It may be noted that the size of C-4 is 260 columns by 9 rows in a time frame of 125 μs.

Formation of VC-4

Container-4 formed above is added with control information known as path overhead (POH) of 9 bytes (one column by nine rows). The container plus POH form VC-4. The container has 260 bytes and one byte of POH, making a total of 261 which matches the total number of payload columns in STM-1 frame.

It may be noted that the size of VC-4 is 261 columns by 9 rows in a time frame of $125\mu s$

Formation of AU-4

A pointer located in the 4th row of the SOH area is associated with the VC whose value indicates the position of the beginning of the VC with respect of the STM-1 frame. This process is known as SDH aligning. The pointer value can be incremented or decremented as necessary to accommodate movements of the position of the VC. Hence, for VC-4 this pointer is called Administrative Unit-4 (AU-4).

Formation of AUG

Since it is the case of forming STM frame from 14 Mbps data stream, the VC-4 is formed in one go, which with a pointer forms the AU-4 (Administrative Unit-4). The AU-4 plus AU-4 pointer make the AUG.

Formation of STM-1 frame by addition of SOH

The first nine bytes of SOH are still to be added to make the total columns 270 for the STM frame. The SOH bytes are added to AUG. Among these SOH bytes, RSOH bytes are terminated at regenerators and MSOH bytes are terminated where AUGs are assembled and disassembled.

The MSOH bytes maintain transparency through regenerators. As already discussed in Section 29.4, SOH bytes include bytes for block framing, error performance checking, order-wire, communication facility for operation and maintenance, protection switching, etc.

All the steps involved in forming STM-1 frame from C-4 are illustrated in a block diagram in Fig. 29.15.

29.11 Synchronous Multiplexing of 34 Mbps Signals

STM-1 frame can contain three 34 Mbps signals. The steps involved in forming STM-1 frame from three 34 Mbps PDH signals are given below.

- Formation of container C-3
- Formation of virtual container VC-3
- Formation of tributary unit TU-3
- Formation of tributary unit group TUG-3

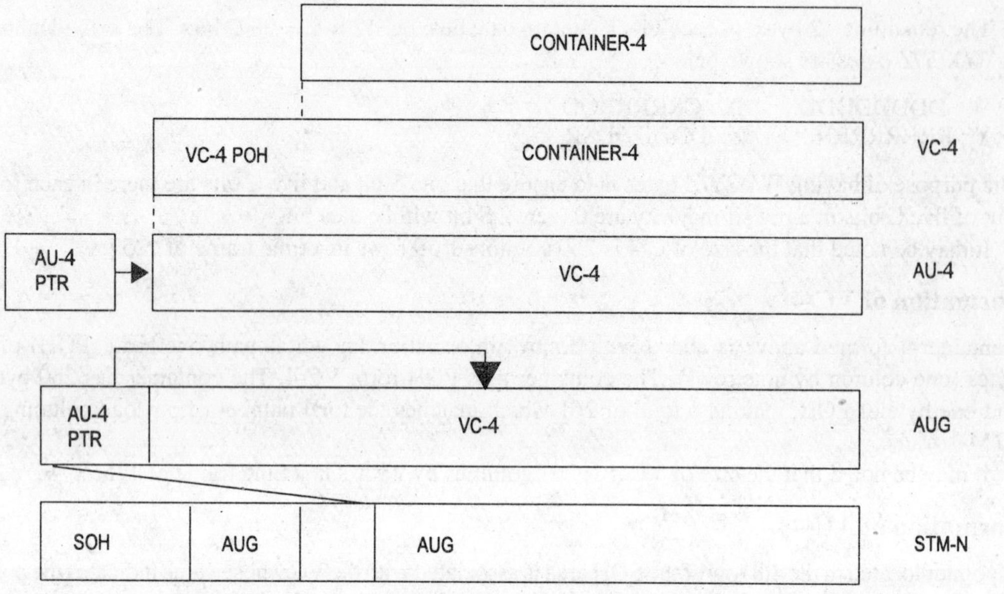

Fig. 29.15 *Formation of STM-1 frame from container-4.*

- Multiplexing TUG-3 into VC-4
- Formation of AU-4
- Formation of AUG
- Formation of STM-1 by addition of SOH

Formation of container C-3

Size of C-3 is 84 columns by 9 rows in a time frame of 125 μs. Unlike C-4, C-3 is filled up by dividing the payload into three sub-frames of three rows each. Since there are 84 columns, we have 84 bytes in each row. The three sub-frames are designated T1, T2, T3.

In each row the 85th byte is added to form VC-3, i.e., VC-3 has 85 columns by 9 rows whereas C-3 has 84 columns by 9 rows.

Bit maps of VC-3 and one sub-frame of C-3 are given in Fig. 29.16. The following observations can be made while looking at the bit map of a sub-frame of C-3.

- It has 84 bytes per row and 3 such rows.
- For every 3 bytes, a regular stuff byte REG exists. The composition of REG is 8 R bits.
- Two sets of five justification control bits (C1, C2) totaling to ten are packed into two bytes along with R bits. These two bytes are designated as CON bytes.
- Two bits S1 and S2, which are justification opportunity bits, are packed together with R bits into two bytes designated as OPR bytes.
- In a sub frame, thus, there are:

  1431 D-bits,

  10 C-bits,

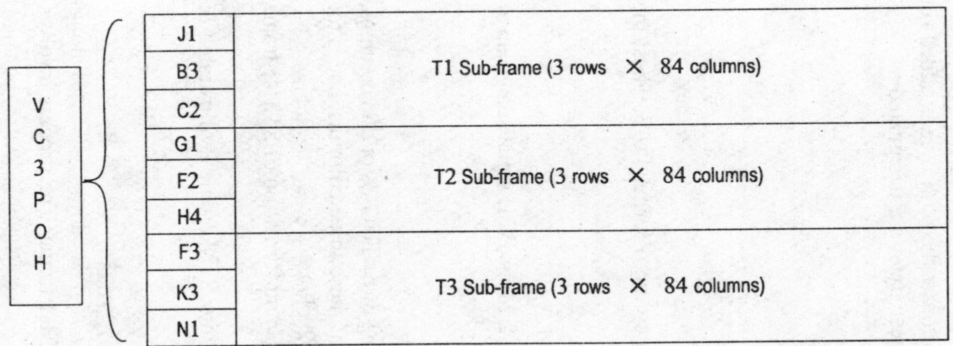

X	3 × 8I	X	3 × 8I	X	3 × 8I	X	3 × 8I	X	X	3 × 8I	X	3 × 8I	X	3 × 8I	X	3 × 8I	X	3 × 8I	X	con
X	3 × 8I	X	3 × 8I	X	3 × 8I	X	3 × 8I	X	X	3 × 8I	X	3 × 8I	X	3 × 8I	X	3 × 8I	X	3 × 8I	X	con
X	3 × 8I	X	3 × 8I	X	3 × 8I	X	3 × 8I	X	X	3 × 8I	X	3 × 8I	X	3 × 8I	X	3 × 8I	X	3 × 8I	X	con
1	3	1	3	1	3	1	3	1	1	3	1	3	1	3	1	3	1	3	1	1

3 × 8I	X	3 × 8I	X	3 × 8I	X	3 × 8I	X	3 × 8I	X	X	3 × 8I	X	3 × 8I	X	3 × 8I	X	3 × 8I	X	3 × 8I	X	con	3 × 8I
3 × 8I	X	3 × 8I	X	3 × 8I	X	3 × 8I	X	3 × 8I	X	X	3 × 8I	X	3 × 8I	X	3 × 8I	X	3 × 8I	X	3 × 8I	X	con	3 × 8I
3 × 8I	X	3 × 8I	X	3 × 8I	X	3 × 8I	X	3 × 8I	X	X	3 × 8I	X	3 × 8I	X	3 × 8I	X	3 × 8I	X	3 × 8I	X	con	q1q28I
3	1	3	1	3	1	3	1	3	1	1	3	1	3	1	3	1	3	1	3	1	1	3

Note : Each sub-frame T1, T2, T3 has 3 rows and 84 columns. For want of space, 84 columns are shown in a split-up fashion. 3 × 8I implies 3 bytes of 8 information bits; X implies one byte 'RRRRRRRR'; 'con' implies one-byte 'RRRRRRC1C2'; q1 implies one-byte 'RRRRRRRS1' and q2 implies 'S2OOOOOOO'; the number of bytes in each set are given by the digits' 1 or' 3 ; the sum of 1s and 3s is 84 which is equal to the number of columns; D-Data bit, R-Fixed stuff bit; O – Overhead bit; S-Justification opportunity bit; C1, C 2 – Justfication control bits

Fig. 29.16 *Mapping of 34 Mbps tributaries into VC-3.*

2 S-bits, and

573 R-bits.

- C1 bits control S1 and C2 bits control S2. Majority vote among C1 indicates the status of S1. In a similar way, majority among C2 bits indicates the status of S2.
- The wastage of 573 bits for R-bits seems unnecessary. But these are unavoidable since STM-1 is designed to accommodate 3 Nos. of 45 Mbps PDH signal of American standards. Thus, because of this wastage, we are able to accommodate only 3 Nos. of 34 Mbps PDH signals in STM-1 frame.

Formation of VC-3

While discussing the formation of C-3, it has been stated that in each row the 85th byte is added (in fact it is the first byte rather, the next 84 being container-3 bytes). These bytes are designated as

J1, B3, CE for the 3 rows of T-1 sub-frame,

G1, F2, H4 for the 3 rows of T-2 sub-frame,

F3, K3, N1 for the 3 rows of T-3 sub-frame.

These are all POH bytes of VC-3.

Formation of TU-3

VC-3 together with a pointer located in an additional column (86th column) forms TU-3. Thus, the size of TU-3 is 86 columns by 9 rows.

Formation of TUG-3

One TU-3 forms one TUG-3 without any change. But when handling 2 Mbps PDH signals, a group of TUs form TUG-3.

Multiplexing TUG-3 into VC-4

Each TUG-3 has a size of 86 columns by 9 rows. After accommodating three TUG-3s or 258 columns, we have three more columns in the 261 column space of VC-4. Out of these three columns, one is for POH. The remaining two are filled with fixed stuff bits (refer to Fig. 29.17).

The next three steps viz. the formation of AU-4, formation of AUG and formation of STM-1 frame by addition of SOH are the same as in Section 29.10.

The entire process of multiplexing three 34 Mbps PDH signals into STM frame is illustrated in Fig. 29.18.

29.12 Synchronous Multiplexing of 2 Mbps Signal

Sixty-three numbers of 2 Mbps PDH signals can be taken in one STM-1 frame. The process involves the following steps :

- Formation of container (C-1)
- Formation of virtual container (VC-12)
- Formation of tributary unit (TU-12)
- Multiplexing of TU-12s to form TUG-3

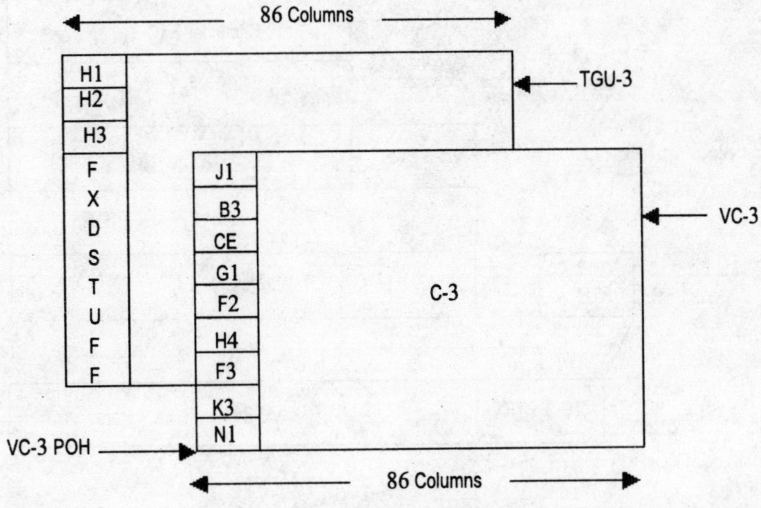

Fig. 29.17(a) *Formation of TU-3 (synchronous multiplexing of 34 Mbps).*

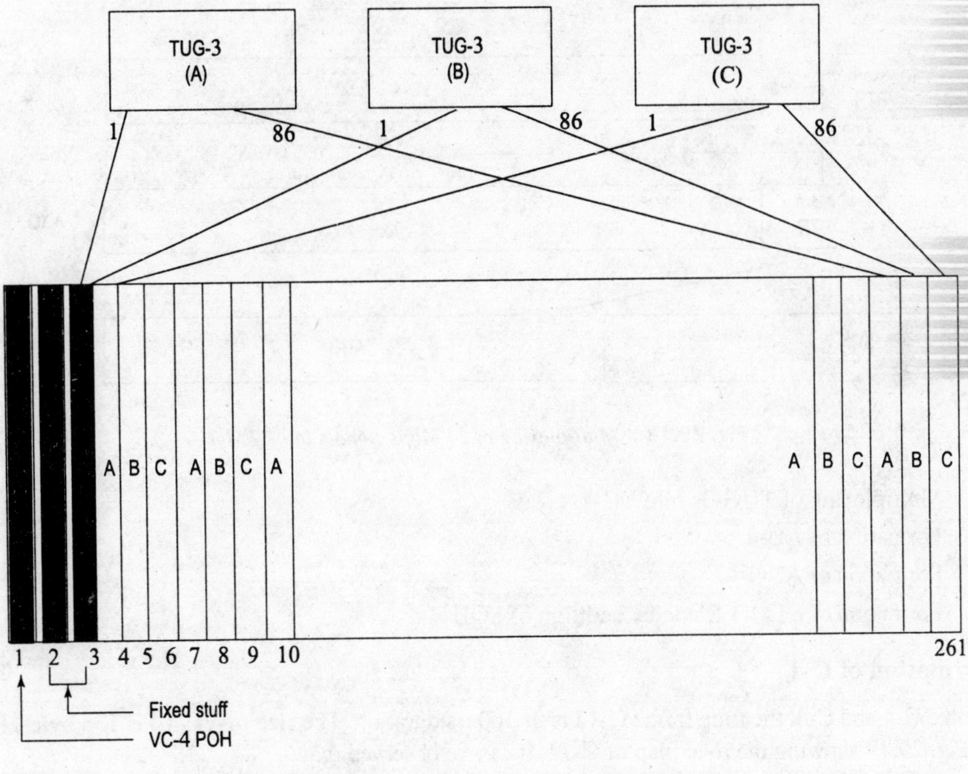

Fig. 29.17(b) *Multiplexing of TUG-3s into VC-4.*

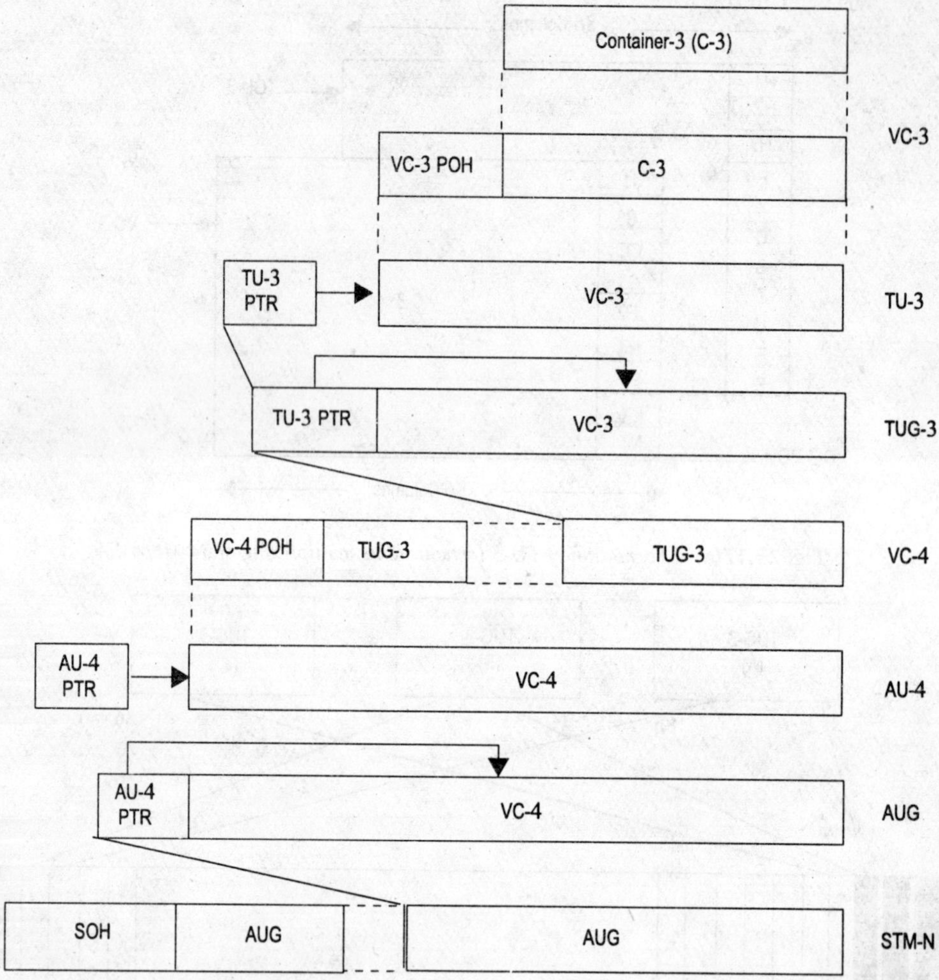

Fig. 29.18 *Multiplexing of 34 Mbps signals into STM-1.*

- Multiplexing of TUG-3s into VC-4
- Formation of AU-4
- Formation of AUG
- Formation of STM-1 frame by addition of SOH

Formation of C-1

Unlike C-4 and C-3, the time frame C-12 is of 500 μs duration. The size of VC-12 is 136 bytes. Refer to Fig. 29.19 showing the byte map of C-12. It can be observed that :

- C-12 has 1023 D-bits
- Five REG bytes with all R-bits

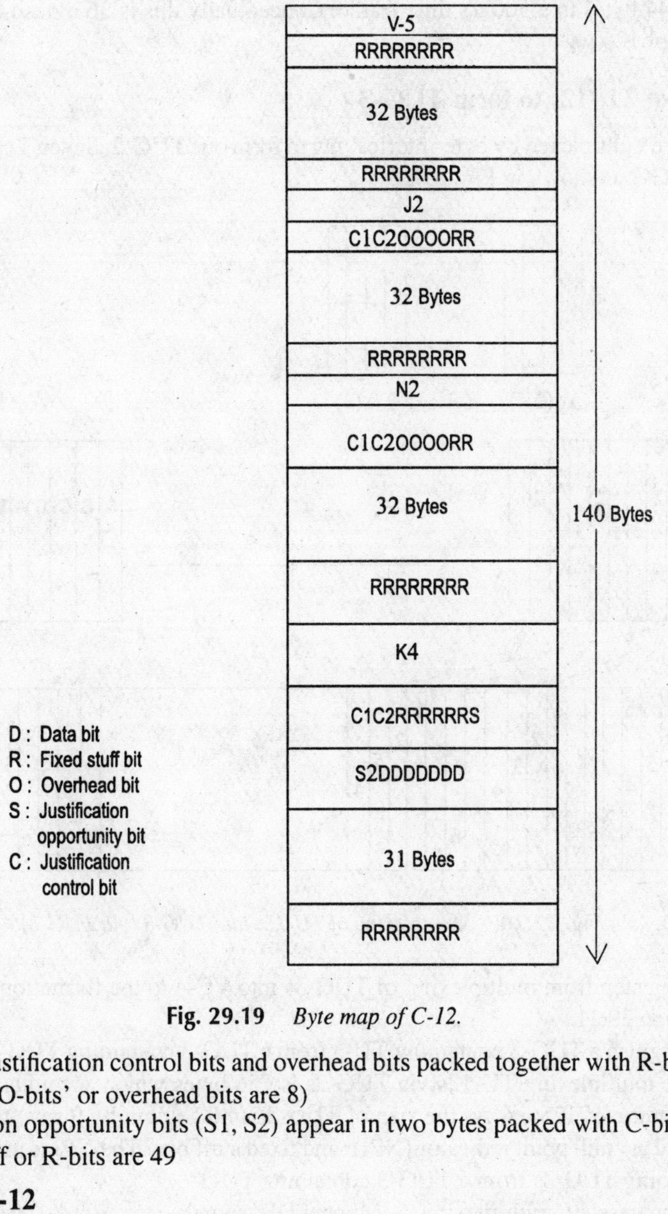

Fig. 29.19 *Byte map of C-12.*

- Two bytes of justification control bits and overhead bits packed together with R-bits (total C1, C2 bits are 6 and 'O-bits' or overhead bits are 8)
- Two justification opportunity bits (S1, S2) appear in two bytes packed with C-bits and D-bits
- Total fixed stuff or R-bits are 49

Formation of VC-12

VC-12 is formed by adding 4 bytes, viz. K4, N2, J2, and V5 to C-12. Please refer to Fig. 29.19. Total bytes in VC-12 are 140.

Formation of TU-12

VC-12, together with the pointer, forms TU-12. The size of TU-12 is 144 bytes, i.e., 4 bytes get added to VC-12 to form TU-12. However, out of 4, only two bytes (V1, V2) are utilized to indicate the location of the first byte (V5) of VC-12.

TU-12 has 144 bytes in a 500 μs time frame. Conceptually this is 36 bytes (4 columns by 9 rows) in a time frame of 125 μs.

Multiplexing of TU-12s to form TUG-3

Three TU-12s are multiplexed by byte-interleaving to form one TUG-2. Seven TUG-2s are multiplexed to form one TUG-3 as shown in Fig. 29.20

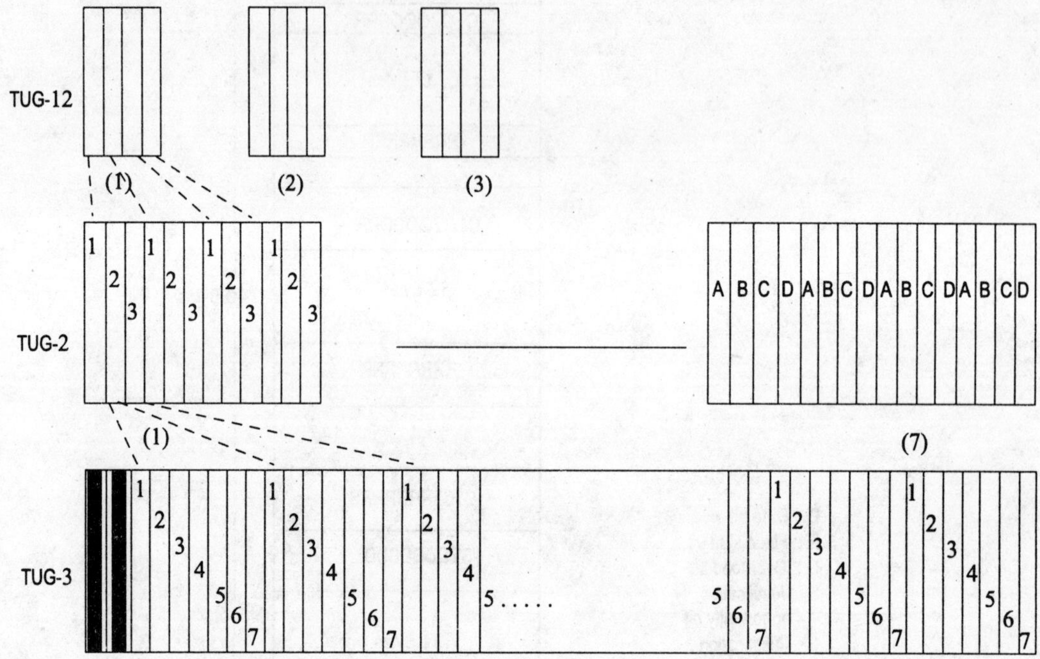

Fig. 29.20 *Multiplexing of TU-12s into TUG-3 (via TUG-2).*

The remaining step from multiplexing of TUG-3s into VC-4 to the formation of STM-1 frame are same as in Section 29.11.

How to distinguish a TUG-3 containing TU-3 from a TUG-3 containing TUG-2s? The payload size of TUG-3 while multiplexing TU-12s via TUG-2s is 756 bytes which accounts for 84 columns by 9 rows in a time frame of 125 μs. As the size of TUG-3 is 86 column by 9 row, the bytes in extra two columns are used as 'null point indicator (NPI)' and fixed stuff bits. The NPI is useful in distinguishing a TUG-3 containing TUG-2s from a TUG-3 containing TU-3.

The entire process of multiplexing 2 Mbps PDH signals into STM-1 frame is illustrated in Fig. 29.21.

29.13 Formation of Higher Order STM Frames

STM-N frame is formed by byte interleaving of STM-1 frames. Figure 29.22 shows the arrangement of N-AUGs multiplexed into the STM-N.

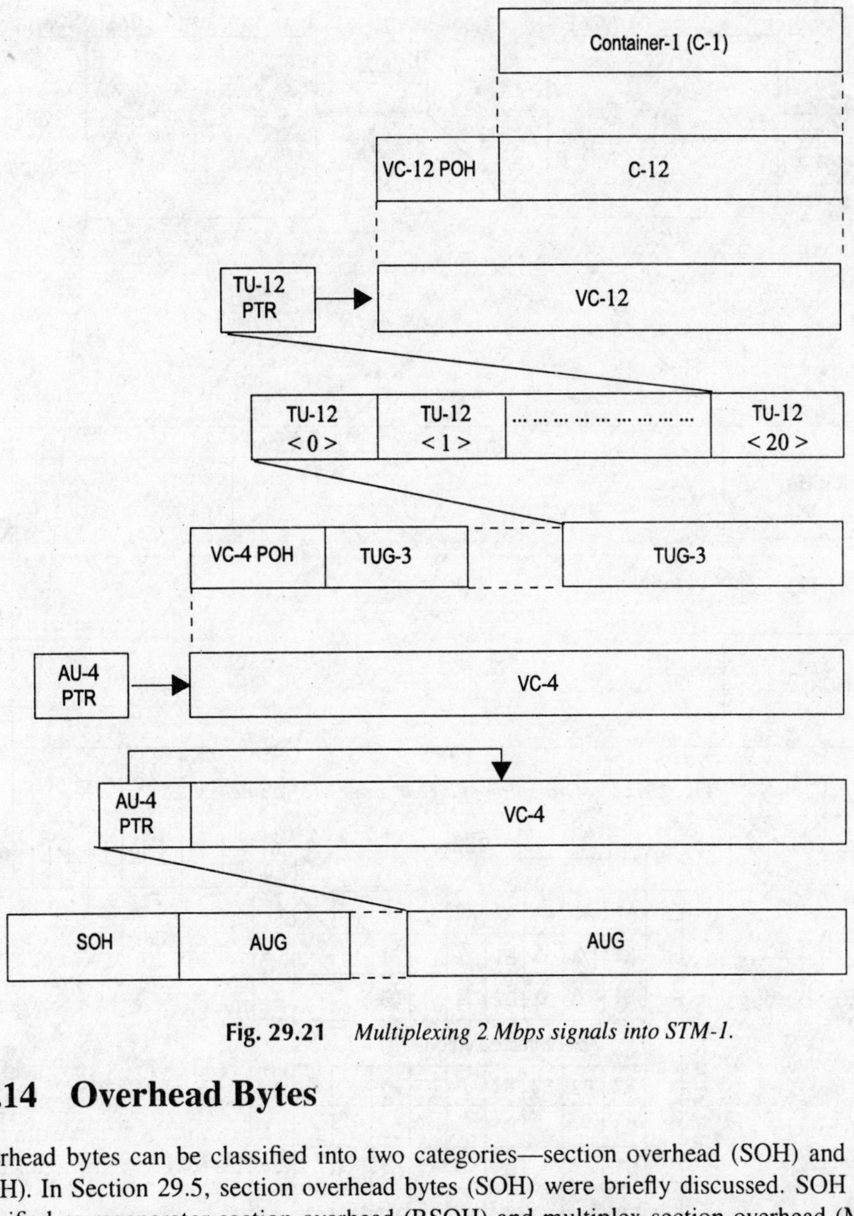

Fig. 29.21 *Multiplexing 2 Mbps signals into STM-1.*

29.14 Overhead Bytes

Overhead bytes can be classified into two categories—section overhead (SOH) and path overhead (POH). In Section 29.5, section overhead bytes (SOH) were briefly discussed. SOH can be further classified as regenerator section overhead (RSOH) and multiplex section overhead (MSOH). These overheads bytes are explained in detail in Section 29.14.1. The other overhead bytes attached to C-n while forming VC-n, viz. path overhead bytes (POH), are explained in detail in Section 29.14.2.

29.14.1 SOH description

Figure 29.23 shows the SOH bytes of STM-1 (refer to Fig. 29.3 also). SOH of STM-N will be N- times that of STM-1.

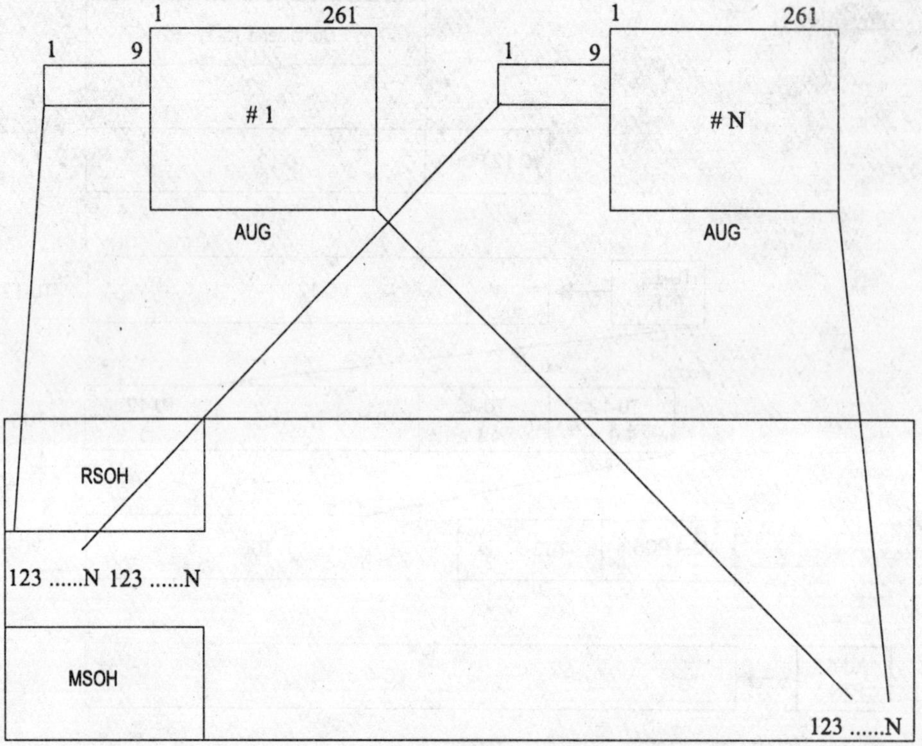

Fig. 29.22 *Formation of higher order STM frame.*

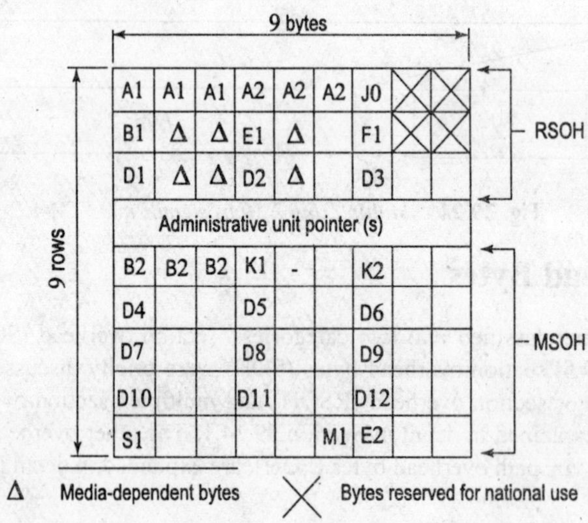

Fig. 29.23 *SOH bytes of STM-1 frame.*

1. Forming A1, A2: Two types of bytes are defined for framing :
   - A1 : 11110110
   - A2 : 00101000

   The frame alignment work of an STM-N frame is composed of 3 × NA1 bytes followed by 3 × NA2 bytes.
2. Regenerator section trace J0: The J0 byte located at S (1, 7, 1) in an STM-N is allocated to a regenerator section trace. This byte is used to transmit repetitively a section access point identifier so that a section receiver can verify its continued connection to the intended transmitter.
3. Spare Z0: These bytes, which are located at positions S(1, 7, 2) to S (1, 7, N) are reserved for future international standardisation.
4. One byte is allocated for regenerator section error monitoring. Its function is to generate bit-interleaved parity 8 (BIP-8) code using even parity. The BIP-8 is computed over all bits of the previous STM-N frame after scrambling and is placed in byte B1 of the current frame before scrambling.
5. Orderwire E1, E2: These two bytes may be used to provide orderwire channels for voice communication. E1 is part of the RSOH and may be accessed at regenerators. E2 is part of the MSOH and may be accessed at multiplex section terminations.
6. User channel F1: This byte is reserved for user purposes (e.g. to provide temporary data/voice channel connections for special maintenance purposes).
7. Data communication channel (DCC) D1-D12: A 192 kbps channel is defined using bytes D1, D2 and D3 as a regenerator section DCC. A 576 kbps channel is defined using bytes D4 to D12 as a multiplex section DCC.
8. BIP-N×24—(B2): The B2 bytes are allocated for a multiplex section error monitoring function. Bit interleaved parity N × 24 is computed over all bits of the previous STM-N frame except for the first three rows of SOH, and is placed in bytes B2 of the current frame before scrambling.
9. Automatic protection switching (APS) channel K1, K2 (b1–b5): Two bytes are allocated for APS signalling for the protection of the multiplex section.
10. MS-RDI K2 (b6 - b8): The multiplex section remote defect indication (MS-RDI) is used to return an indication to the transmit end that the received end has detected an incoming section defect or is receiving MS-AIS.
11. Synchronisation status S1 (b5–b8): Bits 5 to 8 of byte located at S (19, 1, 1) are allocated for synchronisation status messages.
12. MS-REI — (M1): One byte is allocated for use as a multiplex section REI (remote error indication).

29.14.2 POH descriptions

POH bytes are different for VC-4/VC-3 and VC-12, hence these are covered separately.

VC-4 Xc/VC-4/VC-3 POH

The VC-4-Xc/VC-4/VC-3 POH consists of 9 bytes denoted as J1, B3, C2, G1, F2, H4, F3, K3 and N1. These bytes are classified as follows :

- Bytes or bits used for end-to-end communication with independent payload function: J1, B3, C2, G1, K3 (b1 to b4).

- Payload type specific bytes: H4, F2, F3.
- Bits reserved for future international standardisation: K3 (b5 to b8)
- Byte which can be overwritten in an operator domain (without affecting the end-to-end performance monitoring facility of the byte B3) : N1.

1. Path trace : J1 This is the first byte in the virtual container; its location is indicated by the associated AU-4 or TU-3 pointer. This byte is used to transmit repetitively a path access point identifier so that a path receiving terminal can verify its continued connection to the intended transmitter.

2. Path BIP-8 : B3 This byte is allocated for a path error monitoring function. This shall be a BIP-8 code using even parity. The path BIP-8 calculated over all bits of the previous VC-4-Xc/VC-4/VC-3 before scrambling is placed in the B3 byte of the current VC-4/Xc/VC-4/VC-3 before scrambling.

3. Signal label : C2 This byte is allocated to indicate the composition or the maintenance status of the VC-4-Xc/VC-4/VC-3.

4. Path status : G1 This byte is allocated to convey the path status and performance back to a VC-4-Xc/VC-4/VC-3 trail termination source as detected by a trail termination sink. This feature permits the status and performance of the complete duplex trail to be monitored at either end, or at any point along that trail. Bit allocations are shown in Fig. 29.24.

REI				RDI	RESERVED		SPARE
1	2	3	4	5	6	7	8

Fig. 29.24 *Bit map of path status byte (G1).*

5. Path user channels : F2, F3 These bytes are allocated for user communication purposes between path elements and are payload dependent.

6. Position indicator : H4 This byte provides a generalised position indicator for payloads and can be payload-specific (e.g. H4 can be used as a multiframe position indicator for VC-1).

7. Automatic protection switching (APS) channel : K3 (b1-b4) These bits are allocated for APS signalling for protection at the VC-4/3 path levels. K3 (b5 to b8) are spare, meant for future use.

8. Network operator byte : N1 This byte is allocated to provide a tandem connection monitoring (TCM) function.

29.14.3 VC-12 POH

The bytes V5, J2, N2 and K4 are allocated to VC-1 POH.

1. V5 byte : The V5 byte is the first byte of the multiframe and its position is indicated by the TU-1 pointer. It provides functions of error checking, signal label and path status of the VC-1 paths. Bit assignments of the V5 byte are given in Fig. 29.25.

BIP-2	BIP-2	REI	RFI	Signal Label			RDI
1	2	3	4	5	6	7	8

Fig. 29.25 *Bit map of V5 byte.*

Bits 1 and 2 are used for error performance monitoring. A bit interleaved parity (BIP-2) scheme is followed. Bit 1 is set such that parity of all odd number bits (1, 3, 5 and 7) in all the bytes in the previous VC-1 is even and bit 2 is set similarly for the even number bits (2, 3, 6 and 8).

Bit 3 is a VC-1 path remote error indication (REI) that is set to one and sent back towards VC-1 path originator if one or more errors are detected by the BIP-2, it is otherwise set to zero.

Bit 4 is a VC-1 path remote failure indication (RFI). This bit is set to one if a failure is declared, otherwise it is set to zero.

Bits 5 through 7 provide a VC-1 signal label.

Bit 8 is set to 1 to indicate a VC-1 path remote defect indication (RDI), otherwise it is set to zero. The VC-1 path RDI is sent back towards the trail termination source if either a TU-1 server signal failure or trail signal failure condition is detected by the trail termination sink.

2. *Path trace : J2* Byte J2 is used to transmit repetitively a low order path access point identifier so that the path receiving terminal can verify its continued connection to the intended transmitter.

3. *Network operators byte : N2* This byte is allocated to provide a tandem connection monitoring (TCM) function.

4. *Automatic protection switching (APS) channel : K4 (b1–b4)* These bits are allocated for APS signalling for protection at the lower order path level. K4 (b5–b8) is spare, meant for future use.

29.15 Pointers

As discussed in Section 29.9 , the pointer provides a method for allowing flexible and dynamic alignment of the VC-n within the AU-n/TU-n frame. Dynamic alignment means that the VC-n is allowed to "float" within the AU-n/TU-n frame. In this section we will discuss pointers in more detail.

29.15.1 AU-4 pointer

The AU-4 pointer is contained in bytes H1, H2 and H3 as shown in Fig. 29.26. The pointer contained in H1 and H2 designates the location of the byte where the VC-n begins. These two bytes can be viewed as one word. The last ten bits (bits 7–16) of the pointer word carry the pointer value. The AU-4 pointer value is a binary number with a range of 0 to 782 which indicates the offset in the three byte increments between the pointer and the first byte of VC-4. A pointer value of 0 indicates that the VC-4 starts in the byte location that immediately follows the last H3 byte, whereas an offset of 87 indicates that the VC-4 starts three bytes after the K2 byte.

If there is a frequency offset between the frame rate of the AUG and that of the VC-4, then the pointer value will be incremented or decremented as needed, accompanied by a corresponding positive or negative justification byte or bytes. Consecutive pointer operations must be separated by at least three frames in which the pointer value remains constant.

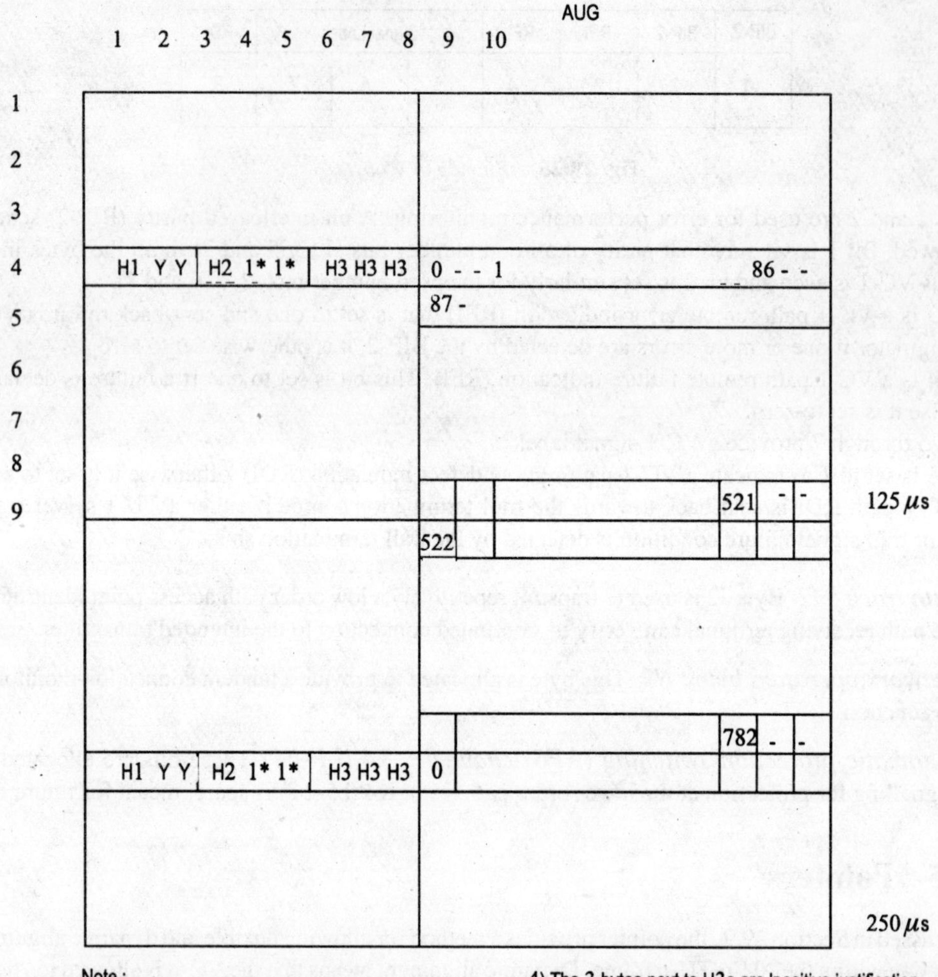

Note :

1) Number of columns in AUG is 270.

2) Out of these, 1 to 9 contain Au-4 pointer.

3) In row-4 of AU-4 pointer, the 3 H3 bytes are 'negative justification opportunity bytes'.

4) The 3 bytes next to H3s are 'positive justification opportunity bytes'.

5) If the frame above contains two sequential AUGs, then, row 4 is repeated.

6) Total duration of the frame is 125 × 2, i.e., 250 μs.

Fig. 29.26 *Au-4 pointer's byte map.*

If the frame rate of the VC-4 is too slow with respect to that of the AUG, then the alignment of the VC-n must periodically slip back in time and the pointer value must be incremented by one. This operation is indicated by inverting the 1-bits (bits 7, 9, 11, 12 and 15). Three positive justification bytes appear immediately after the last H3 byte in the Au-4 frame. Subsequent pointers will contain the new offset.

If the frame rate of the VC-n is too fast with respect to that of AUG, then the alignment of the VC-n must periodically be advanced in time and the pointer value must be decremented by one. This operation is indicated by inverting D-bits (bits 8, 10, 12, 14 and 16). Three negative justification bytes appear in the H3 bytes. Subsequent pointers will contain the new offset.

Bits 1-4 (N-bits) of the pointer work carry a new data flag (NSF) which allows an arbitrary change of the pointer value if that change is due to a change in the payload.

29.15.2 TU-3 pointer

The TU-3 pointer provides a method of allowing flexible and dynamic alignment of VC-3 within the TU-3 frame. Three individual TU-3 pointers are contained in three separate H1, H2 and H3 bytes as shown in Fig. 29.27. The TU-3 pointer value contained in H1 and H2 designates the location of the byte where the VC-3 begins. These two bytes can be viewed as one word. The last ten bits (bits 7–16) of the pointer word carry the pointer value. The TU-3 pointer value is a binary number with a range of 0–764 which indicates the offset between the pointer and the first byte of the VC-3.

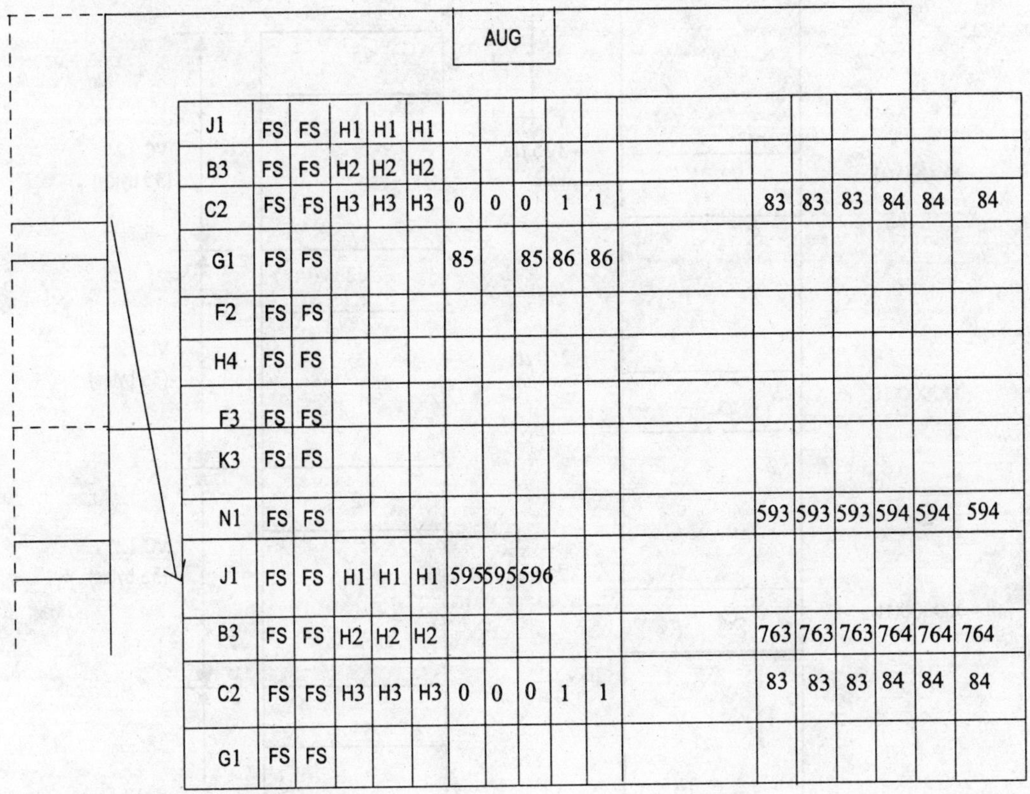

Note :

1) AUG has columns 1 to 270.
2) Out of these, 1 to 9 contain AU-4 pointer.
3) Byte map for columns 10 to 270 shown above.
4) FS in columns 12 and 13 stands for fixed stuff.

Fig. 29.27 *TU-3 pointer's byte map.*

If there is a frequency offset between the TU-3 frame rate and that of the VC-3, then the pointer value will be incremented or decremented as needed, accompanied by a corresponding positive or negative justification byte. Consecutive pointer operations must be separated by at least three frames in which the pointer value remains constant.

If the frame rate of the VC-3 is too slow with respect to that of the TU-3 frame rate, then the alignment of the VC-3 must periodically slip back in time and the pointer must be incremented by one. This operation is indicated by inverting of the 1-bits (bits 7, 9, 11, 13 and 15). A positive justification byte appears immediately after the individual H3 byte in the TU-3 frame. Subsequent TU-3 pointers will contain the new offset.

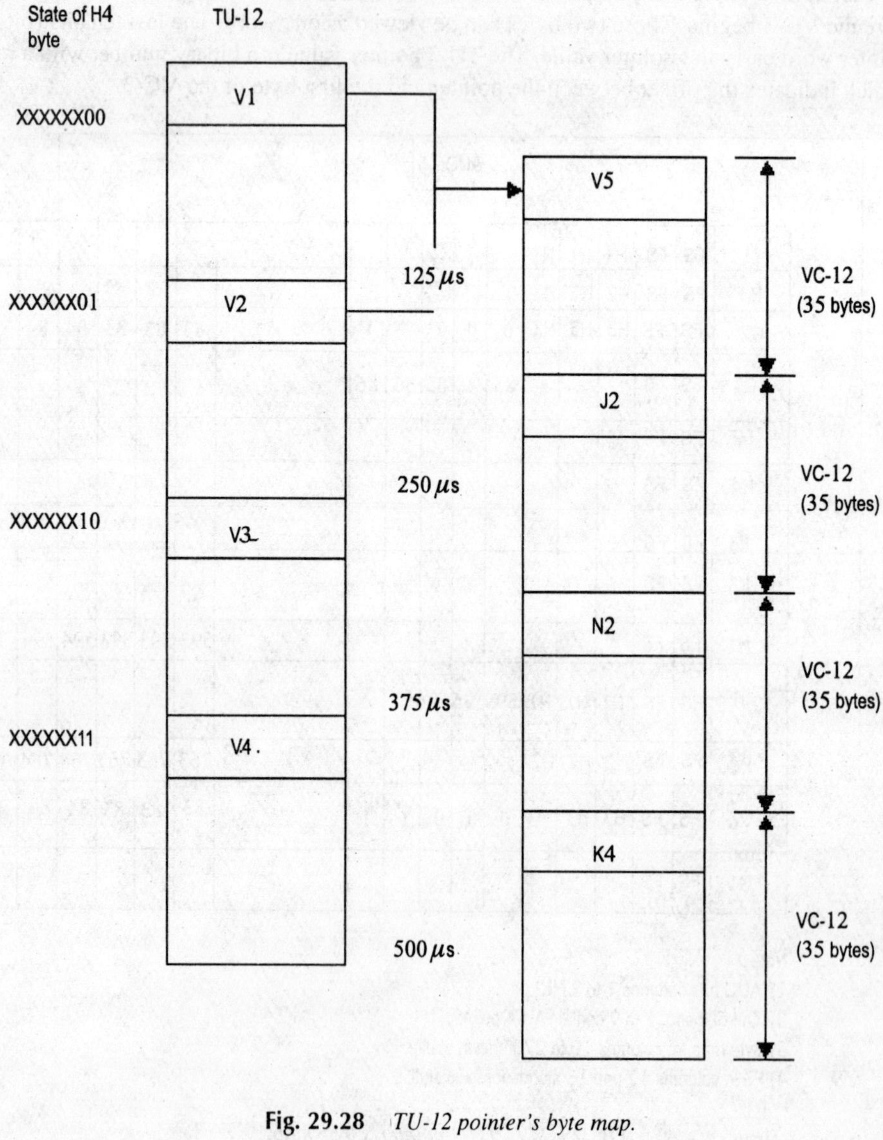

Fig. 29.28 *TU-12 pointer's byte map.*

If the frame rate of the VC-3 is too fast with respect to that of the TU-3 frame rate, then the alignment of the VC-3 must be periodically advanced in time and the pointer must be decremented by one. This operation is indicated by inverting the D-bits (bits 8, 10, 12, 14 and 16). A negative justification byte appears in the individual H3 byte in the TU-3 frame. Subsequent TU-3 pointers will contain the new offset.

Bits 1–4 (N-bits) of the pointer word carry an NDF (new data flag) which allows an arbitrary change of the value of the pointer if that change is due to a change in the VC-3.

29.15.3 TU-1 pointer

The TU-1 pointers provide flexible and dynamic alignment of the VC-12 within the TU-12 multiframes. The TU-1 pointers are contained in the V1 and V2 bytes as illustrated in Fig. 29.28. As shown in Fig. 29.29, two S bits (bits 5 and 6) indicate the tributary unit type. The pointer value (bits 7–16) is a binary number which indicates the offset from V2 to the first byte of VC-12. The pointer bytes are not counted in the offset.

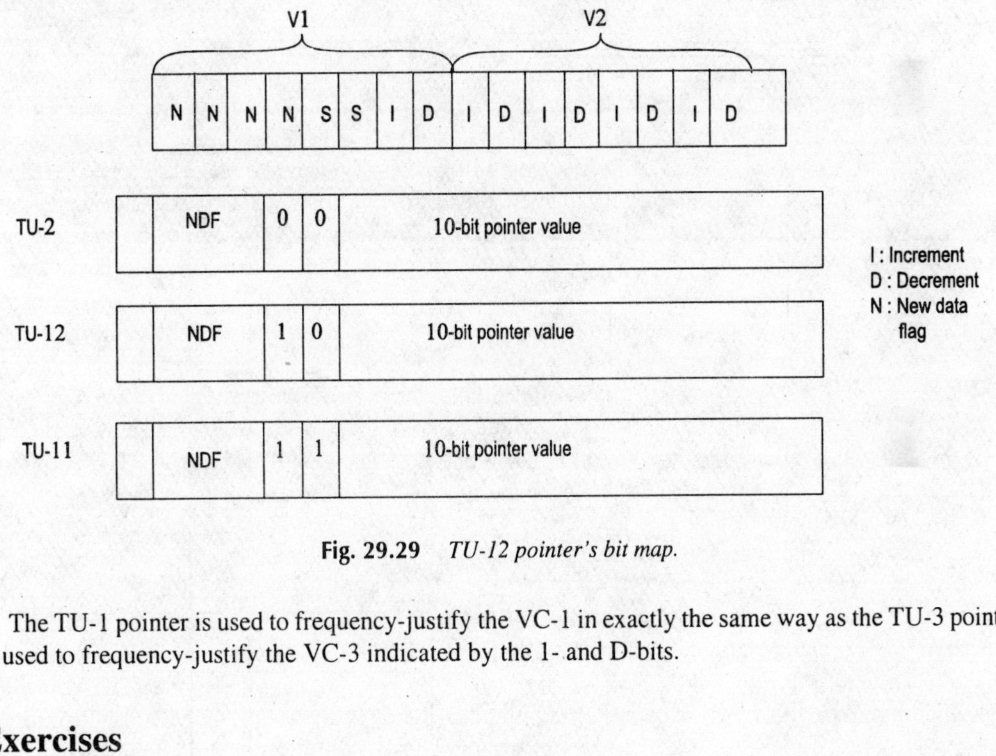

Fig. 29.29 *TU-12 pointer's bit map.*

The TU-1 pointer is used to frequency-justify the VC-1 in exactly the same way as the TU-3 pointer is used to frequency-justify the VC-3 indicated by the 1- and D-bits.

Exercises

1. What are the currently available values of N in STM-N standard for SDH? What is the maximum value of N?
2. In SDH we use the term container whereas in plesiochronous multiplexing we use the term tributary. Why?
3. Derive the time rates for STM-1, STM-4 and STM-16.

References

[1] American National Standards Institute (ANSI), New York-ANSI T1. 105-1995, Telecommunications: SONET — Basic description including multiplex structures, rates and formats, Oct.1995.

[2] International Telecommunication Union - Telecommunication Standardization Sector (ITU-T), http://www.itu.int; ITU-T recommendations G.957 — Optical interfaces for equipment and systems relating to the SDH.

Appendix

A1. Error Function and Complementary Error Function

Error function, denoted by erf (u) is defined as

$$\text{erf}(u) = \frac{2}{\sqrt{\pi}} \int_0^u \exp(-z^2)dz.$$

It has two properties :

1. Symmetry : $\text{erf}(-u) = \text{erf}\quad(u)$
2. Asymptotic to unity : As $u \to \infty \ldots \text{erf}(u) \to 1$

$$\text{i.e.,}\ \frac{2}{\sqrt{\pi}} \int_0^\infty \exp(-z^2)\, dz = 1.$$

(Actually, $\text{erf}(u) = 0.999998$ for $u = 3.3$.)
Complementary error function is defined by

$$\text{erfc}\ (u) = \frac{2}{\sqrt{\pi}} \int_u^\infty \exp(-z^2)dz$$

$$= \frac{2}{\sqrt{\pi}} \int_0^\infty \exp(-z^2)dz - \frac{2}{\sqrt{\pi}} \int_0^u \exp(-z^2)dz = 1 - \text{erf}(u).$$

Values of erf (u) for $u = 0$ to 3.3 (in steps of 0.05) are given below.

Table A1. *The error function*

u	erf(u)	u	erf(u)
0.00	0.00000	1.10	0.88021
0.05	0.05637	1.15	0.89612
1.10	0.11246	1.20	0.91031
1.15	0.16800	1.25	0.92290
0.20	0.22270	1.30	0.93401
0.25	0.27633	1.35	0.94376
0.30	0.32863	1.40	0.95229
0.35	0.37938	1.45	0.95970
0.40	0.42830	1.50	0.96611
0.45	0.47548	1.55	0.97162
0.50	0.52050	1.60	0.97635
0.55	0.56332	1.65	0.98038
0.60	0.60386	1.70	0.98379
0.65	0.64203	1.75	0.98667
0.70	0.67780	1,80	0.98909
0.75	0.71116	1.85	0.99111
0.80	0.74210	1.90	0.99279
0.85	0.77067	1.95	0.99418
0.90	0.79691	2.00	0.99532
0.95	0.82089	2.50	0.99959
1.00	0.84270	3.00	0.99998
1.05	0.86244	3.30	0.999998

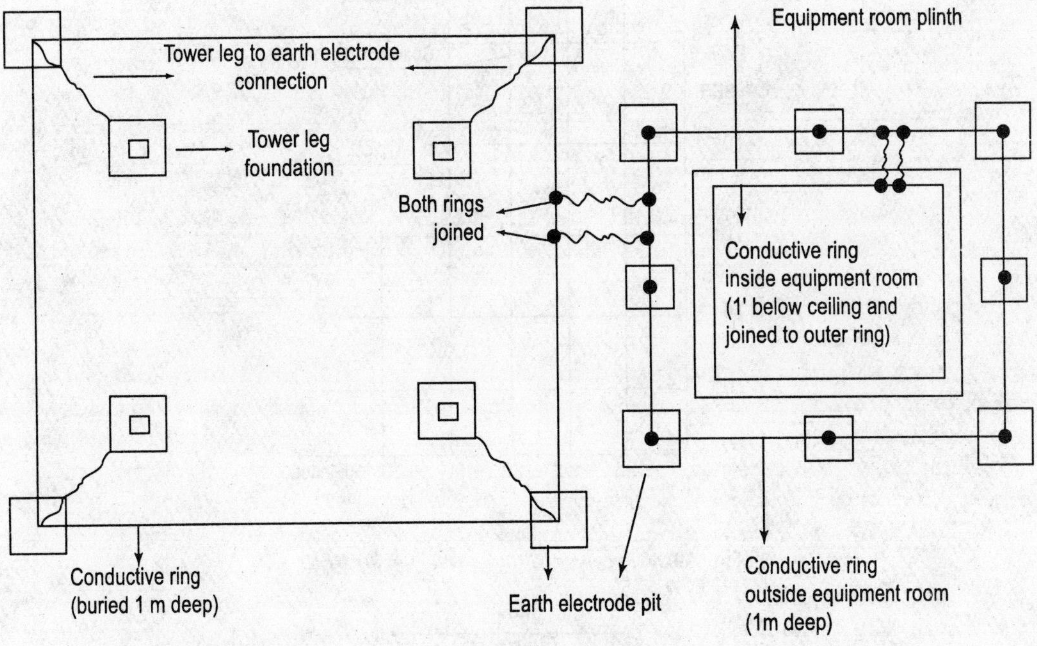

Fig. A2. *Ring earth arrangement at MW station.*

Table A3. *Jitter acceptance test specifications for III-order MUX*

At 2Mb port		At 34 Mb port	
Frequency	Specified UI	Frequency	Specified UI
20 Hz	1.5	100 Hz	1.5
100 Hz	1.5	500 Hz	1.5
500 Hz	1.5	1 kHz	1.5
1 kHz	1.5	3 kHz	0.5
2.4 kHz	1.5	5 kHz	0.3
5 kHz	0.72	10 kHz	0.15
10 kHz	0.36	50 kHz	0.15
18 kHz	0.20	100 kHz	0.15
40 kHz	0.20	200 kHz	0.15
100 kHz	0.20	699 kHz	0.15

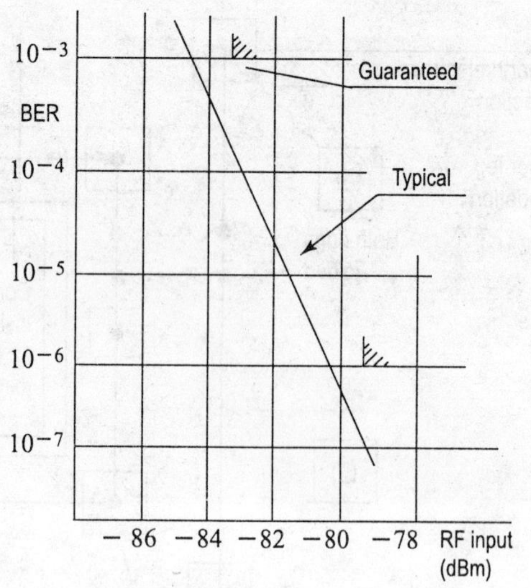

Fig. A4. *Receiver RF input level versus BER.*

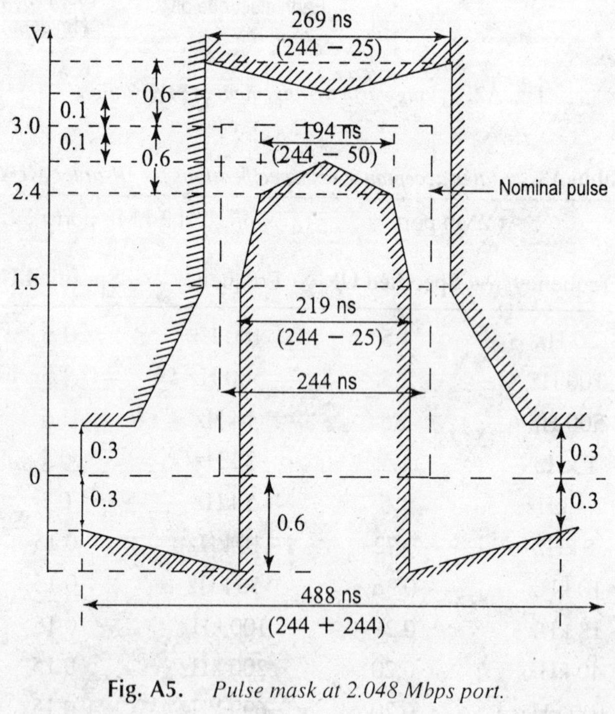

Fig. A5. *Pulse mask at 2.048 Mbps port.*

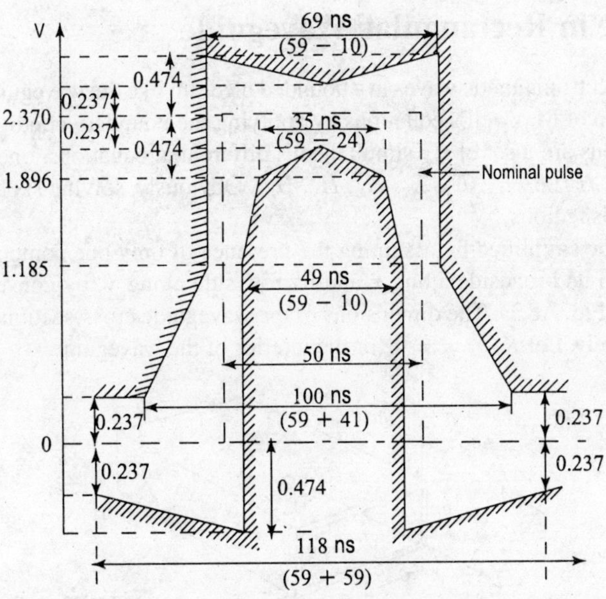

Fig. A6. *Pulse mask at 8.448 Mbps port.*

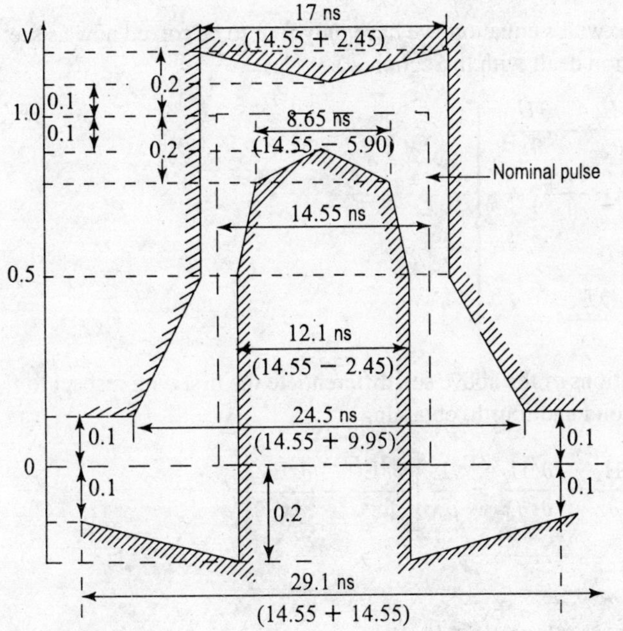

Fig. A7. *Pulse mask at 34.368 Mbps port.*

A8. H_{10} Wave in Rectangular Waveguide

The propagation of electromagnetic waves in a bounded medium like the waveguide can be investigated by finding the solution of Maxwell's equations by applying the boundary conditions.

Maxwell's equations are a set of six simultaneous differential equations, one for each of the three components of E and H (i.e., E_x, E_y, E_z, H_x, H_y, H_z). Obviously, solving six equations by applying boundary conditions is tedious.

The problem can be simplified by assuming the presence of only one component of electric field. For a waveguide with its broadside along x and narrow side along y, by convention we assume the presence of E_y only (Fig. A8.1). The dimensions of the waveguide cross-section are a and b (along x- and y- axis) respectively. Let $K_r = \mu = 1$ for the interior of the waveguide.

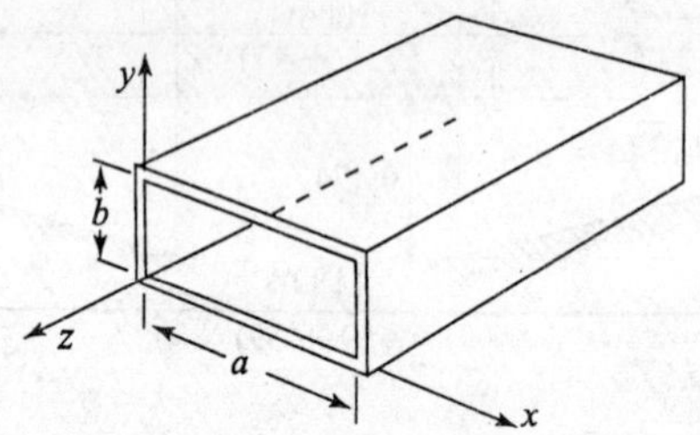

Fig. A8.1 *Coordinate system for rectangular waveguide.*

Out of the six Maxwell's equation, we have only four to be solved now as seen from the rectangular waveguide propagation dealt with in Section 20.3.

$$\left.\begin{aligned} \kappa_0 \frac{\partial E_y}{\partial t} &= \frac{\partial H_x}{\partial z} - \frac{\partial H_z}{\partial x} \\ -\mu_0 \frac{\partial H_x}{\partial t} &= -\frac{\partial E_y}{\partial z} \\ -\mu_0 \frac{\partial H_y}{\partial t} &= 0 \\ -\mu_0 \frac{\partial H_z}{\partial t} &= \frac{\partial E_y}{\partial x} \end{aligned}\right\}. \tag{A8.1}$$

To solve the equations of the above set, differentiate the first with respect to t, and substitute for H_x and H_z from the second and fourth, obtaining

$$\kappa_0 \frac{\partial^2 E_y}{\partial t^2} = \frac{\partial^2 H_x}{\partial t \partial z} - \frac{\partial^2 H_z}{\partial t \partial x} = \frac{1}{\mu_0}\left[\frac{\partial^2 E_y}{\partial z^2} + \frac{\partial^2 E_y}{\partial x^2}\right].$$

Hence

$$\frac{\partial^2 E_y}{\partial x^2} + \frac{\partial^2 E_y}{\partial z^2} - \kappa_0 \mu_0 \frac{\partial^2 E_y}{\partial t^2} = 0,$$

or

$$\frac{\partial^2 E_y}{\partial x^2} + \frac{\partial^2 E_y}{\partial z^2} - \frac{1}{c^2}\frac{\partial^2 E_y}{\partial t^2} = 0. \tag{A8.2}$$

This is the differential equation to be solved for E_y, and similar equations hold for H_x and H_z. It is a wave equation but for the addition of one more term.

Consider the expression

$$E_y = A \sin kx \sin(\omega t - \beta z), \tag{A8.3}$$

where β is the propagation constant, k is a constant dependent on the boundary encountered in propagation, and A is a constant defining the amplitude of E_y. Substitution of this in the differntial equation shows that it is a solution, provided that β, instead of being equal to ω/c as in the plane wave, satisfies the equation

$$\beta^2 = (\omega/c)^2 - k^2. \tag{A8.4}$$

The constant k is easily determined from boundary conditions. The electric field must be zero at those side walls to which it is tangential, that is,

$$E_y = 0 \quad \text{at} \quad x = 0 \quad \text{and} \quad x = a.$$

The form of equation (A8.3) satisfies this condition automatically for $x = 0$, but for $x = a$ it implies

$$\sin ka = 0,$$

i.e., a number of possible values of k as determined by $ka = \pi, 2\pi, 3\pi, \ldots.$

The first value of k gives the field of what we have called the H_{10} mode. For this value of k, (A8.3) becomes,

$$E_y = A \sin(\pi x/a) \sin(\omega t - \beta z), \tag{A8.5}$$

where β is given by $\beta^2 = \left[\frac{\omega}{c}\right]^2 - \left[\frac{\pi}{a}\right]^2.$ (A8-6)

This value of the phase constant β in the waveguide defines the guide wavelength λ_g by the relation $\beta = 2\pi/\lambda_g$ just as the free-space value of β defines the free-space wavelength λ. By substituting this in equation (A8.6) and dividing by $4\pi^2$ we get the fundamental relation between λ and λ_g for the H_{10} mode.

$$\frac{1}{\lambda_g^2} = \frac{1}{\lambda^2} - \left[\frac{m}{2a}\right]^2. \tag{A8.7}$$

This is same as the result obtained in section 20.4

Substitution of equation (A8.5) into (A8.1) enables us to find the following expressions for the field components:

$$\left.\begin{aligned}
E_y &= A \sin\frac{\pi x}{a} \sin(\omega t - \beta z),\\
H_x &= -A\frac{\beta}{\omega\mu_0} \sin\frac{\pi x}{a} \sin(\omega t - \beta z),\\
H_z &= A\frac{\pi/a}{\omega\mu_0} \cos\frac{\pi x}{a} \cos(\omega t - \beta z),\\
E_x &= E_z = H_y = 0,
\end{aligned}\right\} \tag{A8.8}$$

where the phase constant β is given by equation (A8.6).

A solution in $\cos(\omega t - \beta z)$ could equally well have been used in place of the sine term in equation (A8.5), and the two solutions can then be combined in the form

$$E_y = A \sin \frac{\pi x}{a} \varepsilon^{j(\omega t - \beta z)} \tag{A8.9}$$

in the same way as was done in the equations for the plane wave equation. This form of the solution is more convenient for some purposes. It may also be written as

$$E_y = A \sin \frac{\pi x}{a} \varepsilon^{j\omega t} \varepsilon^{-j\beta z}. \tag{A8.10}$$

The lines of force plotted from the field components give the accurate field distribution whose approximate form was deduced from elementary considerations in Section 20.2. The phase and group velocities can now be readily derived. The form $(\omega t - \beta z)$ represents a wave travelling in the positive direction of the z-axis.

By definition

$$\upsilon_p = \frac{\omega}{\beta} = \frac{2\pi c/\lambda}{2\pi/\lambda_g} = c\frac{\lambda_g}{\lambda}, \quad \text{(same as the result in Section 20.4)}$$

$$\upsilon_g = \frac{d\omega}{d\beta} = \frac{c^2 \beta}{\omega} = \frac{c^2}{\upsilon_p} = c\frac{\lambda}{\lambda_g}. \quad \text{(same as the result in Section 20.4)}$$

A more general solution would be found by taking $ka = m\pi$ (m any integer) as the solution of $\sin ka = 0$ in place of the particular value $ka = \pi$ which gave the H_{10} wave. This gives (as the equation of λ_g)

$$\frac{1}{\lambda_g^2} = \frac{1}{\lambda^2} - \left[\frac{m}{2a}\right]^2, \tag{A8.11}$$

and this value corresponds to the more general H_{m0} wave.

A9. General Solution for Rectangular Waveguide

For the most general form of H wave the E_x and E_y components are present, but since the wave is transverse electric, E_z is zero. The wave equation for any of the field components is the general one given in equation

$$\mu k \frac{j^2 E_x}{j t^2} = \frac{j^2 E_x}{j x^2} + \frac{j^2 E_x}{j z^2}. \tag{A9.1}$$

Consider $E_x = A \cos k_1 x \sin k_2 y \sin(\omega t - \beta z)$. For equation (A9.1) it is seen to be a solution (for the case where $k_r = \mu_r = 1$) provided that β satisfies the equation

$$\beta^2 = (\omega/c)^2 - k_1^2 - k_2^2. \tag{A9.2}$$

Using continuity equation,

$$\text{i.e.,} \quad \frac{\partial E_x}{\partial x} + \frac{\partial E_y}{\partial y} + \frac{\partial E_z}{\partial z} = 0,$$

the component E_y can be found to be proportional to $\sin k_1 x \cos k_2 y \sin(\omega t - \beta z)$. This time the boundary conditions $E_x = 0$ at $y = b$, and $E_y = 0$ at $x = 0, x = a$, involve both the dimensions a and b, the values of k_1 and k_2 being thereby determined as $k_1 a = m\pi$, $k_2 b = n\pi$, with integral values for m and n. This gives, as the general expression for λ_g in an H_{mn} wave

$$\frac{1}{\lambda_g^2} = \frac{1}{\lambda^2} - \left[\frac{m}{2a}\right]^2 - \left[\frac{n}{2b}\right]^2, \tag{A9.3}$$

and hence for the critical wavelength λ_c

$$\frac{1}{\lambda_c} = \left[\frac{m}{2a}\right]^2 + \left[\frac{n}{2b}\right]^2. \tag{A9.4}$$

The treatment for the E_{mn} wave is similar, and gives rise to the same equations (A9.3 and A9.4) for λ_g and λ_c respectively.

If the expressions for the field components are derived, it will be found that for the H wave either m or n may be made zero, but if both are made zero, the field vanishes. For the E wave, on the other hand, the field vanishes if either m or n is made zero. Thus, while the fundamental H mode is H_{10} or H_{01} (depending on whether $a>b$ or $a<b$), the fundamental E mode is the E_{11}.

A10. Attenuation of H_{10} Wave in Rectangular Waveguide

Since the power carried by a wave is proportional to the product of E and H, if each of these is attenuated exponentially with an attenuation constant α, and the power will decrease according to the relation: $P_z = P_0 e^{-2\alpha_z}$.

Hence the rate of power decrease or loss of power per unit length is equal to:

$$-dP_{z/dz} = 2\alpha P_0 \exp(-2\alpha z) = 2\alpha P_z.$$

Thus, if P_{loss} is the power lost in a unit length of guide through which the power flow is P, the attenuation constant is given by $\alpha = P_{\text{loss}}/2P$.

By the boundary condition, the magnitude of the surface current density is equal to that of the tangential magnetic field at the surface. The current density in the lower surface of the guide is due to the magnetic field components, H_x and H_z, and its mean-square value is given by:

$$I^2 = \frac{1}{2}\left|H\right|^2 = \frac{1}{2}\left[\left|H_z\right|^2 + \left|H_x\right|^2\right]_{y=0}$$

Multiplication of I^2 by the surface resistivity R_s gives the mean power loss per unit area. Inserting the values for H_z and H_x from equation of H_{10} wave and integrating over the width from $x = 0$ to $x = a$ we get the mean power loss per unit length as

$$\frac{1}{2}R_s \int_0^a \left\{\left[\frac{A\pi/a}{\omega\mu_0}\right]^2 \cos^2\frac{\pi x}{a} + \left[\frac{A\beta}{\omega\mu_0}\right]^2 \sin^2\frac{\pi x}{a}\right\} dx$$

which reduces to $(1/2)(A/120\pi)^2(a/2)$.

A similar calculation gives the mean power loss per unit length for the side wall $x = 0$ as

$$\frac{1}{2}R_s\left[(A/120\pi)(\lambda/\lambda c)\right]^2 b.$$

Doubling the sum of these two losses gives the total loss per unit length for all four walls. Then, using equations for losses and putting $\lambda_c = 2a$, we can calculate the attenuation constant as:

$$\alpha = \frac{R_s}{120\pi b}\left\{1 + \frac{2b}{a}\left[\frac{\lambda}{2a}\right]^2\right\}\frac{1}{\sqrt{\{1 - (\lambda/2a)^2\}}} \quad \text{nepers per metre.}$$

A11. Theory of Circular Waveguides

The wave equation in x, y, z components applies equally to propagation in a circular waveguide, but in that form it is difficult to satisfy the boundary conditions. The equation must be transformed into cylindrical coordinates so that a point instead of being represented by the coordinates (x, y, z), is represented by (r, θ, z), where r is the radius vector and θ the polar angle related to x and y as shown in Fig A11.1.

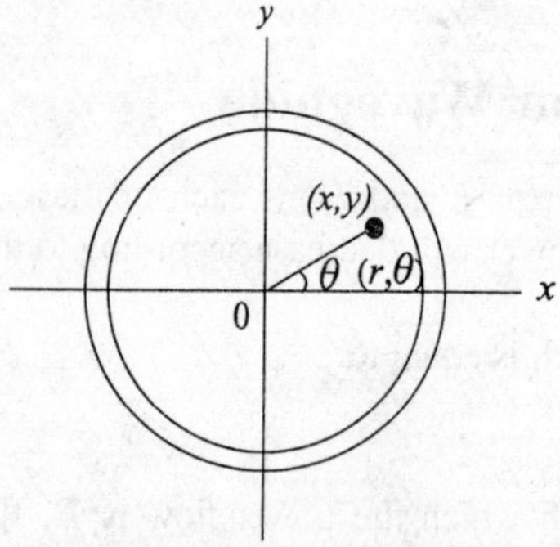

Fig. A11.1 *Cylindrical coordinate system for circular waveguide.*

$$x = r\cos\theta, \quad y = r\sin\theta.$$

The transformation can be carried out most simply by regarding r and θ as two independent variables, in terms of which x and y are defined by the above equations. By partial differention, we get:

$$\frac{\partial}{\partial r} = \frac{\partial}{\partial x}\frac{\partial x}{\partial r} + \frac{\partial}{\partial y}\frac{\partial y}{\partial r}$$
$$= \frac{\partial}{\partial x}\cos\theta + \frac{\partial}{\partial y}\sin\theta.$$

Multiplication by r and differentiation with respect to r gives:

$$\frac{\partial}{\partial r}\left[r\frac{\partial}{\partial r}\right] = \left[\frac{\partial}{\partial x} + x\frac{\partial^2}{\partial x^2} + y\frac{\partial^2}{\partial x\partial y}\right]\frac{\partial x}{\partial r} + \left[\frac{\partial}{\partial y} + y\frac{\partial^2}{\partial y^2} + x\frac{\partial^2}{\partial x\partial y}\right]\frac{\partial y}{\partial r},$$

which reduces to:

$$r\frac{\partial}{\partial r}\left[r\frac{\partial}{\partial r}\right] = x^2\frac{\partial^2}{\partial x^2} + y^2\frac{\partial^2}{\partial y^2} + 2xy\frac{\partial^2}{\partial x \partial y} + x\frac{\partial}{\partial x} + y\frac{\partial}{\partial y}. \tag{A11.1}$$

Similarly,

$$\begin{aligned}\frac{\partial}{\partial \theta} &= \frac{\partial}{\partial x}\frac{\partial x}{\partial \theta} + \frac{\partial}{\partial y}\frac{\partial y}{\partial \theta}\\ &= -y\frac{\partial}{\partial x} + x\frac{\partial}{\partial y}.\end{aligned}$$

A second differentiation gives

$$\frac{\partial^2}{\partial \theta^2} = y^2\frac{\partial^2}{\partial x^2} + x^2\frac{\partial^2}{\partial y^2} - 2xy\frac{\partial^2}{\partial x \partial y} - x\frac{\partial}{\partial x} - y\frac{\partial}{\partial y}\ldots \tag{A11.2}$$

Adding (A11.1) and (A11.2),

$$\begin{aligned}r\frac{\partial}{\partial r}\left[r\frac{\partial}{\partial r}\right] + \frac{\partial^2}{\partial \theta^2} &= (x^2+y^2)\frac{\partial^2}{\partial x^2} + (x^2+y^2)\frac{\partial^2}{\partial y^2}\\ &= r^2\left[\frac{\partial^2}{\partial x^2} + \frac{\partial^2}{\partial y^2}\right].\end{aligned}$$

Substitution of this result transforms the wave equation into its polar form

$$\frac{1}{r}\frac{\partial}{\partial r}\left[r\frac{\partial E_z}{\partial r}\right] + \frac{1}{r_2}\frac{\partial^2 E_z}{\partial \theta^2} + \frac{\partial^2 E_z}{\partial z^2} - \frac{1}{c^2}\frac{\partial^2 E_z}{\partial t^2} = 0, \tag{A11.3}$$

k_r and μ_r being taken equal to unity. We should expect the part of the solution of this equation involving z and t to be of the same form as in the previous wave solutions, that is, $\cos(\omega t - bz)$ or $e^{i(\omega t - bz)}$. If the complete solution be written as $E_z = Fe^{j(\omega t - bz)}$ (or $F\cos(\omega t - bz)$, where F is that part of the solution involving r and θ, substitution of this in (A11.3) gives:

$$\frac{1}{r}\frac{\partial}{\partial r}\left[r\frac{\partial F}{\partial r}\right] + \frac{1}{r^2}\frac{\partial^2 F}{\partial \theta^2} + \left[-\beta^2 + \frac{\omega^2}{c^2}\right]F = 0,$$

which can be written as

$$\frac{1}{r}\frac{\partial}{\partial r}\left[r\frac{\partial F}{\partial r}\right] + \frac{1}{r^2}\frac{\partial^2 F}{\partial \theta^2} + k^2 F = 0. \tag{A11.4}$$

Let us consider first the simpler case where E_z and hence F are independent of θ so that (A11.4) reduces to

$$\frac{1}{r}\frac{d}{dr}\left[r\frac{dF}{dr}\right] + k^2 F = 0,$$

$$\frac{d^2 F}{dr^2} + \frac{1}{r}\frac{dF}{dr} + k^2 F = 0.$$

This equation, which is a special case of what is called Bessel's equation, can only be solved in terms of simple functions by means of an infinite series. Substitution of the series

$$F = 1 - \frac{(kr)^2}{2^2} + \frac{(kr)^4}{(2.4)^2} - \frac{(kr)^6}{(2.4.6)^2} + \ldots, \tag{A11.5}$$

will show that this satisfies the equation. It is convenient however, to introduce as the solution a new function which will be defined by this series. The function is called a Bessel function and the usual symbol for it is $J_0(kr)$. In this particular case it is of zeroth order, indicated by the subscript 0. The full expression for E_z is then

$$E_z = AJ_0(kr)\varepsilon^{j(\omega t - \beta z)}, \tag{A11.6}$$

A being as before an arbitrary amplitude constant.

So far, k and β are undetermined. But they are readily found by application of a boundary condition. At $r = a$, if a is the radius of the tube, E_z is tangential to the tube wall and must therefore be zero. Hence we have the boundary condition that

$$J_0(ka) = 0.$$

Numerical values of $J_0(s)$ for a variable s can be calculated from the series in equation (A11.5). When plotted against s these values give the curve shown in Fig. A11.2. From it we see that $J_0(s)$ is zero for a series of values of s, of which the first which has the value 2.405, is the only one to be considered here. Thus k is determined as 2.405/a. The propagation constant β, which is given by $k^2 = (\omega/c)^2 - \beta^2$, can now be found to be

$$\beta = \left\{\left[\frac{\omega}{c}\right]^2 - \left[\frac{2.405}{a}\right]^2\right\}^{1/2}. \tag{A11.7}$$

From this the guide wavelength λ_g is determined as

$$\frac{1}{\lambda_g} = \frac{\beta}{2\pi} = \left\{\frac{1}{\lambda^2} - \left[\frac{2.405}{2\pi a}\right]^2\right\}^{1/2}, \tag{A11.8}$$

and then the critical wavelength as $\lambda_c = 2\pi a/2.405$. This is an E mode and is known as the E_{01}, the subscript 0 from the Bessel function, and the subscript 1 because the first root of $J_0(s)$ was taken.

Higher order modes involve solutions of (A11.4) in which F is not independent of θ. The variation of F with θ must be of period 2π so that there be no discontinuity in the field. The simplest variation of this type is a sinusoidal one, F being written as $F = R\sin\theta$, where R is a function of r. Substitution of this in (A11.4) shows that R must satisfy the equation

$$\frac{d^2R}{dr^2} + \frac{1}{r}\frac{dR}{dr} + \left(k^2 - \frac{1}{r^2}\right)R = 0,$$

which again has a series solution, this time denoted by the Bessel function of the first order $J_1(kr)$. The form of this function is shown in Fig. A11.2.

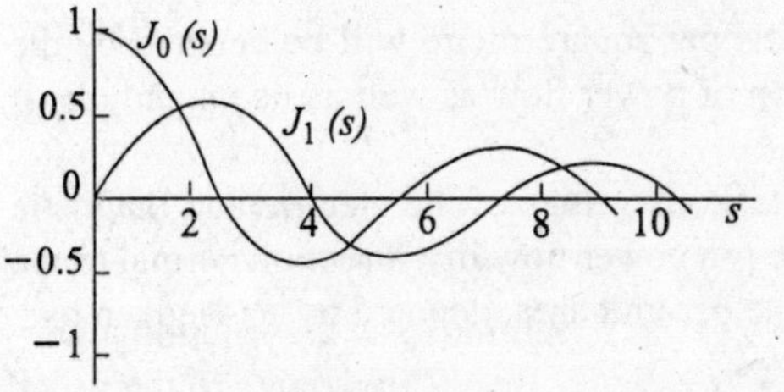

Fig. A11.2 *Graph of Bessel function $J_0(S)$ and $J_1(S)$.*

Since the complete expression for E_z in this case is given by

$$E_z = AJ_1(kr)\sin\theta e^{j(wt-bz)}, \tag{A11.9}$$

the boundary condition will require that

$$J_1(ka) = 0.$$

The smallest value of ka to satisfy this is, from Fig. A11.2, $ka = 3.832$. This defines an E_{11} wave, which will not be considered here in any more detail.

For H waves the field components also satisfy equation (A11.3). The two simplest modes, the H_{01} and H_{11} have their H_z components given by the same expression as those for E_z in equations (A11.6) and (A11.7). With H waves, however, the boundary conditions are different since H_z is not required to vanish at the boundary. There is no E_z, but there is another electric component E_0 tangential to the walls. It can be shown from Maxwell's equations that E_0 is proportional to H_z/r, and thus the boundary condition for the H_{01} wave is

$$\left[\frac{d}{dr}J_0(kr)\right]_{r=a} = 0.$$

Values of k satisfying this evidently occur at places where the slope of the $J_0(s)$ curve is zero. The first such value of s is 3.832, of which the approximate value can be seen from Fig. A11.2 (excluding $s = 0$ which makes the field vanish), hence k is given by $k = 3.832/a$.

In the case of the H_{11} wave the corresponding boundary condition is

$$\left[\frac{d}{dr}J_1(kr)\right]_{r=a} = 0,$$

whence the first solution is $k = 1.841/a$.

From these values of k the values of b and λ_g for H_{01} and H_{11} waves can be found as before. The expressions will be the same as those of equations (A11.7) and (A11.8) with 2.405 replaced by 3.832 and 1.841 respectively.

A12. Power and the Poynting Vector

The energy carried by an electromagnetic wave as a plane wave and as a guided wave (in rectangular wave guide) is calculated here based on **E** and **H**. Consider an imaginary plane surface in a space chosen perpendicular to the direction of motion of a plane wave. There will be a flow of power through this surface, and the power passing through a unit area in watts per square metre will be denoted by the symbol **P**. This is a vector quantity, indicating the direction of power flow as well as its magnitude. It is called the Poynting vector.

Poynting's theorem states that, if E and H are instantaneous values of the electric and magnetic vectors at a point and θ the angle between them, then there is a power flow in a direction normal to the plane containing E and H of which the instantaneous value per unit area, denoted by P_a is given by

$$P_a = EH \sin\theta \quad \text{webers/m}^2 \quad \left[\boldsymbol{P} = \boldsymbol{E} \times \boldsymbol{H}\right]$$

The direction of $\boldsymbol{P}$ relative to the fields is given by the 'corkscrew rule' and is as shown in Fig. A12.1. A plane wave can be represented as

$$E_x = A_e^{j(\omega t - \beta z)},$$

$$H_y = A\sqrt{[k_0/\mu_0]}e^{j(\omega t - \beta z)}.$$

Poynting vector through a surface coinciding with the plane $z = 0$ at the instant $t = 0$ is:

$$P_a = A^2(k_0/\mu_0)^{1/2} = A^2/120\pi,$$

and its direction is along the positive z-axis (refer to Fig. A12.1.)

In dealing with propagation in a waveguide, the power flow is along the guide axis, and so our surface is chosen perpendicular to the axis of the guide. The possible field components in this cross-section are E_x, E_y, E_z, H_x and H_y. There will then be two components of $\boldsymbol{P}$, viz., $E_x H_y$ and $E_y H_x$, but the direction of the second is, by the corkscrew rule, along the negative z-axis. Thus the value of $\boldsymbol{P}$ at the point (x, y) in the cross-section is given by equation

$$P_a = E_x H_y - E_y H_x.$$

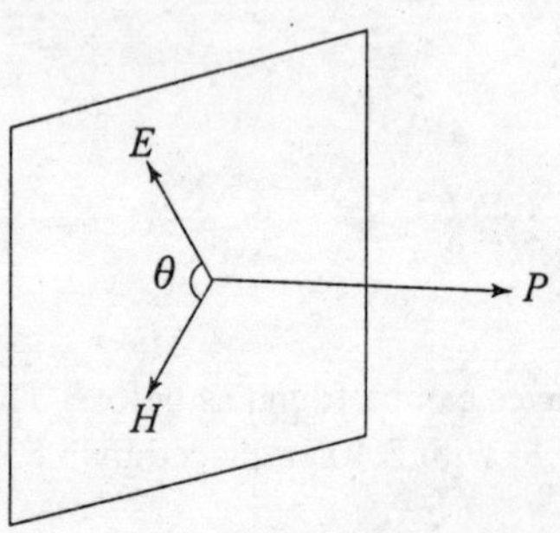

Fig. A12.1 *Vector directions in Poynting's theorem.*

Application of this formula to the equations of the H_{10} wave given in Appendix A8 in a rectangular guide at a point in the plane $z = 0$ gives

$$P_a = \frac{A^2}{120\pi}\frac{\lambda}{\lambda_g}\sin^2\frac{\pi x}{a}\sin^2\omega t.$$

This is the instantaneous power flow at one point in the cross-section. To obtain the instantaneous total power P_{inst}, this expression must be integrated over the whole area of the waveguide cross-section, and becomes

$$P_{inst} = \frac{A^2}{120\pi}\frac{\lambda}{\lambda_g}\int_0^a \sin^2\frac{\pi x}{a}dx\int_0^b dy \sin^2\omega t$$

$$= \frac{A^2}{120\pi}\frac{\lambda}{\lambda g}\frac{ab}{2}\sin^2\omega t.$$

Usually the mean value P over an interval of time is required, so that $\sin^2\omega t$ is replaced by its mean value 1/2, and the mean power is:

$$P = \frac{A^2}{120\pi}\frac{\lambda}{\lambda_g}\frac{ab}{4} \quad \text{watts.}$$

This expression is used in calculating attenuation in a waveguide.

Since the maximum electric field occurs half-way accross the guide at $x = a/2$, and since the maximum value is equal to A, P can be written as

$$P = \left[E_{max}\right]^2 \frac{ab}{480\pi}\frac{\lambda}{\lambda_g} \quad \text{watts.}$$

A13. Signalling System Number 7

A13.1 Introduction

The transition of network control signalling from in-channel signalling to common-channel signalling has finally culminated in signalling system number 7 (SS7). SS7 is a general purpose common channel signalling system that can be used over a variety of digital and circuit-switched networks as well as ISDN. SS7 specifications were first issued by CCITT in 1980 and revised in 1984, 1988, 1992.

Though SS7 with its mechanisms for internal control and network intelligence is specifically designed for ISDNs, it can be used over a variety of digital circuit-switched networks (GSM being one such network) because of its open-ended design.

The objectives of SS7 are to provide an internationally standardized general purpose common-channel signalling system with the following characteristics:

- Optimized for use in digital telecom networks in conjunction with digital stored program-control exchanges utilizing 64 kbps digital channels
- Able to meet information transfer requirements for call control, remote control, management and maintenance
- Able to provide for transfer of information in correct sequence without loss or duplication
- Suitable for operation over analog channels at speeds below 64 kbps
- Suitable for use on point-to-point terrestrial and satellite links

A13.2 ITU-T Recommendations on SS7

The ITU-T recommendations that comprise SS7 are listed below.

Sl. No.	Rec. No.	Title and date
1.	Q.700	Introduction to CCITT SS7, 1993
2.	Q.701	Functional description of the message transfer part (MTP) of SS7, 1993
3.	Q.702	Signalling data link, 1988
4.	Q.703	Signalling link, 1996
5.	Q.704	Signalling network functions and messages, 1996
6.	Q.705	Signalling network structure, 1993
7.	Q.706	Message transfer part signalling performance, 1993
8.	Q.707	Testing and maintenance, 1988
9.	Q.708	Number of international signalling point codes, 1993
10.	Q.709	Hypothetical signalling reference connection, 1993
11.	Q.710	Simplified MTP version for small systems, 1988
12.	Q.711	Functional description of the signalling connection control part (SCCP), 1996
13.	Q.712	Definition and functions of SCCP messages, 1996
14.	Q.713	SCCP formats and codes, 1996
15.	Q.714	SCCP procedures, 1996
16.	Q.715	SCCP user guide, 1996
17.	Q.716	SCCP performances, 1993
18.	Q.721	Functional description of the SS7 telephone user part (TUP), 1988
19.	Q.722	General function of telephone messages and signals, 1988
20.	Q.723	Formats and codes, 1988
21.	Q.724	Signalling procedures, 1988
22.	Q.725	Signalling performance in the telephone application, 1993
23.	Q.730	ISDN supplementary service, 1993
24.	Q.731.1	Direct dialing-in, 1996
25.	Q.731.3	Calling line identification presentation, 1993
26.	Q.731.8	Sub-addressing, 1992
27.	Q.732.2	Call offering supplementary services, 1996
28.	Q.732.7	Explicit call transfer, 1996
29.	Q.733.1	Call waiting, 1992
30.	Q.733.2	Call hold, 1993
31.	Q.734.1	Conference calling, 1993
32.	Q.734.2	Three party service, 1996
33.	Q.735.1	Closed user group, 1993

34.	Q.735.6	Global virtual network service,1996
35.	Q.736.1	International telecommunication charge card, 1995
36.	Q.736.3	Reverse charging, 1995
37.	Q.737.1	User-to-user signalling, 1993
38.	Q.750	Overview of SS7 management, 1993
39.	Q.751.1	Network element management information model for MTP, 1995
40.	Q.752	Monitoring and measurements for SS7 networks, 1997
41.	Q.753	SS7 management functions MRVT, SRVT & CVT and definition of the OMASE-User, 1997
42.	Q.754	Operations, maintenance & administration part (OMAP), 1997
43.	Q.755	SS7 protocol tests, 1993
44.	Q.756	Guide book to OMAP, 1997
45.	Q.761	Functional description of the ISDN user part of SS7, 1993
46.	Q.762	General function of messages and signals, 1993
47.	Q.763	Formats and codes, 1993
48.	Q.764	Signalling procedures, 1993
49.	Q.766	Performance objectives in the ISDN application, 1993
50.	Q.767	Application of ISUP for international ISDN interconnections, 1991
51.	Q.768	Signalling interface between an international switching center and an ISDN satellite sub-network, 1995
52.	Q.771	Functional description of transaction capabilities, 1993
53.	Q.772	Transaction capabilities information element definitions, 1993
54.	Q.773	Transaction capabilities formats and encoding, 1993
55.	Q.774	Transaction capabilities procedures, 1993
56.	Q.775	Guidelines for using transaction capabilities, 1997
57.	Q.780	SS7 test specification general description, 1995
58.	Q.781	MTP level 2 test specification, 1996
59.	Q.782	MTP level 3 test specification, 1996
60.	Q.783	TUP test specification, 1988
61.	Q.784	ISUP basic call test specification, 1991
62.	Q.785	ISUP protocol test specification for supplementary services, 1991
63.	Q.786	SCCP test specification, 1992
64.	Q.787	Transaction capabilities test specification, 1993
65.	Q.788	User network interface to user network interface compatibility test specifications, 1997

A13.3 SS7 Architecture

A13.3.1 Overlay of packet-switched network on circuit-switched network

In the system of common-channel signalling, control messages are routed through the network to perform call management and network management functions. These are packets that get routed through the network. Thus, a packet-switched network is overlaid on a circuit-switched network to operate and control the circuit-switched network.

However, for this overlay, SS7 does not dictate any hardware implementation. All SS7 functions can be implemented in the circuit switching nodes as additional functions or as separate switching points that carry only the control packets. In the latter case, some portions of SS7 also need to be implemented on circuit switching nodes as they have to receive the control signals.

A13.3.2 Functional entities in SS7

SS7 defines the following three functional entities namely,

- signalling point (SP)
- signal transformer point (STP)
- signalling link (SL).

Signalling point (SP) is any point in the signalling network capable of handling SS7 control messages. It can be the end-point for control messages (circuit switching nodes) or the network control centre. The SP may not be capable of processing messages not directly addressed to itself.

Signal transfer point (STP) is a signalling point which is capable of routing control messages. It receives a message on one signalling link and transfers it to another. STP can be a pure routing node, or can also include functions of end points.

Signalling link (SL) is a data link that connects signalling points (SPs).

A13.3.3 Control plane and information plane in SS7

The distinction between packet switching signalling function and circuit switching information transfer function can well be understood through the concept of 'control plane' and 'information plane'.

The control plane is responsible for establishing and managing connections. These connections are requested by the user over the D-channel using Q.931 (dialogue between user and exchange). For this purpose, the local exchange acts as a signalling point (SP)—it translates the dialogue with user (Q.931) into control messages inside the network that perform the user-requested functions and vice versa. Internal to the network, SS7 is used to establish and maintain a connection involving one or more SPs and STPs.

Once a connection is set up, information is transferred end-to-end in the 'information plane'. For this to happen, a circuit is set up from the local exchange of one user to that of another (routed through one or more other circuit switching nodes which act as transit centres, but are SPs essentially).

A13.4 Signalling Network Structure in SS7

Signalling network in SS7 has SPs and STPs organized in hierarchical form with SPs at the lower rung. The STPs at higher level can have several rungs.

A13.4.1 Parameters to be considered in the design of signalling network

The following parameters are to be considered in the designing of signalling network for SS7 and the number of levels thereof :

- STP capacities
  - Number of SLs that can be handled.
  - Signalling message transfer time
  - Message throughput capacity
- Network performance
  - Number of SPs
  - Signalling delays
- Availability and reliability requirements

A13.4.2 ITU-T guidelines for signalling network structure

ITU-T issued the following guidelines for signalling network design:

- Each SP that is not an STP is at the same time connected to at least two STPs.
- The meshing of STPs is as complete as possible.

A13.5 Protocol Architecture

SS7 protocols are layered into four levels. Their equivalence with OSI model is depicted in Fig. A13.1.

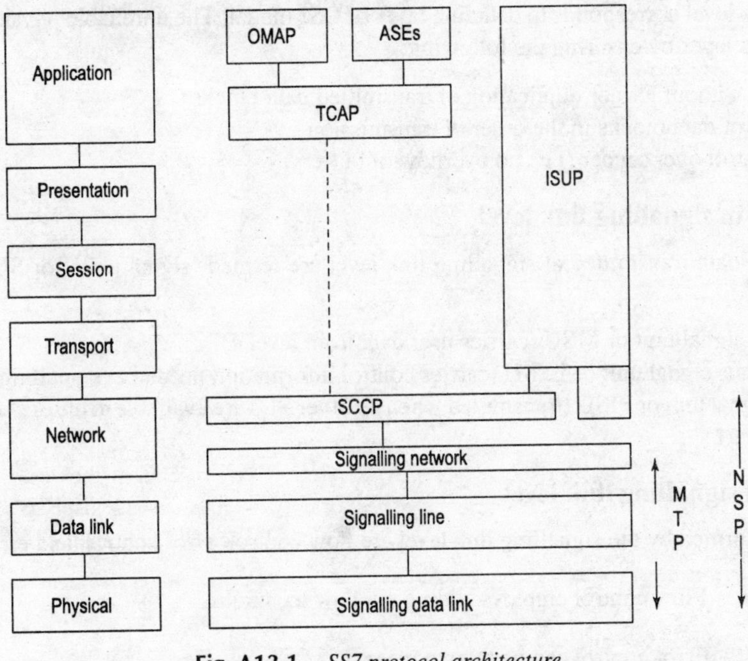

Fig. A13.1 *SS7 protocol architecture*

OMAP: Operations, Mntc & Administration; ASE: Application service element; TCAP: Transactions capabilities appl. Part; ISUP: ISDN user part; SCCP: Signalling connection control part; MTP: Message transfer part; NSP: Network service part.

The lowest three levels of SS7 are referred to as MTP (manage transfer point). MTP provides a 'datagram' style of service for routing messages through the SS7 network. MTP was developed prior to SCCP and was tailored to the real-tine needs of telephony applications. SCCP was designed later, to meet the requirements of ISDN such as expanded addressing capability and reliable message transfer.

The split-up of SS7 into MTP and SCCP facilitates the usage of SCCP (overhead) only when required.

The three levels in MTP, viz.,

- signalling data link level,
- signalling link level, and
- signalling network level

shall be described now.

A13.5.1 Signalling data link level

Signalling data link level which is the equivalent of physical layer of OSI model, provides full duplex physical link dedicated to SS7 traffic optimized for 64 kbps. However, lower speed connections and analog links with modems are also allowed. Routing via a satellite is also considered.

A13.5.2 Signalling link level

Basic purpose

Signalling link level corresponds to data link layer of OSI model. The unreliable physical link is made reliable by this layer by ensuring the following:

- Delivery without loss or duplication of transmitted data blocks
- Delivery of data blocks in the order of transmission
- Flow control over sender (i.e., no overflow of buffers)

Signal units in signalling link level

The blocks of data transmitted at signalling link level are termed 'signal units' or SUs. There are 3 types of SUs :

- Message signal unit or MSU (carries user data from level 4)
- Level status signal unit or LSSU (carries control information needed at signalling link level)
- Fill-in signal unit or FISU (transmitted when no other SUs are available to ensure continuous error monitoring).

Functions of signalling link level

Functions performed by the signalling link level are flow control, error control, and error monitoring.

Flow control: Flow control employs sliding window technique.

Error control: Error control can be by two means:

- Basic method of go-back-N ARQ.
- Preventive cyclic retransmission (PCR)

The latter is applicable for signalling links with one-way propagation delay greater than or equal to 15 ms, i.e., links established via satellite.

For a link with relatively long propagation delay, if each message unit is comparatively short, then the link may be idle for most of the time. Hence it is not good practice to wait for negative ACK before retransmitting. Instead, a node (whenever it has no MSUs pending) automatically retransmits unacknowledged MSUs without waiting for positive or negative ACK.

However, this has an associated problem. Since only positive ACK's are sent by the other side, there is a danger that a unit in error may go undetected for a considerable time, more so when the flow of traffic is heavy. In such situations, one side is so busy sending new units that it rarely performs voluntary retransmission.

The way out of this problem is 'forced retransmission procedure'. In this procedure, when a predetermined number of outstanding unacknowledged signal units exist, the transmission of new units is interrupted and the retained signal units are retransmitted cyclically until the number of unacknowledged signal unit is reduced.

Error monitoring: Two types of error monitoring are supported by signalling link level:

1. Signalling unit rate error monitor: This is for a signal link in service for deciding when to take it out of service. A counter initialized to zero is maintained and manipulated based on two parameters namely (i) T—threshold above which error is signalled to layer-3, (ii) $1/D$—lowest error rate that would cause signalling of error in layer-3. For each signalling unit (SU) received in error, the counter is incremented by 1 and decremented by D for every sequence of received SUs (error /error free). When the count equals T, link is declared unreliable. For 64 kbps links $T = 64$ and $D = 256$ $(1/D = 0.004)$, this algorithm is termed 'Leaky Bucket Algorithm'.

2. Alignment error rate monitoring: This is employed when the signal link is being initialized and aligned. This provides a criteria for rejecting a link for service due to excessive error rate. A counter is used for this purpose which is initialized to zero. It is incremented by one for each signal unit received in error. If the counter exceeds a threshold before the end of an initial 'proving period', the proving period is aborted. In the event of failure, the proving period procedure may be tried up to five times—if all five successive attempts fail, the signal link is rejected.

A13.5.3 Signalling network level

This level includes functions related to network management and message handling.

Message handling functions

These are performed at every SP including STP. These are based on the 'routing label' and 'service info octet' parts of MSU.

There are three categories of message handling functions:

Discrimination: By analyzing the destination code in the routing label, it is determined whether the message is at its destination or to be relayed further.

Routing: This determines the SL to be used in forwarding a message. The message may be from discrimination functions or level-4. For a given source–destination pair, several alternate routes may be possible; the value of SLS field specifies which particular route is to be followed.

With a 4-bit SLS field, 16 different routes are possible. These are virtual circuits in the network. The control signals associated with a single call follow the same route usually. The user part varies the route selection from one call to the next so that the MTP can distribute traffic uniformly.

Distribution: By analyzing the 'service indicator' portion of 'service indicator octet', it is determined to which user part the message is to be delivered.

Network management functions

The objective of this set of functions is to overcome link degradation (failures or congestion). The unavailability shall not be more than 10 minutes per year for any route. Such a lofty goal is achievable through redundancy of links and dynamic routing.

SS7 lays strong emphasis on network management functions because

- the performance of networks control signalling architecture affects all subscribers to the network,
- degradation of a nation's signalling system will have repercussions beyond the borders,
- recovery and restoration actions involve multiple networks. Without SS7 involvement in this field, every public network would be forced to enter into several agreements/protocols with other networks which would be very inconvenient.

Signalling network management functions fall into three catagories namely,

(*a*) Signalling traffic management for diverting traffic to prevent congestion/ loss of packets,

(*b*) signalling link management which restores failed SLs, activated new SLs, and deactivated aligned SLs, and

(*c*) signalling route management which distributes information about network stations to block or unblock signalling routes.

A13.6 Signalling Connection Control Part (SCCP)

SCCP was developed as a higher layer above MTP for the following reasons:

(*a*) The addressing capabilities of MTP are limited to delivery of messages to a specified node

(*b*) MTP cannot use 'global titles' for routing

(*c*) MTP uses a 4-bit service indicator to distribute messages within a node—this is a very limited way of doing message transfer

(*d*) MTP does not have all the routing capabilities laid down by OSI model

The above limitations are overcome by SCCP in the following way:

- Limitation in (*a*) are overcome by providing addressing that uses destination point codes (DPCs) plus subsystem numbers (SSNs), i.e., more explicit addressing. SSN is the local addressing information that identifies each of the SCCP users at a node.
- Limitations in (*b*) are overcome by addressing messages with global titles such as dialed digit that are not in a form usable by MTP for routing. This is done by mapping global titles to the form DCP + SSN.

- Limitations in (*c*) are overcome by providing enhanced message transfer services, two of them connectionless, two connection-oriented (a total of 4). The four classes of message transfer services provided by SCCP are

  0 – basic connectionless
  1 – sequenced (MTP) connectionless class
  2 – basic connection-oriented
  3 – flow-control—connection-oriented

  In class '0' service, user provides block of data called 'NSDU' (network service data unit) to SCCP for delivery to a user at another node. NSDUs are transported independently and may be out of sequence. Thus, we have the 'datagram' service.

  In class '1' service, a particular stream of NSDUs can be specified to be delivered in sequence.

  Class-2 service provides basic connection-oriented service, i.e., a virtual circuit through a signalling network.

  Class-3 service provides for flow control over logical connection, and for detection of message loss and wrong sequence.

  (In class-2 and class-3 services, each logical connection is given a unique signalling link code.)

- SCCP overcomes the limitations under (*d*) with its in-built 'routing control' mechanism. Upon receipt of a message from MTP or a frame of connection-oriented control or connectionless control, SCCP forwards it to MTP for transfer, or passes it to connection-oriented control or connectionless control.

References

[1] Black, U., *ISDN and SS7: Architecture for Digital Signaling Networks*, Prentice Hall, NJ, 1997
[2] Russell, R., *SS7*, McGraw-Hill, NY, 1995
[3] Jabbari, B., Common channel signaling system number 7 for ISDN and intelligent networks, *Proceedings of IEEE*, Special Issue on SS7, April 1992.
[4] Andrew S. Tanenbaum, *Computer Networks*, Third edition, Chapter 3, PHI.

List of Abbreviations used in the Book

5B6B	5 bit to 6 bit code
5-BPF	5-stage band-pass filter
ACSR	Aluminum conductor steel reinforced
ADM	Add–drop multiplexer
AGC	Automatic gain control
AGCH	Access grant channel
AIS	Absence of input signal (All 1's sent)
AIUR	Air interface user rate
ALC	Automatic level control
ALBO	Automatic line build out
ALM	Alarm signal
AM	Amplitude modulation
AMI	Alternate mark inversion
AMPS	Advanced mobile phone system
APC	Automatic phase control
ARQ	Automatic repeat request
ASC	Analog service channel
ASCII	American standard code for information interchange
ATM	Asynchronous transfer mode
AuC	Authentication center

AUG	Administrative unit group
AWGN	Additive white Gaussian noise
BCCH	Broadcast control channel
BER	Bit error rate
BIT-COMB	Bit combiner
BPF	Bandpass filter
BRZ	Bipolar return to zero
BS	Base site equipment
BSC	Base station controller
BSS	Base station subsystem
BSSGP	Base station subsystem GPRS protocol
BTS	Base transceiver
B-U CONV	Binary-to-unipolar converter
BW	Bandwidth
CAMEL	Customized application for mobile enhanced logic
CCCH	Common control channel
CCH	Control channel
CCIR	Consultative Committee for International Radio
CCITT	Consultative Committee for International Telegraphy and Telephony
CCK	Common cipher key
CCS	Call control service
CDMA	Code division multiple access
CEPT	Conference of European Posts and Telegraphs
CKSN	Ciphering key sequence number
CLMS	Connectionless message service
CMI	Coded mark inversion
CN	Core network
CNMI	Central network management interface
CODEC	Coder–decoder
COMP	Comparator
COMS	Connection-oriented message service
CRF	Call receive filter
CSFM	Compelled sequence multi-frequency
CSU	Channel service unit
CTF	Call transmission filter
CTS	Clear to send
CW	Continuous wave

D-AMPS	Digital advanced mobile phone service
DCCH	Dedicated control channel
DCE	Data communications equipment
DCK	Derived cipher key
DCS 1800	GSM in 1800 MHz band
DCT	Discrete cosine transform
DECT	Digital enhanced cordless telecommunications
DEL-EQL	Delay equalizer
DM	Direct mode
DM-DW	Direct mode dual watch
DMR 770	Digital Microwave Radio (7 GHz band) Model 770
DM-REP	Direct mode repeater
DPCM	Differentiated pulse code modulation
DPDCH	Dedicated physical control channel
DPSK	Differential phase shift keying
DPU	Digital processor unit
DSC	Digital service channel
DSU	Data service units
DTAX	Digital trunk automatic exchange
DTE	Data terminal equipment
DTMF	Dual tone multiple frequency
DXC	Digital cross connects
EAROM	Electrically alterable ROM
ECMA	European Computer Manufacturers Association
EDGE	Enhanced data rates for global evolution
EFI	Error free interval
EHF	Extra high frequency (30 GHz to 300 GHz range)
EIA	Electronic Industries Alliance
EIR	Equipment identity register
ELPD	Erasable programmable logic device
ETSI	European Telecommunications Standards Institute
FACCH	Fast associated dedicated control channel
FAS	Frame alignment signal
FAW	Frame alignment word
FCCH	Frequency correction channel
FD	Frequency diversity
FDDI	Fibre distributed digital interface
FDM	Frequency division multiplexing

FDMA	Frequency division multiple access
FEC	Forward error correcting code
FISU	Fill-in signal unit
FM	Frequency modulation
FMA	Frequency multiplexing by analog filtering
FPLMTS	Future Public Land Mobile Telecommunication System
F-PLS	Frame pulse
FR	Frame relay
FS	Frame sync signal
FS1, FS2	Frame pattern signals
F-SYNC	Frame synchronizer
FT	Fixed radio terminal
GCK	Group cipher key
GEO	Geosynchronous earth orbit
GGSN	Gateway GPRS support node
GMAP processor	Group mapping processor
GMRI	Gyromagnetic resonance interaction
GMSC	Gateway MSC
GMSK	Gaussian-filtered minimum shift keying
GPRS	General packet radio system
GSM	Global system for mobile communications
GSN	GPRS support nodes
GTP	GPRS tunneling protocol
HDB	High density bipolar
HDLC	High level data link protocol
HEO	High elliptical orbit
HF	High frequency
HL	Hotline
HLR	Home location register
HRF alarm	Higher data rate (34 Mbps) receive failure
HS	Handshake
HSCSD	High speed circuit switched data
HTF alarm	Higher data rate (34 Mbps) transmit failure
HYB	Hybrid
I/O	Input / Output
I/P	Input
ICL	Inserted correction loss
IDB	Internal distribution board

IF	Intermediate frequency
IMEI	International mobile equipment identity
IMSI	International mobile subscriber identity
IMT-2000	International Mobile Telecommunications-Year 2000 Program
INTFC	Interface
ISDN	Integrated services digital network
ISI	Inter symbol interference
ISO	International Standards Organization
ISUP	ISDN user part
ITSI	Individual TETRA subscriber identity
ITU	International Telecommunication Union
KSG	Key stream generator
KSS	Key stream sequence
LA	Location area
LAI	Location area identification
LAN	Local area network
LAPD	Link access procedure for D-channel
LAPDm	LAPD mobile version
LEO	Low earth orbit
LL alarm	Loss of link alarm (no tributary received from far end)
LLC	Logical link control
LO	Local oscillator
LOL	Loss of lock of the PLL
LOTC	Loss of transmit clock
LPF	Low-pass filter
LSI	Line station interface
LSSU	Logical status signal unit
MAC	Medium access control
MAP	Maximum aposteriori probability
MBS	Manual busy indication
MCC	Mobile country code
MEO	Medium earth orbit
MGCK	Modified group cipher key
MM	Mobility management
MNC	Mobile network code
MPFAIL	Microprocessor fail
MS	Mobile station
MSC	Mobile service switching center

MSC	Mobile services switching centre
MSISDN	Mobile subscriber ISDN number
MSRN	Mobile station roaming number
MSU	Message signal unit
MT	Mobile terminal
MTC	Mobile terminated call
MTIE	Maximum time error interval
MULDEX	Multiplexer–demultiplexer
MUX	Multiplexer
MW	Microwave
NFB	No-fuse breaker
NFW	Non-frame word
NMI	Network management interface
NMS	Network monitoring system
NMS	Network management system
NNI	Network node interface
NRZ	Non-return to zero
NSS	Network and switching subsystem
NSS	Network subsystem
O/P	Output
OFC	Optical fibre cable
OLTE	Optical line terminating equipment
OMC	Operations and maintenance center
OSI model	Operation System Interface – 7layer reference model
OSS	Operation subsystem
OTAR	Over the air re-keying
PAD	Packet assemblers / disassemblers
PAM	Pulse amplitude modulation
PCD	Program control and display teminal
PCH	Paging control channel
PCM	Pulse code modulation
PCS	Personal communication system
PDCH	Packet data channel identifier
PDH	Plesiochronous digital hierarchy
PDM	Pulse duration modulation
PDO	Packet data optimized
PDTCH	Packet data traffic channels
PEI	Peripheral equipment interface

PH-DEM	Phase demodulator
PH-ERROR-DET	Phase error detector
PH-MOD	Phase modulator
PIN	Personal identity number PIN
PISN	Private integrated service network
PLL	Phase-locked loop
PLMN	Public land mobile networks
PLS-DET	Pulse detector
PM	Pulse modulation
POH	Path overhead
PSU	Power supply unit
PPM	Pulse position modulation
PRACH	Physical random access channel
PRBS	Pseudo random binary sequence
PROT	Protection channel
PS	Power supply
P-S-CONV	Parallel-to-serial converter
PSK	Phase shift keying
PSPDN	Packet switched public data network
PSTN	Public switched telephone network
PSU	Power supply unit
PT	Portable radio terminal
PTM	Point-to-multi-point
PTP	Point to point
PTP-CLNS	PTP connectionless network service
PTP-CONS	Connection-oriented network service
PTS	Proceed to send
PUK	PIN-unlocking key
PWM	Pulse width modulation
QAM	Quadrature amplitude modulation
QoS	Quality of service
QPRS	Quadrature partial response signalling
QPSK	Quadrature phase shift keying
RACH	Random access channel
RALM	Receive alarm
RCF	Receive channel filter
RDSTF	Receive de-stuff signal
REG	Regular channel

R -AUTO-PROT	Regular-auto-protection mode selection
RF	Radio frequency
RGNR	Regenerator
RGPCLK	Receive gap clock
RLP	Radio link protocol
RMT	Remote terminal failure
RPE-LPC	Regular pulse excited linear predictive decoder
RSS	Radio subsystem
RSTFDT	Receive stuffing detect
RTS	Ready to send
RTT	Round trip time
Rx	Receiver
SACCH	Slow associated dedicated control channel
SACFA	Standing Advisory Committee on Frequency Allocation
SAGE	Security Algorithm Group of Experts
SAP	Service access point
SC-AMP	Service channel amplifier
SC-TEL DISP	Service channel telephone display
SCH	Synchronization channel
SC-INTFC-D	Service channel interface unit for DSC
SCK	Static cipher key
SCRB	Scrambler
SDCCH	Stand alone dedicated control channel
SDH	Synchronous digital hierarchy
SDU	Service data unit
SG	Supergroup
SGSN	Serving GPRS support node
SHF	Super high frequency (3 to 30 GHz)
SIM	Subscriber identity module
SL	Signalling link
SLIC	Subscriber loop interface circuit
SMS	Short message service
SN	Subscriber number
SNDCP	Sub network dependent convergence protocol
SNR	S/N ratio
SOH	Section overhead
SP	Signalling point
SRES	Signed response

SS	Supplementary service
SS7	Signalling system number 7
STM	Synchronous transport module
STP	Signal transformer point
STR	Symbol timing (clock) recovery
STUFF-CONT	Stuff controller
STUFF-PLS	Stuff pulse
SwMI	Switching management infrastructure
SWO	Switch-over
SWO-CONT	Switch-over control
SYS alarm	System alarm
TAX	Trunk automatic exchange
TCF	Transmission channel filter
TCH	Traffic channels
TCH/F	Full rate TCH
TCH/H	Half rate TCH
TD-CDMA	Time division CDMA
TDD	Time division duplex
TDM	Time division multiplexing
TDMA	Time division multiple access
TE	Terminal equipment
TEI	Terminal equipment identity
TETRA	Terrestrial trunked radio
TIC	Trunk interface IC
TIF	Tributary input failure
TL alarm	Tributary failure alarm (no tributary input, i.e., local)
TLLI	Temporary logical link identity
TM	Trunked mode
TMN	Telecommunication Management Network
TMSI	Temporary mobile subscriber identity number
T-MUX	Trans multiplexer
TRSV-EQL	Transversal equalizer
TTS	Tone transmission service
TU	Tributary unit
TUG	Tributary unit group
TVP	Time variant parameter
Tx	Transmitter
UAK	User authentication key

UDP	User datagram protocol
UHF	Ultra high frequency
UMTS	Universal mobile telecommunication system
UPT	Universal personal telecommunications
UTRA	UMTS terrestrial radio access
V + D	Voice plus data
VAD	Voice activated detection
VC	Virtual container
VCC	Visitor country code
VCO	Voltage controlled oscillator
VCXO	Voltage controlled crystal oscillator
VDB	Visitor database
VF	Voice frequency
VHE	Virtual home environment
VHF	Very high frequency
VLF	Very low frequency
VLR	Visitor location register
VNDC	Visitor national destination code
W-CDMA	Wideband CDMA
WN	Weighting network
WPC	Wireless planning and coordination cell
WS	Wayside
WS-INTFC	Wayside interface unit
WS-SWO	Wayside switchover
ZXP WS/DSC	Wayside and digital service channel traffic section

Index